McKAY BOOKS & CDs
Engineering
9.00
CB0159971 02G
DESPAULT ROBERT

I1006772

WATER RESOURCE SYSTEMS PLANNING AND ANALYSIS

Daniel P. Loucks
Jery R. Stedinger
Douglas A. Haith

Cornell University

Prentice-Hall, Inc.
Englewood Cliffs, New Jersey 07632

Library of Congress Cataloging in Publication Data

Loucks, Daniel P
Water resource systems planning and analysis.

Includes bibliographies and index.
1. Water resources development—Planning.
2. Water resources development—Mathematical models. I. Stedinger, Jery R., joint author.
II. Haith, Douglas A., joint author. III. Title.
TC409.L883 333.91′15′0724 80-18721
ISBN 0-13-945923-5

Editorial/production supervision
and interior design by Lori Opre
Manufacturing buyer: Anthony Caruso

Printed in the United States of America

10 9 8 7 6 5 4 3 2 1

Prentice-Hall International, Inc., *London*
Prentice-Hall of Australia Pty. Limited, *Sydney*
Prentice-Hall of Canada, Ltd., *Toronto*
Prentice-Hall of India Private Limited, *New Delhi*
Prentice-Hall of Japan, Inc., *Tokyo*
Prentice-Hall of Southeast Asia Pte. Ltd., *Singapore*
Whitehall Books Limited, *Wellington, New Zealand*

To our teachers, colleagues,
and students throughout the world
who have taught us not only what we know,
but also how much we have yet to learn.

Contents

Preface

The constantly increasing demand for a sufficient quantity and quality of water, properly distributed in time and space, has forced engineers and planners to contemplate and propose ever more comprehensive, complex, and ambitious plans for water resource systems. Such plans include the regulation of natural water supplies and the transportation of water between watersheds, river basins, nations, and as it has been suggested, even continents. These projects are undertaken in an attempt to provide water of an adequate quality and quantity at the times and places where it is of sufficient value to justify the effort.

Which of the infinite variety and variations of possible water resource systems and policies should be implemented? Information that can help answer this question is what water resource systems analysts try to provide. Water resource systems analysts use mathematical models and methods to aid engineers, planners, economists, and the public to sort through the myriad of schemes which are and could be proposed. Systems analysis methods can help identify plans or sets of plans and policies that achieve to the greatest extent possible the needs, goals, and aims of those who plan, pay for, and make use of, or are affected by water resource facilities and management plans.

The application of systems methods such as mathematical optimization and simulation can significantly aid in the definition, evaluation, and selection of water resources investments, designs, and policies. There is increasing use and documentation of systems analysis methods in water resources planning, and courses in water resource systems analysis have become commonplace in graduate and undergraduate programs in engineering schools. This book serves as an introductory text for undergraduate and graduate students in such courses and to practitioners involved in, or responsible for, the planning and management of water resource systems.

Although we have attempted to incorporate into each chapter the current thinking in, and approaches to, water resource systems planning and analysis, this book is not intended to be a review of the literature. Rather, it is intended to introduce readers to the methods useful in water resource systems planning. We have tried to organize the discussion about various quantitative methods for evaluating and comparing alternative water resources projects and plans in a form useful for teaching and self-study. We have generally selected modeling techniques that have seen successful field applications. The tone of the work reflects our conviction that the most useful models for planning are often the simplest, chiefly because they are easy to understand, to explain, and to use. This does not imply that highly sophisticated complex models are less useful. Rather, the analyst should realize that the appropriate model complexity is a compromise between detail and a potential for increased accuracy on the one hand and a savings in model development time, computational requirements, and model simplicity on the other.

Unfortunately, it is difficult to actually teach the practical art of water resource systems analysis in a book such as this and we have declined the temptation to try. In practice, a successful study requires that the system engineers and planners grasp the issues and concerns that are important to those in the planning process, those who will determine the decisions that are made, and those who are directly affected by such decisions. In addition, system engineers and planners must understand the operation of the water system under study including its peculiarities and the alternative courses of action. Then by skillful application and use of models and techniques such as those presented here, the systems engineer can determine and illustrate the important consequences of alternative plans. He or she can identify plans that provide a reasonable compromise among the possibly conflicting objectives of the numerous groups of individuals who are, or will be, involved in or affected by the project. The importance of the art in systems analysis can be grasped from reviewing the various case studies contained in references given throughout the text. However, in practice the art is most likely realized only when diligent, skillful, and open-minded systems engineers ply their trade with honesty, dedication, and the broadest viewpoint possible.

This book requires only an introductory background in calculus, matrix algebra, and probability theory. Chapters 2, 3, and 4 introduce the quantitative methods that form the basis of water resource systems analysis: mathematical programming and simulation, probability and statistics, and benefit-cost analysis and multiobjective modeling. These concepts and methods are discussed in the context of water resources design and management problems.

The application of these quantitative methods form the principal portion of the book. We have made a somewhat artificial but common division of river basin models into water *quantity* and water *quality* models. The modeling approaches to the two problems have significant differences, and at least from a pedagogical viewpoint, there are advantages to separate discussions of water quantity and quality problems when this is possible.

Surface water quantity management problems and models are discussed in four chapters. Chapter 5 treats river basin phenomena as deterministic: future quantities such as streamflows, water requirements, and prices are considered known and certain. Although this is a very simplified view of real water problems, the resulting models are conceptually simple and are adequate for some preliminary planning applications.

Stochastic water quantity models are presented in Chapters 6 and 7. Chapter 6 introduces a variety of synthetic streamflow generating models that can be used to assist in comparing and evaluating proposed or existing water resource systems. Chapter 7 presents a range of stochastic optimization models useful for planning both the design and operation of reservoirs and for solving river basin water allocation problems. These models provide a more realistic representation of the variability of hydrologic elements of water resource systems than do the models in Chapter 5. Stochastic models constitute a more "advanced" topic and require a reasonable understanding of deterministic models and of probability. Finally, Chapter 8 presents some further detail on the water quantity aspects of irrigation planning and operation, an increasingly important activity of water resources planners.

In one sense, water quality management can be viewed as just one of the many purposes to be achieved in river basin planning. However, in many areas of the world, and particularly in developed regions, water pollution control has a very high priority for water resources planning. Chapters 9 and 10 introduce a variety of simulation and optimization models for surface water quality planning. Water quality models are based on mass conservation and descriptions of the biochemical and physical phenomena associated with water pollution. The discussion in these two chapters illustrates how the costs of water quality management alternatives, such as wastewater treatment, flow augmentation, and land disposal of wastewaters, can be incorporated into the models that describe and predict various biochemical and physical phenomena. This provides a means of comparing and evaluating the economic as well as the environmental impacts of various water management alternatives.

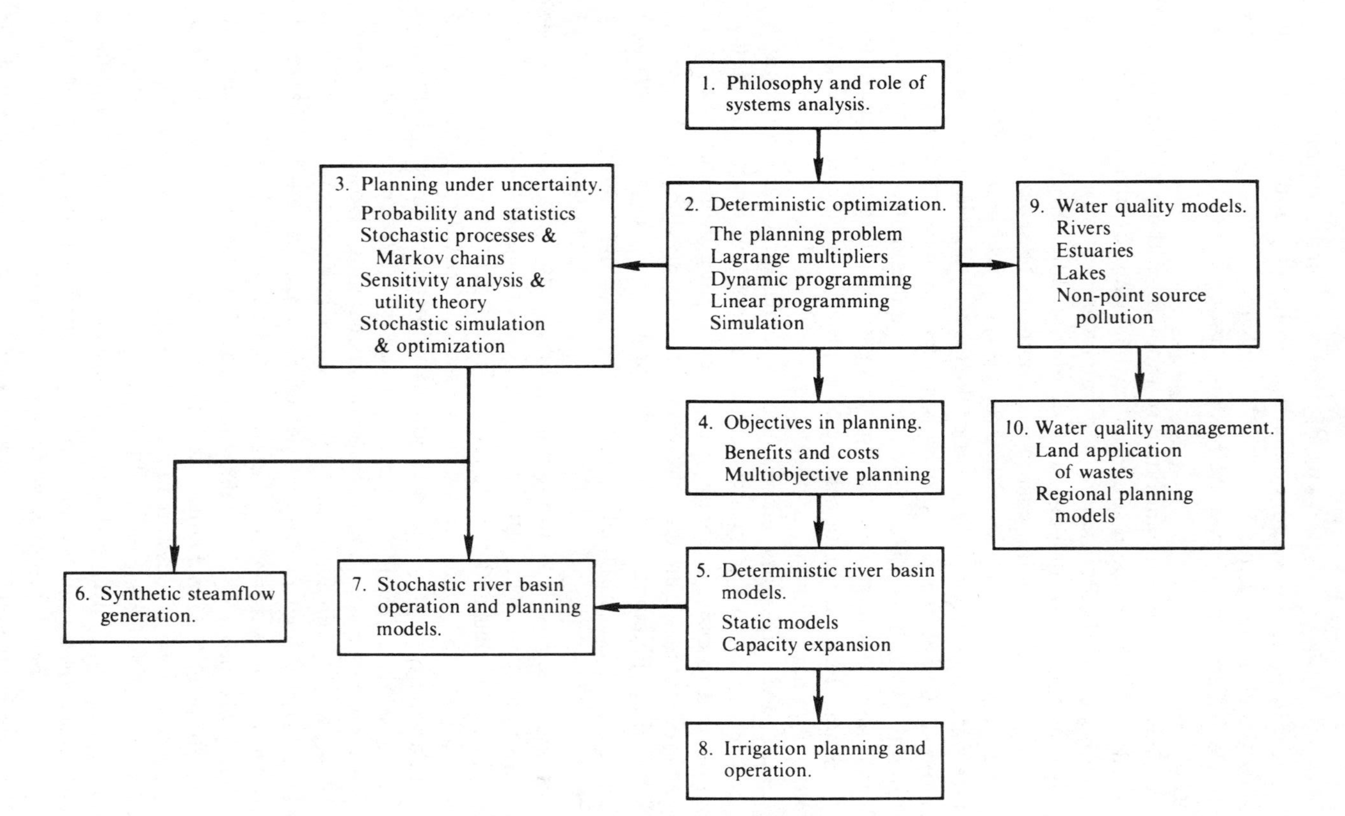

EXHIBIT A. Topics and relationships of chapters.

Exhibit A illustrates the major topics of each chapter. Arrows indicate when the material in one chapter builds upon the material in another chapter. The material may be taught and organized in several ways to meet the needs and interests of an instructor or a reader. We have taught a two-semester course on water resource systems planning. The first semester emphasized deterministic models and methods (Chapters 1, 2, 4, 5, and 8). The second semester addressed stochastic models (Chapters 3, 6, and 7) and water quality problems (Chapters 9 and 10). For those interested primarily in water quality management, Chapters 1, 2, 4, 9, and 10 plus at least Section 5 of Chapter 3 would constitute a good course of study. For instructors whose students have had advanced training in either mathematical programming and optimization or in probability, statistics, and time series analysis, large parts of Chapters 2 and 3 may be covered rapidly. However, we have generally found that going over this material, if only briefly, is useful to the student.

Many individuals helped the authors in the preparation of this book. The authors give special thanks to Helga van der Leeden and Patty Apgar who diligently typed and helped edit many drafts of this book. We must also thank Professor Ir. L.J. Mostertman, Director of the International Institute for Hydraulic and Environmental Engineering, Delft, the Netherlands, and Professors Donald Harleman, David Marks, and Frank Perkins at the Massachusetts Institute of Technology for their financial support and for the opportunity to teach from these notes at their institutes. It is, of course, impossible to recognize and acknowledge the invaluable help of all of our teachers, colleagues, and students at numerous universities and institutes who through their comments and their work have contributed to the production of this book. We particularly thank Stephen Burges, Rolf Deininger, John Dracup, Donald Erlenkotter, Louis Falkson, Warren Hall, James Heaney, Perry McCarty, Gerald Orlob, Sergio Rinaldi, Charles Scherer, Christine Shoemaker, Rodolfo Soncini-Sessa, and Harald Stehfest for their careful reading and criticism of all or parts of earlier drafts. Any remaining errors of fact or opinion are of course solely the responsibility of the other two authors of this book! We would welcome comments from anyone that will help us improve any future editions.

Ithaca, New York

D.P. LOUCKS

J.R. STEDINGER

D.A. HAITH

PART I

Overview

This book on water resource systems planning and analysis is concerned with the development and application of quantitative mathematical modeling methods to problems of water management. Quantitative modeling methods are typically discussed in books or courses on operations research or systems analysis. In contrast to such texts on systems methodology, the primary focus of this book is on the application of these methods to problems of water management. The introductory chapter that follows provides a perspective on the types of water problems that are most readily analyzed using systems methods and explains why such quantitative methods can be, and have been, helpful during the planning process.

CHAPTER 1

Planning and Analysis of Water Resource Systems

1.1 INTRODUCTION TO WATER RESOURCES PLANNING

Water: too much, too little, too dirty. Throughout the world, these are the conditions that prompt water resources planning. To meet the demands for the desired quantity and quality of water at particular locations and times, engineers—together with economists, political scientists, lawyers, planners and conservationists—have gained considerable experience in designing, constructing, removing, and operating structures and implementing nonstructural measures that will permit improved management of natural water supplies.

The incentive to plan for increased control of any water resource often follows a major disaster, such as a flood, a drought, intolerable water quality conditions, or a waterborne disease epidemic. Following the crises that often trigger water resources planning, citizens' review committees, planning boards, advisory groups, and public hearings may all help sustain the momentum needed to carry plans through to implementation. Just as rapidly as public support develops for investments in engineering structures for controlling and managing water, environmental and preservation interest groups emerge to critically question the wisdom of such investments. The concern of these groups is not limited to just the conservation and preservation of scenic areas and wild rivers, but often includes the broad regional impacts that could result from changes in air and water quality, noise levels, land use, and transportation corridors. Even if the issue is the improvement of water quality, generally considered a desirable goal, the opposition will certainly include some of those who must pay to achieve it. The issues that create conflict in water resources planning are real and are not necessarily the result of human stubbornness, self-interest, or inflexibility. Reasonable, rational, and informed individuals will frequently reach opposing conclusions concerning the wisdom of proposed projects because of the differences in their assessment of the value of the costs and benefits of the projects.

Water resources planning must take into account multiple users, multiple purposes, and multiple objectives. Different people have different goals, perspectives, and values. Planning for maximum net economic benefits "to whomsoever they accure" is not sufficient: it matters *who* pays and *who* benefits. Issues of equity, risk, redistribution of national wealth, environmental quality, and social welfare are as important as economic efficiency. It is clearly impossible to develop a single objective that satisfies all interests, all adversaries, and all political and social viewpoints.

Increased public involvement in water resources planning has changed the manner in which engineers approach this task. It has forced tradition-bound planners and designers to broaden their perspective and to examine a wider range of alternative plans. It has shifted more of the responsibility for making choices from the engineers or planners to the politicians or public officials. Even though engineers and planners play an important part in the decision-making process, it is elected politicians who are, or at least should be, responsible and accountable for decisions that involve compromises among various public interests and concerns.

Water resources engineers and planners should develop a number of reasonable alternatives for public officials to consider; they should also evaluate the economic, environmental, political, and social impacts that might result from each alternative. Today we study river basins, lakes, estuaries, and other water resource systems with a greater awareness of their complexity, their sensitivity and adaptability to exogenous forces, and of our own limitations in understanding their behavior. We realize the general irreversibility of the impact of large projects on natural ecosystems and on our own social and economic organization. Construction of large reservoirs commits valuable land and water resources to a limited range of uses and thus forecloses society's options to use these resources in very different ways. This awareness is humbling, but it is also a challenge. It has stimulated the development, over the last several decades, of improved analytical tools and methodologies for defining and evaluating alternatives for managing such systems so that the best possible decisions will be made.

Those who are involved in the development of water resource systems methodology know that the use of these tools cannot guarantee development of optimal plans for water resources development and management. Given the competing and changing objectives and priorities of different interest groups, it is unclear how useful the concept of an "optimal plan" really is. What system methodology can do, however, is to help define and evaluate, in a rather detailed manner, numerous alternatives that represent various possible compromises among conflicting groups, values, and management objectives. In particular, a rigorous and objective analysis should help to identify the possible trade-offs between quantifiable objectives so that further debate and analysis can be more informed. The art of systems analysis is to identify those issues and concerns which are important and significant and to structure the analysis to shed light on these issues.

Although the systems approach to water resources planning is not restricted to mathematical modeling, models do exemplify the approach. They can represent in a fairly structured and ordered manner the important interdependencies and interactions among the various control structures and users of a water resource system. Models permit an evaluation of the economic and physical consequences of alternative engineering structures, of various operating and allocating policies, and of different assumptions regarding future flows, technology, costs, and social and legal requirements. Although this systems methodology cannot define the best objectives or assumptions, it can identify good decisions, given those objectives and assumptions.

In acknowledging the role of systems methodology in the water resource planning process, one should recognize the inherent limitation of models as representations of any real problem. The input data, including objectives and other assumptions, may be controversial or uncertain. Of course, these inputs affect the output. Future events are not known with certainty, and our knowledge concerning any water resource system is always limited. Moreover, the results of any quantitative analysis of alternatives are often only a small part of the input to the overall planning and decision-making process. Equally important may be purely qualitative factors, including subjective inferences drawn from the quantitative analysis.

We should not expect, therefore, to have the precise results of any quantitative systems study accepted and implemented. A measure of the success of any systems study resides in the answer to the following questions: Did the study have a beneficial impact in the planning and decision-making process? Did the results of such studies make the debate over the proper choice of alternatives more informed? Did it introduce competitive alternatives which otherwise could not have been considered?

There seems to be no end of challenging water resource systems planning problems facing water resources planners. How one models any specific water resource problem depends on (1) the objectives of the analysis; (2) the data required to evaluate the projects; (3) the time, data, money, and computational facilities available for the analysis; and (4) the modelers' knowledge and skill. Model development is an art, requiring judgment in abstracting from the real world the components that are important to the decision to be made and that can be illuminated by quantitative methods, and judgment in expressing those components and their interrelationships mathematically in the form of a model.

1.2 WATER RESOURCE SYSTEMS ANALYSTS, ENGINEERS, AND POLICYMAKERS

To engage in a successful water resource systems study, the systems analyst must possess not only the requisite mathematical and systems methodology skills, but also an understanding of the environmental engineering, economic, political, cultural, and social aspects of water resources planning problems. For example, to study the impact of a large land development plan, the analyst should be able to predict how the proposed plan would affect runoff and, in turn, the quantity and quality of surface waters and groundwaters, and how the development would affect flood flows and conversely, how flood flows would affect the planned development. The analyst must have an understanding also of the biological, chemical, and physical processes that are influenced or caused by various nutrients, biodegradable wastes, chemicals, and other constituents that may be discharged into and transported by water bodies as a result of the proposed development.

A reasonable knowledge of economic theory is just as important as an understanding of hydraulic, hydrologic, and environmental engineering disciplines. Economics has always had, and will continue to have, a significant role in the planning of water resources investments. It is obvious that the results of most water resources management decisions have a direct impact on people and their relationships. Hence inputs from those having a knowledge of law, regional planning, and political science are also needed during the comprehensive planning of water resource systems, especially during the development and evaluation of the results of various planning models.

Early water resource systems studies were often undertaken with a naive view of the appropriate role and impact of models and modelers in the policymaking process. The policymaker could foresee the need to make a decision. He or she would tell the systems group to study the problem. They would then model it, identifying feasible solutions and their consequences, and recommend one or at most a few solutions. The policymaker, after waiting patiently for these recommendations, would then make a yes or no decision. Experience to date suggests the following:

1. A final solution to a water resources planning problem rarely exists; plans and projects are dynamic and change and evolve over time as facilities are added and modified and the uses and demands placed on the facilities change.
2. For every major decision there are many minor decisions, made by different agencies or management organizations responsible for different aspects of a project; plans evolve.
3. The time normally available to study a water resources problem is shorter than the time necessary for an adequate state-of-the-art mathematical modeling study, or if there is sufficient time, the objectives of the original study will have significantly shifted by the time the study is completed.

This experience emphasizes not only some of the limitations and difficulties that any water resource systems study may encounter, but more important, it emphasizes the need for constant communication among the analysts, engineers responsible for the systems operations, and policymakers. The success or failure of many past water resource studies is in a large part attributable to the efforts expended or not expended in ensuring adequate and meaningful communication among systems planners, professional engineers responsible for system operation and design, and public officials responsible for major decisions and setting general policies. It is these engineers and public officials, after all, who need the information that can be derived from various models and analyses, and they must have it in a form useful and meaningful to them. At the beginning of any study, objectives are usually poorly defined. As more is learned about what can be gotten, people are better able to identify what they want. Close communication among analysts, engineers, and public officials throughout the modeling process is essential if systems studies are to make their greatest contribution to the planning process.

Furthermore, those who will use the models, and present the information derived from the models to those responsible for making decisions, must be intimately involved with model development, solution, and analysis. Only then can they appreciate the assumptions upon which any particular model is based, and hence adequately evaluate the reliability of the results. Any water resource systems study that involves only outside consultants, and minimal communication between consultants and planners within a responsible management agency, is unlikely to succeed in having a significant impact on the planning process. Models that are useful are alive, constantly being modified and applied by those groups which are responsible for plan preparation, evaluation, and implementation.

1.3 CHARACTERISTICS OF SYSTEMS ANALYSIS APPLICATIONS

Successful systems analysis applications exhibit a number of common characteristics. These are reviewed here because they provide insight into whether a systems study of a particular problem may be worthwhile. If the planners' objectives are very unclear, few alternative courses of action exist, or there is little scientific understanding of the issues involved, then mathematical modeling and sophisticated methodologies are frequently of little use. Successful applications of systems analysis are characterized by:

1. A systems focus or orientation: Attention needs to be and is devoted to the interaction of elements within the system as a whole as well as to the elements themselves.
2. The use of interdisciplinary teams: In many complex and nontraditional problems it is not at all clear from the start what disciplinary viewpoints will turn out to be most appropriate. It is essential that the participants in such work—coming from different established disciplines—become familiar with the techniques, vocabulary, and concepts of the other disciplines. It might be said that participation in interdisciplinary research requires a willingness to make mistakes at the fringes of one's technical competence.
3. The use of formal mathematical models: The overwhelming preference of most systems analysts is to use mathematical models to assist in system description and evaluation and to provide an unambiguous record of the assumptions and data used in the analysis.

Not all water resources planning problems are suitable candidates for study using systems analysis methods. The systems approach is most appropriate when:

4. The system's objectives are reasonably well defined and organizations and individuals can be identified who have the necessary authority and power to implement possible decisions.
5. There are many alternative decisions that may satisfy the stated objectives and the best decision is not obvious.

This text on quantitative planning techniques will be particularly useful for water resources problems which have two additional characteristics:

6. The alternative solutions and the objectives of the system being analyzed are describable by a reasonably tractable mathematical representation.
7. The parameters of the model are estimatable from readily obtainable data.

These last four conditions are, of course, rarely met in practice. In such cases, systems analysis studies may still help in providing new insights and understanding to the problem of concern, but will probably be less successful than would otherwise be the case.

EXERCISES

1-1. What are the characteristics of water resources planning or management problems that are most suitable for analysis using quantitative systems analysis techniques?

1-2. Identify some specific water resource systems planning problems and for each problem specify in words possible objectives, the unknown decision variables whose values need to be determined, and the constraints or relationships that must be met by any solution of the problem.

1-3. From a review of the recent issues of various journals pertaining to water resources and the appropriate areas of engineering, economics, planning and operations research, identify those journals that contain articles on water resource systems planning and analysis, and the topics or problems currently being discussed.

1-4. Many water resource systems planning problems involve considerations that are very difficult if not impossible to quantify, and hence they cannot easily be incorporated into any mathematical model for defining and evaluating various alternative solutions. Briefly discuss what value these admittedly incomplete quantitative models may have in the planning process when nonquantifiable aspects are also important. Can you identify some planning problems that have such intangible objectives?

BOOKS ON WATER RESOURCE SYSTEMS ANALYSIS

1. BISWAS, A. K. (ed.), *Systems Approach to Water Management*, McGraw-Hill Book Company, New York, 1976.
2. BUGLIARELLO, G., and F. GUNTHER, *Computer Systems and Water Resources*, Elsevier, Amsterdam, 1974.
3. BURAS, N., *Scientific Allocation of Water Resources*, American Elsevier Publishing Co., Inc., New York, 1972.

4. BURKE, R., III, and J. P. HEANEY, *Collective Decision Making in Water Resources Planning*, Lexington Books, D. C. Heath & Company, Lexington, Mass., 1975.

5. CIRIANI, T. A., U. MAIONE, and J. R. WALLIS (eds.), *Mathematical Models for Surface Water Hydrology*, John Wiley & Sons Ltd., London, 1977.

6. DAVIS, R. K., *The Range of Choice in Water Management*, Johns Hopkins Press, Baltimore, Md., 1968.

7. DEININGER, R. A. (ed.), *Models for Environmental Pollution Control*, Ann Arbor Science Publishers, Inc., Ann Arbor, Mich., 1973.

8. DE NEUFVILLE, R., and D. H. MARKS, *Systems Planning and Design: Case Studies in Modelling, Optimization and Evaluation*, Prentice-Hall, Inc., Englewood Cliffs, N.J., 1974.

9. DORFMAN, R., H. D. JACOBY, and H. A. THOMAS, JR. (eds.), *Models for Managing Regional Water Quality*, Harvard University Press, Cambridge, Mass., 1972.

10. FIERING, M. B, *Streamflow Synthesis*, Harvard University Press, Cambridge, Mass., 1967.

11. FIERING, M. B, and B. JACKSON, *Synthetic Streamflows*, Water Resources Monograph 1, American Geophysical Union, Washington, D.C., 1971.

12. FLEMING, G., *Computer Simulation Techniques in Hydrology*, American Elsevier Publishing Co., Inc., New York, 1975.

13. HAIMES, Y. Y., *Hierarchical Analysis of Water Resources Systems*, McGraw-Hill Book Company, New York, 1977.

14. HAIMES, Y. Y., W. A. HALL, and H. T. FREEDMAN, *Multiobjective Optimization in Water Resources Systems: The Surrogate Worth Trade-off Method*, Elsevier, Amsterdam, 1975.

15. HALL, W. A., and J. A. DRACUP, *Water Resources Systems Engineering*, McGraw-Hill Book Company, New York, 1970.

16. HAMILTON, H. R., S. W. GOLDSTONE, J. W. MILLIMAN, A. L. PUGH, E. G. ROBERTS, and A. ZELLNER, *Systems Simulation for Regional Analysis: An Application to River Basin Planning*, The MIT Press, Cambridge, Mass., 1969.

17. HOWE, C. W., *Benefit-Cost Analysis for Water System Planning*, Water Resources Monograph 2, American Geophysical Union, Washington, D.C., 1971.

18. HUFSCHMIDT, M. M., and M. B FIERING, *Simulation Techniques for Design of Water-Resources Systems*, Harvard University Press, Cambridge, Mass., 1966.

19. JAMES, A. (ed.), *Mathematical Models in Water Pollution Control*, John Wiley & Sons Ltd., Chickester, Great Britain, 1978.

20. KNEESE, A. V., and S. C. SMITH (eds.), *Water Research*, Johns Hopkins Press, Baltimore, Md., 1966.

21. KNEESE, A. V., and B. T. BOWER, *Managing Water Quality: Economics, Technology, and Institutions*, Johns Hopkins Press, Baltimore, Md., 1968.

22. KNETSCH, J. L., *Outdoor Recreation and Water Resources Planning*, Water Resources Monograph 3, American Geophysical Union, Washington, D.C., 1974.

23. MAASS, A., M. M. HUFSCHMIDT, R. DORFMAN, H. A. THOMAS, JR., S. A. MARGLIN, and G. M. FAIR, *Design of Water-Resource Systems*, Harvard University Press, Cambridge, Mass., 1962.

24. MAJOR, D. C., *Multiobjective Water Resource Planning*, Water Resources Monograph 4, American Geophysical Union, Washington, D.C., 1977.

25. MAJOR, D. C., and R. L. LENTON, *Applied Water Resources Systems Planning*, Prentice-Hall, Inc., Englewood Cliffs, N.J., 1979.

26. META Systems, Inc., *Systems Analysis in Water Resources Planning*, Water Information Center, Inc., Port Washington, N.Y., 1975.

27. O'LAOGHAIRE, D. T., and D. M. HIMMELBLAU, *Optimal Expansion of a Water Resources System*, Academic Press, Inc., New York, 1974.

28. OVERTON, D. E. and M. E. MEADOWS, *Stormwater Modeling*, Academic Press, Inc., New York, 1976.

29. RINALDI, S., R. SONCINI-SESSA, H. STEHFEST, and H. TAMURA, *Modeling and Control of River Water Quality*, McGraw-Hill Book Company, New York, 1979.

30. RUSSELL, C. S., D. G. AREY, and R. W. KATES, *Drought and Water Supply*, Johns Hopkins Press, Baltimore, Md., 1970.

31. THOMANN, R. V., *Systems Analysis and Water Quality Management*, McGraw-Hill Book Company, New York, 1972.

32. THRALL, R. N., et al., *Economic Modeling for Water Policy Evaluation*, North-Holland Publishing Co., Inc., Amsterdam, 1976.

PART II

Methods of Analysis

Quantitative methods for defining and evaluating alternative water resources plans include a variety of mathematical techniques. These techniques are drawn from the subject area that has been labeled systems analysis, systems engineering, operations research, or management science. For most purposes these terms are synonymous. It is the mastery of these analytical planning methods, together with a knowledge of economics, environmental engineering, hydraulics, hydrology, and water resources engineering, that distinguishes a water resource systems analyst.

It is not clear how well one must master these quantitative systems analysis methods to be an effective water resource systems analyst. Some analysts have found postgraduate study in operations research and economics highly useful. Certainly anyone expecting to participate in water resource systems research will find advanced graduate courses helpful. On the other hand, many of the more theoretical and highly mathematical aspects of operations research or economics are likely to be of only marginal value in the actual practice of water resource systems planning.

The following three chapters review the operations research methods most commonly used in water resource systems planning. Unlike basic texts in these subjects, this review will be in the context of water resource systems planning. Still the emphasis of this review will be on the *methods* of analysis; the water planning problems that serve to illustrate these methods will be relatively simple. For those engineers and planners who have been introduced to systems analysis methods, this review should serve to refresh their knowledge. For those who have not previously studied these methods, this review is intended to introduce analytical planning methods in a manner requiring only an elementary knowledge of calculus, matrix algebra, and economics. The remainder of the book will emphasize river basin planning and how various systems analysis methods discussed in this section can be used in that planning process.

CHAPTER 2

Identification and Evaluation of Water Management Plans

2.1 INTRODUCTION

The task of water resource system planners may be broadly described as the identification or development of possible water resource system designs or management plans and evaluation of their economic, ecological, environmental, and social impacts. An important and quantifiable criterion for plan evaluation is the economic benefits and costs a plan would entail were it implemented. Most proposed regional water resource plans involve large investments in land, reservoirs, pipes, canals, and/or pumping and electric generating facilities. As a result of these capital investments, economic returns may be obtained for decades. In both an economic and ecological sense these investments are often irreversible: once the forest of a valley is cleared and replaced by the lake behind a dam, it is almost impossible ecologically and economically to destroy the dam and regain the forest.

The emphasis of this book is on mathematical models that can be used to identify those management plans which best meet society's objectives. For this purpose economic and other objectives are expressed mathematically. Management plans involve the selection of many engineering design and operating variables. In the mathematical models used to describe a water resource system, these design and operating variables are called *decision variables*, for it is the best values of these variables which are to be determined. They may include, for example, the capacities of various reservoirs and pipelines of a water supply distribution system, or the allocation of land and water for various crops at an irrigation site, or the location and capacity of various flood control reservoirs and levees along a developed river. The value of each decision variable may affect both the costs and benefits associated with any particular plan. Plan formulation can be thought of as assigning particular values to each of the relevant decision variables.

Assume that there are P plans, each designated by the letter p. For each plan, there exist n_p decision variables x_j^p. Together these variables and their values, expressed by the vector $\mathbf{X}_p$, define the pth plan. The index j distinguishes one decision variable from another, and the index p distinguishes one plan from another. The goal of the analyst, in this case, may be to find the particular plan p, defined by $\mathbf{X}_p$, that maximizes the plan's net benefits or some other objective. This objective can be expressed mathematically as

$$\underset{p \in P}{\text{maximize}} \; \text{NB}(\mathbf{X}_p) \tag{2.1}$$

As written above, the economic objective requires an evaluation of the net benefits derived from each plan p and the selection of the particular plan in the set of all plans whose net benefits are a maximum. When there is a large number of possible system configurations and possible values of the decision variables, it may no longer be feasible to consider all possible combinations, as is suggested in equation 2.1. In this case it is often convenient not to consider discrete plans, but to consider all possible values of continuous decision variables x_j in a search for a plan or plans that maximize net benefits and other objectives. This can be expressed as

$$\text{maximize} \; \text{NB}(\mathbf{X}) \tag{2.2}$$

such that $\mathbf{X}$ is contained in the set of all feasible or possible plans.

The foregoing objective is the continuous equivalent of the discrete objective function 2.1. While maximization of equation 2.1 requires comparison of $\text{NB}(\mathbf{X}_p)$ for every p, maximization of equation 2.2 by complete enumeration is now impossible because of the infinity of possible plans. Maximization of equation 2.2 falls in the domain of mathematical optimization methods which identify optimal combinations of the decision variables without complete enumeration of the alternatives. These methods are the topic of most of this chapter. However, before proceeding to that discussion, a review of some engineering economics is provided. Methods of engineering economics are used to compare plans that yield different time streams of net benefits.

2.2 EVALUATION OF TIME STREAMS OF BENEFITS AND COSTS

Alternative plans may involve construction of different projects with different economic lives, and hence they result in different *time streams* of future benefits and costs. The net benefit generated in time period t by plan p, $\text{NB}_t(\mathbf{X}_p)$, will be written simply as NB_t^p. Each plan is characterized by the time stream or vector of net benefits it generates over its planning period T_p.

$$\{\text{NB}_1^p, \text{NB}_2^p, \text{NB}_3^p, \ldots, \text{NB}_{T_p}^p\} \tag{2.3}$$

Clearly, if in any time period t the benefits exceed the costs, then $\text{NB}_t^p > 0$; and if costs exceed benefits, $\text{NB}_t^p < 0$. This section presents the standard methods used to compare the different net-benefit time streams produced by different plans perhaps having planning periods ending in different years T_p.

Fundamental to the evaluation of the value of different time streams is the concept of the *time value of money*. From time to time, individuals, private corporations, and governments all borrow money. Two transactions characterize this activity: (1) the amount borrowed is repayed, and (2) an additional amount called *interest* is paid to the lender for the priviledge of having had their money at one's disposal. In the private sector the interest rate is often identified as the marginal rate of return on capital. If individuals have a certain amount of money, called *capital*, at their disposal, they can either use it themselves to earn more money or they can lend it to others and receive the prevailing market interest rate. Assuming that people with capital allocate their money to where it yields the largest returns, most investors should be receiving at least the prevailing interest rate as the return on their capital.

Consider now the problem of determining the value of the time stream of benefits in equation 2.3, where the prevailing interest rate in each period is r. The meaning of r is that if one lends an amount PV at the beginning of period 1, then at the end of period 1 one should receive the principal PV plus the interest rPV. Hence the value of one's assets at the end of period 1 is $(1 + r)$PV.

If V_1 is the value of one's assets at the end of period 1, this relationship may be written

$$V_1 = (1 + r)\text{PV} \tag{2.4}$$

If one immediately reinvests these assets at the end of period 1, one could lend $(1 + r)$PV and hence would have at the end of period 2

$$V_2 = (1 + r)[(1 + r)\text{PV}] = (1 + r)^2\text{PV} \tag{2.5}$$

Repeating this argument, one can show that by lending PV at the beginning of period 1 and always reinvesting, one would have assets at the end of period t equal to

$$V_t = (1 + r)^t\text{PV} \tag{2.6}$$

A rewording of this argument would show that equation 2.6 is also valid if PV is negative. Hence if one borrows an amount $|\mathrm{PV}|$ at the beginning of period 1, he or she then has an obligation at the end of time period t equal to $(1 + r)^t |\mathrm{PV}|$.

Equation 2.6 is the basic relation needed to determine the value at the beginning of period 1 of net benefits NB_t to accrue during time period t. Thus if V_t is equated to the net benefits NB_t received at the end of time period t, the amount one would need to have on hand (or owe) such that by lending or borrowing they would obtain resources NB_t at the end of time period t is called the present value PV of NB_t. Using equation 2.6 yields

$$\mathrm{NB}_t = (1 + r)^t \mathrm{PV}$$

or (2.7)

$$\mathrm{PV} = (1 + r)^{-t} \mathrm{NB}_t$$

The total present value of the net benefits generated by plan p, denoted PV_p, is the sum of the present values of the net benefits accrued at the end of each time period.

$$\mathrm{PV}_p = \sum_{t=1}^{T_p} (1 + r)^{-t} \mathrm{NB}_t \qquad (2.8)$$

The present value of the net benefits achieved by two or more plans having the same economic planning horizons T_p can be used as an economic basis for plan selection. If the economic lives of projects differ, the present value of the plans is an inappropriate measure for comparison and plan selection. If plans have very different economic lives, they are essentially solving different problems, or at least providing very different solutions to the same problem. A more reasonable comparison of alternative plans can often be obtained by extending all plans to cover the longest planning period required to adequately evaluate all the plans.

Rather than extending the economic life of each project to a common future year in order to compare their present values, an alternative procedure is to convert each time stream of net benefits to an equivalent average annual net benefit, and then compare these annual net benefits. This is done in two steps. First, one computes the present value of the time stream of net benefits, using equation 2.6. These present net benefits then can be converted to the corresponding average annual benefits of the project. That is, the total benefits provided by project p can be described by their present value PV_p or as an average annual benefit stream of NB each year.

Net benefit	NB	NB	NB	...	NB
Year	1	2	3	...	T_p

To obtain the value of the average annual benefits NB corresponding to any present value PV and time horizon T, one equates

$$\mathrm{PV} = \sum_{t=1}^{T} (1 + r)^{-t}\mathrm{NB} = \left[\frac{(1 + r)^T - 1}{r(1 + r)^T}\right]\mathrm{NB} \tag{2.9a}$$

and thus

$$\mathrm{NB} = \left[\frac{r(1 + r)^T}{(1 + r)^T - 1}\right]\mathrm{PV} \tag{2.9b}$$

The *capital recovery factor* CRF_T is the factor in equation 2.9b which converts a fixed payment or present value at the beginning of a project's life into an equivalent fixed periodic payment

$$\mathrm{CRF}_T = \frac{r(1 + r)^T}{(1 + r)^T - 1} \tag{2.10}$$

This factor is often used to compute the annual cost of engineering structures that have a fixed initial construction cost C_0 and a constant annual operation, maintenance, and repair (OMR) cost. The total annual cost TAC is simply the initial cost times the capital recovery factor plus the annual OMR costs.

$$\mathrm{TAC} = (\mathrm{CRF}_T)C_0 + \mathrm{OMR} \tag{2.11}$$

For private investments requiring borrowed capital, interest rates are usually established, and hence fixed, at the time of borrowing. However, benefits may be affected by changing interest rates, which are not easily predicted. For the purposes of this discussion interest rates were assumed constant, as is done in practice. For a discussion on the effects of inflation, see Hanke et al. [9].

This discussion has justified interest rates based on the possible behavior of a private firm or an individual. In the economic evaluation of public-sector investments, the same relationships are used even though government agencies are not generally free to loan or borrow funds on private money markets. In the case of public-sector investments, the appropriate interest rate is a matter of public policy. It specifies the rate at which the government is willing to collect taxes or forego current benefits to its citizens to provide larger benefits in future time periods. It can be viewed as the government's estimate of the time value of public monies or the marginal rate of return to be achieved by public investments.

These are the definitions and concepts of engineering economics that will be required in the remainder of the text. More detailed discussions of the application of engineering economics to water resources planning are contained in refs. 8, 11, 14, and 16. The exercises at the end of this chapter extend this discussion.

2.3 PLAN FORMULATION

The previous section discussed the comparison and evaluation of predefined plans. The plan yielding the largest value of the objective function could be found simply by comparing the value of the objective achieved by each plan and selecting that plan which achieved the largest value. It is certainly possible that a change in the value of one or more decision variables included in a particular plan (thus defining a new plan) might produce an even greater value of the objective. Methods of plan "optimization" are the subject of this section. In mathematical terms, such methods must identify the values of the decision variables x_j, contained in $\mathbf{X}$, that maximize the appropriate net benefit function NB($\mathbf{X}$) or other objectives.

The values that the decision variables may assume are rarely unrestricted. Usually, various functional relationships among these variables must be satisfied. A quantity of water completely consumed by one user cannot be simultaneously or subsequently allocated to another. This is an example of a physical constraint that may restrict the values of variables representing the allocation of water to alternative users in different time periods. Technological restrictions limit the capacities and sizes to pipes, generators, and pumps to those commercially available. The minimum allowable quality of water in a river or lake may be specified by law, an example of a legal restriction. There may also be limited funds available to finance water resources development projects, a financial constraint.

Physical, technical, legal, financial, and other restrictions on the values of the decision variables can generally be expressed by equations of the form $g_i(\mathbf{X}) = b_i$, commonly called *constraints*. The subscript i denotes one of, say, m such constraints, where b_i is a constant, possibly 0. From a mathematical point of view, these constraints define the water resource system, its components and their interrelationships, and the permissible values of the decision variables.

Plan formulation involves finding values of the decision variables that satisfy all m constraint equations. Typically, there exist many more decision variables than constraints, and hence, if any solution exists, there may be many solutions that satisfy all the constraints. The existence of alternative plans is a characteristic of most water resource systems planning problems. Any solution or plan that satisfies all these constraint equations is termed a *feasible* plan. The particular feasible solution or plan that maximizes the net benefit function NB($\mathbf{X}$) or another objective $F(\mathbf{X})$ is called the *optimal* plan (from this limited point of view).

The plan formulation process can be expressed by the following model or problem:

$$\text{maximize } F(\mathbf{X}) \tag{2.12}$$

subject to

$$g_i(\mathbf{X}) = b_i \qquad i = 1, 2, \ldots, m \tag{2.13}$$

Equation 2.12 of the planning model is called an *objective function.* An objective function is the function that is to be "optimized" (i.e., either maximized or minimized).

The problem is to find the values of each decision variable x_j in $\mathbf{X}$ that, in this case, maximize the objective function $F(\mathbf{X})$ while satisfying each of the constraint equations 2.13. Equations 2.12 and 2.13 together form a planning or management model. Since all the functions in this model are completely general, the foregoing planning model is also an example of a completely general constrained optimization model. Methods for solving constrained optimization models such as this are termed *mathematical programming algorithms.*

The type of solution procedure (or algorithm) most appropriate for any particular constrained optimization (or mathematical programming) model depends on the particular mathematical form of the objective function and of the constraint equations. There is no universal solution procedure that will solve efficiently all constrained optimization models. Hence model builders tend to model physical–economic water resource systems by mathematical expressions that are of a form compatible with one or more known efficient solution procedures. This may be a possible source of error if convenient mathematical expressions do not provide the best possible description of the system. Another source of error is our inability to quantify and express mathematically all the planning objectives; the technical, economic, and political uncertainties; and other important considerations that will play a role in the decision-making process. Hence at best any mathematical planning model is only an approximate description of the real water resource system. The optimal solution of the model is optimal only with respect to the particular model, not necessarily with respect to the real problem. It is important to realize this limited meaning of the word "optimal," a word commonly used by water resources and other systems analysts.

2.4 PLANNING MODELS AND SOLUTION PROCEDURES

There are two basic approaches for solving planning models: *simulation* and *optimization.* Simulation relies on trial-and-error to identify near-optimal solutions. The value of each decision variable is set, and the resulting objective values are evaluated. The difficulty with the simulation approach is that there is often a frustratingly large number of feasible solutions or plans. Even when combined with efficient techniques for selecting the values of each decision variable, an enormous computational effort may lead to a solution that is still far from the best possible.

To their credit, simulation methods are able to solve water resource systems planning models with highly nonlinear relationships and constraints. Constrained optimization procedures are seldom able to deal with all the complexities and nonlinearities which are easily incorporated in a simulation model. Still, when an optimization procedure can be constructed to efficiently solve an adequate approximation to the real problem, they can greatly narrow down the search with simulation for a global optimum by identifying plans that may be close to the optimum.

Constrained optimization algorithms include a diverse set of techniques that use calculus and matrix algebra. Optimization techniques include Lagrange multipliers, linear programming, dynamic programming, quadratic programming, and geometric programming. Each of these and other solution procedures are highly dependent on the mathematical structure of the management model. This chapter presents three constrained optimization techniques: classical constrained optimization using Lagrange multipliers, dynamic programming, and linear programming. These three methods, particulary the latter two, are the most commonly used in water resource systems planning; they are used throughout the book. Some aspects of simulation modeling are also reviewed.

2.5 OBJECTIVE FUNCTIONS AND CONSTRAINT EQUATIONS

Typical planning models generally include at least one objective function that is either to be maximized or minimized and which serves to rank the alternative solutions or plans. In virtually every case the objective function is a scalar function; that is, no matter how many terms are included, the dimensions of each term must be homogeneous. One cannot consider as a single objective function the simultaneous maximization of irrigation water (cubic meters/year) and hydropower production (megawatt-hours/year).

In addition to an objective, planning problems incorporate a number of requirements which are formulated as constraints. It is important to distinguish the different roles played by the objective function and the constraints. The optimal solution of the planning problem is a plan that achieves the largest (or smallest) value of the objective while satisfying all the constraints. Constraints can be of two kinds. One type of constraint expresses an actual physical limitation that cannot be violated at any cost. Such limitations may include the conservation of mass, the magnitude of fixed resources, or the capacity of existing or proposed facilities.

The second type of constraint is in some sense an implicit objective or goal which in fact could be violated, although the cost of such a violation may be high. Such constraints include restrictions on minimum streamflows to maintain water quality, schedules of water deliveries, and budgetary limitations. When goals are formulated as constraints, all feasible solutions must satisfy these goals. There is no explicit incentive for overfulfilling these goals, nor are the goals permitted to be reduced if the cost of meeting them is too high. These adjustments are often made after an examination of the optimal solution of the planning problem as initially formulated. Decision makers want to know how much money can be saved, for example, if some requirement is relaxed or what the cost may be of providing more of some service. Hence, the solution to any particular planning problem provides decision makers and analysts with estimates of what is possible. They can use this information as a point of departure in their search for more desirable alternatives.

To clarify these concepts, consider irrigation-water requirements. If these requirements are truly constraints, the implication is that there would be an enormously large penalty sustained if the targeted quantity of water were not supplied, and further, no additional benefits would accrue if more water were supplied. On the other hand, irrigation-water requirements may be considered an objective in that there is probably an increase in benefits if some water beyond that which is targeted is provided, and there is a loss in benefits associated with any deficit allocation or with large excess allocations.

Sometimes the planning problem is stated in the form of maximizing an objective function subject to probabilistic constraints. One statement representing this class of constraints is: The probability that the power produced is greater than or equal to the "power requirement" must be no less than 0.95. Such a probabilistic constraint might be written

$$\Pr[\text{power produced} \geq \text{power requirement}] \geq 0.95 \qquad (2.14)$$

where 0.95 is the lower limit on the fraction of time in which the power requirement is to be met.

Occasionally, deciding whether a requirement should be a constraint or an objective is difficult, simply because objectives are not well defined at the beginning of the plan formulation process. In some cases there is little distinction between requirements stated as objectives and requirements stated as constraints.

2.6 LAGRANGE MULTIPLIERS

Consider a quantity of water Q that can be allocated to three water users, denoted by the index $j = 1, 2$, and 3. The problem is to determine the allocation x_j to each user j that maximizes the total net benefits. In this example the net benefit resulting from an allocation of x_j to user j is approximated by

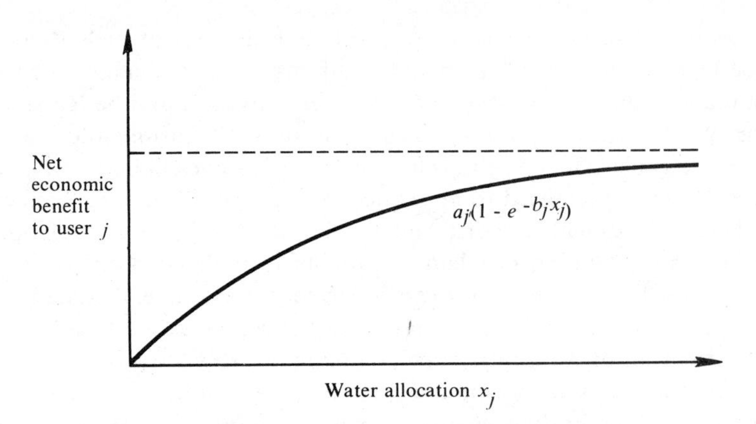

FIGURE 2.1. Net-benefit function for user *j*.

the function $a_j[1 - \exp(-b_j x_j)]$, where a_j and b_j are known positive constants. These net-benefit functions are of the form illustrated in Figure 2.1.

The three allocation variables x_j are unknown decision variables. The values that these variables can assume are restricted between 0 (since a negative allocation is impossible) and values whose sum, $x_1 + x_2 + x_3$, does not exceed the available supply of water Q. Since the more water that is allocated the greater the net benefits, the maximum total net benefits are obtained when all the water is allocated. Therefore, the constraint equations can be written as

$$\sum_{j=1}^{3} x_j = Q \tag{2.15}$$

$$x_j \geq 0 \qquad j = 1, 2, 3 \tag{2.16}$$

These constraint equations and inequalities, together with the objective function

$$\text{maximize NB}(\mathbf{X}) = \sum_{j=1}^{3} a_j[1 - \exp(-b_j x_j)] \tag{2.17}$$

define the planning or management problem.

One of the most direct procedures for finding the optimal solution to this problem is through the use of *Lagrange multipliers*. For the moment ignore the inequality constraints given in equation 2.16. The first step in this procedure is to multiply the constraint equation written as

$$\sum_{j=1}^{3} x_j - Q = 0 \tag{2.18}$$

by a variable or *multiplier* λ and subtract the product from the original objective function. For all feasible values of **X** this term does not change the objective function's value because for feasible **X**'s the equality in equation

2.18 must apply. For the example problem, the augmented objective function is

$$L(\mathbf{X}, \lambda) = \sum_{j=1}^{3} a_j[1 - \exp(-b_j x_j)] - \lambda\left(\sum_{j=1}^{3} x_j - Q\right) \tag{2.19}$$

Since there is only one constraint equation in this problem, there is only one multiplier λ in equation 2.19. If there were other constraints, each constraint i would be multiplied by a different multiplier λ_i.

In this problem there are now three unknown allocations x_j and one unknown multiplier λ. Solutions to the original problem can be obtained by setting the partial derivatives of the function L with respect to each of the unknown variables equal to zero. This results in the following four equations:

$$\begin{aligned}
\frac{\partial L}{\partial x_1} &= 0 = a_1 b_1 \exp(-b_1 x_1) - \lambda \\
\frac{\partial L}{\partial x_2} &= 0 = a_2 b_2 \exp(-b_2 x_2) - \lambda \\
\frac{\partial L}{\partial x_3} &= 0 = a_3 b_3 \exp(-b_3 x_3) - \lambda \\
\frac{\partial L}{\partial \lambda} &= 0 = x_1 + x_2 + x_3 - Q
\end{aligned} \tag{2.20}$$

Equations 2.20 are necessary conditions for a local maximum or minimum ignoring the nonnegativity conditions ($x_j \geq 0$). Since the objective function 2.17 involves the maximization of the sum of concave functions (functions whose slopes are decreasing), any local optima will also be the global maxima. (Similarly, a global minimum can be obtained from the minimization of the sum of convex functions, i.e., functions whose slopes are increasing.)

The optimal solution to this problem is found by solving for each x_j in terms of λ.

$$x_j = \frac{1}{b_j}\left[-\ell n\left(\frac{\lambda}{a_j b_j}\right)\right] \tag{2.21}$$

Insisting that the total allotment $x_1 + x_2 + x_3$ must equal Q in the optimal solution, one can solve for λ by substituting equation 2.21 into constraint equation 2.18.

$$Q = \sum_{j=1}^{3} \frac{-\ell n\,[\lambda/(a_j b_j)]}{b_j} = -\sum_{j=1}^{3} \ell n\left[\left(\frac{\lambda}{a_j b_j}\right)^{1/b_j}\right]$$

or

$$\lambda = \left[e^{-Q}\left(\prod_{j=1}^{3} (a_j b_j)^{1/b_j}\right)\right]^{1/[(1/b_1)+(1/b_2)+(1/b_3)]} \tag{2.22}$$

Hence knowing Q and the a_j's and b_j's, equation 2.22 determines the value of λ. Substitution of this value in equation 2.21 yields the optimal allotments provided all the x_j's are nonnegative. Although not necessary for this particular problem, one must sometimes check the second-order conditions and solve the model with various x_j set equal to 0, the boundary conditions, to verify whether a global maximum or minimum was obtained.

For this example problem, the first-order conditions, equations 2.20, are solvable and the number of special conditions is sufficiently small so that the optimal solution can be obtained without undue computational effort. For most water resources planning problems, this is not the case. What typically may happen is that the equations resulting from equating the partial derivatives to zero are so nonlinear as to make their simultaneous solution impossible algebraically and extremely difficult using numeric techniques.

2.6.1 Meaning of Multiplier λ

The value of λ has an important economic meaning. For the reader to appreciate its value, it is useful to review the general Lagrange multiplier solution procedure. Consider a general constrained optimization problem containing n decision variables x_j and m constraint equations ($m < n$).

$$\text{maximize (or minimize) } F(\mathbf{X}) \tag{2.23}$$

subject to constraints

$$g_i(\mathbf{X}) = b_i \qquad i = 1, 2, 3, \ldots, m \tag{2.24}$$

where $\mathbf{X}$ is the vector of all x_j. The Lagrangian $L(\mathbf{X}, \lambda)$ is formed by combining equations 2.24 with the objective function 2.23.

$$L(\mathbf{X}, \lambda) = F(\mathbf{X}) - \sum_{i=1}^{m} \lambda_i (g_i(\mathbf{X}) - b_i) \tag{2.25}$$

Solutions of the equations

$$\begin{aligned} \frac{\partial L}{\partial x_j} &= 0 \qquad \text{for all decision variables } j \\ \frac{\partial L}{\partial \lambda_i} &= 0 \qquad \text{for all constraints } i \end{aligned} \tag{2.26}$$

are possible local optima.

The meaning of the values of the multipliers λ_i at the optimum can be seen by manipulation of equations 2.26. Multiplying $\partial L/\partial x_j$ by dx_j for each j and summing the resulting equations over all j yields

$$\sum_{j=1}^{n} \frac{\partial F}{\partial x_j} dx_j - \sum_{i=1}^{m} \lambda_i \left(\sum_{j=1}^{n} \frac{\partial g_i}{\partial x_j} dx_j \right) = 0 \tag{2.27}$$

where all of the partial derivatives are evaluated at the optimal solution. The first term in equation 2.27 is the marginal change dF in the objective $F(\mathbf{X})$ which results from small changes $dx_1, \ldots, dx_n$ in the decision variables. At the optimum, a small change in the decision variables will also effect the constraints in equations 2.24, where

$$dg_i = \sum_{j=1}^{n} \frac{\partial g_i}{\partial x_j} dx_j = db_i \qquad (2.28)$$

Equation 2.28 relates the value of small changes in the decision variables to the resulting change dg_i in the constraint functions; if the constraints are to remain satisfied, any change dg_i in the constraint function must equal the specified change db_i in b_i. Substituting equation 2.28 into equation 2.27 yields the relationship between a small change in the value of the objective dF and small changes db_i in each constant b_i

$$dF = \sum_{i=1}^{m} \lambda_i \, db_i \qquad (2.29)$$

For the original problem every db_i must equal zero so that equation 2.29 specifies that, to first order, every incremental change in the decision variables $dx_1, \ldots, dx_n$ that continues to satisfy the constraints ($db_i = 0$ in equation 2.28) results in no change in the objective function ($dF = 0$). These are the first-order conditions for a constrained maximum or minimum to the problem in equations 2.23 and 2.24.

Equation 2.29 can also be used to determine the change in the optimal value of the objective function $F(\mathbf{X})$ that would result from a small change db_i in the right-hand side of constraint i. If the right-hand side b_i of constraint i is increased by an amount db_i, then at the optimum dF is increased by $\lambda_i \, db_i$. In shorthand notation,

$$\left. \frac{\partial F}{\partial b_i} \right|_{\mathbf{X}=\mathbf{X}^*} = \lambda_i \qquad (2.30)$$

Equation 2.30 points out that each multiplier λ_i is the rate of change in the optimal value of the original objective function, 2.23, with respect to a change in the value (the right-hand side) of the corresponding ith constraint. If a constraint is inactive or redundant (i.e., is satisfied whether or not it is included in the model), then the associated λ_i will equal 0. These λ_i's are called *dual variables* or *shadow prices.* They are important economic parameters which are often as useful as the optimal values of the decision variables. Use of the shadow prices is discussed in more detail in Chapter 4.

2.6.2 Slack and Surplus Variables

The planning problem as formulated so far has only considered equality constraints. If inequality constraints exist, nonnegative *slack* or *surplus* variables can be introduced to create equality constraints. For example, an inequality constraint of the form $g_i(\mathbf{X}) \leq b_i$ can be converted to $g_i(\mathbf{X}) +$

$s_i = b_i$ if $s_i \geq 0$. Likewise, $g_i(\mathbf{X}) \geq b_i$ can be converted to $g_i(\mathbf{X}) - s_i = b_i$ as long as each $s_i \geq 0$.

To ensure that a slack or surplus variable is nonnegative, the square s_i^2 of an unrestricted slack or surplus variable s_i may be used. This results in constraint equations of the form

$$g_i(\mathbf{X}) \pm s_i^2 - b_i = 0 \tag{2.31}$$

Differentiating a Lagrangian containing terms $\lambda_i(g_i(\mathbf{X}) \pm s_i^2 - b_i)$ with respect to s_i, and setting the derivative equal to zero, results in equations

$$\lambda_i s_i = 0 \tag{2.32}$$

This requires that either λ_i or s_i equal 0.

If λ_i equals 0, then the ith constraint is inactive or redundant and does not constrain or affect the optimal solution. Thus a change in b_i will not change the optimal value of the objective function. A change in b_i will be counterbalanced by a corresponding change in s_i^2.

If s_i is 0, then the ith constraint may affect the optimal solution. A change in b_i will generally change the optimal value of the objective function, and hence λ_i will be either positive or negative, although in special cases it could be zero.

The solution of the set of partial differential equations 2.26 and equations 2.32 involves a trial-and-error process, equating to zero a λ_i or a s_i for each inequality constraint and solving the remaining equations, if possible. This tedious procedure, along with the need to check boundary solutions when nonnegativity conditions are imposed, detracts from the utility of classical Lagrange multiplier methods for solving all but relatively simple water resources planning problems.

2.7 DYNAMIC PROGRAMMING

The water supply allocation problem in the previous section considered a single net-benefit function for each water user. Now assume that the *gross* benefits are defined by the function $a_j[1 - \exp(-b_j x_j)]$ for each use j and the costs are defined by the concave functions $c_j x_j^{d_j}$, where c_j and d_j are known positive constants and $d_j < 1$. This cost function has decreasing marginal or average costs as the quantity allocated x_j increases. In other words, each user's cost exhibits economies of scale.

Assuming that the objective remains one of maximizing the total net benefits, the new planning model is

$$\text{maximize} \sum_{j=1}^{3} \{a_j[1 - \exp(-b_j x_j)] - c_j x_j^{d_j}\} \tag{2.33}$$

subject to

$$\sum_{j=1}^{3} x_j \leq Q \tag{2.34}$$

and the nonnegativity condition

$$x_j \geq 0 \quad \text{for each use } j \tag{2.35}$$

This modified problem is difficult to solve using Lagrange multipliers since the equations resulting from equating the partial derivatives of the Lagrangian to zero (equations 2.26 and 2.27) cannot be solved explicitly as before. In addition, the new objective function 2.33 may not be concave over the range of possible values of x_j. The model is, however, readily solvable using an optimizing procedure called dynamic programming [2, 19].

A first step in the dynamic programming procedure is to structure this allocation problem as a sequential allocation process or a multistage decision-making procedure, as illustrated in Figure 2.2. The allocation to each user is considered a decision *stage* in a sequence of decisions. When portion x_j of the total water supply Q is allocated at stage j, this results in net benefits $R_j(x_j) = a_j[1 - \exp(-b_j x_j)] - c_j x_j^{d_j}$. A *state* variable s_j is defined as the amount of water available to the remaining $(4 - j)$ users or stages. Finally, a state transformation function $s_{j+1} = s_j - x_j$ defines the state in the next stage as a function of the current state and the current allocation or decision.

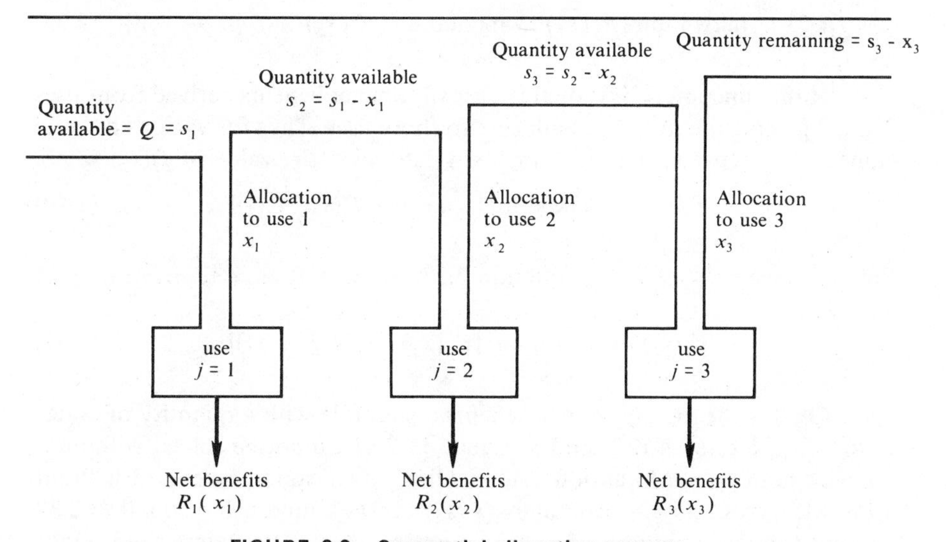

FIGURE 2.2. Sequential allocation process.

2.7.1 Recursive Equations

With these definitions, it is possible to write the allocation problem defined by equations 2.33 to 2.35 as

$$f_1(Q) = \underset{\substack{x_1+x_2+x_3 \leq Q \\ x_1, x_2, x_3 \geq 0}}{\text{maximum}} [R_1(x_1) + R_2(x_2) + R_3(x_3)] \tag{2.36}$$

The function $f_1(Q)$ is the maximum net benefits that can be obtained from allocation of a quantity of water Q to users 1, 2, and 3. Each allocation x_1, x_2, and x_3 cannot be negative and their sum cannot exceed Q. Equation 2.36 represents a problem with three decision variables. It can be transformed to three problems, each having only one decision variable.

Equation 2.36 can be written

$$f_1(Q) = \underset{0 \leq x_1 \leq Q}{\underset{x_1}{\text{maximum}}} \{R_1(x_1) + \underset{0 \leq x_2 \leq Q - x_1 = s_2}{\underset{x_2}{\text{maximum}}} [R_2(x_2) + \underset{0 \leq x_3 \leq s_2 - x_2 = s_3}{\underset{x_3}{\text{maximum}}} R_3(x_3)]\} \tag{2.37}$$

Let function $f_3(s_3)$ equal the maximum net benefits derived from use 3 given a quantity s_3 available for allocation to that use. Hence for various discrete values of s_3 between 0 and Q, one can determine the value of $f_3(s_3)$ where

$$f_3(s_3) = \underset{0 \leq x_3 \leq s_3}{\underset{x_3}{\text{maximum}}} [R_3(x_3)] \tag{2.38}$$

Since $s_3 = s_2 - x_2$, equation 2.37 can now be rewritten in terms of only x_1, x_2, and s_2:

$$f_1(Q) = \underset{0 \leq x_1 \leq Q}{\underset{x_1}{\text{maximum}}} \{R_1(x_1) + \underset{0 \leq x_2 \leq s_2}{\underset{x_2}{\text{maximum}}} [R_2(x_2) + f_3(s_2 - x_2)]\} \tag{2.39}$$

Now let the function $f_2(s_2)$ equal the maximum net benefits derived from uses 2 and 3 given a quantity s_2 to allocate to those uses. Thus for various discrete values of s_2 between 0 and Q, one can determine the value of $f_2(s_2)$ where

$$f_2(s_2) = \underset{0 \leq x_2 \leq s_2}{\underset{x_2}{\text{maximum}}} [R_2(x_2) + f_3(s_2 - x_2)] \tag{2.40}$$

Finally, since $s_2 = Q - x_1$, equation 2.37 can be written in terms of only x_1 and Q:

$$f_1(Q) = \underset{0 \leq x_1 \leq Q}{\underset{x_1}{\text{maximum}}} [R_1(x_1) + f_2(Q - x_1)] \tag{2.41}$$

Here $f_1(Q)$ is the maximum net benefits achievable with a quantity of water Q to allocate to uses 1, 2, and 3. Equation 2.41 cannot be solved without a knowledge of $f_2(s_2)$. Equation 2.40 yielding $f_2(s_2)$ cannot be solved without a knowledge of $f_3(s_3)$. Fortunately, $f_3(s_3)$ can be found using equation 2.38 without reference to any other maximum net benefit function $f_j(s_j)$. Once the value of $f_3(s_3)$ is determined, the value of $f_2(s_2)$ can be computed, which will allow determination of $f_1(Q)$, the quantity of interest.

Since equations 2.38, 2.40, and 2.41 must be solved in sequence, they are known as *recursive equations*. Recursion involves the successive solution of a series of equations, each one dependent on a knowledge of the values derived from the previous equations. Knowledge of $f_3(s_3)$ from equation 2.38 provides the information needed to solve for $f_2(s_2)$ using equation 2.40; this in turn permits the solution of $f_1(Q)$, the original objective, using equation 2.41.

Recursive sets of equations are fundamental to dynamic programming. It is sometimes easier and quicker to solve numerous single variable problems than a single multivariable problem. Each recursive equation represents a stage at which a decision is required, hence the term "multistage decision-making procedure."

When the state variables or quantity of water available s_j at stage j and the decision variables or allocations x_j to use j are allowed to take on only a finite set of discrete values, the problem is a discrete dynamic programming problem. The solution will always be a global maximum (or minimum) regardless of the concavity, convexity, or even the continuity of the functions $R_j(x_j)$. Obviously, the smaller the difference or interval between each discrete value of each state and decision variable, the greater will be the mathematical accuracy of the solution when the x_j are actually continuous decision variables.

Solving discrete dynamic programming problems to find the value of the objective function, and also the values of the decision variables that maximize or minimize the objective function, is best done through the use of tables, one for each stage of the decision-making process. Table 2.1 gives the values

TABLE 2.1. Values of User Net-Benefit Functions $R_j(x_j)$

x_j	$R_1(x_1)$	$R_2(x_2)$	$R_3(x_3)$
0	0	0	0
1	−0.5	6.5	−6.9
2	3.0	10.1	0
3	6.6	10.9	6.3
4	10.0	9.6	11.5
5	13.1	7.0	15.6

of the user net-benefit functions for our example. The computations associated with the first stage, defined by equation 2.38, are presented in Table 2.2a. Computations for the intermediate stage, defined by recursive equation 2.40 having the general form

$$f_j(s_j) = \underset{\substack{x_j \\ 0 \leq x_j \leq s_j}}{\text{maximum}} [R_j(x_j) + f_{j+1}(s_j - x_j)] \tag{2.42}$$

for various values of s_j from 0 to Q, are given in Table 2.2b.

Table 2.2c contains the calculations for the final recursive equation 2.41. The numbers in the tables are based on the assumption that $Q = 5$ and $a_j =$ 100, 50, 100; $b_j = 0.1, 0.4, 0.2$; $c_j = 10, 10, 25$; and $d_j = 0.6, 0.8, 0.4$ for $j = 1, 2, 3$, respectively.

TABLE 2.2a. Calculation of $f_3(s_3)$ from Equation 2.38

State	$R_3(x_3)$								
s_3	x_3:	0	1	2	3	4	5	$f_3(s_3)$	x_3^*
0		0						0	0
1		0	−6.9					0	0
2		0	−6.9	0				0	0, 2
3		0	−6.9	0	6.3			6.3	3
4		0	−6.9	0	6.3	11.5		11.5	4
5		0	−6.9	0	6.3	11.5	15.6	15.6	5

TABLE 2.2b. Calculation of $f_2(s_2)$ from Equation 2.40

State	$R_2(x_2) + f_3(s_2 - x_2)$								
s_2	x_2:	0	1	2	3	4	5	$f_2(s_2)$	x_2^*
0		0						0	0
1		0	6.5					6.5	1
2		0	6.5	10.1				10.1	2
3		6.3	6.5	10.1	10.9			10.9	3
4		11.5	12.8	10.1	10.9	9.6		12.8	1
5		15.6	18.0	16.4	10.9	9.6	7.0	18.0	1

TABLE 2.2c. Calculation of $f_1(Q)$ from Equation 2.41

State	$R_1(x_1) + f_2(Q - x_1)$								
Q	x_1:	0	1	2	3	4	5	$f_1(Q)$	x_1^*
5		18.0	12.3	13.9	16.7	16.5	13.1	18.0	0

Keeping a record of the optimal allocation x_j^* associated with each state-variable value makes it possible to backtrack through each successive table to find the optimal values of each decision variable. From Table 2.2c the maximum net benefits $f_1(Q)$ equal 18, and the allocation x_1^* that resulted in this maximum is 0. Hence the optimal s_2, the quantity available to allocate to uses 2 and 3, is $Q - x_1 = 5 - 0 = 5$. From Table 2.2b the optimal allocation x_2^*, to use 2 given $s_2 = 5$, is 1. Hence the optimal s_3, the quantity available to allocate to use 3, is $s_2 - x_2 = 5 - 1 = 4$. From Table 2.2a the optimal allocation x_3^*, to use 3, is 4. So the optimal allocations are 0, 1, and 4 to uses 1, 2, and 3, respectively. The sum of the allocations are, in this case, equal to Q. All available water is allocated.

2.7.2 Principle of Optimality

To illustrate the principle upon which dynamic programming is based, it is useful to represent this allocation problem by the network shown in Figure 2.3. The nodes of the network represent the states or quantities of water available to allocate to that and all following uses or stages. The links represent the possible or feasible decisions given the state-variable value and stage. Associated with each link or allocation decision is a net benefit $R_j(x_j)$.

Beginning at use or stage 3, for each feasible discrete value of state s_3, the maximum of all $R_3(x_3)$ values on the links entering each node s_3 will define

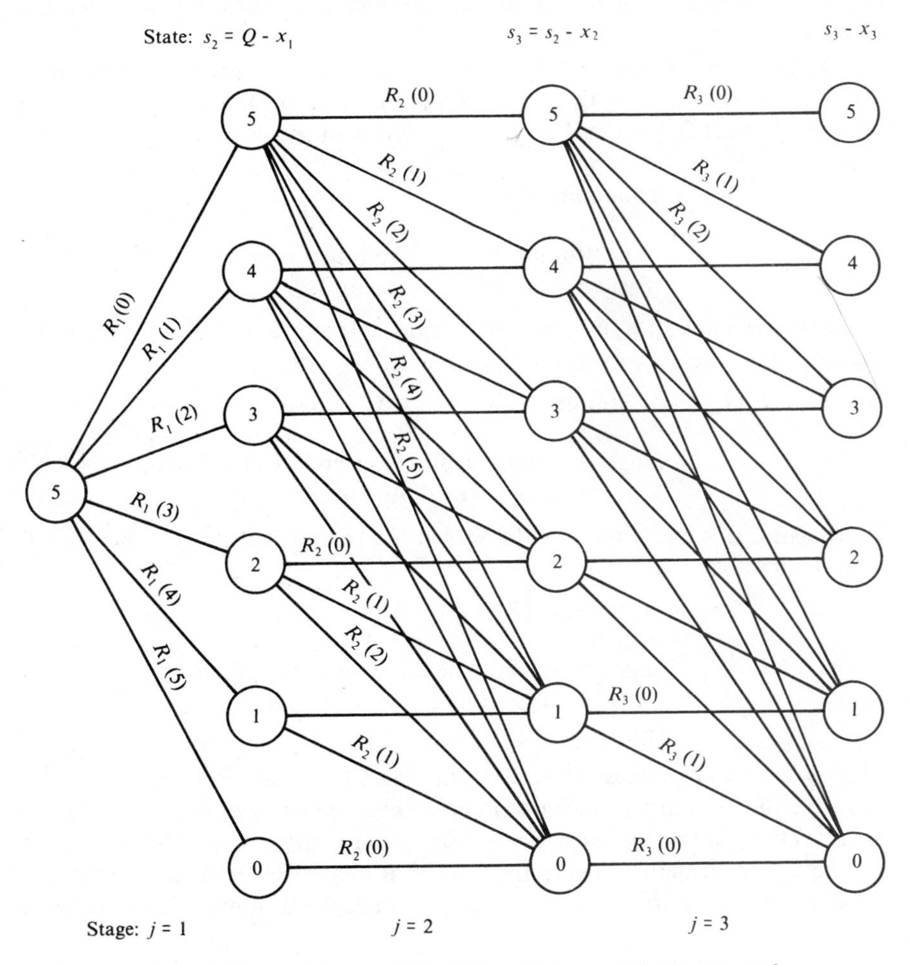

FIGURE 2.3. Network of feasible allocation decisions x_j and net benefits $R_j(x_j)$.

$f_3(s_3)$. Having these values for each value of s_3 permits a shift to the next stage, stage 2, to compute, for each state s_2, $f_2(s_2)$, the maximum benefit from uses 2 and 3 given s_2 resources to allocate. Finally, with each $f_2(s_2)$ and the values on the links from stage 1 to stage 2, one can move to stage 1 and compute $f_1(Q)$, the maximum benefit from uses 1, 2, and 3 given a quantity $Q = 5$ of water to allocate. Each $f_j(s_j)$ is the maximum of all $R_j(s_j - s_{j+1}) + f_{j+1}(s_{j+1})$ on those links entering node s_j from nodes s_{j+1}.

The procedure just described is a process that moves backward through the network from stage 3 to stage 1 to obtain a solution. It is based on the principle that *no matter in what state of what stage one may be, in order for a policy to be optimal, one must proceed from that state and stage in an optimal manner.*

The procedure could just as well begin at stage 1 and proceed forward through the network. In this case a function $f'_j(s_j)$ would be defined as the total net benefit from uses 1 to j given s_j units of water to allocate to those uses.

$$\begin{aligned} f'_1(s_1) &= \underset{x_1 \leq s_1}{\underset{x_1}{\text{maximum}}} \, [R_1(x_1)] \\ &= \text{maximum net benefits from use 1 with } s_1 \text{ units} \\ &\quad \text{of water available to allocate} \end{aligned} \tag{2.43}$$

Since the optimal s_1 is unknown, equation 2.43 must be solved for various discrete values of x_1 between 0 to Q. Next,

$$\begin{aligned} f'_2(s_2) &= \underset{x_2 \leq s_2}{\underset{x_2}{\text{maximum}}} \, [R_2(x_2) + f'_1(s_2 - x_2)] \\ &= \text{maximum net benefits from uses 1 and 2 with} \\ &\quad s_2 \text{ units of water available to allocate} \end{aligned} \tag{2.44}$$

Once again, this equation must be solved for various values of s_2 between 0 and Q. Finally,

$$\begin{aligned} f'_3(s_3) &= \underset{x_3 \leq s_3 = Q}{\underset{x_3}{\text{maximum}}} \, [R_3(x_3) + f'_2(s_3 - x_3)] \\ &= \text{maximum net benefits from uses 1, 2, and 3} \\ &\quad \text{with } s_3 = Q \text{ units of water available to} \\ &\quad \text{allocate} \end{aligned} \tag{2.45}$$

This process that moves forward from stage 1 to stage 3 is based on the principle that *no matter in what state of what stage one may be, in order for a policy to be optimal, one had to get to that state and stage in an optimal manner.*

These two principles paraphrase what Richard Bellman [2] termed the *principle of optimality*. It is a very simple concept but quite often elusive to apply.

As shown above, this particular allocation problem is just as easily solved proceeding backward from use or stage 3 to use or stage 1, or proceeding forward from use or stage 1 to use or stage 3. While some problems can be solved equally well by either backward- or forward-moving procedures, other problems may lend themselves to one or the other approach, but not to both. In either case, there must be a starting or ending point that does not depend on other stages in order to be able to define the first of the recursive equations.

Unlike other constrained optimization procedures, discrete dynamic programming methods are often simplified by the addition of constraints. The addition of lower and upper limits on each of the allocations x_j, for example, could narrow the number of discrete values of x_j that have to be considered.

2.7.3 Multiple State Variables

There are certain types of multistage problems that do cause some computational difficulties. Suppose that in the preceeding example allocation problem, the water was used for three different irrigated crops. In addition to water, land is also required. Assume that A units of land are available for all three crops and that u_j units of water are required for each unit of irrigated land containing crop j. The management or planning problem is now to determine the allocations x_j of water and x_j/u_j units of land that maximize the total net benefits subject to the added restriction on land resources:

$$\text{maximize} \sum_{j=1}^{3} R_j(x_j) \tag{2.46}$$

subject to

$$\sum_{j=1}^{3} x_j \leq Q \tag{2.47}$$

$$\sum_{j=1}^{3} \frac{x_j}{u_j} \leq A \tag{2.48}$$

$$x_j \geq 0 \qquad j = 1, 2, 3 \tag{2.49}$$

Unlike the original problem (equations 2.33 to 2.35), there are now two allocations to make at each stage: water and land. Hence an additional state variable r_j is required to indicate the amount of land available for allocation to the remaining $4 - j$ crops. The general recursive relation becomes

$$f_j(s_j, r_j) = \underset{\substack{0 \leq x_j \leq s_j \\ x_j \leq r_j u_j}}{\text{maximum}} \left[R_j(x_j) + f_{j+1}\left(s_j - x_j, r_j - \frac{x_j}{u_j}\right)\right] \tag{2.50}$$

which must be solved for various discrete values of both state variables s_j and r_j $(0 \leq s_j \leq Q, 0 \leq r_j \leq A)$.

Although the addition of a second state variable causes no conceptual difficulties, it does increase the required computational effort. The larger the number of state variables, the more combinations of discrete states that must be examined at each stage. If done on a computer, this added dimensionality requires more computer time and storage capacity. The existence of more than three state variables can exceed the computational capacity of a computer if many discrete values of each state variable are required. This occurs because of the exponential increase in the total number of discrete states that have to be considered as the number of state variables increases. This phenomenon is termed the "curse of dimensionality" of multiple-state-variable dynamic programming problems.

2.7.4 Additional Applications

Three common dynamic programming applications in water resources planning concern water allocation, capacity expansion, and reservoir operation. The previous three-user water allocation problem illustrates the first type of application. The other two applications are presented below.

Capacity Expansion. Consider a municipality that must plan for the future expansion of its water supply system or some component of that system, such as a reservoir, aqueduct, or treatment plant. The capacity needed at the end of each period t has been estimated to be D_t. The cost, $C_t(s_t, x_t)$, of adding capacity x_t in each period t is a function of that added capacity as well as of the existing capacity s_t at the beginning of the period. The planning problem is to find that time sequence of capacity expansions that minimizes the present value of total future costs and meets the projected requirements. Hence the planning model can be written as

$$\text{minimize} \sum_{t=1}^{T} \alpha_t C_t(s_t, x_t) \tag{2.51}$$

where α_t is the discount factor $(1 + r)^{-(t-1)}$. This discount factor assumes an interest rate of r in each period t and that the construction costs are incurred at the beginning of each period. The constraints define each period's final capacity, or equivalently the next period's initial capacity, s_{t+1} as a function of the known existing capacity s_1 and each expansion x_t up through period t.

$$s_{t+1} = s_1 + \sum_{\tau=1}^{t} x_\tau \qquad \text{for } t = 1, 2, \ldots, T \tag{2.52}$$

This may be simply expressed by a series of continuity relationships,

$$s_{t+1} = s_t + x_t \qquad \text{for } t = 1, 2, \ldots, T \tag{2.53}$$

In this problem, the constraints must also ensure that the actual capacity s_{t+1} at the end of each future period t is no less than the capacity required D_t at the end of that period.

$$s_{t+1} \geq D_t \qquad \text{for } t = 1, 2, 3, \ldots, T \tag{2.54}$$

There may also be constraints on the possible expansions in each period defined by a set Ω_t of feasible capacity additions in each period t:

$$x_t \in \Omega_t \tag{2.55}$$

Figure 2.4a illustrates this type of capacity expansion problem. The cost functions $C_t(s_t, x_t)$ typically exhibit fixed costs and economies of scale, as illustrated in Figure 2.4b.

The constrained optimization model defined by equations 2.51 and 2.53 to 2.55 can be restructured as a multistage decision-making process and solved using either a forward- or a backward-moving dynamic programming solution procedure. The stages of the model will be the periods in which

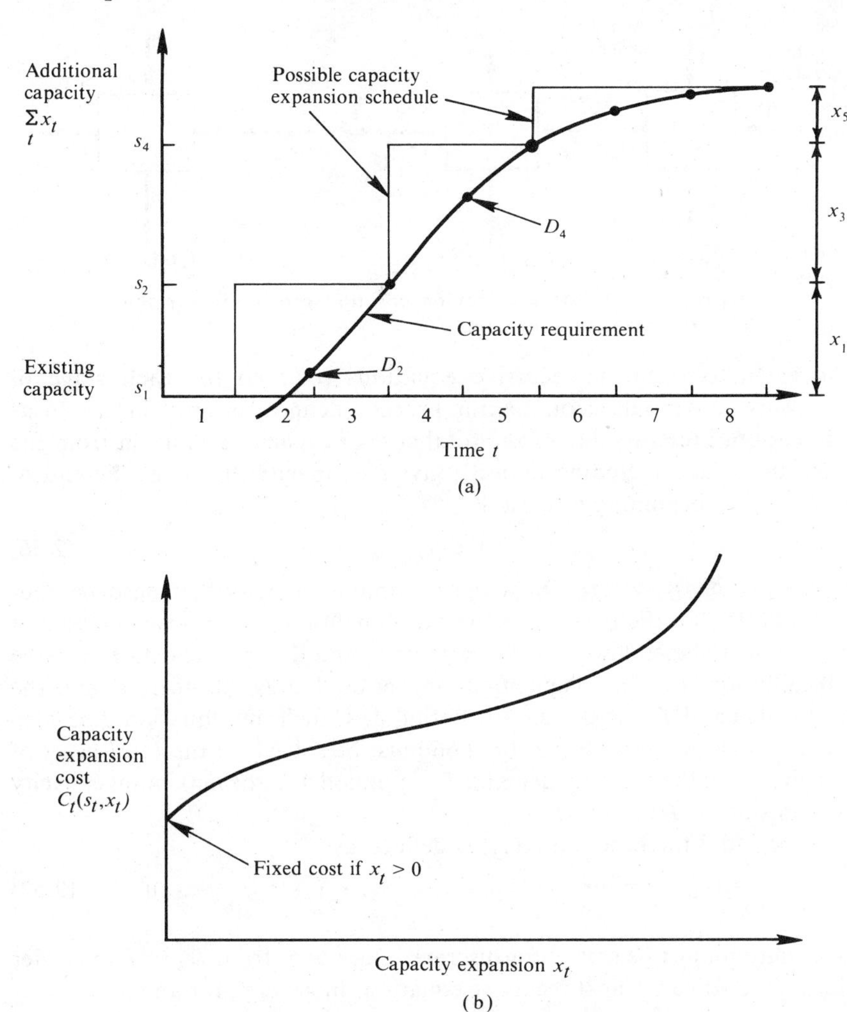

FIGURE 2.4. (a) Example capacity expansion program. (b) Example capacity expansion cost function assuming an existing capacity s_t in period t.

capacity expansion decisions are made. The states will be either the capacity s_{t+1} at the end of a stage or period t if a forward-moving solution procedure is adopted, or the capacity s_t at the beginning of a stage or period t if a backward-moving solution procedure is used. Another dynamic programming approach to the capacity expansion problem is given in Chapter 5 for the case when one has to select capacity increments from among a set of fixed size alternatives. A forward-moving solution procedure, illustrated by Figure 2.5, is used for the problem considered here.

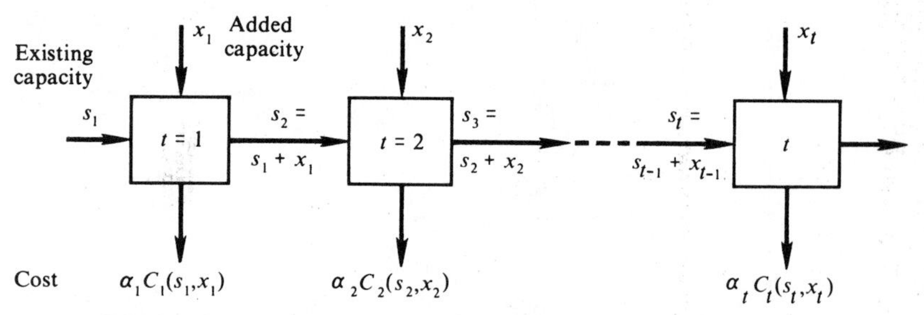

FIGURE 2.5. A forward-moving solution procedure for the capacity expansion program.

To begin to write the recursive equations required for each stage of the forward-moving decision-making process, define the function $f_t(s_{t+1})$ as the minimum present value of the total cost of capacity expansion from the present up to and including period t given a capacity of s_{t+1} at the end of period t. Thus beginning in period 1,

$$f_1(s_2) = \alpha_1 C_1(s_1, s_2 - s_1) \tag{2.56}$$

for $D_1 \leq s_2 \leq D_T$, where D_T is the maximum capacity that need be considered. Note that the term x_1 in the cost function $C_1(s_1, x_1)$ is expressed in terms of the assumed known state variable s_1 and the specified state variable s_2. In addition, the final capacity s_2 in period 1 must be no less than the required demand D_1 at the end of period 1. Constraint equation 2.55 also applies to $x_1 = s_2 - s_1$. Equation 2.56 must be solved for discrete values of s_2 ranging from the capacity demand D_1 in period 1 to the maximum capacity to be considered $D^{\max}$.

For period 2 the function $f_2(s_3)$ is defined as

$$f_2(s_3) = \underset{\substack{x_2 \\ x_2 \in \Omega_2}}{\text{minimum}} [\alpha_2 C_2(s_3 - x_2, x_2) + f_1(s_3 - x_2)] \tag{2.57}$$

This equation must be solved for discrete values of s_3 from D_2 to D_T in order to be able to solve the next recursive equation. In general, for any period $t > 1$, the recursive equation is

$$f_t(s_{t+1}) = \underset{\substack{x_t \\ x_t \in \Omega_t}}{\text{minimum}} [\alpha_t C_t(s_{t+1} - x_t, x_t) + f_{t-1}(s_{t+1} - x_t)] \tag{2.58}$$

This equation must be solved for all $D_t \leq s_{t+1} \leq D_T$. The recursive equation for the last construction period, $t = T$, needs to be solved for the value of s_{T+1} equal to D_T which minimizes the total cost

$$f_T(D_T) = \underset{\substack{x_T \\ x_T \in \Omega_T}}{\text{minimum}} [\alpha_T C_T(D_T - x_T, x_T) + f_{T-1}(D_T - x_T)] \qquad (2.59)$$

A backward-moving solution procedure can be formulated using a very similar notation. The functions $f'_t(s_t)$ will define the minimum present value of the total cost of capacity expansion in periods t through T given an initial capacity in period t of s_t. Note the difference in the definitions of $f_t(s_{t+1})$ used in the recursive equations just developed for a forward-moving solution procedure, and of $f'_t(s_t)$ just defined for a backward-moving solution procedure.

For period T, the last period in which capacity expansion is considered in order to meet a demand of D_T,

$$f'_T(s_T) = \underset{\substack{x_T \\ x_T \in \Omega_T}}{\text{minimum}} [\alpha_T C_T(s_T, x_T)] \qquad (2.60)$$

This equation must be solved for all discrete values of s_T from D_{T-1} to D_T. For all periods t between 2 and $T - 1$,

$$f'_t(s_t) = \underset{\substack{x_t \\ x_t \in \Omega_t}}{\text{minimum}} [\alpha_t C_t(s_t, x_t) + f'_{t+1}(s_t + x_t)] \qquad (2.61)$$

which must be solved for all discrete values of s_t from D_{t-1} to D_T.

Finally, in period $t = 1$, the value

$$f'_1(s_1) = \underset{\substack{x_1 \\ x_1 \in \Omega_1}}{\text{minimum}} [\alpha_1 C_1(s_1, x_1) + f'_2(s_1 + x_1)] \qquad (2.62)$$

for the actual value of s_1 will equal $f_T(D_T)$ found by equation 2.59, namely the minimum present value of the total cost of the capacity expansion program.

At this point some comments are appropriate on how the solution from such models should be used. Clearly, as time advances, better data regarding future costs and capacity demands (and perhaps better methods of analysis, too) will become available. Hence the solution from such capacity expansion models is not intended to be a guide for capacity expansion up to the end of some future period T, but only a guide as to what decision to make *this current period* ($t = 1$). The final period T is selected far enough into the future so as not to influence the value of x_1, the expansion in the current period. At the end of the current period, forecasts of future cost and demands can be updated, the planning horizon extended, and the model resolved to obtain a better estimate of x_2, the expansion desired in period 2. Thus the models are used in a sequential process, as well as being solved by a sequential or recursive process.

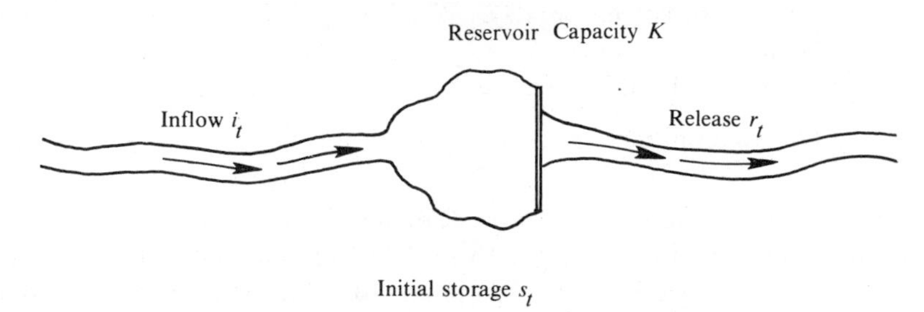

FIGURE 2.6. Reservoir operation variables for each period *t*.

Reservoir Operation. Figure 2.6 illustrates a single reservoir having inflows i_t and making releases r_t in each period t. In deterministic problems such as this one, the sequence of inflows i_t is assumed known, and the sequence of releases r_t is to be determined. Given a known reservoir storage capacity of K, the reservoir operating problem involves finding the sequence of releases r_t that maximizes total net benefits. These net benefits may be a function of the storage volume as well as of the release.

Let s_t be the initial storage volume in period t. Assume that the net benefits in each period t can be defined as functions of the initial and final storage volumes (s_t and s_{t+1}) and release (r_t), and can be denoted by $\text{NB}_t(s_t, s_{t+1}, r_t)$. Also assume that these net benefit functions for each period t will be the same from one year to the next, at least for the foreseeable future. Benefits associated with storage might stem from lake recreation, hydropower, and the protection of various species of wildlife and their habitats. Release benefits could result from navigation, water supplies, and wastewater dilution.

Let i_t represent the inflow in a period t. Assume there are T periods in a year. A management objective might be to maximize the total annual benefits.

$$\text{maximize} \sum_{t=1}^{T} \text{NB}_t(s_t, s_{t+1}, r_t) \tag{2.63}$$

The constraints include a mass balance of inflows and outflows or releases in each period t. There are many ways to express this mass balance. Assuming no significant evaporation or seepage losses, one approach is to equate the final storage volume s_{t+1} in period t (which is the same as the initial storage volume in period $t + 1$) to the initial storage volume s_t plus inflow i_t minus release r_t.

$$s_{t+1} = s_t + i_t - r_t \qquad \text{for each period } t \tag{2.64}$$

Note that if period t is T, then $T + 1$ is equal to 1, the initial period in the year. The constraints must also include the capacity restriction K on each storage volume s_t:

$$s_t \leq K \qquad \text{for each period } t \tag{2.65}$$

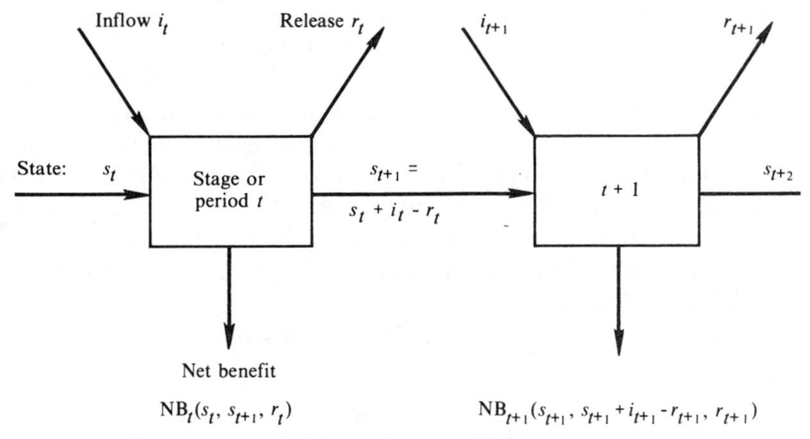

FIGURE 2.7. Sequential reservoir operation process.

In this example, evaporation and seepage, flood control storage, and dead storage are ignored. These aspects of reservoir design and operation will be discussed in more detail in Chapters 5 and 7.

The reservoir operating problem defined by equations 2.63 to 2.65, can be viewed as a multistage decision-making process as illustrated in Figure 2.7. The stages are the time periods, and the states are the storage volumes. Again, either a forward- or backward-moving sequence of recursive equations can be formulated, one for each stage of the process. Proceeding backward, a particular period is selected after which it is assumed the reservoir will no longer be operated. This can be any period in any year, because the eventual "steady-state" optimal policy $\{r_t\}$ derived from the model will be independent of this arbitrary assumption provided that both the average inflows i_t and net benefit functions $NB_t(\cdot)$ in each period t do not change from one year to the next.

Let the arbitrary terminal period be period T. Only one period of operation remains, which is the period on the far right of the time shown in Figure 2.8. Next define a function $f'_T(s_T)$ that is the maximum net benefit derived from operating the reservoir in the last period of that last year, given an initial storage volume of s_T,

$$f_T^1(s_T) = \underset{\substack{r_T \geq 0 \\ r_T \leq s_T + i_T \\ r_T \geq s_T + i_T - K}}{\text{maximum}} [NB_T(s_T, s_T + i_T - r_T, r_T)] \tag{2.66}$$

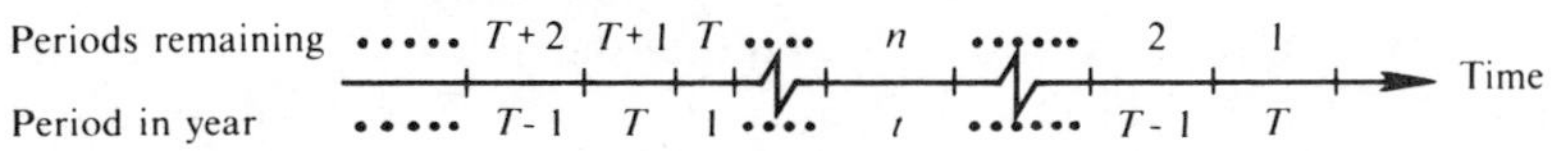

Periods remaining ••••• $T+2$ $T+1$ T •••• n •••••• 2 1

Time

Period in year ••••• $T-1$ T 1 •••• t •••••• $T-1$ T

FIGURE 2.8. Relationship between periods t and n at each stage of reservoir operating problem.

The constraints on the release r_T limit it to the water available, and force a spill if the available water exceeds the reservoir capacity K. Equation 2.66 must be solved for discrete values of s_T from 0 (or some minimum allowable storage volume in that period) to the maximum possible storage volume K. These values of $f_T^1(s_T)$ will be needed to solve the next recursive equation.

Moving backward in time (from right to left in Figure 2.8), the next stage is the previous period, $T-1$. There are now two periods remaining for reservoir operation. In this case the function $f_{T-1}^2(s_{T-1})$ represents the maximum total net benefit with two periods to go, given an initial storage of s_{T-1} in period $T-1$. Since $s_T = s_{T-1} + i_{T-1} - r_{T-1}$, $f_T^1(s_T)$ can be expressed in terms of the state variable s_{T-1}, the decision variable r_{T-1}, and the known average inflow i_{T-1}.

$$f_{T-1}^2(s_{T-1}) = \underset{\substack{r_{T-1} \geq 0 \\ r_{T-1} \leq s_{T-1}+i_{T-1} \\ r_{T-1} \geq s_{T-1}+i_{T-1}-K}}{\text{maximum}} [\text{NB}_{T-1}(s_{T-1}, s_{T-1} + i_{T-1} - r_{T-1}, r_{T-1}) + f_T^1(s_{T-1} + i_{T-1} - r_{T-1})] \tag{2.67}$$

Again, this must be solved for all discrete values of s_{T-1} between 0 and K.

Continuing backward in time, the general recursive equation for each period t with n ($n > 1$) periods remaining can be written

$$f_t^n(s_t) = \underset{\substack{r_t \geq 0 \\ r_t \leq s_t+i_t \\ r_t \geq s_t+i_t-K}}{\text{maximum}} [\text{NB}_t(s_t, s_t + i_t - r_t, r_t) + f_{t+1}^{n-1}(s_t + i_t - r_t)] \tag{2.68}$$

where the index n proceeds from 1 and increases at each successive stage and the index t cycles backward from period T to 1 and then to period T again. The relationship between periods t and the index n can be seen in Figure 2.8.

Now the question remains, how many recursive equations 2.68 must be solved to obtain the optimal release policy r_t for each period t associated with each discrete value of the initial storage volume s_t? Usually after proceeding through only three to four years, the optimal release r_t associated with each initial storage volume s_t will be the same as the corresponding r_t and s_t in the previous year. This is called a *stationary solution*. The maximum annual net benefit resulting from this policy will equal $f_t^{n+T}(s_t) - f_t^n(s_t)$ for any value of s_t and t. One can recognize that indeed the stationary policy has been identified when the values $f_t^{n+T}(s_t) - f_t^n(s_t)$ are independent of s_t and t.

A simple example may help to illustrate this reservoir operating model and its solution. Consider a reservoir site at which it is desirable to maintain a constant storage volume of 20 and a constant release of 25. Assume that the capacity K of the reservoir is 30 and that the inflows i_t are 10, 50, and 20 in three distinct seasons ($t = 1, 2, 3$), respectively. Desired is an operating policy that minimizes the annual sum of squared deviations from these desired storage volume and release values or targets. Hence $\text{NB}_t(s_t, s_{t+1}, r_t)$ in equation 2.63 will equal $[(20 - s_t)^2 + (25 - r_t)^2]$.

Letting s_t take on the discrete values 0, 10, 20, and 30, and r_t the discrete values of 10, 20, 30, and 40, Table 2.3 summarizes the solution of equations 2.66 to 2.68, using expanded tables similar to those shown in Tables 2.2.

In this example the optimal reservoir releases r_t^* derived after the first stage (period 3 with only $n = 1$ period remaining) defined the stationary sequential release policy. Not only are the releases r_t^* the same in each succeeding year but also the difference $f_t^{n+3}(s_t) - f_t^n(s_t)$ is a constant 275 for all s_t and t for $n > 1$. This value is the minimum annual sum of squared deviations that can be obtained by following the derived sequential operating policy (indicated in the lower right corner of Table 2.3).

By simulating the reservoir's operation with this policy and the deterministic sequence of inflows, the corresponding optimal sequence of storage volumes and releases can be determined. With the stationary operating policy

TABLE 2.3. Solution of Reservoir Operation Example Problem (Equations 2.66 to 2.68), Indicating Minimum Sum of Squared Deviations, $f_t^n(s_t)$, and Optimal Reservoir Releases, r_t^*, in Each Period t and Stage n

Initial Storage, s_t	*n = 1*		*n = 2*		*n = 3*	
	$f_3^1(s_3)$	r_3^*	$f_2^2(s_2)$	r_2^*	$f_1^3(s_1)$	r_1^*
0	425	20	450	30	1075	10
10	125	20, 30	250	30	575	10, 20
20	25	20, 30	350	40	275	20
30	125	20, 30	—	[a]	375	30

s_t	*n = 4*		*n = 5*		*n = 6*	
	$f_3^4(s_3)$	r_3^*	$f_2^5(s_2)$	r_2^*	$f_1^6(s_1)$	r_1^*
0	1200	10	725	30	1350	10
10	600	10	525	30	850	10, 20
20	300	20	625	40	550	20
30	400	30	—	[a]	650	30

s_t	*n = 7*		STATIONARY POLICY s_t	OPTIMAL RELEASES r_1^*	r_2^*	r_3^*
	$f_3^7(s_3)$	r_3^*				
0	1475	10	0	10	30	10
10	875	10	10	10, 20	30	10
20	575	20	20	20	40	20
30	675	30	30	30	—	30

[a] Requires $r_t \geq 50$, hence infeasible.

and the deterministic and cyclic inflows, the same sequence of storage volumes and releases will occur year after year, once any of the stationary storage volumes are observed beginning from any assumed initial storage volume. Readers can verify that for the example problem and its sequential operating policy defined in Table 2.3, the stationary storage volumes are 20, 10, and 30 for periods 1, 2, and 3, respectively. The other possible storage volumes and their associated releases are transient and will not be observed once the reservoir operation achieves a stationary condition.

Before ending this discussion of dynamic programming methods applied to water resources planning problems, it is appropriate to examine a major assumption that has been made in each of the problems presented. The first is that the net benefits or costs resulting from each decision at each stage of the problem are dependent only on the state variables and are otherwise independent of decisions made at other stages. If the returns at any stage are dependent on the decisions made at other stages in a way not captured by the state variables, then dynamic programming is not an appropriate solution technique, except perhaps as a rough approximation. For example, dynamic programming is not suited for determining the optimal capacity of a reservoir or the optimal target release or storage volumes along with its operating policy because capacity and target decisions affect the constraints on system operation and the net benefit function in not just one, but every time period or stage. For such planning problems other methods, such as linear optimization models, are more successful. A special method for solving linear optimization models, namely linear programming, is the topic of the next section.

2.8 LINEAR PROGRAMMING

The water planning models discussed in Section 2.7 have a nonlinear objective function and linear constraints. If the objective function, as well as the constraints, are linear, then very efficient procedures for solving these constrained linear optimization problems, called *linear programming*, may be used. This section is devoted to illustrating some linear water planning problems and reviewing, rather briefly, these linear programming techniques.

Unlike most other optimization techniques, linear programming packages are available at most of the major scientific computer facilities throughout the world. Hence applied systems analysts need only a knowledge of how to use the computer programs that are available and an understanding of the meaning of the output of such programs. To use linear programming it is not necessary to understand all the details of the linear programming solution procedures. This is a distinct advantage over most other types of optimization. The fact that linear programming solution procedures are readily available has created the incentive to structure many nonlinear as well as linear water planning problems as linear optimization models. Various methods for converting nonlinear relationships to linear ones will be discussed following the development of some simple linear water planning models. Appendix 3A provides a discussion of the use of a widely used and commercially available linear programming package.

2.8.1 Reservoir Storage-Yield Models

First, consider the development of storage-yield functions for a single reservoir. A reservoir storage-yield function defines the maximum fixed or constant reservoir release or yield that will be available, at a given level of reliability, during each period of operation. Given a series of known inflows, the yield from any reservoir will depend on its active storage capacity. Figure 2.9 illustrates two typical storage-yield functions for a single reservoir.

There are many methods available for deriving storage-yield functions. A few of these are discussed in Chapter 5. One very versatile method uses linear optimization. It involves structuring and solving a model having a linear objective function and linear constraints.

For this deterministic example, consider a single reservoir that must provide a uniform release or yield r in each period t for which a record of known (historical or synthetic) streamflows is available. The problem is to find the maximum uniform yield r obtainable with a given capacity. The objective is

$$\text{maximize } r \tag{2.69}$$

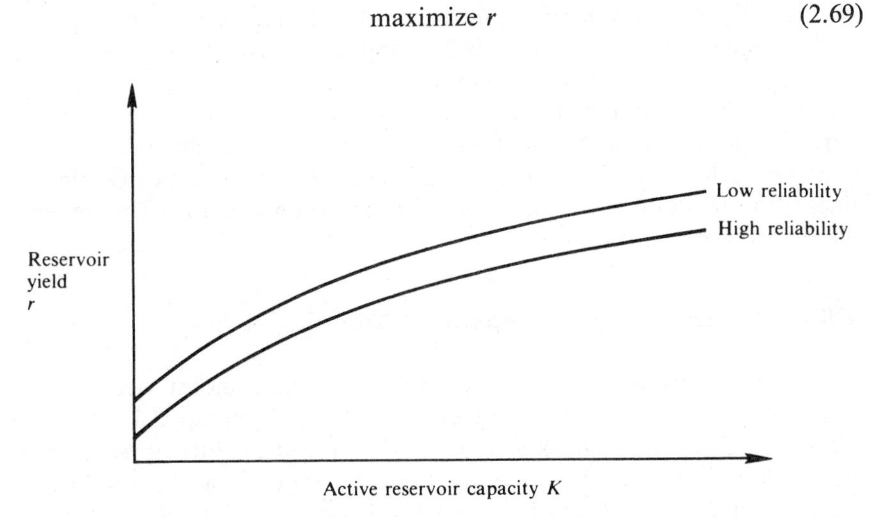

FIGURE 2.9. Reservoir storage-yield functions.

This maximum yield is constrained by the water available in each period and by the reservoir capacity.

As has been discussed previously, two sets of constraints are needed to define the relationships among the inflows, the reservoir storage volumes, the yields, any excess release, and the reservoir capacity. The first is the set of continuity equations that equate the unknown final reservoir storage volume s_{t+1} in period t to the unknown initial reservoir storage volume s_t plus the known inflow i_t, minus the yield r and any excess release r_t in period t.

$$s_{t+1} = s_t + i_t - r - r_t \qquad t = 1, 2, \ldots, T \qquad T + 1 = 1 \tag{2.70}$$

If, as indicated in equation 2.70, one assumes that period 1 follows period T, it is not necessary to specify the value of the initial storage volume s_1 and/or final storage volume s_{T+1}. The resulting "steady-state" solution is essentially based on the assumption that the entire inflow sequence will repeat itself again and again. Obviously, this is very unlikely, but for this deterministic example it is as good as the alternative assumptions that could be made. This assumption should have only a minor effect on the model's behavior if the streamflow sequence is relatively long.

The second set of required constraints ensures that the reservoir storage volumes s_t in each period t are no greater than the active reservoir capacity K.

$$s_t \leq K \qquad t = 1, 2, 3, \ldots, T \tag{2.71}$$

To derive a storage-yield function, the model defined by equations 2.69 to 2.71 must be solved for various values of capacity K. Clearly, an upper bound on the yield in this example would equal the mean flow $\sum_t^T i_t/T$.

Note the similarity of equations 2.70 and 2.71 to equations 2.64 and 2.65. The only difference between equations 2.64 and 2.70 is the inclusion of a uniform release r in equation 2.70. The addition of that variable r in each period t precludes the efficient use of dynamic programming for solving equations 2.69 to 2.71. However, since the objective and all constraints are linear, the model can be solved using linear programming. This model is discussed further in Chapter 5.

2.8.2 Water Quality Management Models

Linear programming solution techniques will be demonstrated using the simple water quality management example shown in Figure 2.10. A stream receives waste from sources located at sites 1 and 2. Without some waste treatment at these sites, the water quality indicator (such as dissolved oxygen concentration), q_i mg/ℓ, at sites 2 and 3 will continue to be below the desired concentration Q_i. The problem is to find the level of wastewater treatment

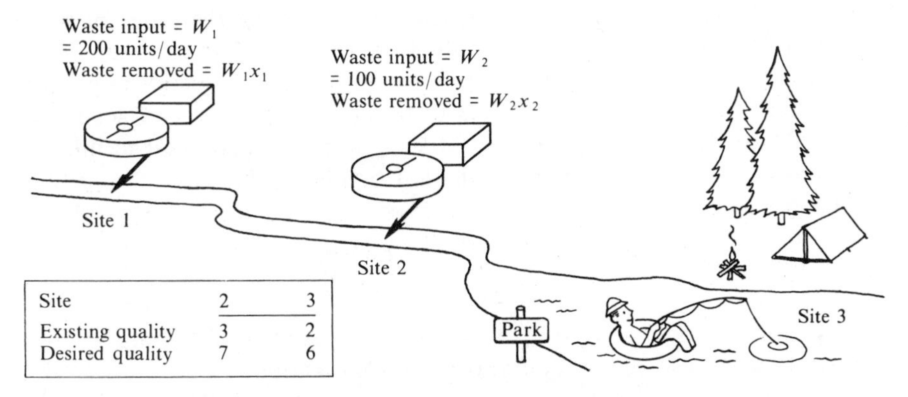

FIGURE 2.10. Water quality management problem.

(waste removed) at sites 1 and 2 required to achieve the desired concentrations at sites 2 and 3 at a minimum total cost. This may be a naive objective for an actual water quality management problem. Still it serves to illustrate the use of another linear model and some linear programming procedures that can be used for its solution.

Assume that for each unit of waste removed (not discharged into the stream) at site 1 the quality index at site 2 improves by 0.025 mg/ℓ, and the quality index at site 3 improves by 0.0125 mg/ℓ. For each unit of waste removed at site 2 the quality index at site 3 improves by 0.025 mg/ℓ. In this example the water quality at site 2 is measured just upstream of the point of waste water discharge; hence waste discharged from site 2 affects the quality only at site 3. Denote these transfer coefficients as a_{ij} (the improvement in the quality index at site j per unit of waste removed at site i), W_i as the amount of waste to be treated at site i, and x_i as the fraction of waste removed by treatment at site i. Then the quality improvement at site j due to W_ix_i units of waste removed at site i is $(a_{ij})(W_ix_i)$ mg/ℓ. For all values of x_1 between 0 and 1, the quality index at site 2 will equal the current concentration q_2 plus the improvement, $(a_{12})(W_1x_1)$, due to treatment at site 1. The quality index at site 3 will equal the current concentration q_3 plus the improvement $(a_{13})(W_1x_1) + (a_{23})(W_2x_2)$ due to treatment at sites 1 and 2.

The cost of treatment $C_i(x_i)$ at each site i will be a function of the fraction of waste removed x_i. The objective in this example problem is to find the values of the removal fractions, x_1 and x_2, that minimize the total cost,

$$\text{minimize } C_1(x_1) + C_2(x_2) \tag{2.72}$$

while meeting the desired quality standards Q_j at sites $j = 2$ and 3:

$$q_2 + a_{12}W_1x_1 \geq Q_2 \tag{2.73}$$

$$q_3 + a_{13}W_1x_1 + a_{23}W_2x_2 \geq Q_3 \tag{2.74}$$

To complete the planning model, constraints confining the range of waste removal fractions to their feasible values are required. For this example, at least 30% removal will be required at both sites. This corresponds to primary treatment of municipal waste waters to prevent the discharge of suspended and floating solids. An upper limit of 95% will reflect the best technology available without resorting to distillation or piping the wastes to another location.

$$x_i \geq 0.30 \qquad i = 1, 2 \tag{2.75}$$

$$x_i \leq 0.95 \qquad i = 1, 2 \tag{2.76}$$

The planning model defined by equations 2.73 through 2.76 contains linear constraints and possibly a nonlinear objective function.

In many actual planning situations each cost function $C_i(x_i)$ will be unknown. At best only an approximate estimate may be available without additional design and cost studies. In this example each cost function will be assumed to be unknown. The challenge will be to try to solve the planning problem, that is, find the least-cost combination of x_1 and x_2 without requiring an expensive design and cost study to obtain improved treatment cost data.

The solution of this problem without complete knowledge of the cost functions in the objective will illustrate how modeling studies can assist in planning data collection programs. Very often planning exercises are divided into to distinct phases, data collection followed by data analysis. The two phases should be integrated from the beginning. Model builders must be aware of the data that are available, or that can be obtained at a reasonable cost. Data collection programs should be geared to the need for various data and the accuracy required. There is no advantage to spending money or time collecting data or improving the accuracy of such efforts if such data have little influence on the final solution or decision. Models can be used to assess the sensitivity of possible solutions to changes in the values of various parameters or assumptions. This use of models for such *sensitivity analyses* is often of more value to planners and decision makers than is their use for identifying possible solutions.

Returning to the water quality management problem, substitution of the values for each of the known coefficients (q_i, a_{ij}, W_i, and Q_i) in equations 2.73 to 2.76 results in the following sets of constraints:

1. The desired water quality at site 2:

$$q_2 + a_{12}W_1x_1 \geq Q_2$$

which yields

$$3 + (0.025)(200)x_1 \geq 7$$

or

$$x_1 \geq 0.8 \tag{2.77}$$

2. The desired water quality at site 3:

$$q_3 + a_{13}W_1x_1 + a_{23}W_2x_2 \geq Q_3$$

which yields

$$2 + (0.0125)(200)x_1 + (0.025)(100)x_2 \geq 6$$

or

$$x_1 + x_2 \geq 1.6 \qquad (2.78)$$

3. Limits on the fraction of waste removed:

$$x_1 \geq 0.3$$
$$x_2 \geq 0.3 \qquad (2.79)$$

and

$$x_1 \leq 0.95$$
$$x_2 \leq 0.95 \qquad (2.80)$$

2.8.3 Graphical Solution and Analysis

Since this is a problem involving only two unknown variables, all combinations of x_1 and x_2 that satisfy the constraints 2.77 through 2.80 can be determined graphically as shown in Figure 2.11. Once these *feasible* solutions

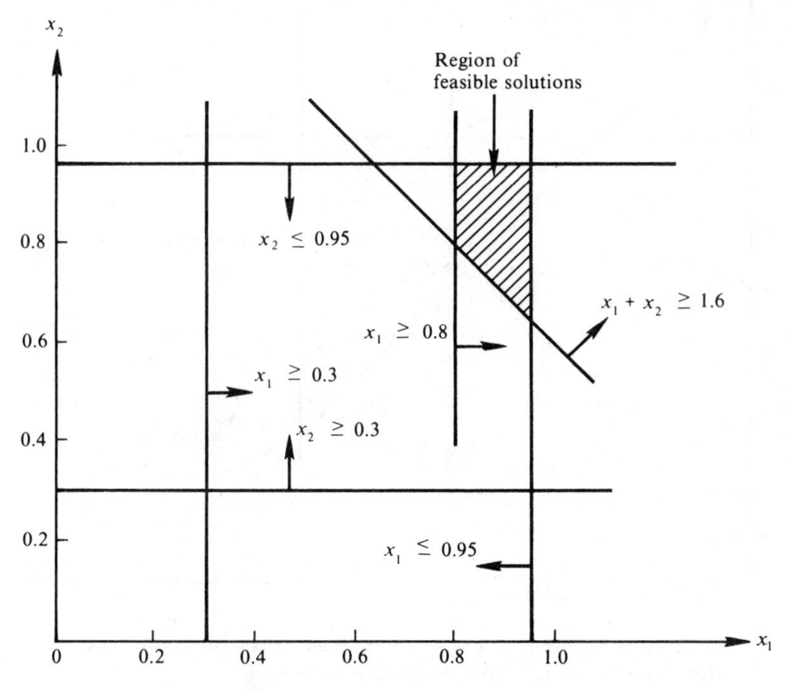

FIGURE 2.11. Plot of water quality planning model constraints.

have been identified, it will be necessary to identify those that might be optimal (i.e., that might result in a minimum total cost).

The shaded area in Figure 2.11 represents combinations of x_1 and x_2 that satisfy each constraint. Constraints that do not border this *feasible region*, such as equations 2.79, are called *redundant constraints*. For example, if x_1 is to be no less than 0.8 (equation 2.77), this implies that x_1 is certainly greater than 0.3 (equation 2.79).

To find the minimum-cost solution within the feasible region identified in Figure 2.11, assume that the objective function 2.72 can be expressed as a linear function,

$$\text{minimize } C = c_1x_1 + c_2x_2 \tag{2.81}$$

in which each c_i represents the cost per unit fraction of waste removal at site i. Assigning reasonable values to the total cost C and to each marginal cost c_i in equation 2.81 permits one to display in Figure 2.11 the set of feasible combinations of x_1 and x_2 that result in the specified total cost. Two such plots are shown in Figure 2.12 assuming a value for the total cost, C, of 2.7.

If the arbitrarily assigned value of the total cost C used in Figure 2.12 is reduced, the corresponding set of feasible values of x_1 and x_2 will shift toward the origin of the graph (i.e., to the left). Hence to find the minimum

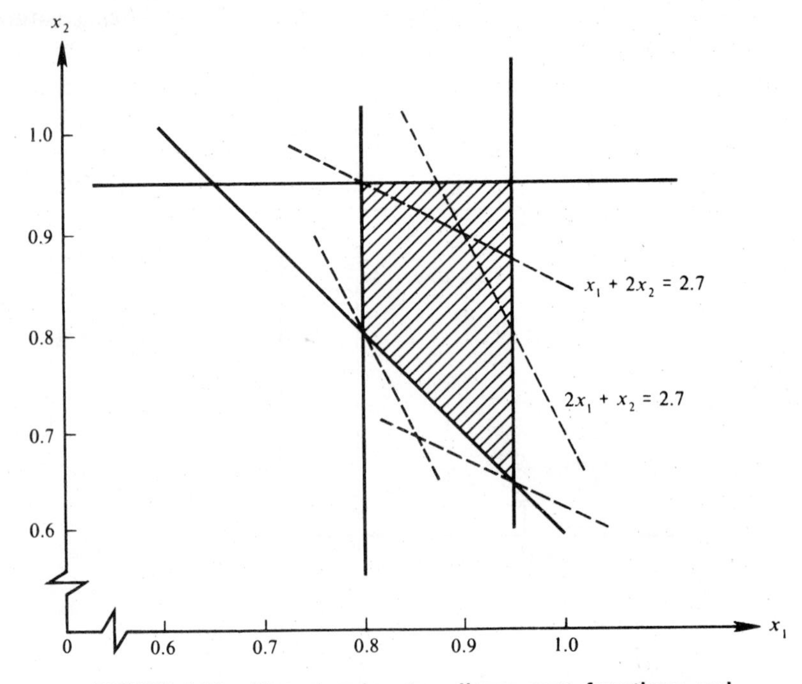

FIGURE 2.12. Plot showing two linear cost functions and their parallel shift to minimum-cost solutions.

total cost, the line $C = c_1x_1 + c_2x_2$ should be moved as close to the origin as is possible while still intersecting the feasible region. When this is accomplished, the point at which the line intersects the boundary of the feasible region will represent the values of x_1 and x_2 that achieve the minimum possible total cost. This parallel shift is indicated by the smaller dashed lines in Figure 2.12 for both cost functions.

It is also possible to examine the effect of changes in the values of the marginal costs c_i. Note that for all values of c_1 greater than c_2, the least-cost solution will be $x_1 = x_2 = 0.8$. If, on the other hand, c_1 is less than c_2, the least-cost solution will be $x_1 = 0.95$ and $x_2 = 0.65$. Finally, in the unlikely event that $c_1 = c_2$, there are many solutions which achieve the minimum cost.

Of course, the actual minimum cost will depend on the actual values of c_1 and c_2, but the planner responsible for studying the problem without any additional information has at least an estimate of the treatment efficiencies that will result in least-cost solutions. If after an examination of each treatment site it is obvious that the marginal cost of treating the waste at site 1 will always be greater than that for treating the waste at site 2, simply because of the larger required hydraulic capacity, then it is unnecessary to spend more money to find the exact cost functions at the two sites.

There are other interesting questions that could arise. These include project sequencing or staging issues, that is, determining which treatment facility to build first given a budget constraint that prevents constructing both treatment facilities simultaneously. One can question the assumptions made with regard to future waste loads W_i and the transfer coefficients a_{ij}. Some of this sensitivity information can be derived from the dual variables or shadow prices associated with each constraint.

2.8.4 Dual Variables—Shadow Prices

The dual variables for this problem specify the rate of change in the total cost per unit change in the quantity $(Q_j - q_j)/a_{ij}W_i$ at sites $j = 2$ or 3 and with a unit change in the lower or upper limits of the fractions of waste removed at sites 1 and 2. Obviously, the value of each nonzero dual variable will depend on the values of the marginal costs c_1 and c_2. Assume that $c_1 = 10$ and $c_2 = 6$. Recall that the right-hand side of equation 2.77 is $(Q_2 - q_2)/a_{12}W_1$. Denote this expression as b_2. If b_2 is increased from 0.8 to 0.9, the optimal solution, as can be determined from Figure 2.12, changes to $x_1 = 0.9$ and $x_2 = 0.7$; the minimum total cost C increases from 12.8 to 13.2. Hence $\Delta C/\Delta b_2 = 0.4/0.1 = 4.0$. Similarly, if the right-hand side of equation 2.78, $(Q_3 - q_3)/a_{13}W_1 = (Q_3 - q_3)/a_{23}W_2$, denoted as b_3, is decreased from 1.6 to 1.5, a redrawing of Figure 2.12 will show that the optimal solution becomes $x_1 = 0.8$ and $x_2 = 0.7$; the minimum total cost C

decreases from 12.8 to 12.2. Hence $\Delta C/\Delta b_3 = 0.6/0.1 = 6.0$. These quantities, $\Delta C/\Delta b_2 = 4.0$ and $\Delta C/\Delta b_3 = 6.0$, are the dual variables of those two water quality constraints.

Perhaps of more interest to those responsible for water quality planning is the change in the total cost associated with a change in the quality deficit $Q_j - q_j$, or quality standard Q_j at sites $j = 2$ or 3. These values can be obtained graphically, if the coefficients $a_{ij}W_i$ of the variables x_i in the quality constraints 2.77 and 2.78 are kept on the left-hand side of the equations, or directly from the dual variables $\Delta C/\Delta b_j$. Since $\Delta C/\Delta b_2 = \Delta C/\Delta[Q_2 - q_2)/a_{12}W_1] = 4.0$, it follows that $\Delta C/\Delta(Q_2 - q_2) = 4.0/a_{12}W_1$, which equals 0.8. Hence, if Q_2 is increased (or decreased) by one unit and $a_{12}W_1$ remains constant, the minimum cost will increase (or decrease) by 0.8. Similarly, if Q_3 is increased (or decreased) by a unit, the minimum total cost will increase (or decrease) by $\Delta C/\Delta(Q_3 - q_3) = 6.0/a_{13}W_1 = 6.0/a_{23}W_2 = 2.4$. The point to be made here is that the dual variable associated with each constraint indicates the change in the objective function per unit change in the entire right-hand side of that constraint, not necessarily per unit change in some component of that right-hand side.

From Figures 2.11 and 2.12 it is clear that equations 2.79 are redundant, and that equations 2.80 also do not affect the least-cost solution. Hence the dual variables of those sets of constraints are zero. They represent the cost savings associated with a reduction in the lower or upper bounds of the waste removal efficiencies.

2.8.5 Dual Models

The dual variables associated with each constraint can also be found by structuring and solving what is termed the *dual model* of the original *primal model.*

If the original linear programming model is of the form

$$\text{minimize} \sum_{i=1}^{m} c_i x_i \tag{2.82}$$

subject to

$$\sum_{i=1}^{m} a_{ij} x_i \geq b_j \qquad j = 1, 2, \ldots, n \tag{2.83}$$

$$x_i \geq 0 \qquad i = 1, 2, \ldots, m \tag{2.84}$$

then its dual is

$$\text{maximize} \sum_{j=1}^{n} b_j y_j \tag{2.85}$$

subject to

$$\sum_{j=1}^{n} a_{ij} y_j \leq c_i \qquad i = 1, 2, \ldots, m \tag{2.86}$$

$$y_j \geq 0 \qquad j = 1, 2, \ldots, n \tag{2.87}$$

The dual model contains a variable y_j for each constraint of the original primal model. The right-hand sides b_j of the primal model are the coefficients of the objective function in the dual model. The objective coefficients c_i of the primal are the right-hand sides of the dual model. The direction of the inequalities are reversed and minimization is changed to maximization, and vice versa.

Equality constraints in the primal can be converted to two inequality constraints, one a less-than-or-equal-to constraint, and the other a greater-than-or-equal-to constraint, which forces an equality. The direction of an inequality can be reversed by multiplying every term by -1.

Consider the example water quality management problem in which $c_1 = 10$ and $c_2 = 6$. The nonredundant and active portion of the original primal model is

$$\text{minimize } 10x_1 + 6x_2 \tag{2.88}$$

subject to

$$x_1 \geq 0.8 \tag{2.89}$$

$$x_1 + x_2 \geq 1.6 \tag{2.90}$$

$$x_1 \leq 0.95 \tag{2.91}$$

$$x_2 \leq 0.95 \tag{2.92}$$

The dual of this model can be derived more easily if all the constraints are greater-than-or-equal-to inequalities. This can be done by multiplying inequalities 2.91 and 2.92 by -1. Assigning the variables y_1, y_2, y_3, and y_4 to constraint equations 2.89 through 2.92, respectively, the dual can be written as

$$\text{maximize } 0.8y_1 + 1.6y_2 - 0.95y_3 - 0.95y_4 \tag{2.93}$$

subject to

$$y_1 + y_2 - y_3 \leq 10 \tag{2.94}$$

$$y_2 - y_4 \leq 6 \tag{2.95}$$

From Figure 2.12 one can see that the last two constraints of the primal model, equations 2.91 and 2.92, do not affect the least-cost solution and hence could be omitted from the model. This implies that y_3 and y_4 equal zero, which reduces the dual model to another two-variable problem that can be solved graphically. The graphical solution is shown in Figure 2.13.

The maximum value of the objective function 2.93 is obtained at $y_1 = 4$ and $y_2 = 6$, as illustrated in Figure 2.13. These are the exact values of the dual variables associated with constraints 2.89 and 2.90. The objective value 12.8 is the same as that obtained from solving the primal problem defined by equations 2.88 through 2.92. Using graphical techniques, readers can verify that the values of the dual variables associated with constraints 2.94 and 2.95 of the dual model will equal the optimal values of the x_1 and x_2 variables of the primal problem. Hence the variables y_j of the dual model are the dual variables of the primal model, and the variables x_i of the primal model are

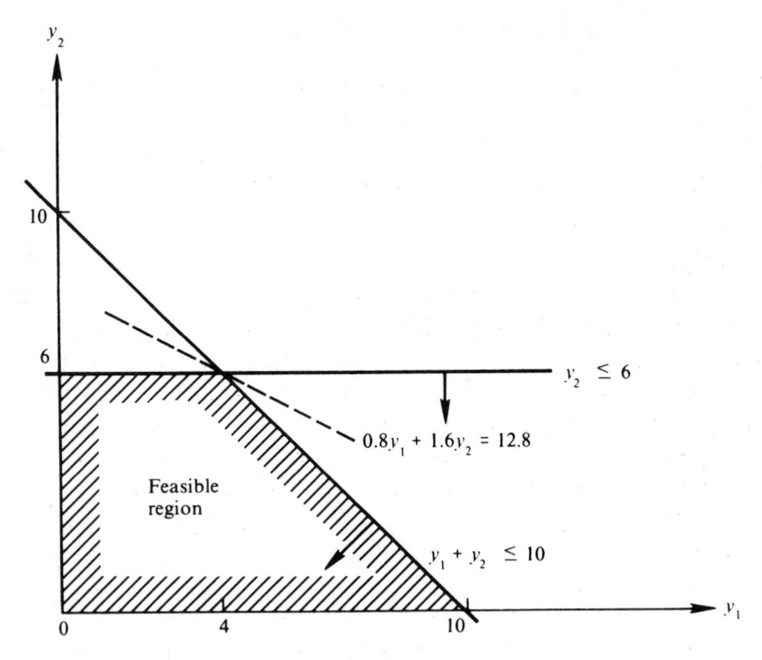

FIGURE 2.13. Graphical solution of dual model.

the dual variables of the dual model. The dual of the dual model is the original primal model.

This dual model has a physical interpretation which can be seen somewhat clearer if in the primal problem the terms $W_i x_i$ are combined into a single decision variable, namely the quantity of waste to be removed at each site i. Then the c_i terms in the objective function 2.81 represent the cost per unit of waste removal at site i. In its general form the primal model can be written as

$$\text{minimize} \sum_i \quad c_i \quad \cdot \quad (W_i x_i) \quad = \quad C$$

$$\begin{bmatrix}\text{cost per unit}\\ \text{waste removed}\\ \text{at site } i\end{bmatrix} \cdot \begin{bmatrix}\text{quantity of}\\ \text{waste removed}\\ \text{at site } i\end{bmatrix} = [\text{total cost}]$$

subject to

$$\sum_i \quad a_{ij} \quad \cdot \quad (W_i x_i) \quad \geq \quad Q_j - q_j$$

$$\begin{bmatrix}\text{quantity improvement}\\ \text{at site } j \text{ per unit}\\ \text{waste removed at site } i\end{bmatrix} \cdot \begin{bmatrix}\text{quantity of waste}\\ \text{removed at site } i\end{bmatrix} \geq \begin{bmatrix}\text{quality}\\ \text{deficit at}\\ \text{site } j\end{bmatrix}$$

$$W_i x_i \quad \leq \quad 0.95 W_i \quad \text{for all } i$$

$$\begin{bmatrix}\text{quantity of waste}\\ \text{removed at site } i\end{bmatrix} \leq \begin{bmatrix}\text{maximum waste}\\ \text{removal at site } i\end{bmatrix}$$

$$W_i x_i \geq 0.3W_i \quad \text{for all } i$$

$$\begin{bmatrix}\text{quantity of waste}\\ \text{removed at site } i\end{bmatrix} \geq \begin{bmatrix}\text{minimum waste}\\ \text{removal at site } i\end{bmatrix}$$

Let y_j be the dual variable associated with the quality constraints at site j and y_i^u and y_i^l be the dual variables associated with the upper and lower bounds, respectively, of the waste removal at site i. The general dual model can be written

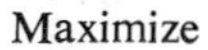

Maximize

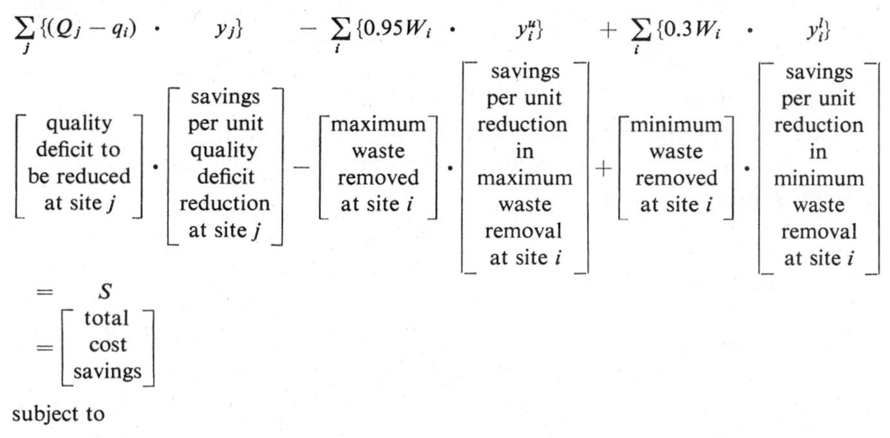

$$\sum_j \{(Q_j - q_i) \cdot y_j\} - \sum_i \{0.95W_i \cdot y_i^u\} + \sum_i \{0.3W_i \cdot y_i^l\}$$

$$\begin{bmatrix}\text{quality}\\ \text{deficit to}\\ \text{be reduced}\\ \text{at site } j\end{bmatrix} \cdot \begin{bmatrix}\text{savings}\\ \text{per unit}\\ \text{quality}\\ \text{deficit}\\ \text{reduction}\\ \text{at site } j\end{bmatrix} - \begin{bmatrix}\text{maximum}\\ \text{waste}\\ \text{removed}\\ \text{at site } i\end{bmatrix} \cdot \begin{bmatrix}\text{savings}\\ \text{per unit}\\ \text{reduction}\\ \text{in}\\ \text{maximum}\\ \text{waste}\\ \text{removal}\\ \text{at site } i\end{bmatrix} + \begin{bmatrix}\text{minimum}\\ \text{waste}\\ \text{removed}\\ \text{at site } i\end{bmatrix} \cdot \begin{bmatrix}\text{savings}\\ \text{per unit}\\ \text{reduction}\\ \text{in}\\ \text{minimum}\\ \text{waste}\\ \text{removal}\\ \text{at site } i\end{bmatrix}$$

$$= S$$

$$= \begin{bmatrix}\text{total}\\ \text{cost}\\ \text{savings}\end{bmatrix}$$

subject to

$$\sum_j \{a_{ij} \cdot y_j\} - y_i^u + y_i^l \leq c_i \quad \text{for all } j$$

$$\begin{bmatrix}\text{quality}\\ \text{deficit}\\ \text{reduction}\\ \text{at site } j\\ \text{per unit}\\ \text{waste}\\ \text{removal}\\ \text{at site } i\end{bmatrix} \cdot \begin{bmatrix}\text{savings}\\ \text{per unit}\\ \text{quality}\\ \text{deficit}\\ \text{reduction}\\ \text{at site } j\end{bmatrix} - \begin{bmatrix}\text{saving}\\ \text{per unit}\\ \text{reduction}\\ \text{in}\\ \text{maximum}\\ \text{waste}\\ \text{removal}\\ \text{at site } i\end{bmatrix} + \begin{bmatrix}\text{savings}\\ \text{per unit}\\ \text{reduction}\\ \text{in}\\ \text{minimum}\\ \text{waste}\\ \text{removal}\\ \text{at site } i\end{bmatrix} \leq \begin{bmatrix}\text{cost}\\ \text{per unit}\\ \text{waste}\\ \text{removal}\\ \text{at site } i\end{bmatrix}$$

$$y_j, y_i^u, y_i^l \geq \text{for all } j \text{ and } i$$

The dual constraints reflect the fact that the optimal total cost savings per unit of waste removed at each waste source site i (resulting from quality deficit reductions at each quality site j or changes in the maximum or minimum waste removals at site i) cannot exceed the total cost per unit of waste removal at each site i. Any change in the primal or original model that further restricts or constrains the problem will never result in an improvement in the value of the objective function. Unless the change is redundant and hence leaves the optimal solution unchanged, the objective function value will indicate increased costs or reduced benefits. This fact is reflected in the dual model, which shows that any cost saving y_i^u associated with a unit reduction in the maximum allowable waste removal at each source site will actually be negative. If any upper limit is binding (nonredundant) in the primal model (which for the example problem will occur for site 1 if $c_1 < c_2$), then any decrease in this limit will increase the minimum total cost C and hence the maximum cost savings S.

Models having more than two decision variables cannot be solved easily using graphical techniques. For n-variable problems ($n > 0$), algebraic methods can be used to define the region of feasible solutions and the optimal solution, if it exists. Using these algebraic solution techniques, the more constraints a problem has, the more difficult or time consuming is the solution procedure. The addition of more variables may not significantly increase computation time. Hence solving the dual of a heavily constrained primal problem may be more efficient than solving the primal itself. In either case the values of both the decision and dual variables will be available when the optimal solution to either problem has been found.

2.8.6 Algebraic Solution Procedures

Anyone who has studied linear programming methods will have been introduced to one or more versions of the *simplex method* used to solve constrained linear optimization problems. It is a very efficient search procedure for finding a feasible solution and then an optimal solution, if one exists. Since the initial development of the simplex method by Dantzig in 1947 [5], he and many others, including Charnes [4] and Dorfman [6], have made linear programming one of the most versatile and universally used constrained optimization methods for economic-engineering planning.

The simplex procedure is based on the realization that one or more corner points, defined by the simultaneous solution of the constraint equations that form the boundary of the feasible region, will be an optimal solution, if an optimal solution exists. The simplex procedure identifies a sequence of corner points, each one an improvement over the last, until no further improvement in the objective function's value is possible. It is at this corner point that the solution is optimal. If there is no feasible solution, or if the feasible region is unbounded such that one or more decision variables assume an infinite value, then no optimal solution exists.

In texts on simplex procedures, corner-point solutions are computed using a sequence of coefficient matrices called a *simplex tableaux*. These represent sets of simultaneous constraint equations that determine the corner points of the feasible region. Since most water resource systems analysts will use computer programs already available for solving linear optimization problems, there is no need to go into the detail of the simplex procedure. What the water resource systems engineer needs to know are the techniques available to model as linear functions some of the nonlinear relationships found in water planning problems.

2.8.7 Linearization Techniques

Linear programming problems require that all decision variables be nonnegative. Negative values for allocations, reservoir releases, treatment plant efficiencies, and the like, are usually meaningless. Occasionally, it is appropriate to define a variable unrestricted in sign. For example, if in the water quality planning problem in Section 2.8.2 there was a requirement that the maximum difference between waste removal fractions x_1 and x_2 were some known constant Δ, then a constraint could be written which would force the absolute value of the difference to be no greater than Δ.

$$|x_1 - x_2| \leq \Delta \tag{2.96}$$

To convert constraint 2.96 to a linear one, one can define two nonnegative variables u_1 and u_2 and let

$$x_1 - x_2 = u_1 - u_2 \tag{2.97}$$

If the difference $x_1 - x_2$ is negative, $-u_2$ will equal that difference if $u_1 = 0$. If the difference is positive, u_1 will equal the difference if $u_2 = 0$. Ensuring that the sum of u_1 and u_2 does not exceed Δ,

$$u_1 + u_2 \leq \Delta \tag{2.98}$$

will force u_1 or u_2 to equal 0 if the constraint $u_1 + u_2 = \Delta$ is binding. Clearly, any unrestricted variable x can be replaced by the difference of two nonnegative variables $u_1 - u_2$.

Another typical problem in water planning studies involves minimizing the maximum of a series of unknown decision variables. For example, the active capacity K of a reservoir is the maximum of all reservoir storage volumes s_t required in each period of operation. Referring to the storage-yield model, in which only active storage capacity is being considered, equations 2.69 to 2.71 could just as easily be written as

$$\text{minimize } [\underset{t}{\text{maximum}}\,(s_t)] \tag{2.99}$$

subject to the continuity equations

$$s_{t+1} = s_t + i_t - r - r_t \qquad \text{for all periods } t \tag{2.100}$$

where the inflows i_t and uniform yield r are specified. By introducing an unknown capacity variable K, the problem can be represented as a constrained linear optimization model:

$$\text{minimize } K \tag{2.101}$$

subject to the continuity equations 2.100 and capacity limits

$$s_t \leq K \quad \text{for each period } t \tag{2.102}$$

Solving equations 2.100 through 2.102 for various values of the yield r will result in the same storage-yield functions that would be derived from solving equations 2.69 to 2.71 for various values of the capacity K.

Linear programming can be used to maximize concave functions, or minimize convex functions, once they are replaced by piecewise-linear functions. Two common methods of doing this are illustrated (methods 1 and 2) in Figure 2.14. Note that each technique involves the addition of both variables and constraints. A third method allows the minimization of concave functions or the maximization of convex functions. This method requires the use of linear mixed-integer optimization solution procedures in which some variables are forced to assume only integer values. Linear mixed-integer optimization programs are available at most scientific computer facilities.

Method 1 in Figure 2.14 requires the addition of a variable x_j and (except for the last segment) a constraint bounding that variable, for each segment j of the function. As the number of segments increases, so will the accuracy of the linear approximation of $f(x)$, but so also will the required number of variables and constraints. Method 2 requires only two additional constraints regardless of the number of variables w_j. Convex functions can be minimized using methods 1 or 2 just as they can be used to maximize concave functions.

If the functions that are to be minimized are convex, or if the functions that are to be maximized are concave, then methods 1 and 2 will result in accurate representations of the piecewise-linear approximation. For method 1, each variable x_j will equal its upper limit before x_{j+1} will be greater than zero. The requirement that the variables x_j fill up in order will be fulfilled, since any other situation would be nonoptimal. This is because the slopes of each successive linear segment are decreasing for concave functions being maximized, and are increasing for convex functions being minimized.

To obtain an accurate approximation using method 2, at most two unknown variables w_j may be positive, and those two positive variables must be adjacent to one another (i.e., $w_j + w_{j+1}$ must equal 1 for some segment j). Like method 1, method 2 provides a linear interpolation between two adjacent end points of a segment of the nonlinear function [20].

Methods 1 and 2 can be used with special separable linear programming packages to minimize a concave function or maximize a convex function. Separable programming ensures that the segment variables x_j assume positive values in the correct sequence, that is, x_j will equal its maximum value if $x_{j+1} > 0$ or that only two adjacent w_j's are nonzero. This special feature of some linear programming does not guarantee global optimality. However, if the nonlinear functions are relatively "smooth," globally optimal solutions may result.

Method 3 is a means of maximizing or minimizing concave or convex functions. It requires a linear mixed-integer optimization procedure. Such procedures are generally available but are less efficient and are more costly than noninteger linear programming. Using method 3, any separable function, whether convex, concave, or a combination of these shapes, can be approximated for inclusion in a linear model. Except for strictly convex functions that are to be maximized, or concave functions that are to be minimized, a combination of methods 1 or 2 and 3 will be preferred to method 3 alone, since fewer integer variables may be required.

Method 3 with some modification is also appropriate for fixed-cost or fixed-charge problems. For example, assume that the cost of capacity expansion x_t in period t equals

$$C_t(x_t) = \begin{cases} 0 & \text{if } x_t = 0 \\ a_t + b_t(x_t) & \text{if } x_t > 0 \end{cases} \tag{2.103}$$

where a_t is the fixed cost and $b_t(x_t)$ is the variable cost. Introducing integer variables z_t, the cost component of the objective, and its associated constraints, would be written

$$\text{minimize} \sum_t [a_t z_t + b_t(x_t)] \tag{2.104}$$

subject to

$$\begin{aligned} z_t &\leq 1 \text{ and integer} \quad && \text{for all } t \\ x_t &\leq M_t z_t && \text{for all } t \end{aligned} \tag{2.105}$$

Each constant M_t in equations 2.105 is a specified upper bound for x_t. These constraints force each z_t to be positive, and therefore equal to 1, whenever x_t is positive. Of course, the variable cost functions $b_t(x_t)$ must also be made piecewise linear prior to using linear mixed-integer programming solution methods.

If linear mixed-integer solution methods are not readily available, it may be possible to use conventional (noninteger) linear programming, assuming various integer values of the integer variables if there are not too many of these variables. Rounding off continuous variables to the nearest integer solution may not yield the optimal solution, or even a feasible solution.

A typical problem encountered in structuring linear water planning models is the need to include nonlinear separable functions; a separable function of n variables $f(x_1, x_2, \ldots, x_n)$ is one that can be written as the sum of n individual functions

$$f_1(x_1) + f_2(x_2) + f_3(x_3) + \ldots + f_n(x_n)$$

each of which depends on only one of the decision variables. Separable functions may arise in either the objective or the constraint set. To include such separable functions in linear or mixed integer programming models, one can simply use methods 1, 2, or 3 to approximate each separate function. The

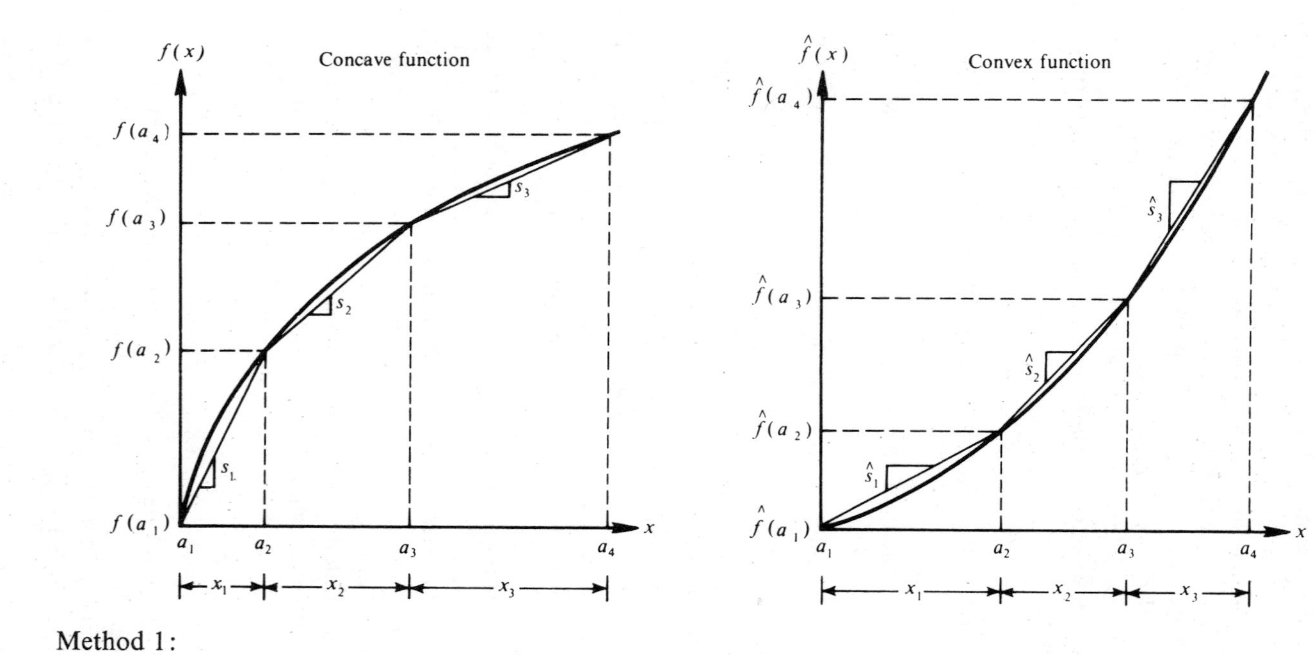

Method 1:

$$\text{maximize } f(x) \simeq s_1x_1 + s_2x_2 + s_3x_3 + \ldots = \sum_j s_jx_j$$

or

$$\text{minimize } \hat{f}(x) \simeq \hat{s}_1x_1 + \hat{s}_2x_2 + \hat{s}_3x_3 + \ldots = \sum_j \hat{s}_jx_j$$

subject to

$$a_1 + x_1 + x_2 + x_3 + \ldots = x$$

$$x_j \leq a_{j+1} - a_j \quad \text{for all segments } j$$

Method 2:

$$\text{maximize } f(x) \simeq f(a_1)w_1 + f(a_2)w_2 + f(a_3)w_3 + \ldots = \sum_j f(a_j)w_j$$

or

$$\text{minimize } \hat{f}(x) \simeq \hat{f}(a_1)w_1 + \hat{f}(a_2)w_2 + \hat{f}(a_3)w_3 + \ldots = \sum_j \hat{f}(a_j)w_j$$

subject to

$$a_1w_1 + a_2w_2 + a_3w_3 + \ldots = x$$
$$w_1 + w_2 + w_3 + \ldots = 1$$

Method 3:

$$\text{maximize or minimize } f(x) \simeq f(a_1)z_1 + f(a_2)z_2 + f(a_3)z_3 + f(a_4)z_4 + s_1x_1 + s_2x_2 + s_3x_3 = \sum_j [f(a_j)z_j + s_jx_j]$$
$$\hat{f}(x) \simeq \hat{f}(a_1)z_1 + \hat{f}(a_2)z_2 + \hat{f}(a_3)z_3 + \hat{f}(a_4)z_4 + \hat{s}_1x_1 + \hat{s}_2x_2 + \hat{s}_3x_3 = \sum_j [\hat{f}(a_j)z_j + \hat{s}_jx_j]$$

subject to

$$a_1z_1 + a_2z_2 + a_3z_3 + a_4z_4 + \ldots + x_1 + x_2 + x_3 + \ldots = x$$
$$z_1 + z_2 + z_3 + \ldots = 1$$
$$x_j \leq (a_{j+1} - a_j)z_j \quad \text{for all segments } j$$
$$z_j \text{ integer} \quad \text{for all segments } j$$

FIGURE 2.14. Several piecewise linearization methods for nonlinear separable functions.

overall approximation for $f(x_1, x_2, \ldots, x_n)$ is then the sum of the approximations for each $f_i(x_i)$.

2.8.8 Special Structure Problems

Certain linear planning models have a special structure which permits a more efficient solution procedure than the simplex method. One such problem is called the *transportation problem*. This problem involves the transport of goods from m "origins" to n "destinations" at minimumal total cost. For example, consider an irrigation district serviced by several groundwater or surface-water sources. Let the subscripts i indicate a source site and j an irrigation area within the district. Assume a known supply of water a_i is available at each source of supply i and a known demand b_j at each irrigation area j. The problem is to find the quantity of water x_{ij} to pump or transport from source site i to use site j so that the total cost is minimized. Assuming that c_{ij} is the cost of moving a unit of water from origin i to destination j, the problem can be written as:

$$\text{minimize} \sum_{i}^{m} \sum_{j}^{n} c_{ij} x_{ij} \tag{2.106}$$

subject to

1. Amount pumped from each site i cannot exceed the supply:

$$\sum_{j} x_{ij} \leq a_i \quad \text{for each reservoir site } i \tag{2.107}$$

2. Amount pumped to each site j must meet the demand:

$$\sum_{i} x_{ij} \geq b_j \quad \text{for each irrigation area } j \tag{2.108}$$

3. Nonnegativity:

$$x_{ij} \geq 0 \qquad \forall i, j \tag{2.109}$$

The special solution procedures to solve any linear optimization problem having this specific structure, regardless of the meaning of the variables or parameters, are often performed faster by hand than by computer, given the time needed to prepare the problem for computer solution. Again, these procedures can be found in most introductory texts in operations research or linear programming [20].

Similarly, there are efficient solution procedures for a variety of network problems, such as finding the maximum flow through a network, or finding the least-cost path from one point in a network to another. These special methods can also be found in most texts on constrained optimization methods.

2.9 SIMULATION AND SEARCH METHODS

Simulation is perhaps the most widely used method for evaluating alternative water resource systems. The reason for its popularity lies in its mathematical simplicity and versatility. The advent of high-speed computation has enabled planners to write very detailed simulation programs to describe the operation of water resources systems. Some of these include the use of synthetic streamflows and precipitation, and models of flood-wave attenuation, hydroelectric power generation, irrigation, recreation, municipal water supply development, water quality control, navigation, and other purposes of water resource systems. Moreover, economic benefits and costs can be identified for each project and activity in a basin. Hence both the physical and economic responses of various alternatives can be estimated.

Simulation is *not* an optimizing procedure. Rather, for any set of design and operating policy parameter values, it merely provides a rapid means for evaluating the anticipated performance of the system. It is necessary for the analyst to specify the trial design (or, equivalently, to allow the computer to do so in accordance with some algorithm), whereupon the simulation model yields estimates of the economic, environmental, and other responses associated with that trial. Simulation methods do not identify the optimal design and operating policy, but they are an excellent means of evaluating the expected performance resulting from any design and operating policy. Hence they are often used to assist water resources planners in evaluating those designs and operating policies defined by simpler optimization models.

The difficulties encountered during simulation studies are those associated with any computer programming exercise. There are problems associated with writing, debugging, testing, and executing a computer program of substantial length, involving vast amounts of input and output data. There are difficulties associated with how to define the boundaries of the system that is to be simulated and the level of detail within the system that should be modeled. In other words, there are difficulties in identifying just what it is one should be simulating, especially if no preliminary screening of alternatives has been done using, perhaps, relatively simpler optimization models.

Finally, there are difficulties associated with sampling in the multidimensional space which contains the vector of the design and operating decision variables. Questions of the adequacy of a sample, the stability of the simulation and sampling (Monte Carlo) procedure, the accuracy and reliability of the numerical computation procedures, and others haunt the analyst performing a simulation study.

2.9.1 Types of Simulation Models

A simulation model may be *time-sequenced* or *event-sequenced.* In a time-sequenced model a fixed time interval, Δt, is selected and the computer examines the state of the system (flows, storage volumes, demands, etc.) at successive time intervals. Events of interest can sometimes go unnoticed in a time-sequenced study. For example, consider a reservoir which receives precipitation during Δt and from which the same volume of water is lost by evaporation and seepage. On examining the initial and terminal reservoir states corresponding to the beginning and end of the interval Δt, there is no evidence of precipitation or evaporation or leakage. An event-sequenced simulation considers a sequence of events, like floods, when they happen. The time interval between events is a random variable. An event-sequence simulation consults the "clock" whenever an event of interest occurs. While this has the distinct advantage of not missing any significant events, the form in which water resources data are available makes event sequencing very cumbersome for most uses.

A time-sequenced model requires the selection of a characteristic time interval Δt. If Δt is small, the simulation will require relatively more computer time. Conversely, if Δt is large, many of the approximations on which the simulation model is based may be invalid. That is, during a long interval the values of the variables that describe the state of the system might change so much that calculations, based on the initial and final state of the system (at the start and end of the time interval), may not be accurate. In such cases one may have to consider intermediate conditions in the computation or some sort of average. Recreation benefits provide a typical example. If Δt is long, say a full season, it might happen that the change in storage volume within that season is sufficiently great to create a vastly different recreation benefit at the end of Δt than at the beginning. If it is presumed that the change in volume occurs at a uniform rate throughout the season, and if the loss function is a linear transformation of the change, it is acceptable to use the average recreation benefit as representative of the entire season. But such functions are generally nonlinear, and hence defining some sort of average is not trivial.

A simulation may be *deterministic* or *stochastic.* If the system is subject to random input events, or generates them internally, the model is said to be at least partially stochastic. If no random components are involved, the model is deterministic. Certain models can be operated in both modes, switching from one to the other as dictated by the state of the system. For example, in simulating a groundwater regime, evaporation from the groundwater (a stochastic quantity because of its dependence upon stochastic meteorological phenomena such as temperature, wind, humidity, and radiation) is very important when the groundwater table lies close to the surface of the ground.

However, if the groundwater is drawn down an appreciable depth, the stochastic components are damped by the overburden and hence the relevant features of evaporation can be suitably modeled by a deterministic relationship.

A simulation may deal with *steady-state* or *transient* conditions. The study of a water resources system during its initial years, perhaps involving one strategy for filling the reservoirs and one for diverting damaging floods which occur before the structures are ready to receive them, lies in the area of transient analysis. Study of the operation of a water resource system over a relatively long period of time during which no major changes in the system occur would be done with a steady-state analysis.

Simulation is a surrogate for asking "What if?". Although not directly identifying optimal designs or operating policies, by incorporating the large number of variables required to define a system, simulation indicates which design questions might fruitfully be asked and which data are most urgently needed. Despite the complexity of major river basin simulation efforts, even the most formidable computer program is built piece by piece, subroutine by subroutine, until finally the whole program is complete.

2.9.2 Sampling and Search Procedures

One of the major problems inherent in simulation modeling is the determination of the number of sets of design and operating policy values that need to be simulated. Each simulation will result in a description of the system's performance. Many such performance pictures, each one corresponding to a specific set of the design and operating parameter values, define a performance response surface. Assuming only two design or operating parameters x_1 and x_2, the response surface $f(x_1, x_2)$ generated from simulations of all values of x_1 and x_2 might appear as in Figure 2.15. The next few paragraphs outline some methods for sampling such response surfaces.

The *uniform grid* sampling approach consists of an examination of specific uniformly spaced values of the decision variables. In large problems with many decision variables, typical of water resources management problems, an evaluation of the system's response on even a course grid requires a very large number of simulations. The uniform grid sampling procedure is depicted in Figure 2.15.

The *random sampling* approach consists of randomly selecting feasible values of the decision variables. This is not a trivial exercise, but if it can be done, the analyst can make quantitative statements about the reliability of the results. Consider a vector $\mathbf{X}$ of k decision variables $(x_1, x_2, \ldots, x_k)$ and a single objective or performance function $B(\mathbf{X})$. Consider the set of all feasible decision vectors or plans $\mathbf{X}$. Assume that the performance functions $B(\mathbf{X})$ for

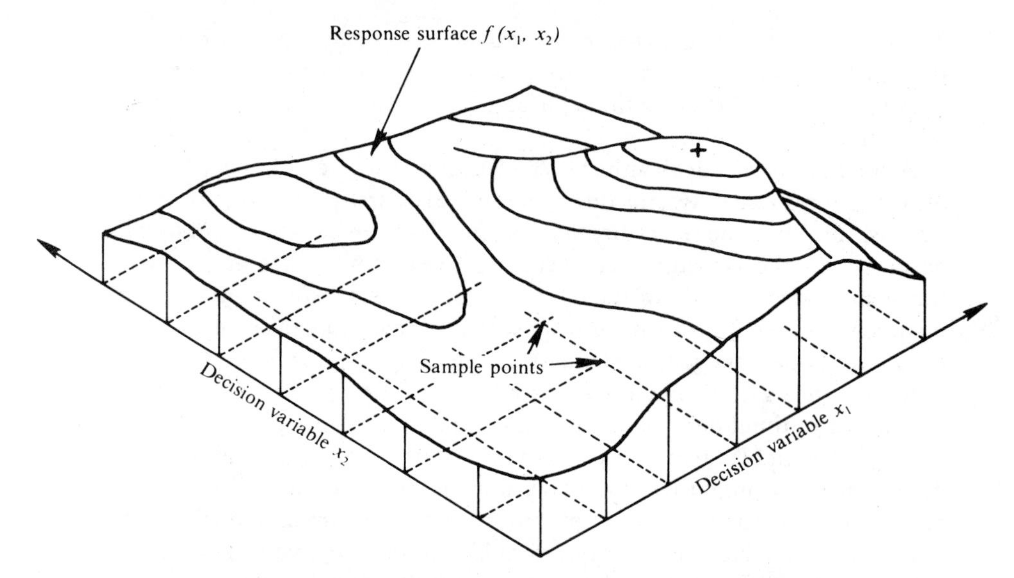

FIGURE 2.15. Uniform grid sampling search procedure.

100p % of these plans is less than some value B_p. One could randomly select n feasible plans from the set of all possible plans and calculate, via simulation, the value of their performance functions $B(\mathbf{X}_1)$, $B(\mathbf{X}_2)$, . . . , $B(\mathbf{X}_n)$. The probability that each $B(\mathbf{X}_i)$ is less than B_p is just p. Hence, the probability that all n performance values $B(\mathbf{X}_1)$, . . . , $B(\mathbf{X}_n)$ are less than B_p is p^n. Thus the probability that at least one $B(\mathbf{X}_i)$ exceeds B_p is $1 - p^n$.

This analysis allows one to determine the probability that $B(\mathbf{X})$ for the best of n randomly selected plans is in the upper $100(1 - p)$ percent of the objective values achieved by all feasible plans. For example, if 30 randomly selected plans are examined, the probability that the best of the 30 performance values falls in the upper 10% of the values yielded by all feasible plans is $1 - 0.9^{30}$, or 0.957. This result is obtained without knowing the value of B_p. While the probability that the best of 30 trial designs is in the upper 10% of all designs is 96%, one does not know how large the absolute difference between the maximum possible value of B and the best of those observed may be.

The nature of the results (i.e., the shape of the response surface) is a useful indicator of how many samples should be taken. Fortunately, most response surfaces for water resources problems are relatively smooth and flat near the optimum. A random search of a response surface is illustrated by Figure 2.16.

A general characteristic of both uniform grid and random sampling is that no information from previous trials is used to determine the values of the decision variables in subsequent trials. *Sequential search procedures* utilize previous results to calculate what adjustments might result in "better" performance; better is defined with respect to the objective function.

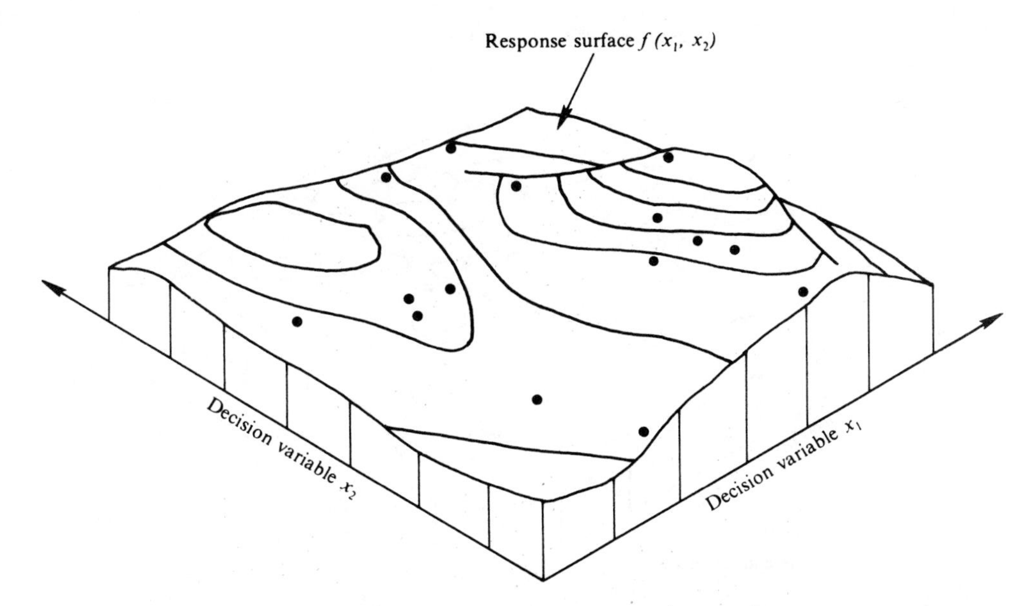

FIGURE 2.16. Random sampling search procedure.

In water resources analyses the most frequently used sequential search procedure is *trial and error*. This is used because it is simple and because the problems are too large to be helped much by formal sequential search procedures. The trial-and-error method moves in the direction the user "feels" will result in the greatest change of the performance function. A user's ability to do this depends on their understanding of the water resource system being simulated.

Another sequential search approach is to use the partial derivatives of the objective function to identify the direction in which to change the decision vector **X** so as to maximize the rate of increase of the objective. Lacking a functional form of the response surface so that derivatives can be obtained analytically to indicate the path of *steepest ascent*, a decision variable x_j can be marginally adjusted and the simulation model can be run. The difference in performance values for the marginal change in x_j can be used as an approximation to the partial derivative so that

$$\frac{\partial B}{\partial x_j} \cong \frac{\Delta B}{x_j - (x_j - \Delta x_j)} \qquad j = 1, \ldots, k \tag{2.110}$$

where B is the value of performance function and x_j a decision variable. This type of search is demonstrated in Figure 2.17. A procedure that successively adjusts each variable individually is not very efficient when the number of variables is large.

Each of the search and sampling methods has its own advantages. It follows that the most used search procedure often involves a combination of

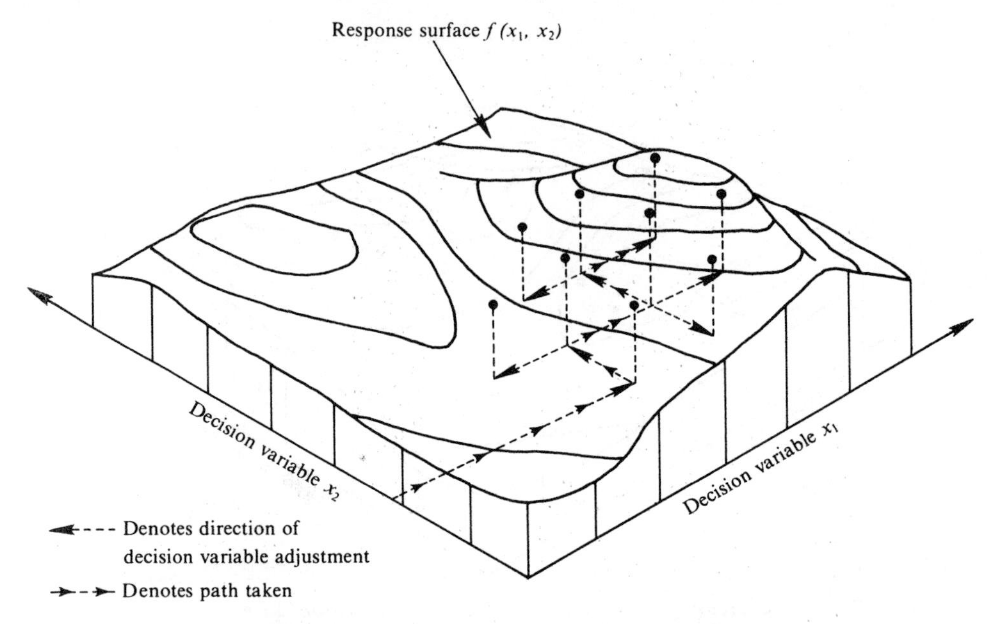

FIGURE 2.17. Trial-and-error search procedure.

all these methods. This "hybrid" approach could make use of random or grid sampling to locate initial, "interesting" regions of the response surface followed by a directed steepest ascent search to locate the optimum. There is no guarantee of global optimality in any of these or more sophisticated sequential search procedures unless the problem satisfies a number of very restrictive mathematical conditions.

When investigating alternative designs by simulation in search of a good solution, it is important to be able to determine when a sufficient search has been made. Aside from complete enumeration, no criteria for determining optimal stopping time for a search of a complex response surface have appeared in the literature. An important consideration is the problem of local and global optimality. Witness the problem of locating the locally optimum peak B instead of the global peak A in Figure 2.18. The usual procedure employing successive runs of the simulation model has been to balance the cost of each successive action against the likelihood of increased information about the system response surface in an intuitive learning procedure.

2.10 CONCLUSION

This chapter provides a brief review of the methods commonly used by water resource systems planners to help define and evaluate alternative water management investments and operating policies. These methods have

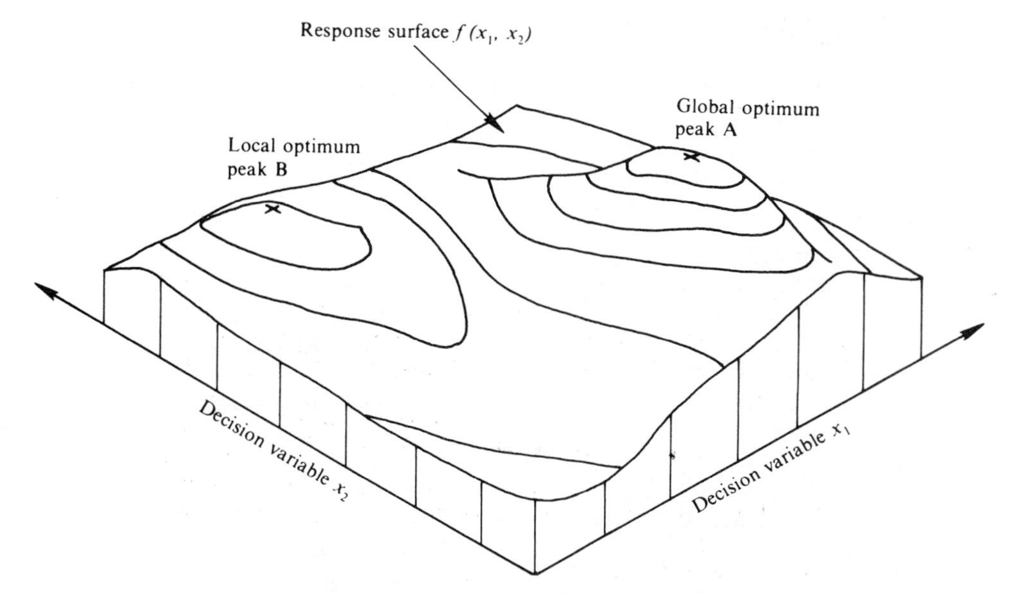

FIGURE 2.18. Problem of local optimality.

included engineering economics, mathematical optimization, and simulation techniques. The emphasis has been on the application of these methods to some simplified water resource planning problems. Additional detail on the methods themselves can be found in many of the references listed at the end of this chapter. More realistic applications are presented in later chapters.

APPENDIX 2A

COMPUTER PROGRAMS FOR LINEAR PROGRAMMING

This appendix discusses the basic input requirements and capabilities of IBM's Mathematical Programming Systems MPS and MPSX [13, 17]. This serves to illustrate the basic features of such general-purpose programs in that the capabilities and input format of programs available on other lines of computers are similar.

Linear programming computer programs have two stages: compilation and execution. Each stage has required input. The input for the compilation stage is a program—a list of statements—specifying exactly what operations the computer is to perform. The input for the execution stage is comprised of the data defining the particular linear optimization problem to be solved. These two input streams are discussed in the following sections.

2A.1 Program Statements

There are many possible program statements that can be written. A simple sequence of program statements that will solve any standard linear optimization problem is the following:

PROGRAM	
INITIALZ	
MOVE(XDATA,'data set name')	(identifies name of problem to be solved)
MOVE(XPBNAME,'PBFILE')	(lists problem statistics)
CONVERT('SCRATCH','SUMMARY')	('SCRATCH' may not be needed in some cases)
BCDOUT	(optional, prints out input)
TITLE('any title desired')	(optional)
SETUP('BOUND','bounds name','MAX')	[or just SETUP or SETUP('MAX') for maximization]
MOVE(XOBJ,'objective name')	(identifies name of objective function)
MOVE(XRHS,'right hand side name')	(identifies name of right-hand-side vector)
PICTURE	(optional, prints out coded matrix of coefficients)
CRASH	(optional, finds feasible solution if possible)
PRIMAL	(finds optimal solution, if one exists)
SOLUTION	(prints out solution)
EXIT	
PEND	

For the IBM MPS or MPSX program, program statements are punched on cards starting in *column 10*. This particular program always minimizes the objective value, unless the argument 'MAX' is included in the SETUP statement, as indicated above. Alternatively, for a maximization problem the objective coefficients in the data input stream could be multiplied by -1. The data set name, the title name (if any), the single-variable bounds vector name (if any), the objective name, and the right-hand-side vector name must be placed within single quotes, and they must correspond *exactly* to the names specified in the data cards.

2A.2 Input Data

All variable and data set names may be composed of from *one to eight* characters drawn in any order from A through Z and from 0 through 9. Numerical input values may have at most 12 input characters, including a decimal point. For negative numbers a minus sign must also be indicated. It is recommended that the decimal point always be in the same column for all numeric data.

The first card of the data stream is a NAME card. It has the following format:

Columns:	1–4	15–22
	NAME	Data set name

The data set name must correspond *exactly* to the data set name specified in the program.

The next card in the data stream is a ROWS card which has the format

Columns:	1–4
	ROWS

The cards following the ROWS card specify the name of each row of the problem (including the objective row) and the relationship of the row value to the right-hand side. These cards have the format

Columns:	2–3	5–12
	Relationship	Row name

The possible relationships are:

N—no relationship (the objective row is not constrained)
G—greater than or equal to the corresponding right-hand-side value
L—less than or equal to the corresponding right-hand-side value
E—equal to the corresponding right-hand-side value

The row names given to the objective row must correspond *exactly* to those given in the program.

After all the rows have been specified, the following card is the COLUMNS card. It has the format

Columns:	1–7
	COLUMNS

The cards that follow it specify the coefficients of each of the columns or variables of the problem. They have the format

Columns:	5–12	15–22	25–36	40–47	50–61
	Column name	Row name	Entry value	Row name (optional)	Entry value (optional)

Entry values need be specified for only the nonzero entries in the matrix, and they must conform with the convention for numerical inputs. Use of the option for specifying two entries per card can reduce the number of input cards for a job. The column cards must be ordered so that the cards for each column (variable) name are together.

After all the columns have been named and their entries specified, the next card in the data stream is the right-hand-side or RHS card. It has the format

Columns:	1–3
	RHS

The cards following the RHS card specify the nonzero entries of the RHS vector. They have the format

Columns:	5–21	15–22	25–26
	Right-hand-side name	Row name	Entry value

The right-hand-side name must correspond *exactly* with the name given in the program. The name is the same for all right-hand-side values that are to be included in the optimization problem.

After the right-hand-side cards is the bounds section, if desired and if applicable. In this section values of variables may be fixed, or upper and/or lower limits may be placed on variables. The card formats are:

Columns:	1–6
	BOUNDS

Columns:	2–3	5–12	15–22	25–36
	Relationship	Bounds name	Column name	Entry value

Possible relationships are:

FX—fixed value—value of column name is equal to entry value
UP—upper bound—value of column name is not greater than entry value
LO—lower bound—value of column name is not less than entry value
FR—free range—variable can range from $-\infty$ to ∞

The bounds name must correspond *exactly* to that given in the control program (SETUP card). The name is the same for all bounds that are to be considered together in the same optimization problem.

After the last of the bounds cards comes the ENDATA card, which is the last card of the input stream. It has the format

Columns:	1–6
	ENDATA

2A.3 Example

To clarify the preceding sections, consider the following problem:

$$\text{maximize } 2x_1 + 3x_2 + 4x_3$$

subject to

$$\begin{aligned}
x_1 + x_2 &\leq 2 \\
x_1 + 2x_2 &\geq 1 \\
x_1 - 0.5x_3 &\geq 0 \\
x_2 &\leq 3 \\
x_2 &\geq 1 \\
x_3 &\leq 2 \\
x_1, \quad x_2, \quad x_3 &\geq 0
\end{aligned}$$

Before preparing the input for the computer, it is helpful to construct, at least mentally if not actually, a matrix that defines each row name, row type, column or variable name and its coefficient for each row, and the right-hand-side value of each row. Such a matrix for the example problem is illustrated in Table 2A.1. The last two constraint rows will be included in the bounds section of the input data.

TABLE 2A.1. Data Matrix for Example Problem

ROWS	COLUMNS:	x_1	x_2	x_3	RHS
Row Name	*Row Type*	*Coefficient Matrix*			*RHS1*
ROBJ	N	−2	−3	−4	
ROW1	L	1	1		2
ROW2	G	1	2		1
ROW3	G	1		−0.5	
ROW4	L		1		3
	BOUNDS				
Name	*Type*				
BN1	LO		1		
BN1	UP			2	

The names given to each row and variable or column, as well as the names given to the RHS and BOUNDS vectors, are completely arbitrary, as long as they do not exceed eight characters. This matrix can assist in preparing the data cards required for computer solution.

The listing in Table 2A.2 is the entire deck needed to solve this problem with the exclusion of the first several system control cards that may vary for

TABLE 2A.2. MPSX Input for Sample Problem

```
//(first system control card = job account card)            ┌These cards      ┐
//(second system control card calling MPS or MPSX)          │may vary at      │
//(third system control card)                               │each computer    │
        PROGRAM                                             └facility         ┘
        INITIALZ
        MOVE(XDATA,'SAMPROB')
        MOVE(XPBNAME,'PBFILE')
        CONVERT('SUMMARY')
        BCDOUT
        SETUP('BOUND','BN1')
        MOVE (XOBJ,'ROBJ')
        MOVE(XRHS, 'RHS1')
        PICTURE
        CRASH
        PRIMAL
        SOLUTION
        EXIT
        PEND
/*                    ┌These system control cards may also vary┐
//EXEC.SYSIN DD*      └at each computer facility               ┘
NAME          SAMPROB
ROWS
 N   ROBJ
 L   ROW1
 G   ROW2
 G   ROW3
 L   ROW3
COLUMNS
      X1      ROBJ       -2.0      ROW1       1.0
      X1      ROW2        1.0      ROW3       1.0
      X2      ROBJ       -3.0      ROW1       1.0
      X2      ROW2        2.0      ROW4       1.0
      X3      ROBJ       -4.0      ROW3      -0.5
RHS
      RHS1    ROW1        2.0
      RHS1    ROW2        1.0
      RHS1    ROW4        3.0
BOUNDS
 LO BN1       X2          1.0
 UP BN1       X3          2.0
ENDATA
/*
//
```

different computer facilities. Note that the nonnegativity constraints ($x_j \geq 0$) need not be included in the input. The simplex procedure assumes that, unless otherwise specified, all variables will be nonnegative. Also, the two bounds could have been included in the ROWS, COLUMNS, and RHS just as are the other constraints. The program is more efficient, however, if the bounds section is used when single variables are bounded as two are in this example problem.

Table 2A.3 lists most of the resulting output in the format used by MPSX.

2A.4 Output Data

The output for this problem shown in Table 2A.3 includes the list of program statements and a summary of program statistics. This summary, together with the listed input data and picture, a coded version of Table 2A.4, are very useful in detecting possible punching errors in the input data cards. Note that for this example, in which each coefficient in the objective function was multiplied by -1 for minimization, the optimal solution was obtained after four iterations, and equals -13.

In the rows section of the solution, the ACTIVITY is the value of the objective function or constraint. The negative of each dual variable is listed in the last column, labeled DUAL ACTIVITY. For example, if the right-hand side of the first constraint row were increased from 2 to 3, the objective function would decrease by 3, from -13 to -16. This would result from a change in the optimal value of x_2 from 1 to 2. Similarly, if the right-hand side of the third constraint were decreased from 0 to -1, the optimal solution would be $x_1 = 0$, $x_2 = 2$, and $x_3 = 2$. The objective function value would equal -14, an increase of -1. Note that the dual-variable value would not apply if the right-hand side were increased from 0 to 1 or if the first constraint row right-hand side were decreased from 2 to 1. In both cases the problem would be infeasible. This illustrates the fact that while in general the dual variable indicates the change in the objective function associated with either an increase or decrease in the right-hand side, this may not always be the case. How this can be determined will be discussed shortly.

The optimal value of each unknown variable is included under ACTIVITY in the columns section of the solution. The last column of that section, labeled REDUCED COST, indicates the change in the objective function associated with a unit change in the variables currently at their upper or lower limits, indicated by a UL or LL in the columns section. Again, the extent to which these variables can change is not specified. In this example problem, if x_3 is increased by one unit, from 2 to 3, the new solution would be $x_1 = 1.5$, $x_2 = 0.5$, and $x_3 = 3$, which changes the objective function value from -13 to -16.5, a difference of -3.5.

TABLE 2A.3. MPSX Output for Sample Problem

Program Statements:

```
0001    PROGRAM
0002    INITIALZ
0096    MOVE(XDATA,'SAMPROB')
0097    MOVE(XPBNAME,'PBFILE')
0098    CONVERT('SUMMARY')
0099    BCDOUT
0100    SETUP('BOUND','BN1')
0101    MOVE(XOBJ,'ROBJ')
0102    MOVE(XRHS,'RHS1')
0103    PICTURE
0104    CRASH
0105    PRIMAL
0106    SOLUTION
0107    EXIT
0108    PEND
```

Statistics:

```
SUMMARY
1 - ROWS SECTION.
    0 MINOR ERROR(S) - 0 MAJOR ERROR(S).
2 - COLUMNS SECTION.
    0 MINOR ERROR(S) - 0 MAJOR ERROR(S).
3 - RHS'S SECTION.
    RHS1
    0 MINOR ERROR(S) - 0 MAJOR ERROR(S).
5 - BOUNDS SECTION.
    BN1
    0 MINOR ERROR(S) - 0 MAJOR ERROR(S).
NUMBER OF ELEMENTS BY COLUMN ORDER
    6 X1      .....4    X2      .....4    X3      .....2
NUMBER OF ELEMENTS BY ROW ORDER, EXCLUDING RHS'S, INCLUDING
                                                  SLACK ELEMENT
    1 N ROBJ .....4   L ROW1 .....3   G ROW2 .....3   G ROW3
                                      .....3  L ROW4   .....2
PROBLEM STATISTICS
    5 LP ROWS,     8 VARIABLES,     15 LP ELEMENTS, DENSITY = 37.50
THESE STATISTICS CONTAIN ONE SLACK VARIABLE FOR EACH ROW
    0 MINOR ERRORS,     0 MAJOR ERRORS.
```

Input Data:

```
BCDOUT - USING PBFILE
  NAME          SAMPROB
  ROWS
   N   ROBJ
   L   ROW1
   G   ROW2
   G   ROW3
   L   ROW4
```

TABLE 2A.3. (Continued)

```
COLUMNS
    X1          ROBJ    -   2.00000        ROW1          1.00000
    X1          ROW2        1.00000        ROW3          1.00000
    X2          ROBJ    -   3.00000        ROW1          1.00000
    X2          ROW2        2.00000        ROW4          1.00000
    X3          ROBJ    -   4.00000        ROW3    -      .50000
RHS
    RHS1        ROW1        2.00000        ROW2          1.00000
    RHS1        ROW4        3.00000
BOUNDS
 LO BN1         X2          1.00000
 UP BN1         X3          2.00000
ENDATA
```

Picture:

```
LOWER BOUND       1
UPPER BOUND          A
                        R
                        H
                  X  X  X  S
                  1  2  3  1
ROBJ       N     -A -A -A
ROW1       L      1  1     A
ROW2       G      1  A     1
ROW3       G      1    -T
ROW4       L         1     A
```

SUMMARY OF MATRIX

SYMBOL	RANGE			COUNT (INCL.RHS)
Z		LESS THAN	.000001	
Y	.000001	THRU	.000009	
X	.000010		.000099	
W	.000100		.000999	
V	.001000		.009999	
U	.010000		.099999	
T	.100000		.999999	1
1	1.000000		1.000000	6
A	1.000001		10.000000	6
B	10.000001		100.000000	
C	100.000001		1.000.000000	
D	1.000.000001		10.000.000000	
E	10.000.000001		100.000.000000	
F	100.000.000001		1.000.000.000000	
G		GREATER THAN	1.000.000.000000	

MINIMUM = .500000E+00 MAXIMUM = .400000E+01

TABLE 2A.3. (Continued)

Solution:

SOLUTION (OPTIMAL)

... NAME ...	... ACTIVITY ...	DEFINED AS
FUNCTIONAL	13.00000–	ROBJ
RESTRAINTS		RHS1
BOUNDS		BN1

SECTION 1 – ROWS

NUMBER	... ROW ...	AT	... ACTIVITY ...	SLACK ACTIVITY	.. LOWER LIMIT .	.. UPPER LIMIT .	. DUAL ACTIVITY
1	ROBJ	BS	13.00000–	13.00000	NONE	NONE	1.00000
2	ROW1	UL	2.00000		NONE	2.00000	3.00000
3	ROW2	BS	3.00000	2.00000–	1.00000	NONE	.
4	ROW3	LL	.	.	.	NONE	1.00000–
5	ROW4	BS	1.00000	2.00000	NONE	3.00000	.

SECTION 2 – COLUMNS

NUMBER	. COLUMN .	AT	... ACTIVITY ...	.. INPUT COST ..	.. LOWER LIMIT .	.. UPPER LIMIT .	. REDUCED COST .
6	X1	BS	1.00000	2.00000–	.	NONE	.
7	X2	BS	1.00000	3.00000–	1.00000	NONE	.
8	X3	UL	2.00000	4.00000–	.	2.00000	3.50000–

There is an alternative method for including constraints that are both upper and lower bounded. When both upper and lower limits are desired, one of these limits is defined in the normal manner (i.e., by the row type and the RHS value) and the other limit is specified by the difference, or range, in a separate RANGES section of the input data. In the example problem the fourth and fifth constraints together define such a situation. In this case the constraint includes only one variable, x_2, and its value must be no less than 1.0 and no greater than 3.0. The difference, or range, is 2.0. Hence, $1 \leq$ ROW4 ≤ 3 implies that $1 \leq x_2 \leq 3$.

The method involves defining a single name for the vector of ranges of each applicable constraint in a given problem, and specifying that name in the program statements and in the input data. For this problem the name RN1 is used. Then a separate RANGES (column 1 through 6) section is defined in input data. Within this section the range name (columns 5 through 12), the constraint name (columns 15 through 22), and the value of the range (columns 25 through 36) are specified.

Table 2A.4 illustrates the revised data matrix and Table 2A.5 includes a listing of the program statements, data input, and the resulting solution. Compare this solution with that given for the same problem in Table 2A.3 The solution is exactly the same, except that the upper and lower bounds on x_2 in Table 2A.3 have been replaced by the same upper and lower bounds on ROW4 in Table 2A.5.

TABLE 2A.4. Data Maxtrix for Modified Example Problem

ROWS	COLUMNS:	x_1	x_2	x_3	RHS	RANGE
Row Name	*Row Type*	*Coefficient Matrix*			*RHS1*	*RN1*
ROBJ	N	−2.0	−3.0	−4.0		
ROW1	L	1.0	1.0		2.0	
ROW2	G	1.0	2.0		1.0	
ROW3	G	1.0		−0.5		
ROW4	L		1.0		3.0	2.0
		BOUNDS				
Name	*Type*					
BN1	UP			2.0		

2A.5 Sensitivity Analysis

Also in this modified problem the program statement RANGE has been inserted after the SOLUTION program statement. This provides additional information on the sensitivity of the optimal solution to changes in the values

TABLE 2A.5. MPSX Output for Modified Example Problem

Program Statements:

```
PROGRAM
INITIALZ
MOVE(XDATA,'SAMPROB')
MOVE(XPBNAME,'PBFILE')
CONVERT('SUMMARY')
BCDOUT
SETUP('BOUND','BN1','RANGE','RN1')
MOVE(XOBJ,'ROBJ')
MOVE(XRHS,'RHS1')
PICTURE
CRASH
PRIMAL
SOLUTION
RANGE
EXIT
PEND
```

Input Data:

```
NAME          SAMPROB
ROWS
 N  ROBJ
 L  ROW1
 G  ROW2
 G  ROW3
 L  ROW4
COLUMNS
    X1        ROBJ     –   2.00000    ROW1         1.00000
    X1        ROW2         1.00000    ROW3         1.00000
    X2        ROBJ     –   3.00000    ROW1         1.00000
    X2        ROW2         2.00000    ROW4         1.00000
    X3        ROBJ     –   4.00000    ROW3     –    .50000
RHS
    RHS1      ROW1         2.00000    ROW2         1.00000
    RHS1      ROW4         3.00000
RANGES
    RN1       ROW4         2.00000
BOUNDS
 UP BN1       X3           2.00000
ENDATA
```

TABLE 2A.5. (Continued)

Picture:

```
LOWER BOUND
UPPER BOUND       A
                      R
                      A
                      N
                      G
                    R E
                    H
              X X X S
              1 2 3 1
ROBJ     N   -A-A-A
ROW1     L    1 1   A
ROW2     G    1 A   1
ROW3     G    1  -T
ROW4     L      1   A A
```

(see Table 2A.3 for meaning of symbols used in matrix)

Solution:

SOLUTION: (OPTIMAL)

...NAME...	...ACTIVITY...	DEFINED AS
FUNCTIONAL	13.00000–	ROBJ
RESTRAINTS		RHS1
BOUNDS....		BN1
RANGES....		RN1

of the constraints and decision variables. The output resulting from the inclusion of the RANGE program statement in the example problem is presented in Table 2A.6.

Sections 1 and 3 of Table 2A.6 apply to the rows or constraints of the problem. The LOWER and UPPER LIMIT values of the rows are specified by the input data. Any unit change in the value of the constraint between the LOWER and UPPER ACTIVITY values will result in a change in the objective function specified by UNIT COST. For the binding constraints (section 1) these UNIT COST values are the dual variables, since the value of the constraint equals the value of the right-hand side. But having the information presented in this section makes it clear over what ranges of RHS values these dual variables apply. Clearly, the dual variable of -3.0 applies to an increase in the value of ROW1 from 2 to 4, but not to any decrease, as has been discussed previously. Similarly, the dual variable of -1.0 applies to a decrease in the value of ROW3 from 0 to -1 but not to any increase. For those constraints that are not binding (section 3), each UNIT COST value represents the change in the objective function value per unit change in the value of the constraint, but not in a change in the RHS value of that constraint.

TABLE 2A.6. Range Data for Optimal Solution

SECTION 1 – ROWS AT LIMIT LEVEL

NUMBER	ROW	AT	ACTIVITY	SLACK ACTIVITY	LOWER LIMIT UPPER LIMIT	LOWER ACTIVITY UPPER ACTIVITY	UNIT COST UNIT COST	UPPER COST LOWER COST
2	ROW1	UL	2.00000	.	NONE	2.00000	3.00000	
					2.00000	4.00000	3.00000–	
4	ROW3	LL	.	.	.	1.00000–	1.00000–	
					NONE	.	1.00000	

SECTION 2 – COLUMNS AT LIMIT LEVEL

NUMBER	COLUMN	AT	ACTIVITY	INPUT COST	LOWER LIMIT UPPER LIMIT	LOWER ACTIVITY UPPER ACTIVITY	UNIT COST UNIT COST	UPPER COST LOWER COST
8	X3	UL	2.00000	4.00000–	.	.	3.50000	.50000–
					2.00000	2.00000	3.50000–	INFINITY–

SECTION 3 – ROWS AT INTERMEDIATE LEVEL

NUMBER	ROW	AT	ACTIVITY	SLACK ACTIVITY	LOWER LIMIT UPPER LIMIT	LOWER ACTIVITY UPPER ACTIVITY	UNIT COST UNIT COST	UPPER COST LOWER COST
3	ROW2	BS	3.00000	2.00000–	1.00000	3.00000	1.00000	
					NONE	4.00000	7.00000	
5	ROW4	BS	1.00000	2.00000	1.00000	.	1.00000	
					3.00000	2.00000	7.00000	

SECTION 4 – COLUMNS AT INTERMEDIATE LEVEL

NUMBER	COLUMN	AT	ACTIVITY	INPUT COST	LOWER LIMIT UPPER LIMIT	LOWER ACTIVITY UPPER ACTIVITY	UNIT COST UNIT COST	UPPER COST LOWER COST
6	X1	BS	1.00000	2.00000–	.	.	7.00000	5.00000
					NONE	1.00000	1.00000	3.00000–
7	X2	BS	1.00000	3.00000–	.	1.00000	1.00000	2.00000–
					NONE	2.00000	7.00000	10.00000–

In sections 2 and 4 of Table 2A.6, the INPUT COST and the UPPER and LOWER LIMIT values are specified by the input data. The LOWER and UPPER ACTIVITY values indicate the limits in which a unit change in the value of the variable will result in a change in the objective function value specified by the UNIT COST. Note that if x_3 is decreased from 2 to 0, the objective function value will increase by 3.5 per unit change, or 7.0. This is the same as the REDUCED COST specified in the solution section of Table 2A.3. If x_3 equaled 0, the optimal solution would be $x_1 = x_3 = 0$, $x_2 = 2$, and the objective function would equal -6, which is an increase of 7 from the original -13. The variable x_3 cannot be increased without changing the status of the solution. The solution status (or "basis") changes when any nonzero variable becomes zero, and a zero-valued variable becomes nonzero.

The UPPER COST and LOWER COST values in sections 2 and 4 represent the highest and lowest cost coefficients at which the variable would be maintained at its current value. If the cost coefficient were increased above the UPPER COST or decreased below the LOWER COST, the variable value would decrease to the LOWER ACTIVITY or increase to the UPPER ACTIVITY, respectively.

All of this information refers to a single change in a variable or constraint value, all other variables and constraint values remaining constant. If several changes are made simultaneously, the same conclusions hold if the sum of the absolute changes, measured in terms of the percent of the way from the current value to the upper or lower limit, is less than 100%.

One final remark regarding the computer solution of linear optimization problems. During the matrix inversion processes required to solve simultaneous linear equations, numerical errors due to round-off or truncation may result. The probability of encountering such numerical errors in the solution procedure can be reduced by scaling the coefficient matrix (i.e., by reducing the range of the magnitude of the matrix coefficients). If a column is multiplied by any constant, the units of the column variable are changed accordingly. Multiplying a row by any constant does not affect any of the dimensions of the variables, only those of the variable coefficients and right-hand side. For large optimization problems, scaling can be an important factor in reducing possible errors and also in increasing the speed of computation.

EXERCISES

Engineering Economics

2-1. Consider two alternative water resource projects, A and B. Project A will cost \$2,533,000 and return \$1,000,000 at the end of 5 years and \$4,000,000 at the end of 10 years. Project B will cost \$4,000,000 and

will return \$2,000,000 at the end of 5 and 15 years, and another \$3,000,000 at the end of 10 years. Project A has a life of 10 years, and B has a life of 15 years. Assuming an interest rate of 0.10 per year:
(a) What is the present value of each project?
(b) What is each project's annual net benefit?
(c) Would the preferred project differ if the interest rate were 0.05?
(d) Assuming that each of these projects would be replaced with a similar project having the same time stream of costs and returns, show that by extending each series of projects to a common terminal year (e.g., 30 years), the annual net benefits of each series of projects will be the same as found in part (b).

2-2. Prove the identity defined by equation 2.9a.

2-3. Show that if compounding occurs at the end of m equal length periods within a year in which the *nominal* annual interest rate is r, then the *effective* annual interest rate, r', that rate which when used in equation 2.3 will yield the annual interest, is equal to

$$r' = \left(1 + \frac{r}{m}\right)^m - 1$$

2-4. From the equation defining the effective interest rate in Exercise 2-3, show that when compounding is continuous (i.e., when $m \longrightarrow \infty$), the compound interest factor required to convert a present value to a future value in year T is e^{rT}. [*Hint*: Use the fact that $\lim_{k \to \infty} (1 + 1/k)^k = e$, the base of natural logarithms.]

2-5. The term "capitalized cost" refers to the present value PV of an infinite series of end-of-period equal payments, A. Assuming an interest rate of r, show that as the terminal period $T \longrightarrow \infty$, $\text{PV} = A/r$.

2-6. The *internal rate of return* of any project or plan is the interest rate that equates the present value of all receipts or income with the present value of all costs. Referring to Exercise 2-1, show that the internal rate of return of projects A and B are approximately 8 and 6%, respectively. These are the interest rates r, for each project, that essentially satisfy the equation

$$\sum_{t=0}^{T} (R_t - C_t)(1 + r)^{-t} = 0$$

2-7. In the discussion above, maximum annual net benefits were used as an economic criterion for plan selection. The maximum benefit–cost ratio, or annual benefits divided by annual costs, is another criterion. Benefit–cost ratios should be no less than one if the annual benefits are to exceed the annual costs. Consider two projects, I and II:

	PROJECT	
	I	*II*
Annual benefits	20	2
Annual costs	18	1.5
Annual net benefits	2	0.5
Benefit–cost ratio	1.11	1.3

What additional information is needed before one project can be considered preferable to another?

2-8. Bonds are often sold to raise money for water resource project investments. Each bond is a promise to pay a specified amount of interest, usually semiannually, and to pay the face value of the bond at some specified future date. The selling price of a bond may differ from its face value. Since the interest payments are specified in advance, the current market interest rates dictate the purchase price of the bond.

Consider a bond having a face value of $10,000, paying $500 annually for 10 years. The coupon interest rate is 500/10,000, or 5%. If the bond is purchased for $10,000, the actual interest rate will equal the coupon rate. But suppose that one can invest money in similar quality (equal risk) bonds or notes and receive 10% interest. As long as this is possible, the $10,000, 5% bond will not sell in a competitive market. In order to sell it, its purchase price has to be such that the actual interest rate will be 10%. In this case, show that the purchase price will be $6927.

The interest paid by some bonds, especially municipal bonds, may be exempt from state and federal income taxes. If an investor is in the 50% income tax bracket, for example, a 5% municipal tax-exempt bond is equivalent to a 10% taxable bond. Clearly, this exemption helps reduce local taxes needed to pay the interest on municipal bonds, as well as providing attractive investment opportunities to individuals in high tax brackets. Note the difference between these municipal, state, or federal bonds and *savings bonds*, in which the purchase price and interest rate are specified.

Lagrange Multipliers

2-9. Replacing the objective function 2.17 with $(12x_1 - x_1^2) + (8x_2 - x_2^2) + (18x_3 - 3x_3^2)$, which is also concave, find the values of each x_j that maximize the new objective when each x_j is unrestricted. Next assume that the sum of all x_j cannot exceed 10. Show that at the optimal solution the marginal values, $\partial F(\mathbf{X})/\partial x_j$, are each equal to the shadow price or dual variable λ associated with the constraint $\sum_{j=1}^{3} x_j = 10$. Finally, solve the problem again with the constraint that $\sum_{j=1}^{3} x_j \leq 15$.

2-10. How would the Lagrange multiplier procedure differ if the function 2.17, or its replacement in Exercise 2-9, were to be minimized?

2-11. Referring to Figure 2.1, assume that the objective was to minimize the squared deviations from the actual allocations x_j and some desired or target allocations T_j. Given a supply of water Q less than the sum of all target allocations T_j, structure a planning model and its corresponding Lagrangian. Will a global minimum be obtained from solving the partial differential equations derived from the Lagrangian? Why?

2-12. Using Lagrange multipliers, prove that the least-cost design of an open-top cylindrical storage basin of any volume $V > 0$ has one-third of its cost in its base and two-thirds of its cost in its side, regardless of the cost per unit area of its base or side. (It is these types of rules that end up in handbooks in engineering design.)

Dynamic Programming

2-13. Solve for the optimal allocations x_1, x_2, and x_3 for the problem defined by equation 2.36 and Table 2.1 for $Q = 3$ and 4. Also solve for the optimal allocation policy if $Q = 7$ and each x_j must not exceed 4.

2-14. Rewrite the dynamic programming model for reservoir operation defined by equations 2.66 to 2.68 using final storage volumes rather than reservoir releases as the decision variables. Using this model, verify the optimal operating policy shown in Table 2.3 for the example problem. Which model do you think is easier to solve? How would either model change if more importance were given to the desired releases than to the desired storage volumes?

2-15. Show that the constraint limiting a reservoir release, r_t, to be no greater than the initial storage volume, s_t, plus inflow, i_t, is redundant to the continuity equation $s_t + i_t - r_t = s_{t+1}$.

2-16. Develop a general recursive equation for a forward-moving dynamic programming solution procedure for a single reservoir operating problem. Define all variables and functions used.

2-17. The following table provides estimates of the discounted costs of additional wastewater treatment plant capacity needed at the end of each 5-year period for the next 20 years. Find the capacity expansion schedule that minimizes the present value of the total future costs. If there is more than one least-cost solution, indicate which one you think is better, and why.

Period	*Years*	DISCOUNTED COST OF ADDITIONAL CAPACITY: UNITS OF ADDITIONAL CAPACITY					*Additional Required Capacity at End of Period*[a]
		2	*4*	*6*	*8*	*10*	
1	1–5	12	15	18	23	26	2
2	6–10	8	11	13	15		6
3	11–15	6	8				8
4	16–20	4					10

[a]This is the total required capacity that must be added to the existing capacity at the beginning of year 1.

Assuming that the cost of expansion in any 5-year period must be paid at the beginning of the period, what is the appropriate discount factor needed to compute the present value of the cost?

2-18. Consider a wastewater treatment plant in which it is possible to include five different treatment processes in series. These treatment processes must together remove at least 90% of the 100 units of influent waste. Assuming that R_i is the amount of waste removed by process i, the following conditions must hold:

$$20 \leq R_1 \leq 30$$
$$0 \leq R_2 \leq 30$$
$$0 \leq R_3 \leq 10$$
$$0 \leq R_4 \leq 20$$
$$0 \leq R_5 \leq 30$$

(a) Write a constrained optimization planning model for finding the least-cost combination of removals R_i that together will remove 90% of the influent waste. The costs of the various discrete sizes of each unit process i are dependent upon the waste entering the process i as well as the amount of waste removed, as indicated in the accompanying table on page 88.

(b) Solve this problem by dynamic programming.

(c) Could the following conditions be included in the model?

(i) $R_4 = 0$ if $R_3 = 0$, or

(ii) $R_3 = 0$ if $R_2 \leq 20$.

(d) Show how this problem can be described by a network in which the nodes are the state variables (the influent quantity, I_i) at each stage (treatment process i) and the links are the decision variables (the quantities of waste removed, R_i) and the associated costs,

PROCESS i:		1	2	3	4	5
Influent, I_i	*Removal,* R_i	*Annual Cost* $= C_i(I_i, R_i)$				
100	20	5				
100	30	10				
80	10		3	3	1	
80	20		9		2	
80	30		13			
70	10		4	5	2	
70	20		10		3	
70	30		15			
60	10			6	2	3
60	20				4	6
60	30					9
50	10			7	3	4
50	20				5	8
50	30					10
40	10			8	5	5
40	20				7	12
40	30					18
30	10				8	8
30	20				10	12
20	10					8

$C_i(I_i, R_i)$. Indicate on the network the calculations required to find the least-cost path from state 100 at stage 1 to state 10 at stage 6 using a forward- and backward-moving dynamic programming solution procedure.

(e) How would this network change if the two additional constraints stated in part (c) were required?

2-19. The city of Eutro Falls is under a court order to reduce the amount of phosphorus which it discharges in its sewage to Lake Algae. The city presently has three wastewater treatment plants. Each plant i currently discharges P_i kg/day of phosphorus into the lake, and must reduce this to a *total* for the three plants of P kg/day.

Let X_i be the *percent* of the phosphorus removed by additional treatment at plant i, and $C_i(X_i)$ the cost of such treatment ($/year) at each plant i. Structure a planning model to determine the least-cost (i.e., a cost-effective) treatment plan for the city. Restructure the model for solution by dynamic programming. Define the stages, states, decision variables, and the recursive equation for each stage.

Linear Programming

2-20. Assume that there are m industries or municipalities adjacent to a river which discharge their wastes into the river. Denote the discharge sites by the subscript i and let W_i be the kg of waste discharged into the river each day at those sites i. To improve the quality downstream, wastewater treatment may be required at each source site i. Let x_i be the fraction of waste W_i removed by treatment at each site i. Develop a model for estimating how much waste removal is needed at each site to maintain acceptable water quality in the river at a minimum total cost. Use the following additional notation:

a_{ij} = decrease in quality at site j per unit of waste discharged at site i

q_j = quality at site j that would result if all controlled upstream discharges were eliminated (i.e., $W_1 = W_2 = 0$)

Q_j = minimum acceptable quality at site j

C_i = cost per unit (fraction) of waste removed at site i

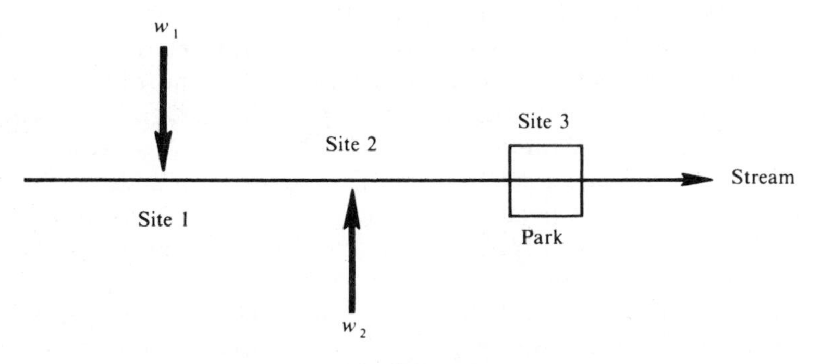

EXERCISE 2-21

2-21. Assume that there are two sites along a stream, $i = 1, 2$, at which waste (BOD) is discharged. Currently, without any wastewater treatment, the quality (DO), q_2 and q_3, at each of sites 2 and 3 is less than the minimum desired, Q_2 and Q_3, respectively. For each unit of waste removed at site i upstream of site j, the quality improves by A_{ij}. How much treatment is required at sites 1 and 2 that meets the standards at a minimum total cost? Following are the necessary data:

C_i = cost per unit of waste treatment at site i (both C_1 and C_2 are unknown but for the same amount of treatment, whatever that amount, $C_1 > C_2$)

R_i = decision variables, unknown waste removal fractions at sites $i = 1, 2$

$$A_{12} = \tfrac{1}{20} \qquad W_1 = 100 \qquad Q_2 = 6$$

$$A_{13} = \tfrac{1}{40} \qquad W_2 = 75 \qquad Q_3 = 4$$

$$A_{23} = \tfrac{1}{30} \qquad q_2 = 3 \qquad q_3 = 1$$

2-22. (a) Develop a linear programming model of the reservoir operating problem presented in Exercise 2-14, using both methods 1 and 2 of Figure 2.14.

(b) Suppose that the reservoir capacity K were unknown and were to be determined for various relative weights w associated with a cost function $C(K)$ similar in shape to Figure 2.4b. How would the modified objective [i.e., minimize $\{(20 - s_t)^2 + (25 - r_t)^2 + w \cdot C(K)\}$] be included in a linear mixed-integer optimization model?

(c) Write a linear programming model to compute the active storage capacity required to minimize the maximum percentage deviation from a known target storage volume and a known target release in each period. How could the solution of the model be used to define a reservoir operating policy?

2-23. Using the network representation of the wastewater treatment plant design problem defined in Exercise 2-18(d), write a linear programming model for finding the least-cost sequence of unit treatment process (i.e., the least-cost path through the network). [*Hint*: Let each decision variable x_{ij} indicate whether or not the link between nodes (or states) i and j connecting two successive stages is on the least-cost or optimal path. The constraints for each node must ensure that if the optimal path enters the node, it must also leave the node.]

2-24. Two types of crops can be grown in a particular irrigation area each year. Each unit quantity of crop A can be sold for a price P_A and requires W_A units of water, L_A units of land, F_A units of fertilizer, and H_A units of labor. Similarly, crop B can be sold at a unit price of P_B and requires W_B, L_B, F_B, and H_B units of water, land, fetilizer, and labor, respectively, per unit of crop. Assume that the available quantities of water, land, fertilizer, and labor are known, and equal W, L, F, and H, respectively.

(a) Structure a linear programming model for estimating the quantities of each of the two crops that should be produced in order to maximize total income.

(b) Solve the problem graphically, using the following data:

	REQUIREMENTS PER UNIT OF:		*Maximum Available*
Resource	*Crop A*	*Crop B*	*Resource*
Water	2	3	60
Land	5	2	80
Fertilizer	3	2	60
Labor	1	2	40
Unit price	30	25	

(c) Define the meaning of the dual variables, and their values, associated with each constraint.

(d) Write the dual model of this problem and interpret its objective and constraints.

(e) Solve the primal and dual models using an existing computer program, and indicate the meaning of all output data.

(f) Assume that one could purchase additional water, land, fertilizer, and labor with capital that could be borrowed from a bank at an annual interest rate r. How would this opportunity alter the linear programming model? The objective continues to be a maximization of net income.

(g) Assume that the unit price P_j of crop j were a decreasing linear function $(P_j^0 - b_j x_j)$ of the quantity, x_j, produced. How could the linear model be restructured so as to identify not only how much of each crop to produce, but also the unit price at which each crop should be sold in order to maximize total income?

2-25. Water resources planning usually involves a series of separate tasks. Let the index i denote each task, and H_i the set of tasks that must precede task i. The duration of each task i is estimated to be d_i. Develop a linear programming model to identify the starting times and the sequence of tasks that minimizes the time, T, required to complete the total planning project.

Simulation

2-26. Develop a flowchart of a model for simulating the operation of a reservoir in order to evaluate alternative reservoir operating policies.

2-27. Show how Exercise 2-13 can be solved using simulation and a technique for searching for the best solution.

2-28. Identify and discuss a water resources planning situation that illustrates the need for a combined optimization–simulation study in order to identify the best alternative solutions and their impacts.

2-29. (a) Using linear programming, derive an annual storage-yield function for a reservoir at a site having the following record of annual flows:

Year y	*Flow* Q_y	*Year y*	*Flow* Q_y
1	5	9	3
2	7	10	6
3	8	11	8
4	4	12	9
5	3	13	3
6	3	14	4
7	2	15	9
8	1		

Find the values of the storage required for yields of 2, 3, 3.5, 4, 4.5, and 5.

(b) Write a flow diagram for computing the maximum yield of water that can be obtained given any value of active reservoir storage capacity, K, using simulation.

2-30. How many different simulations of a water resource system would be required to ensure that there is no less than a 95% chance that the best solution obtained is within the better 5% of all possible solutions that could be obtained? What assumptions must be made in order for your answer to be valid? Can any statement be made comparing the value of the best solution obtained from all the simulations to the value of the truly optimal solution?

2-31. Assume that in a particular river basin 20 development projects are being proposed. Assume that each project has a fixed capacity and operating policy, and it is only a question of which of the 20 projects would maximize the net benefits to the region. Assuming that 5 minutes of computer time is required to simulate and evaluate each combination of projects, show that it would require about 36 days of computer time even if 99% of the alternative combinations could be discarded using "good judgment." What does this suggest about the use of simulation for regional interdependent multiproject water resources planning?

REFERENCES

1. Baumol, W. J., *Economic Theory and Operations Analysis*, 4th ed., Prentice-Hall, Inc., Englewood Cliffs, N.J., 1977.
2. Bellman, R., *Dynamic Programming*, Princeton University Press, Princeton, N.J., 1957.
3. Bradley, S. P., A. C. Hax, and T. L. Magnanti, *Applied Mathematical Programming*, Addison-Wesley Publishing Co., Inc., Reading, Mass., 1977.
4. Charnes, A., and W. W. Cooper, *Management Models and Industrial Applications of Linear Programming*, Vols. 1 and 2, John Wiley & Sons, Inc., New York, 1961.
5. Dantzig, G. B., *Linear Programming and Extensions*, Princeton University Press, Princeton, N.J., 1963.
6. Dorfman, R., P. A. Samuelson, and R. M. Solow, *Linear Programming and Economic Analysis*, McGraw-Hill Book Company, New York, 1958.
7. Eckstein, O., *Water Resource Development: The Economics of Project Evaluation*, Harvard University Press, Cambridge, Mass., 1961.
8. Grant, E. L., W. G. Ireson, and R. S. Leavenworth, *Principles of Engineering Economy*, 6th ed., The Ronald Press Company, New York, 1976.
9. Hanke, S. H., P. H. Carver, and P. Bugg, Project Evaluation during Inflation, *Water Resources Research*, Vol. 11, No. 4, 1975.
10. Hillier, F. S., and G. I. Lieberman, *Operations Research*, 2nd ed., Holden-Day, Inc., San Francisco, 1974.
11. Hirshliefer, J., J. C. DeHaven, and J. W. Milliman, *Water Supply: Economics, Technology and Policy*, University of Chicago Press, Chicago, 1960 (4th impression with new Postscript, 1969).
12. Intrigator, M. D., *Mathematical Optimization and Economic Theory*, Prentice-Hall, Inc., Englewood Cliffs, N.J., 1971.
13. *Introduction to MPSX and Its Optional Features MIP and GUB*, GH20-0849, International Business Machines Corp., White Plains, N.Y., 1971.
14. James, L. D., and R. R. Lee, *Economics of Water Resources Planning*, McGraw-Hill Book Company, New York, 1971.
15. Krutilla, J. V., and O. Eckstein, *Multiple Purpose River Development*, Johns Hopkins Press, Baltimore, Md., 1958.
16. Kuiper, E., *Water Resources Project Economics*, Butterworth & Company Ltd., London, 1971.
17. *Mathematical Programming System-Extended (MPSX), Program Description*, TNL SH20-0968-1, International Business Machines Corp., White Plains, N.Y., 1972.
18. McBean, E. A., and D. P. Loucks, Planning and Analysis of Metropolitan Water Resource Systems, Report prepared for Office of Water Resources Research, USDI, Cornell University Water Resources Center, Technical Report 84, Ithaca, N.Y., June 1974.
19. Nemhauser, G. L., *Introduction to Dynamic Programming*, John Wiley & Sons, Inc., New York, 1966.
20. Wagner, H. M., *Principles of Operations Research*, Prentice-Hall, Inc., Englewood Cliffs, N.J., 1975.

CHAPTER 3

Water Resources Planning under Uncertainty

3.1 INTRODUCTION

Uncertainty is always an element in the planning process. It arises because the values of many factors that affect the performance of water resource systems cannot be known with certainty when a system is planned and constructed. The success and performance of a project depend on future meteorological, demographic, social, technical, and political conditions which influence and determine future costs, benefits, environmental impacts, and social acceptibility. Uncertainty arises due to the stochastic nature of meteorological processes such as evaporation, rainfall, and temperature; similarly, the future populations of cities, per capita water usage rates, irrigation patterns, and priorities for water uses are not known with certainty. This chapter reviews some methods for dealing with uncertainty and illustrates their use in water resources planning.[1]

There are many ways to deal with uncertainty, depending on its severity

[1]Some authors distinguish between *uncertainty*, when the likelihood or probability of some event is not known, and *risk*, when the probability is known. That distinction is not made here.

and how the uncertain quantities or parameters will affect the operation of the system. The simplest approach is to replace the uncertain quantities either by their expected value or some critical (e.g., "worst-case") value and then proceed with a deterministic approach.

Use of the *expected value* or *median value* of an uncertain quantity can be acceptable if the uncertainty or variation in a quantity is reasonably small and does not critically affect the performance of the system. If expected values of uncertain parameters or variables are used in a deterministic model, the planner can then assess the importance of uncertainty with sensitivity analysis, discussed in Section 3.5.1.

Replacement of uncertain quantities by either expected or worst-case values can grossly affect the evaluation of project performance when important parameters are highly variable. Consider the evaluation of the recreation potential of a reservoir. The elevation of the pool varies from year to year depending on the inflow and demand for water. Pool levels and their associated probabilities are given in Table 3.1. The expected use of the recreation facility with different pool levels is also given in the table.

TABLE 3.1. Data for Determining Reservoir Recreation Potential

Possible Pool Levels	*Probability of Each Level*	*Recreation Potential in Visitor-Days per Day for Reservoir with Different Pool Levels*
10	0.10	25
20	0.25	75
30	0.30	100
40	0.25	80
50	0.10	70

The average pool level $\bar{L}$ is simply the sum of each possible pool level times its probability, or

$$\bar{L} = 10(0.10) + 20(0.25) + 30(0.30) + 40(0.25) + 50(0.10) = 30 \tag{3.1}$$

which corresponds to 100 visitor-days per day:

$$\text{VD}(\bar{L}) = 100 \text{ visitor-days per day} \tag{3.2}$$

A worst-case analysis might select a pool level of 10 as a critical value, yielding an estimate of system performance equal to 25 visitor-days per day:

$$\text{VD}(L_{\text{low}}) = \text{VD}(10) = 25 \text{ visitor-days per day} \tag{3.3}$$

Neither of these is a good approximation of the average visitation rate, which is

$$\begin{aligned}\overline{\text{VD}} &= 0.10\text{VD}(10) + 0.25\text{VD}(20) + 0.30\text{VD}(30) \\ &\quad + 0.25\text{VD}(40) + 0.10\text{VD}(50) \\ &= 0.10(25) + 0.25(75) + 0.30(100) + 0.25(80) \\ &\quad + 0.10(70) \\ &= 78.25 \text{ visitor-days per day}\end{aligned} \tag{3.4}$$

Clearly, the average visitation rate $\overline{\text{VD}}$, the visitation rate corresponding to the average pool level $\text{VD}(\bar{L})$, and the worst-case assessment $\text{VD}(L_{\text{low}})$ are very different.

$$\begin{aligned}\overline{\text{VD}} &= 78.25 \\ \text{VD}(\bar{L}) &= 100 \\ \text{VD}(L_{\text{low}}) &= 25\end{aligned} \tag{3.5}$$

Use of only average values in a complex model can produce a poor representation of both the average performance and the range of possible performance of a system. When important quantities are uncertain, a comprehensive analysis needs to evaluate both the expected performance of a project and the risk and possible magnitude of project failures in an economic, ecological, and/or social sense.

This chapter reviews several methodologies for planning when some quantities are uncertain. Section 3.2 is a condensed summary of the important concepts and methods of probability and statistics that are needed in the rest of this chapter and in later chapters.

Section 3.3 presents several probability distributions which are often used to model or describe the distribution of uncertain quantities. The section discusses methods for fitting these distributions using historical information and methods of assessing whether the distributions are an adequate representation of the data.

Section 3.4 presents the basic ideas and concepts of stochastic processes or time series. These are used to model streamflows, rainfall, temperature, or other phenomena whose values change with time. The section also contains a description of Markov chains, a special type of stochastic process used in the stochastic optimization models that appear in Section 3.6 and in Chapter 7.

How one makes a decision when benefits and costs are uncertain and how one measures the cost of making an error are the topic of Section 3.5. Sensitivity analysis is presented as a way of determining the costs of making an error when planning and thus choosing other than the best design. The use of expected benefits, the max-min decision criterion, and utility theory are presented as alternative criteria for evaluating the performance of alternative management plans where benefits and costs are highly variable.

Models useful for describing and optimizing the performance of water resource systems when system performance is affected by random or uncertain quantities are introduced in Section 3.6. These models include river basin simulation, stochastic dynamic programming, and stochastic linear programming.

Chapters 6 and 7 discuss in greater detail how hydrologic uncertainty can be modeled and incorporated into both simulation and optimization models for plan evaluation and formulation. Many topics in Sections 3.2 to 3.4 receive only brief treatment here. Readers desiring additional information should consult applied statistical texts such as Haan [13], Benjamin and Cornell [1], Yevjevich [39, 40], or Kaczmarek [16].

3.2 PROBABILITY CONCEPTS AND METHODS

This section gives an elementary presentation of the basic concepts and definitions of probability and statistics. These concepts are used throughout this chapter and in Chapters 6 and 7.

3.2.1 Random Variables and Distributions

The basic concept in probability theory is that of the *random variable*. A random variable is a function whose value depends on the outcome of a chance event. Examples are (1) the number of years until the flood stage of a river washes away a small bridge, (2) the number of times during a reservoir's life that the level of the pool will drop below a specified level, (3) rainfall in the month of December, and (4) next year's inflow into a reservoir from an unregulated stream. The values of all of these quantities depend on future events whose value cannot be determined before the event has occurred.

The first two examples illustrate *discrete random variables*, random variables that take on values in a discrete set (such as the positive integers). The second two examples illustrate *continuous random variables*. Continuous random variables take on values in a continuous set, such as the real numbers in the two examples. A property of all continuous random variables is that the probability that they equal any single number is zero; thus the probability that the total rainfall in a month is exactly 15.0 cm is zero, while the probability that the total rainfall lies between 14 and 16 cm can be nonzero. Some random variables are combinations of continuous and discrete random variables.

Let X denote a random variable and x a possible value of X. Random variables will be denoted by capital letters and particular values they take on by lowercase letters. For any real-valued random variable X, its *cumulative distribution function* $F_X(x)$, often called simply the distribution function, is the probability that the value of X is less than or equal to x.

$$F_X(x) = \Pr[X \leq x] \tag{3.6}$$

$F_X(x)$ is a nondecreasing function of x because

$$\Pr[X \leq x] \leq \Pr[X \leq x + \delta] \qquad \text{for } \delta > 0 \tag{3.7}$$

In addition,

$$\lim_{x \to +\infty} F_X(x) = 1$$

and

$$\lim_{x \to -\infty} F_X(x) = 0 \tag{3.8}$$

The first limit equals 1 because the probability that X takes on some value must be unity; the second limit is zero because the probability that X takes on no value must be zero.

If X is a real-valued discrete random variable which takes on values x_1, $x_2, \ldots$, the *probability function* $P_X(x_i)$ is defined as the probability X takes on the value x_i.

$$P_X(x_i) = \Pr[X = x_i] \tag{3.9}$$

The value of the cumulative distribution function $F_X(x)$ for a discrete random variable is the sum of the probabilities of all x_i which are less than or equal to x.

$$F_X(x) = \sum_{x_i \leq x} P_X(x_i) \tag{3.10}$$

Figure 3.1 illustrates the probability function $P_X(x_i)$ and the cumulative distribution function of a discrete random variable.

The probability density function $f_X(x)$ for a continuous random variable X is the analogue of the probability function of a discrete random variable. The density function is the derivative of the cumulative distribution function so that

$$f_X(x) = \frac{dF_X(x)}{dx} \geq 0$$

and

$$\int_{-\infty}^{+\infty} f_X(x) = 1 \tag{3.11}$$

Equation 3.11 indicates that the area under the density function is 1. If a and b are any two constants, the cumulative distribution function or the density function may be used to determine the probability that X is greater than a and less than or equal to b because

$$\Pr[a < X \leq b] = F_X(b) - F_X(a) = \int_a^b f_X(x)\, dx \tag{3.12}$$

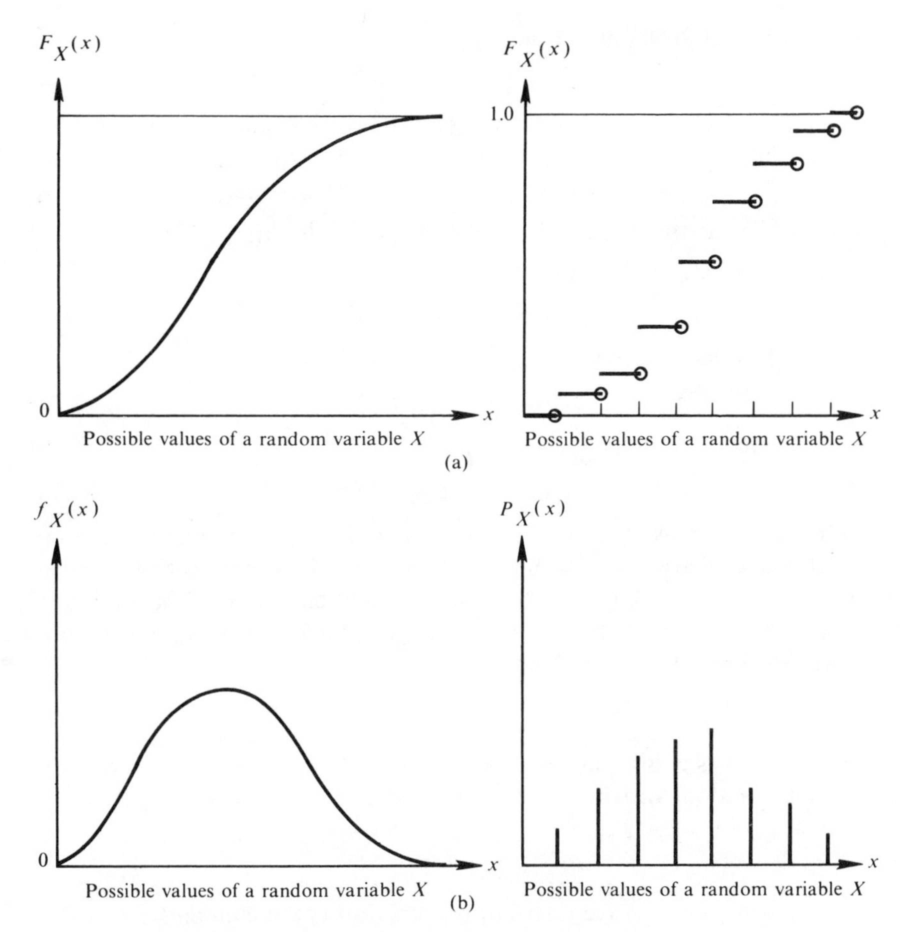

FIGURE 3.1. Cumulative distribution functions and density or probability functions of random variables: (a) continuous distributions; (b) discrete distributions.

The density function specifies the relative frequency with which the value of a continuous random variable falls in different regions.

The joint probability distribution of two or more random variables can also be defined. If X and Y are two continuous real-valued random variables, their joint cumulative distribution function is

$$\begin{aligned} F_{XY}(x, y) &= \Pr[X \leq x \text{ and } Y \leq y] \\ &= \int_{-\infty}^{x} \int_{-\infty}^{y} f_{XY}(u, v)\, du\, dv \end{aligned} \tag{3.13}$$

If two random variables are discrete, then

$$F_X(x, y) = \sum_{x_i \leq x} \sum_{y_i \leq y} P_{XY}(x_i, y_i) \tag{3.14a}$$

where their joint probability function is

$$P_{XY}(x_i, y_i) = \Pr[X = x_i \text{ and } Y = y_i] \tag{3.14b}$$

If X and Y are two random variables and the distribution of X is not influenced by the value taken by Y, and vice versa, the two random variables are said to be *independent*. For independent random variables

$$\Pr[a \leq X \leq b \text{ and } c \leq Y \leq d] = \Pr[a \leq X \leq b]\Pr[c \leq Y \leq d] \tag{3.15}$$

for any a, b, c, and d. As a result,

$$F_{XY}(x, y) = F_X(x)F_Y(y) \tag{3.16}$$

which implies that for continuous random variables

$$f_{XY}(x, y) = f_X(x)f_Y(y) \tag{3.17}$$

while for discrete random variables

$$P_{XY}(x, y) = P_X(x)P_Y(y) \tag{3.18}$$

Other useful concepts are those of the *marginal* and *conditional distributions*. If X and Y are two random variables with a joint cumulative distribution function $F_{XY}(x, y)$, then $F_X(x)$, the marginal cumulative distribution of X, is just the cumulative distribution of X ignoring Y. The marginal cumulative distribution function of X equals

$$F_X(x) = \Pr[X \leq x] = \lim_{y \to \infty} F_{XY}(x, y) \tag{3.19}$$

where the limit is equivalent to letting Y take on any value. If X and Y are continuous random variables, the marginal density of X is given by

$$f_X(x) = \int_{-\infty}^{+\infty} f_{XY}(x, y)\, dy \tag{3.20}$$

The conditional cumulative distribution function is the cumulative distribution function for X given that Y has taken a particular value y. The conditional cumulative distribution function for continuous random variables is given by

$$F_{X|Y}(x \mid y) = \Pr[X \leq x \mid Y = y] = \int_{-\infty}^{x} \frac{f_{XY}(s, y)\, ds}{f_Y(y)} \tag{3.21}$$

The conditional density function is

$$f_{X|Y}(x \mid y) = \frac{f_{XY}(x, y)}{f_Y(y)} \tag{3.22}$$

For discrete random variables, the probability of observing $X = x$, given that $Y = y$ equals

$$P_{X|Y}(x \mid y) = \frac{P_{XY}(x, y)}{P_Y(y)} \tag{3.23}$$

These results can be extended to more than two random variables.

3.2.2 The Expectation Operator

Knowledge of the density function of a continuous random variable or of the probability function of a discrete random variable allows one to calculate the expected value of any function of the random variable. If g is a real-valued function of a continuous random variable X, the expected value of $g(X)$ is

$$E[g(X)] = \int_{-\infty}^{+\infty} g(x) f_X(x)\, dx \tag{3.24}$$

while for a discrete random variable

$$E[g(X)] = \sum_{x_i} g(x_i) P_X(x_i) \tag{3.25}$$

$E[\,\cdot\,]$ is called the *expectation operator*. It has several important properties. In particular, the expectation of a linear function of X is a linear function of the expectation of X. If a and b are constants,

$$E[a + bX] = a + bE[X] \tag{3.26}$$

The expectation of a function of two random variables is given by

$$E[g(X, Y)] = \int_{-\infty}^{+\infty} \int_{-\infty}^{+\infty} g(x, y) f_{XY}(x, y)\, dx\, dy$$

or

$$E[g(X, Y)] = \sum_{x_i} \sum_{y_i} g(x_i, y_i) P_{XY}(x_i, y_i) \tag{3.27}$$

If X and Y are independent, the expectation of the product of a function of X and a function of Y is the product of the expectations

$$E[g(X)h(Y)] = E[g(X)]E[h(Y)] \tag{3.28}$$

3.2.3 Quantiles, Moments, and Their Estimators

While the cumulative distribution function provides a complete specification of the properties of a random variable, it is useful to use some simpler and easily understood measures of the central tendency and range of values that a random variable may assume. Perhaps the simplest approach to describing a random variable's distribution is to report the value of several quantiles. The pth quantile of a random variable X is a value x_p such that

$$\Pr[X < x_p] \le p \le \Pr[X \le x_p] \tag{3.29}$$

If X is a continuous random variable, then in the region where $f_X(x) > 0$, the quantiles are uniquely defined and are obtained by solution of

$$F_X(x_p) = p \tag{3.30}$$

Frequently reported quantiles are the *median* $x_{0.50}$ and the *upper and lower quartiles* $x_{0.25}$ and $x_{0.75}$. The median describes the location or central tendency of the distribution of X because the random variable is, in the continuous case, equally likely to be above as below this value. The interquartile range $[x_{0.25}, x_{0.75}]$ provides an easily understood description of the range of values that the random variable may assume.

In a given application, particularly when safety is of concern, it may be appropriate to use other quantiles. In floodplain management and the design of flood control structures, the 100-year flood $x_{0.99}$ is often the selected design value. In water quality management, a river's minimum 7-day, 10-year low flow is often used as the critical design value.

The natural sample estimate of the median $x_{0.50}$ is the median of the sample. In a sample of size n where $x_{(1)} \leq x_{(2)} \leq \ldots \leq x_{(n)}$ are the observations ordered by magnitude, and for k a non-negative integer, $n = 2k$ (even) or $n = 2k + 1$ (odd), the sample estimate of the median is

$$\hat{x}_{0.50} = \begin{cases} x_{(k+1)} & \text{for } n = 2k + 1 \\ \frac{1}{2}[x_{(k)} + x_{(k+1)}] & \text{for } n = 2k \end{cases} \tag{3.31}$$

Sample estimates of other quantiles may be obtained by using $x_{(i)}$ as an estimate of x_q for $q = i/(n + 1)$ and then interpolating between observations to obtain $\hat{x}_p$ for the desired p. This only works for $1/(n + 1) \leq p \leq n/(n + 1)$ and can yield rather poor estimates of x_p when $(n + 1)p$ is near either 1 or n. An alternative approach is to fit a reasonable distribution function to the observations, as discussed in Section 3.3, and then estimate x_p via equation 3.30, where $F_X(x)$ is the fitted distribution.

Another simple and common approach to describing a distribution's center, spread, and shape is by reporting the moments of a distribution. The first moment about the origin μ_X, is the *mean* of X and is given by

$$\mu_X = E[X] = \int_{-\infty}^{+\infty} x f_X(x)\, dx$$

or

$$\mu_X = E[X] = \sum_{x_i} x_i P_X(x_i) \tag{3.32}$$

Moments other than the first are normally measured about the mean. The second moment measured about the mean is the *variance*, denoted $\text{Var}(X)$ or σ_X^2:

$$\sigma_X^2 = \text{Var}(X) = E[(X - \mu_X)^2] \tag{3.33}$$

The *standard deviation* σ_X is the square root of the variance. While the mean μ_X is a measure of the central value of X, the standard deviation σ_X is a measure of the spread of the distribution of X about μ_X. Another measure of the variability in X is the *coefficient of variation*,

$$\text{CV}_X = \frac{\sigma_X}{\mu_X} \tag{3.34}$$

which expresses the standard deviation as a proportion of the mean. The coefficient of variation is useful for comparing the relative variability of the flow in rivers of different sizes.

The third moment measured about the mean, denoted λ_X, measures the asymmetry or *skewness* of the distribution:

$$\lambda_X = E[(x - \mu_X)^3] \tag{3.35}$$

Typically, the coefficient of skewness γ_X is reported rather than the third moment λ_X. The coefficient of skewness is the third moment rescaled by the standard deviation so as to be dimensionless and hence unaffected by the scale of the random variable.

$$\gamma_X = \frac{\lambda_X}{\sigma_X^3} \tag{3.36}$$

Streamflows and other natural phenomena that are necessarily positive often have distributions with positive skew coefficients, reflecting the asymmetric shape of their distributions.

When the distribution of a random variable is not known but a set of observations $\{x_1, \ldots, x_n\}$ is available, the moments of the distribution of X can be estimated with the sample values as

$$\begin{aligned}
\hat{\mu}_X &= \bar{x} = \frac{1}{n}\sum_{i=1}^{n} x_i \\
\hat{\sigma}_X^2 &= v_X^2 = \frac{1}{n}\sum_{i=1}^{n} (x_i - \bar{x})^2 \\
\hat{\lambda}_X &= \frac{1}{n}\sum_{i=1}^{n} (x_i - \bar{x})^3 \\
\widehat{\mathrm{CV}}_X &= \frac{v_X}{\bar{x}} \\
\hat{\gamma}_X &= \frac{\hat{\lambda}_X}{v_X^3}
\end{aligned} \tag{3.37}$$

The sample estimate of the mean and variance are often denoted $\bar{x}$ and v_X^2. All of these sample estimates must be understood to be only estimates. Unless n is very large, their deviation from the true values of μ_X, σ_X^2, λ_X, CV_X, and γ_X may be large.

When discussing the accuracy of sample estimates, two quantities are often considered, *bias* and *variance*. An estimator $\hat{\theta}$ of a quantity θ is a function of the observed values of the random variable $X_1, \ldots, X_n$; $\hat{\theta}$ may be written $\hat{\theta}[X_1, X_2, \ldots, X_n]$ to emphasize that $\hat{\theta}$ itself is a random variable because its value depends on the observed values of the random variable. An estimator $\hat{\theta}$ of a quantity θ is biased if $E[\hat{\theta}] \neq \theta$ and unbiased if $E[\hat{\theta}] = \theta$. The quantity $\{E[\hat{\theta}] - \theta\}$ is normally called the *bias of the estimator*.

An unbiased estimator has the property that its expected value equals the value of the quantity to be estimated. The sample mean is an unbiased estimate of μ_X because

$$E[\bar{x}] = E\left[\frac{1}{n}\sum_{i=1}^{n} X_i\right] = \frac{1}{n}\sum_{i=1}^{n} E[X_i] = \mu_X \tag{3.38}$$

The estimator v_X^2 is a biased estimator of σ_X^2 where [1]

$$E[v_X^2] = \frac{n-1}{n}\sigma_X^2 \tag{3.39}$$

Hence if many samples of size n are taken in a simulation experiment and the resulting values of v_X^2 are averaged, the average will approach $(n - 1)\sigma_X^2/n$ and not σ_X^2 as might be desired. For this reason, the unbiased estimate of the variance

$$s_X^2 = \frac{nv_X^2}{n-1} \tag{3.40}$$

is often used instead of v_X^2. There is relatively little difference between the two estimators for moderate size n. Both, however, generally produce biased estimates of the standard deviation σ_X [35], although the bias decreases with increasing n.

The second important statistic often used to assess the accuracy of an estimator $\hat{\theta}$ is its variance $\text{Var}(\hat{\theta})$, which equals $E\{(\hat{\theta} - E[\hat{\theta}])^2\}$. For the mean of a set of independent observations, the variance of the sample mean is given by

$$\text{Var}(\bar{x}) = \frac{\sigma_X^2}{n} \tag{3.41}$$

It is common to call $\sigma_X/\sqrt{n}$ the *standard error* of $\bar{x}$ rather than its standard deviation.

The bias measures the difference between the average value of an estimator and the quantity to be estimated. The variance measures the spread or width of the estimator's distribution. Both contribute to the amount by which an estimate deviates from the quantity to be estimated. These two errors are often combined into the *mean square error*, defined as

$$\begin{aligned} \text{MSE}(\hat{\theta}) &= E[(\hat{\theta} - \theta)^2] = \{E[\hat{\theta}] - \theta\}^2 + E\{(\hat{\theta} - E[\hat{\theta}])^2\} \\ &= [\text{Bias}]^2 + \text{Var}(\hat{\theta}) \end{aligned} \tag{3.42}$$

where Bias is $E(\hat{\theta}) - \theta$. The MSE is the expected or average squared deviation of the estimator from the true value of the parameter. It is a convenient measure of how closely $\hat{\theta}$ approximates θ.

Estimation of the coefficient of skewness γ_X provides a good example of the use of the MSE for evaluating the total deviation of an estimate from the true population value. The sample estimate $\hat{\gamma}_X$ of γ_X is often biased, has a large variance, and was shown by Kirby [21] to be bounded so that

$$|\hat{\gamma}_X| \leq \frac{n-2}{(n-1)^{1/2}} \tag{3.43}$$

where n is the sample size. The bounds do not depend on the true skew γ_X. The bias and variance of $\hat{\gamma}_X$ depend on the sample size and the actual distribution of X. Table 3.2 contains the expected value and standard deviation of $\hat{\gamma}_X$ when X has either a normal distribution, for which $\gamma_X = 0$, or a gamma distribution with $\gamma_X = 0.25$, 0.50, 1.00, 2.00, or 3.00.

TABLE 3.2. Sampling Properties of Estimate of Coefficient of Skewness

Expected Value of $\hat{\gamma}_X$

		SAMPLE SIZE			
Distribution of X		*10*	*20*	*50*	*80*
Normal	$\gamma_X = 0$	0.00	0.00	0.00	0.00
Gamma	$\gamma_X = 0.25$	0.13	0.18	0.22	0.23
	$\gamma_X = 0.50$	0.26	0.36	0.44	0.46
	$\gamma_X = 1.00$	0.51	0.70	0.85	0.91
	$\gamma_X = 2.00$	0.97	1.32	1.63	1.74
	$\gamma_X = 3.00$	1.34	1.82	2.25	2.49
Upper bound on skew		2.67	4.13	6.86	8.78

Standard Deviation of $\hat{\gamma}_X$

		SAMPLE SIZE			
Distribution of X		*10*	*20*	*50*	*80*
Normal	$\gamma_X = 0$	0.58	0.47	0.33	0.26
Gamma	$\gamma_X = 0.25$	0.58	0.48	0.34	0.27
	$\gamma_X = 0.50$	0.58	0.49	0.36	0.30
	$\gamma_X = 1.00$	0.59	0.53	0.43	0.37
	$\gamma_X = 2.00$	0.61	0.63	0.60	0.56
	$\gamma_X = 3.00$	0.62	0.70	0.75	0.76

Source: J. R. Wallis, N. C. Matalas, and J. R. Slack, *Just a Moment! Appendix*, National Technical Information Service, PB-231 816, Springfield, Va., 1974.

To illustrate the magnitude of these errors, consider the mean square error of $\hat{\gamma}_X$ calculated from a sample of size 50 when X has a gamma distribution with $\gamma_X = 0.50$, a reasonable value for annual streamflows. The expected value of $\hat{\gamma}_X$ is 0.44; its variance equals $(0.36)^2$, its standard deviation squared. Using equation 3.42, the mean square error of $\hat{\gamma}_X$ is

$$\begin{aligned} \text{MSE}(\hat{\gamma}_X) &= (0.44 - 0.50)^2 + (0.36)^2 = 0.0036 + 0.1296 \\ &= 0.133 \end{aligned} \tag{3.44}$$

An unbiased estimate of γ_X is $(0.50/0.44)\hat{\gamma}_X$; the mean square error of this unbiased estimate of γ_X is

$$\text{MSE}\left(\frac{0.50\hat{\gamma}_X}{0.44}\right) = (0.50 - 0.50)^2 + \left[\left(\frac{0.50}{0.44}\right)(0.36)\right]^2 = 0.167 \qquad (3.45)$$

The mean square error of the unbiased estimate of γ_X is larger than the mean square error of the biased estimate. Unbiasing $\hat{\gamma}_X$ results in a larger mean square error for all the cases listed in Table 3.2 except for the normal distribution and the gamma distribution with $\gamma_X = 3.00$.

As shown here for the skew coefficient, biased estimators often have smaller mean square errors than unbiased estimators; because the mean square error measures the total average deviation of an estimator from the quantity being estimated, this result demonstrates that the strict or unquestioning use of unbiased estimators is not advisable. Additional information on the sampling distribution of quantiles and moments is contained in Appendix 3A.

3.3 DISTRIBUTIONS OF RANDOM EVENTS

A frequent task in water resources planning is the development of a model of some probabilistic or stochastic phenomena such as streamflows, flood flows, rainfall, temperatures, or evaporation. This generally requires that one fit a probability distribution function to a set of observed values of the random variable. Sometimes, one's immediate objective is to estimate a particular quantile of the distribution, such as the 100-year flood or the 7-day, 10-year low flow. Then the fitted distribution will supply an estimate of this quantity. In a stochastic simulation model, fitted distributions are used to generate possible values of the random variable in question.

This section provides a brief introduction to the techniques useful for estimating the parameters of a probability distribution function and determining if the fitted distribution provides a reasonable or acceptable model of the data. Sections are also included on families of distributions based on the normal and gamma distributions. These two families have found frequent use in water resource planning.

3.3.1 Parameter Estimation and Model Adequacy

Given a set of observations to which a distribution is to be fit, one first selects a distribution function to serve as a model of the distribution of the data. The choice of distribution may be based on experience with data of that type, some understanding of the mechanisms giving rise to the data, and/or examination of the observations themselves. One can then proceed to estimate the chosen distribution's parameters and to determine if the fitted distribution provides an acceptable model of the data. A model is generally judged to be unacceptable if it is unlikely that one could have observed the given observations were they actually drawn from the fitted distribution.

In many cases, good estimates of a distribution's parameters are obtained by the *maximum-likelihood-estimation* procedure. Given a set $\{X_1, \ldots, X_n\}$ of independent observations of a continuous random variable X, the joint density of the set is, using equation 3.17,

$$f_{X_1X_2\ldots X_n}(x_1, \ldots, x_n \mid \boldsymbol{\theta}) = f_X(x_1 \mid \boldsymbol{\theta}) \cdot f_X(x_2 \mid \boldsymbol{\theta}) \ldots f_X(x_n \mid \boldsymbol{\theta}) \tag{3.46}$$

where $\boldsymbol{\theta}$ is the vector of the distribution's parameters.

The maximum likelihood estimate of $\boldsymbol{\theta}$ is that vector which maximizes equation 3.46 and thereby makes it as likely as possible to have observed $\{x_1, \ldots, x_n\}$. Considerable work has gone into studying the properties of maximum likelihood parameter estimates. Under rather general conditions, asymptotically the estimated parameters are normally distributed, unbiased, and have the smallest possible variance of any asymptotically unbiased estimator [2]. These, of course, are asymptotic properties, valid for large n. Better estimation procedures, perhaps yielding biased parameter estimates, may exist for small sample sizes. Still, maximum likelihood procedures are to be highly recommended with moderate and large samples, even though the iterative solution of nonlinear equations is sometimes required.

An example of the maximum likelihood procedure for which closed-form expressions for the parameter estimates are obtained is provided by the lognormal distribution. The density function of a lognormally distributed random variable X is

$$f_X(x) = \frac{1}{x\sqrt{2\pi\sigma^2}} \exp\left[-\frac{1}{2\sigma^2}(\ell\text{n}\,(x) - \mu)^2\right] \tag{3.47}$$

Here the parameters μ and σ^2 are the mean and variance of the logarithm of X and not of X itself.

Maximization of the logarithm of the joint density of $\{x_1, \ldots, x_n\}$ is more convenient than maximization of the joint density itself, so the problem becomes the maximization of the *log-likelihood function*

$$\begin{aligned} L &= \ell\text{n} \prod_{i=1}^{n} [f(x_i \mid \mu, \sigma)] \\ &= \sum_{i=1}^{n} \ell\text{n}\, f(x_i \mid \mu, \sigma) \\ &= -\sum_{i=1}^{n} \ell\text{n}\,(x_i\sqrt{2\pi}) - n\,\ell\text{n}\,(\sigma) - \frac{1}{2\sigma^2}\sum_{i=1}^{n} [\ell\text{n}\,(x_i) - \mu]^2 \end{aligned} \tag{3.48}$$

The maximum can be obtained by equating the partial derivatives $\partial L/\partial\sigma$ and $\partial L/\partial\mu$ to zero.

$$0 = \frac{\partial L}{\partial \mu} = \frac{1}{\sigma^2} \sum_{i=1}^{n} [\ell\text{n}\,(x_i) - \mu]$$
$$0 = \frac{\partial L}{\partial \sigma} = -\frac{n}{\sigma} + \frac{1}{\sigma^3} \sum_{i=1}^{n} [\ell\text{n}\,(x_i) - \mu]^2 \tag{3.49}$$

and thereby obtaining

$$\hat{\mu} = \frac{1}{n} \sum_{i=1}^{n} \ell\text{n}\,(x_i)$$
$$\hat{\sigma}^2 = \frac{1}{n} \sum_{i=1}^{n} [\ell\text{n}\,(x_i) - \hat{\mu}]^2 \tag{3.50}$$

The second-order conditions are met and these values do maximize equation 3.48. Note that if one defines a new random variable $Y = \ell\text{n}\ X$, then the maximum likelihood estimates of the parameters μ and σ^2 are the sample estimates of the mean and variance of Y.

$$\hat{\mu} = \bar{y}$$
$$\hat{\sigma}^2 = v_Y^2 \tag{3.51}$$

The second commonly used parameter estimation procedure is the *method of moments*. The method of moments is a quick and simple method for obtaining parameter estimates for many distributions. For a distribution with $m = 1, 2$, or 3 parameters, the first m moments of the postulated distribution are equated to the estimates of the moments calculated using equations 3.37. The resulting nonlinear equations are solved for the unknown parameters. For the lognormal distribution,

$$\mu_X = \exp\,(\mu + \tfrac{1}{2}\sigma^2)$$
$$\sigma_X^2 = \exp\,(2\mu + \sigma^2)[\exp\,(\sigma^2) - 1] \tag{3.52}$$

Substituting $\bar{x}$ and s_X^2 for μ_X and σ_X^2 and solving for μ and σ^2 one obtains

$$\hat{\mu} = \ell\text{n}\left(\frac{\bar{x}}{\sqrt{1 + s_X^2/\bar{x}^2}}\right)$$
$$\hat{\sigma}^2 = \ell\text{n}\,(1 + s_X^2/\bar{x}^2) \tag{3.53}$$

In general, maximum likelihood estimates are to be preferred to those obtained by the moment method. Computer programs for fitting most distributions used in hydrology by both the moments and maximum likelihood methods are contained in Kite [22]; the programs also supply estimates of the standard error of estimated quantiles.

After estimating the parameters of a distribution, it is mandatory that some check of the model's adequacy be made. Such checks vary from simple comparisons of the observations with the fitted model using graphs or tables to rigorous statistical tests. Some of the early and simplest methods of parameter estimation were graphical techniques. Although quantitative techniques are more accurate and precise for parameter estimation, graphical presentations are invaluable for comparing the fitted distribution with the observations for the detection of systematic or unexplained deviations between the two.

The graphical evaluation of the adequacy of a fitted distribution is generally performed by plotting the observations so that they would fall approximately on a straight line if the postulated distribution were the true distribution of the observation. This can be done with the use of special probability papers for some distributions or with the more general technique presented here. Let $\{x_i\}$ denote the observed values and $x_{(i)}$ the ith largest value in the sample so that $x_{(1)} \leq x_{(2)} \leq \ldots \leq x_{(n)}$. Now the random variable $X_{(i)}$ provides a reasonable estimate of the (x_p)th quantile of the true distribution of X for $p = i/(n+1)$. In fact, if one thinks of the cumulative probability U_i associated with the random variable $X_{(i)}$, $U_i = F_X(X_{(i)})$, then if the observations $X_{(i)}$ are independent, the U_i have a beta distribution [12] with density function[1]

$$f_{U_i}(u) = \frac{n!}{(i-1)!(n-i)!} u^{i-1}(1-u)^{n-i} \qquad 0 \leq u \leq 1 \tag{3.54}$$

which has mean and variance

$$\begin{aligned} E[U_i] &= \frac{i}{n+1} \\ \text{Var}(U_i) &= \frac{i(n-i+1)}{(n+1)^2(n+2)} \end{aligned} \tag{3.55}$$

A good graphical check of the adequacy of a fitted distribution $G(x)$ is obtained by plotting the observations $x_{(i)}$ versus $G^{-1}[i/(n+1)]$ [38]. Even if $G(x)$ exactly equaled $F_X(x)$, the true distribution of X, the plotted points will not fall exactly on a 45° line through the origin of the graph; this would only occur if $F_X[x_{(i)}]$ exactly equaled $i/(n+1)$ and therefore $x_{(i)}$ exactly equaled $F_X^{-1}[i/(n+1)]$. An appreciation for how far an individual $x_{(i)}$ can be expected to deviate from $G^{-1}[i/(n+1)]$ can be obtained by plotting $G^{-1}[u_i^{(0.75)}]$ and $G^{-1}[u_i^{(0.25)}]$, where $u_i^{(0.75)}$ and $u_i^{(0.25)}$ are the upper and lower quartiles of the distribution of U_i obtained from integrating the density function in equation 3.54.

Figures 3.2 and 3.3 illustrate the use of this *quantile–quantile plotting technique* by displaying the results of fitting a normal and a lognormal distribution to the annual maximum flows of the Magra River, Italy, at Calamazza for the years 1930–1970. The observations $x_{(i)}$, given in Table 3.3, are plotted on the vertical axis against the quantiles $G^{-1}[i/(n+1)]$ on the horizontal axis.

[1]This distribution can be derived by using that fact that because $X_{(i)}$ is the ith largest observation in a sample of n independent random variables with distribution $F_X(\cdot)$, the distribution of $U_i = F_X[X_{(i)}]$ is that of the ith largest observation in a sample of n uniformly distributed random variables.

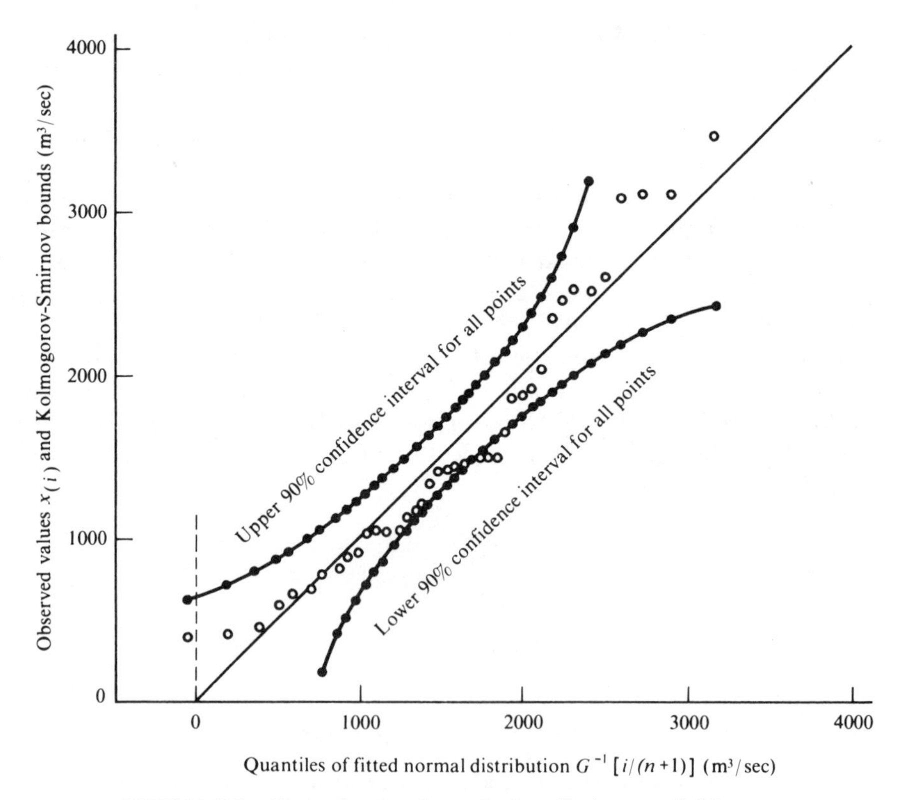

FIGURE 3.2. Plot of annual maximum discharge of Magra River, Italy, versus quantiles of fitted normal distribution.

Rigorous statistical tests are available for trying to determine whether or not it is reasonable to assume that a given set of observations could have been drawn from a particular family of distributions. Although not the most powerful of such tests, the *Kolmogorov–Smirnov test* provides bounds within which *every observation* should lie if the sample is actually drawn from the assumed distribution. In particular, for $G = F_X$, the test specifies that

$$\Pr\left[G^{-1}\left(\frac{i}{n} - C_\alpha\right) \leq X_{(i)} \leq G^{-1}\left(\frac{i-1}{n} + C_\alpha\right) \text{ for every } i\right] = 1 - \alpha \tag{3.56}$$

where C_α is the critical value of the test at significance level α. Formulas for C_α as a function of n are contained in Table 3.4 for three cases: (1) when G is completely specified independent of the sample's values; (2) when G is the normal distribution, and the mean and variance of X are estimated from the sample as $\bar{x}$ and s_X^2; and (3) when G is the exponential distribution and the scale parameter is estimated as $1/(\bar{x})$. For other distributions, the values obtained from Table 3.4 may be used to construct approximate simultaneous confidence intervals for every $x_{(i)}$.

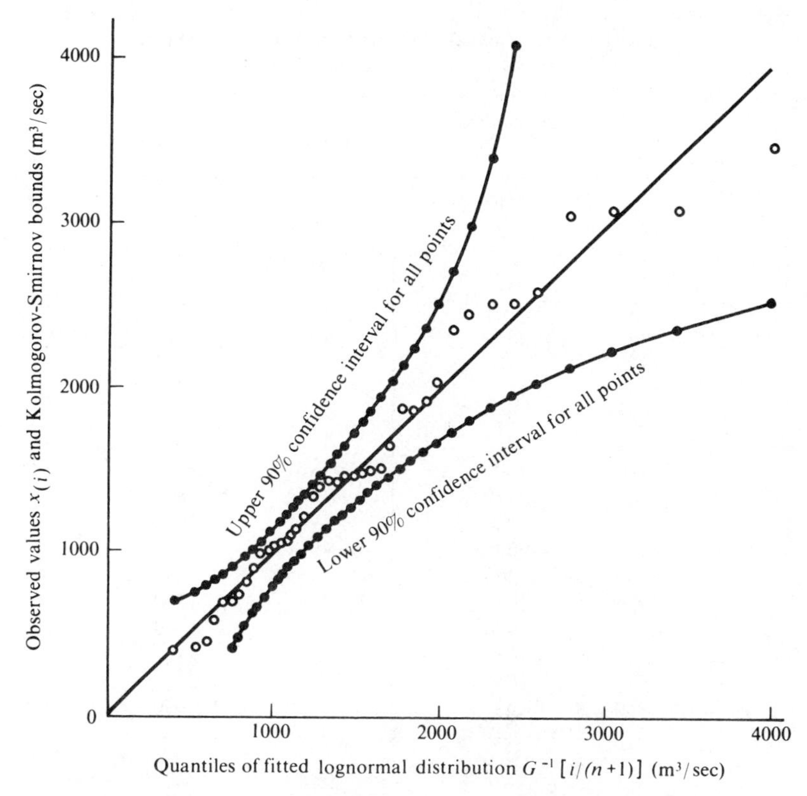

FIGURE 3.3. Plot of annual maximum discharges of Magra River, Italy, versus quantiles of fitted lognormal distribution.

Figures 3.2 and 3.3 contain 90% confidence intervals for the plotted points constructed in this manner. For the normal distribution, the critical value of C_α equals $0.819/(\sqrt{n} - 0.01 + 0.85/\sqrt{n})$, where 0.819 corresponds to $\alpha = 0.10$. For $n = 40$, $C_\alpha = 0.127$. As can be seen in the figures, the annual maximum flows are not consistent with the hypothesis that they were drawn from a normal distribution; three of the observations lie outside the simultaneous 90% confidence intervals for all points, demonstrating a statistically significant lack of fit. The fitted normal distribution underestimates the quantiles corresponding to small and large probabilities while overestimating the quantiles in an intermediate range. In Figure 3.3, deviations between the fitted lognormal distribution and the observations can be attributed to the differences between $F_X(x_{(i)})$ and $i/(n + 1)$. Generally, the points are all near the 45° line through the origin, and no major systematic deviations are apparent.

To actually check if a set of observations or their logarithms are normally distributed, the Shapiro–Wilk test or the nearly as powerful and much simplier correlation test proposed by Filliben [8] should be used (see Exercise

TABLE 3.3. Annual Maximum Discharges on Magra River, Italy, at Calamazza, 1930–1970[a]

Date	*Discharge* (m^3/s)	*Date*	*Discharge* (m^3/s)
1930	410	1951	3070
1931	1150	1952	2360
1932	899	1953	1050
1933	420	1954	1900
1934	3100	1955	1130
1935	2530	1956	674
1936	758	1957	683
1937	1220	1958	1500
1938	1330	1959	2600
1939	1410	1960	3480
1940	3100	1961	1430
1941	2470	1962	809
1942	929	1963	1010
1943	586	1964	1510
1944	450	1965	1650
1946	1040	1966	1880
1947	1470	1967	1470
1948	1070	1968	1920
1949	2050	1969	2530
1950	1430	1970	1490

[a]Value for 1945 is missing.

TABLE 3.4. Critical Values[a] of Kolmogorov–Smirnov Statistic as a Function of Sample Size n

	SIGNIFICANCE LEVEL α				
	0.150	*0.100*	*0.050*	*0.025*	*0.010*
1. F_X completely specified:					
$C_\alpha(\sqrt{n} + 0.12 + 0.11/\sqrt{n})$	1.138	1.224	1.358	1.480	1.628
2. F_X normal with mean and variance estimated as $\bar{x}$ and s_X^2					
$C_\alpha{}'\sqrt{n} - 0.01 + 0.85/\sqrt{n})$	0.775	0.819	0.895	0.995	1.035
3. F_X exponential with scale parameter β estimated as $1/(\bar{x})$					
$(C_\alpha - 0.2/n)(\sqrt{n} + 0.26 + 0.5/\sqrt{n})$	0.926	0.990	1.094	1.190	1.308

[a]Values of C_α are calculated as follows: for case 2, $\alpha = 0.10$, $C_\alpha = 0.819/(\sqrt{n} - 0.01 + 0.85/\sqrt{n})$.

Source: This table is taken from Stephens [30].

3.17). The Shapiro–Wilk test is discussed at length in Hahn and Shapiro [14] and Shapiro and Wilk [28]. These tests are generally more powerful than the Kolmogorov–Smirnov test at detecting departures from normality [8]. The advantage of the Kolmogorov–Smirnov test is that it gives simultaneous confidence intervals for all the observations. These are extremely useful in any graphical evaluation of the adequacy of a fitted distribution.

3.3.2 The Normal and Lognormal Distributions

The normal distribution and its transformation, the lognormal distribution, are certainly the most widely used distributions in science and engineering. The density function of a normal random variable is

$$f_X(x) = \frac{1}{\sqrt{2\pi\sigma^2}} \exp\left[-\frac{1}{2\sigma^2}(x-\mu)^2\right] \qquad -\infty < x < +\infty \tag{3.57}$$

where μ and σ^2 are equivalent to μ_X and σ_X^2, the mean and variance of X. Interestingly, the maximum likelihood estimates of μ and σ^2 are identical to the moment estimates $\bar{x}$ and the biased variance estimate v_X^2.

The normal distribution is symmetric about μ_X and admits values from $-\infty$ to $+\infty$. Thus it is not always satisfactory for modeling physical phenomena such as streamflows, which are necessarily positive and have skewed distributions. A frequently used model for skewed distributions is the lognormal distribution. A random variable X has a lognormal distribution if the natural logarithm of X, $\ell n\ X$, has a normal distribution. If X is lognormally distributed, then by definition $\ell n\ X$ is normally distributed, so that the density function of X is

$$\begin{aligned} f_X(x) &= \frac{1}{\sqrt{2\pi\sigma^2}} \exp\left\{-\frac{1}{2\sigma^2}\left[\ell n\,(x) - \mu^2\right]\right\} \frac{d(\ell n\, x)}{dx} \\ &= \frac{1}{x\sqrt{2\pi\sigma^2}} \exp\left\{-\frac{1}{2\sigma^2}[\ell n\,(x/\mu)]^2\right\} \end{aligned} \tag{3.58}$$

for $x > 0$. A lognormal random variable takes on values in the range $[0, +\infty]$. The parameter μ determines the scale of the X-distribution while σ^2 determines the distribution's shape. The mean and variance of the lognormal distribution are given in equation 3.52. The coefficient of variation and skew are

$$\begin{aligned} \mathrm{CV}_X &= [\exp(\sigma^2) - 1]^{1/2} \\ \gamma_X &= 3\mathrm{CV}_X + \mathrm{CV}_X^3 \end{aligned} \tag{3.59}$$

The maximum likelihood estimates of μ and σ^2 are given in equation 3.50 and the moment estimates in equation 3.53. For reasonable-size samples, the maximum likelihood estimates are generally as good or better than the moment estimates [29].

A useful generalization of the two-parameter lognormal distribution is the shifted lognormal or three-parameter lognormal distribution obtained by letting $\ell n\,(X - \tau)$ be normally distributed, where $X \geq \tau$. Theoretically, τ should be positive if for physical reasons X must be positive; practically, negative values of τ can be allowed when the resulting probability of negative values of X is sufficiently small.

Unfortunately, maximum likelihood estimates of the parameters μ, σ^2, and τ are poorly behaved because of irregularities in the likelihood function [10]. The method of moments does fairly well when the skew of the fitted distribution is reasonably small. A method that does almost as well as the moment method for low-skew distributions, and much better for highly skewed, distributions estimates τ by [29]

$$\hat{\tau} = \frac{x_{(1)}x_{(n)} - \hat{x}_{0.50}^2}{x_{(1)} + x_{(n)} - 2\hat{x}_{0.50}} \tag{3.60}$$

provided that $x_{(1)} + x_{(n)} - 2\hat{x}_{0.50} > 0$, where $x_{(1)}$ and $x_{(n)}$ are the smallest and largest observations and $\hat{x}_{0.50}$ is the sample median; if $x_{(1)} + x_{(n)} - 2\hat{x}_{0.50} < 0$, the sample tends to be negatively skewed and a three-parameter lognormal distribution cannot be fit with this method. Good estimates of μ and σ^2 are then

$$\hat{\mu} = \ell n\left[\frac{\bar{x} - \hat{\tau}}{\sqrt{1 + s_X^2/(\bar{x} - \hat{\tau})^2}}\right] \tag{3.61}$$

$$\hat{\sigma}^2 = \ell n\left[1 + \frac{s_X^2}{(\bar{x} - \hat{\tau})^2}\right] \tag{3.62}$$

3.3.3 The Gamma Distribution and Its Generalizations

The gamma distribution has long been used to model many natural phenomena, including daily, monthly, and annual streamflows as well as flood flows. For a gamma random variable X,

$$\begin{aligned} f_X(x) &= \frac{|\beta|}{\Gamma(\alpha)}(\beta x)^{\alpha-1}e^{-\beta x} \quad \text{for } \beta x \geq 0 \\ \mu_X &= \frac{\alpha}{\beta} \\ \sigma_X^2 &= \frac{\alpha}{\beta^2} \\ \gamma_X &= \frac{2}{\sqrt{\alpha}} = 2\text{CV}_X \end{aligned} \tag{3.63}$$

Here $\Gamma(\alpha)$ is the gamma function; for integer α, $\Gamma(\alpha) = (\alpha - 1)!$ The parameter $\alpha > 0$ determines the shape of the distribution; β is the scale parameter.

The gamma distribution arises naturally in many problems in mathematical statistics and hydrology. It also has a very reasonable shape for such nonnegative random variables as rainfall and streamflow. Unfortunately, its cumulative distribution function is not available in closed form, except for integer α. The gamma family includes two special cases: (1) the exponential distribution when $\alpha = 1$, and (2) the chi-squared distribution when β equals $\frac{1}{2}$ and 2α is an integer.

The gamma distribution has several generalizations. If a constant τ is subtracted from X so that $(X - \tau)$ has a gamma distribution, the distribution of X is a three-parameter gamma. This is also called a *Pearson Type 3 distribution*, because the resulting distribution belongs to the third type of a class of distributions suggested by the statistician Karl Pearson. Another variation is the log Pearson Type 3 distribution obtained by fitting $\ell n\,(X)$ with a Pearson Type 3 distribution. This distribution has found wide use in modeling flood frequencies and has been recommended for that purpose [33]. A discussion of the unusual shapes that this hybrid distribution may take allowing negative values of β is presented by Bobee [4].

The method of moments may be used to estimate the parameters of the gamma distribution. For the three-parameter gamma distribution

$$\hat{\tau} = \bar{x} - 2\left(\frac{s_X}{\hat{\gamma}_X}\right)$$

$$\hat{\alpha} = \frac{4}{(\hat{\gamma}_X)^2} \tag{3.64}$$

$$\hat{\beta} = \frac{2}{s_X \hat{\gamma}_X}$$

where $\bar{x}$, s_X^2, and $\hat{\gamma}_X$ are estimates of the mean, variance, and coefficient of skewness of the distribution of X [5]. For the two-parameter gamma distribution,

$$\hat{\alpha} = \frac{(\bar{x})^2}{s_X^2} \tag{3.65}$$

$$\hat{\beta} = \frac{\bar{x}}{s_X^2} \tag{3.66}$$

Studies by Thom [31] and Matalas and Wallis [25] have shown that maximum likelihood parameter estimates are superior to the moment estimates. Matalas and Wallis [25] present an iteration procedure to obtain parameter estimates for the three-parameter distribution. For the two-parameter gamma distribution, Greenwood and Durand [11, 13, 22] give approximate formulas for the maximum likelihood estimates.

When plotting the observed and fitted quantiles of a gamma distribution, an approximation to the inverse of the distribution function is useful. For $|\gamma| \leq 3$, the Wilson–Hilferty transformation,

$$x_G = \mu + \sigma\left[\frac{2}{\gamma}\left(1 + \frac{\gamma x_N}{6} - \frac{\gamma^2}{36}\right)^3 - \frac{2}{\gamma}\right] \tag{3.67}$$

gives the quantiles x_G of the gamma distribution in terms of x_N, the quantiles of the standard-normal distribution; here μ, σ, and γ are the mean, standard deviation, and coefficient of skewness of x_G. The approximation is discussed by Kirby [20], who provides a modification to make the approximation acceptable for $|\gamma| > 3$.

3.4 STOCHASTIC PROCESSES AND TIME SERIES

Many important random variables in water resources are functions whose value changes with time. Historic records of rainfall or streamflow at a particular site are a sequence of observations called a *time series*. In a time series, the observations are ordered by time, and it is generally the case that the observed value of the random variable at one time influences the distribution of the random variable at later times; the observations are not independent. Time series are conceptualized as being a single observation of a *stochastic process* which is a generalization of the concept of a random variable.

This section has two parts. The first part presents the concept of stationarity and the basic statistics generally used to describe the properties of a stationary stochastic process. The properties of sample estimates of these statistics are discussed in Appendix 3C. The second part of this section presents the definition of a Markov process and the Markov chain model which is used in stochastic optimization models in Section 3.6.2 and in Chapter 7.

3.4.1 Describing Stochastic Processes

A random variable whose value changes through time according to probabilistic laws is called a *stochastic process*. An observed *time series* is said to be one realization of a stochastic process, just as a single observation of a random variable is one possible value the random variable may assume. In the development here, a stochastic process is a sequence of random variables $\{X(t)\}$ ordered by a discrete time variable $t = 1, 2, 3, \ldots$.

The properties of a stochastic process must generally be determined from a single time series or realization. To do this several assumptions are usually made. First, one generally assumes that the process is *stationary*. This means that the probability distribution of the process is not changing over time. In symbols, if $X(t)$ is a stationary stochastic process, then

$$F_{X(t)}[X(t)] = F_X[X(t)] \tag{3.68}$$

In addition, if a process is strictly stationary, the joint distribution of the random variables $X(t_1), \ldots, X(t_n)$ is identical to the joint distribution of $X(t_1 + t), \ldots, X(t_n + t)$ for any t; the joint distribution depends only on the differences $t_i - t_j$ between the times of occurrence of the events.

For a stationary stochastic process, one can write the mean and variance as

$$\mu_X = E[X(t)]$$

and

$$\sigma_X^2 = \text{Var}\,[X(t)] \tag{3.69}$$

independent of time t. The *autocorrelations*, the correlation of X with itself, are given by

$$\rho_X(k) = \frac{\text{Cov}[X(t), X(t+k)]}{\sigma_X^2} \tag{3.70}$$

for any positive integer k. These are the statistics most often used to describe stationary stochastic processes.

When one has available only a single time series, it is necessary to estimate the values of μ_X, σ_X^2, and $\rho_X(k)$ from values of the random variable that one has observed. The mean and variance are generally estimated as they were in equation 3.37,

$$\begin{aligned} \hat{\mu}_X &= \bar{x} = \frac{1}{T}\sum_{t=1}^{T} x_t \\ \hat{\sigma}_X^2 &= v_X^2 = \frac{1}{T}\sum_{t=1}^{T}(x_t - \bar{x})^2 \end{aligned} \tag{3.71}$$

while the autocorrelations $\rho_X(k)$ can be estimated as [6]

$$\hat{\rho}_X(k) = r_k = \frac{\sum_{t=1}^{T-k}(x_{t+k} - \bar{x})(x_t - \bar{x})}{\sum_{t=1}^{T}(x_t - \bar{x})^2} \tag{3.72}$$

The sampling distribution of these estimators depends on the correlation structure of the stochastic process giving rise to the time series. In particular, when the observations are positively correlated as are natural streamflows or annual benefits in a river basin simulation, the variance of $\bar{x}$ and v_X^2 are larger than would be the case if the observations were independent. It is sometimes wise to take this inflation into account. Appendix 3C discusses the sampling distribution of these statistics.

All of this analysis depends on the assumption of stationarity for only then do the quantities defined in equations 3.71 and 3.72 have the intended meaning. Stochastic processes are not always stationary. Urban development, deforestation, agricultural development, climatic shifts, and changes in regional resource management can alter the distribution of rainfall, streamflows, or groundwater levels over time. If a stochastic process is not essentially stationary *over the time span in question*, then statistical techniques that rely on the stationary assumption cannot be employed and the problem generally becomes much more difficult.

3.4.2 Markov Processes and Markov Chains

A common assumption in many stochastic water resources models is that the stochastic process $X(t)$ is a *Markov process*. A Markov process has the property that the dependence of future values of the process on past values is summarized by the current value. In symbols for $k > 0$,

$$F_X[X(t+k) \mid X(t), X(t-1), X(t-2), \ldots] = F_X[X(t+k) \mid X(t)] \quad (3.73)$$

For Markov processes, the current value summarizes the state of the processes. As a consequence, the current value of the process is often referred to as the *state*. This makes physical sense as well when one refers to the state or level of an aquifer or reservoir.

A special kind of Markov process is one whose state $X(t)$ can take on only discrete values. Such a processes is called a *Markov chain*. Often in water resources planning, continuous stochastic processes are approximated by Markov chains. This is done to facilitate the construction of simple stochastic optimization models. The stochastic optimization models in this chapter and in Chapter 7 use this approximation. This section presents the basic notation and properties of Markov chains.

Consider a stream whose annual flow is to be represented by a discrete random variable. Assume that the distribution of streamflows is stationary. In the following development, the continuous random variable representing annual streamflows or some other process is approximated by a discrete random variable Q_y in year y, which takes on values q_i with unconditional probabilities p_i where

$$\sum_{i=1}^{n} p_i = 1 \quad (3.74)$$

It is frequently the case that the value of Q_{y+1} is not independent of Q_y. Such dependence can be modeled by a Markov chain. This requires specification of the *transition probabilities*,

$$p_{ij} = \Pr[Q_{y+1} = q_j \mid Q_y = q_i] \quad (3.75)$$

A transition probability is the conditional probability that the next state is q_j, given that the current state is q_i. The transition probabilities satisfy

$$\sum_{j=1}^{n} p_{ij} = 1 \qquad \text{for all } i \quad (3.76)$$

Figure 3.4 illustrates a possible set of transition probabilities. These probabilities can also be displayed in a matrix, as shown in Figure 3.5. Each element p_{ij} in the matrix is the probability of a transition from streamflow q_i in one year to streamflow q_j in the next. In this example, a low flow tends to be followed by a low flow rather than a high flow, and vice versa.

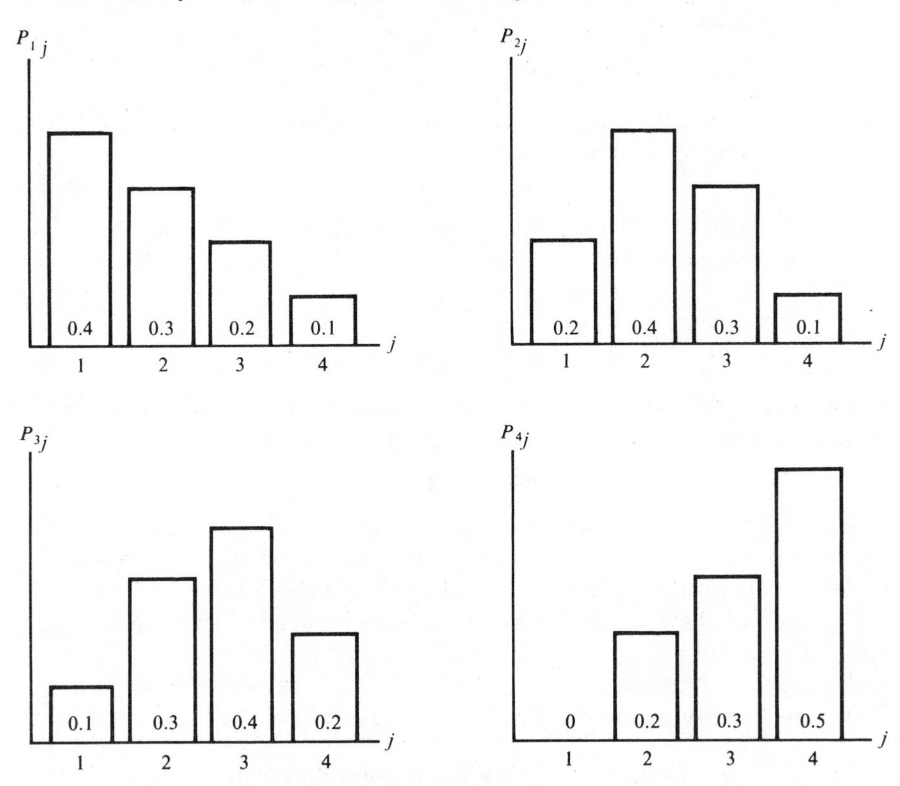

FIGURE 3.4. Conditional probabilities of annual streamflows Q_{y+1} equaling q_j given that Q_y equals q_i for i = 1, 2, 3, 4.

		Probability of streamflow q_j in year $y + 1$			
	i \ j	1	2	3	4
When streamflow was q_i in year y	1	0.4	0.3	0.2	0.1
	2	0.2	0.4	0.3	0.1
	3	0.1	0.3	0.4	0.2
	4	0.0	0.2	0.3	0.5

FIGURE 3.5. Matrix of streamflow transition probabilities p_{ij}.

Let $\mathbf{P}$ be the transition matrix whose elements are p_{ij}. For a Markov chain, the transition matrix contains all the information necessary to describe the behavior of the system. Let p_i^y be the probability that the system resides in state i in year y. Then the probability that $Q_{y+1} = q_j$ is the sum of the probabilities p_i^y that $Q_y = q_i$ times the probability p_{ij} that $Q_{y+1} = q_j$ given that $Q_y = q_i$; in symbols,

$$p_j^{y+1} = p_1^y p_{1j} + p_2^y p_{2j} + \cdots + p_n^y p_{nj} = \sum_{i=1}^{n} p_i^y p_{ij} \tag{3.77}$$

Letting $\mathbf{p}^y$ be the row vector of state resident probabilities $(p_1^y, \ldots, p_n^y)$, this relationship may be written

$$\mathbf{p}^{(y+1)} = \mathbf{p}^{(y)}\mathbf{P} \tag{3.78}$$

To calculate the probabilities of each streamflow state in year $y + 2$, one can use $\mathbf{p}^{(y+1)}$ in equation 3.78 to obtain

$$\mathbf{p}^{(y+2)} = \mathbf{p}^{(y+1)}\mathbf{P} \tag{3.79a}$$

or

$$\mathbf{p}^{(y+2)} = \mathbf{p}^{(y)}\mathbf{P}^2 \tag{3.79b}$$

Continuing in this manner, it is possible to compute the probabilities of each possible streamflow state for years $y + 1, y + 2, y + 3, \ldots, y + k, \ldots$ as

$$\mathbf{p}^{(y+k)} = \mathbf{p}^{(y)}\mathbf{P}^k \tag{3.80}$$

Returning to the four-state example in Figures 3.4 and 3.5, assume that the flow in year y is q_2. Hence in year y the unconditional streamflow probabilities p_i^y are (0, 1, 0, 0). Knowing each p_i^y, the probabilities p_j^{y+1} corresponding to each of the four streamflow states can be determined. From Figure 3.4, the probabilities p_j^{y+1} are 0.2, 0.4, 0.3, and 0.1 for $j = 1, 2, 3$, and 4, respectively.

The probability vectors for nine future years are listed in Table 3.5. Notice

TABLE 3.5. Successive Streamflow Probability Vectors

	STREAMFLOW STATE PROBABILITIES			
Year	p_1^y	p_2^y	p_3^y	p_4^y
y	0.000	1.000	0.000	0.000
$y+1$	0.200	0.400	0.300	0.100
$y+2$	0.190	0.330	0.310	0.170
$y+3$	0.173	0.316	0.312	0.199
$y+4$	0.163	0.312	0.314	0.211
$y+5$	0.159	0.310	0.315	0.216
$y+6$	0.157	0.309	0.316	0.218
$y+7$	0.156	0.309	0.316	0.219
$y+8$	0.156	0.309	0.316	0.219
$y+9$	0.156	0.309	0.316	0.219

that as time progresses, the probabilities reach limiting values. These are the *unconditional* or *steady-state* probabilities. The quantity p_i has been defined as the unconditional probability of q_i. These are the steady-state probabilities which $\mathbf{p}^{(y+k)}$ approaches for large k. It is clear from Table 3.5 that as k becomes larger, equation 3.77 becomes

$$p_j = \sum_{i=1}^{n} p_i p_{ij} \tag{3.81a}$$

or in vector notation equation 3.78 becomes

$$\mathbf{p} = \mathbf{pP} \tag{3.81b}$$

where $\mathbf{p}$ is the row vector of unconditional probabilities $(p_1, \ldots, p_n)$. For this example, $\mathbf{p}$ equals (0.156, 0.309, 0.316, 0.219).

The steady-state probabilities for any Markov chain can be found by solving simultaneous equations 3.81 for all but one of the states j together with the constraint

$$\sum_{i=1}^{n} p_i = 1 \tag{3.74}$$

Annual streamflows are seldom as highly correlated as the flows in this example. However, monthly, weekly, and especially daily streamflows generally have high serial correlations. Assuming that the unconditional steady-state probability distributions for monthly streamflows are stationary, a Markov chain can be defined for each month's streamflow. Since there are 12 months in a year, there would be 12 transition matrices, the elements of which could be denoted as p_{ij}^t. Each defines the probability of a streamflow q_j^{t+1} in month $t + 1$, given a streamflow q_i^t in month t. The steady-state stationary probability vectors for each month can be found by the procedure outlined above, except that now all 12 matrices are used to calculate all 12 steady-state probability vectors.

3.5 PLANNING WITH UNCERTAINTY

As discussed in the introduction, uncertainty is always an element in water resources planning, whether it arises from our inability to predict future population densities and water needs or from the variability of natural sources of water. There are several ways of dealing with uncertainty in water resource systems planning. The next section presents a simple and widely used technique called sensitivity analysis. With this approach, the system is modeled as if the values of uncertain quantities are known. Then the effect of changes in particular parameters and combinations of parameters is assessed. Subsequent sections present more sophisticated techniques of incorporating uncertainty into objective functions and the decision process.

These include the use of expected net benefits, the max-min decision criterion, and utility theory.

3.5.1 Sensitivity Analysis

A simple method for assessing the effect of uncertainty on system performance is to vary the magnitude of the more uncertain parameters or variables (singularly or in combination) and then to examine the results. Such sensitivity analyses provide a means of identifying those parameters or variables to which system performance is particularly sensitive. Whether or not these uncertainties should be reduced by additional study or data collection depends on the cost of the additional information needed to reduce the uncertainty and on the possible increase in benefits such reduction may provide.

A simple capacity expansion model is used here to illustrate the technique of sensitivity analysis and to present three alternative measures of the cost of errors in the model's parameters. This example and the concepts presented are based primarily on the work of H.A. Thomas [32], reproduced in Meta Systems, Inc. [26]. A similar example is provided by Binkley [3].

The classical capacity expansion model for water treatment plants, sewage treatment plants, pipelines, canals, and other structures, which have long lives and show significant economies of scale, can be formulated as follows. The cost of a structure with capacity x is often well approximated by the power function

$$K(x) = ax^b \tag{3.82}$$

where K includes both initial capital costs and the present value of operating and maintenance costs that are related to capacity. The parameter b lies between zero and 1 and reflects the economies of scale. In particular, b is the elasticity of costs defined as

$$b = \frac{x}{K}\frac{dK}{dx} \tag{3.83}$$

As a result, if $b = 0.50$, then a 10% change in capacity will result in only a 5% change in costs. This follows from rearranging equation 3.83 to yield

$$\frac{dK}{K} = b\frac{dx}{x} \tag{3.84}$$

Typical values of b for water treatment and wastewater treatment facilities range from 0.60 to 0.80. The value of b may be as low as 0.35 for some canals and storage dams which have large economies of scale, whereas b is closer to 1 for large systems with many replicate units, such as large municipal wastewater treatment systems [23].

To simplify this example, operating and maintenance costs not related to capacity are assumed to be a negligible fraction of the total cost or else independent of capacity and dependent only on the rate of use of the facilities.

Consider now the problem of scheduling the construction of capacity to meet a growing demand or need. For $b < 1$, it is economically advantageous to build ahead of anticipated demand to capture the economies of scale that result from construction of large facilities. However, these savings must be weighed against the opportunity cost of committing resources to build capacity which may not be needed for many years and possibly not at all.

The problem may be formulated as follows. Assume that (1) the point in time has been reached when capacity is just sufficient to meet demand, (2) the demand for additional services is a linear function $g \cdot t$ of time t, and (3) the demand or need for service must be met with additional capacity. Here the constant linear rate of growth of demand or need g accounts for increases in population and industrial water use, pricing policies and zoning, and other restrictions. Capacity will be added every τ years to meet the anticipated increase in demand $g\tau$ during the subsequent period, as illustrated in Figure 3.6. The optimality of expansion every τ years for some τ is demonstrated by Manne [24], who introduced this model in a more general framework.

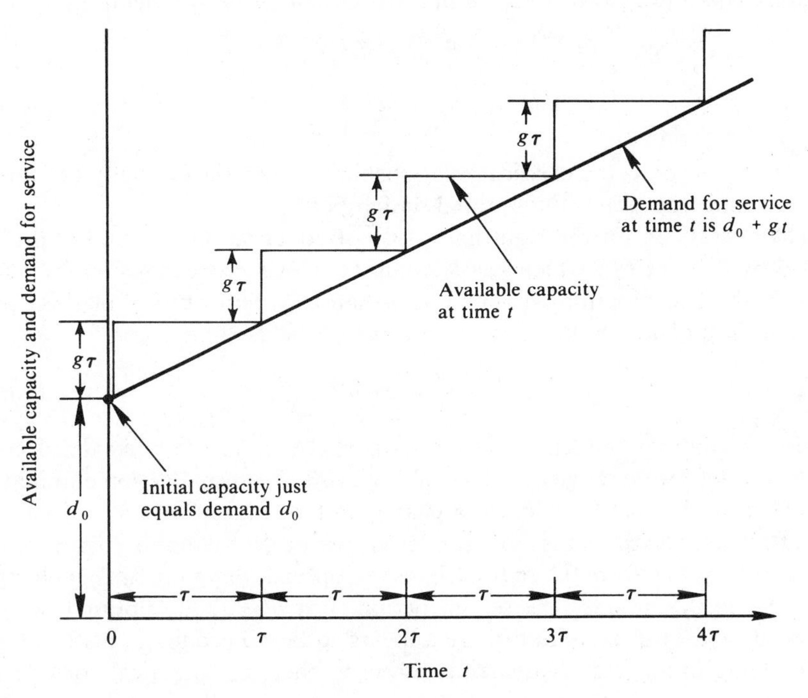

FIGURE 3.6. Future capacity to be added in increments of $g\tau$ every τ years in anticipation of increased demand for service.

With this pattern of expansion, the system will have excess capacity except for every τth year, when capacity will be just sufficient to meet demand.

Using a continuous-time social discount rate r, the discounted present value of the cost of an infinite sequence of capacity additions is

$$C(\tau) = a(g\tau)^b + e^{-r\tau}[a(g\tau^b)] + e^{-2r\tau}[a(g\tau)^b] + \ldots$$

or

$$C(\tau) = \frac{a(g\tau)^b}{1 - e^{-r\tau}} \tag{3.85}$$

Instead of the infinite planning horizon used here, one could use a finite-length planning horizon. Then appropriate compensation would need to be made for those plans having excess capacity in the final period. Minimization of the discounted cost of an infinite sequence of expansions avoids such difficulties.

Rearranging terms in equation 3.85 yields

$$\frac{C(\tau)}{ag^b} = \frac{\tau^b}{1 - e^{-r\tau}} \tag{3.86}$$

Setting the derivative of this expression with respect to the design period length τ equal to zero to find the optimal design period τ^* yields

$$b(\tau^{b-1})(1 - e^{-r\tau}) - \tau^b(re^{-r\tau}) = 0$$

or

$$b = \frac{r\tau^*}{e^{r\tau^*} - 1} \tag{3.87}$$

The last equation is the classical relationship between the elasticity of cost b, the discount rate r, and the optimal design period τ^*.

The expression on the right-hand side of equation 3.87 is well approximated by $(1 - r\tau/6)^3$, which has a similar Taylor series expansion [32]; for $r\tau \leq 2$, the two functions differ by less than 6%. Substitution of this relationship into equation 3.87 and solving for τ^* yields

$$\tau^* = \frac{6}{r}(1 - b^{1/3}) \tag{3.88}$$

Table 3.6 contains the least-cost design period τ^* and the normalized cost $C(\tau^*)/ag^b$ for a wide range of values of r and b. This normalization eliminates the effect on the cost function of a change in the value of a and g which do not affect the length of the optimal design period (see equation 3.86). Note that a change in b from 0.9 to 0.7 triples the optimal design period; a change from 0.7 to 0.5 doubles the design period. For $b = 1$, the optimal design period decreases to zero, and there appears to be no economic justification for building in advance of demand; however, there are practical considerations that make it advantageous to go infrequently through the cycle of planning, financing, and construction of facilities. These considerations

TABLE 3.6. Optimal Design Period τ^* for Different Annual Discount Rates r and Cost Elasticities b

τ^* (*year*)	r (*year*$^{-1}$)	b	$C(\tau^*)/ag^b$
41	0.03	0.5	0.0
22	0.03	0.7	18.0
6.9	0.03	0.9	30.4
25	0.05	0.5	7.0
13.5	0.05	0.7	12.6
4.1	0.05	0.9	19.2
12.4	0.10	0.5	5.0
6.7	0.10	0.7	7.8
2.1	0.10	0.9	10.3
8.3	0.15	0.5	4.0
4.5	0.15	0.7	5.8
1.4	0.15	0.9	7.1

include the social cost of the time required on the part of public officials and others to make and implement a decision.

Our discussion now turns to the question of what is the cost if incorrect values of the parameters a, b, r, and g are used in the analysis. Thus a non-optimal design period length τ could be selected and a plant with a non-optimal capacity x might be built. The optimal design capacity and the true cost of any capacity expansion strategy depends on the parameter values: a, b, r, and g.

The choice of these values results from analysis of demographic, economic, and engineering data. The quality and quantity of such data are limited, as is our ability to forecast the appropriate values for a particular project. The following technique of sensitivity analysis is useful for assessing the cost and the effect of errors in the design parameters. The method is predicated on the assumption of a set of "true" or "correct" parameters denoted by the vector $\mathbf{w}$. Knowledge of the values of the true parameters allows determination of the optimal design capacity $x^* = g\tau^*$. The true parameter values are not really known, so that best available estimates must be used in practice.

The possible effects of errors in the specified parameter values are analyzed with the use of a set of possibly incorrect parameter values $\hat{g}$, $\hat{r}$, $\hat{b}$, and $\hat{a}$, denoted as $\hat{\mathbf{w}}$. Use of these values when determining the appropriate capacity of the facility yields nonoptimal design capacity $\hat{x}$.

Table 3.7 reports the optimal capacity x^* for a set of true parameters and the capacities $\hat{x}$ that would be selected if each of the parameters were incorrectly assumed to have a value 25% too large or too small. The table also

TABLE 3.7. A Simple Approach to Sensitivity Analysis

Case	*Parameter[a] in Error*	*Capacity, $\hat{x}$*	*Cost, $C(\hat{x} \mid \hat{\mathbf{w}})$*
	Overestimation of a Parameter by 25%		
1	None	$6.7 = x^*$	$7.755 = C(x^* \mid \mathbf{w})$
2	$\hat{a} = 1.25$	6.7	9.694
3	$\hat{r} = 0.125$	5.4	6.633
4	$\hat{b} = 0.875$	2.6	10.078
5	$\hat{g} = 1.25$	8.4	9.066
	Underestimation of a Parameter by 25%		
6	$\hat{a} = 0.75$	6.7	5.816
7	$\hat{r} = 0.075$	9.0	9.485
8	$\hat{b} = 0.525$	11.6	5.275
9	$\hat{g} = 0.75$	5.0	6.340

[a]True parameter values are $a = 1, g = 1, r = 0.10$, and $b = 0.70$. In each case, at most one parameter is incorrectly specified.

reports the estimated cost of the nonoptimal designs calculated with the possibly incorrect parameters

$$C(\hat{x} \mid \hat{\mathbf{w}}) = C(\hat{x} \mid \hat{g}, \hat{r}, \hat{b}, \hat{a}) = \frac{\hat{a}(\hat{x})^b}{1 - e^{-\hat{r}\hat{x}/\hat{g}}} \tag{3.89}$$

This is the cost of meeting the projected demand that a designer would calculate if he or she used the incorrect design parameters.

Table 3.7 illustrates the usual approach to sensitivity analysis, which determines how sensitive the values of the objective function and decision variables are to misspecification of the parameters. This type of information is readily obtained when one has an optimization algorithm which, for any set of parameters, determines the optimal solution. Examination of Table 3.7 reveals that except for parameter a, the optimal decision and the minimum cost are quite sensitive to errors in the parameters.

Although this is a deceptively simple approach to doing sensitivity analysis, it provides little help to the planner who wants to know what the costs or consequences are of misspecifying some parameter and thus implementing a nonoptimal design. Table 3.7 gives, for each vector of parameters $\hat{\mathbf{w}}$, only the minimum cost $C(\hat{x} \mid \hat{\mathbf{w}})$ achievable with that $\hat{\mathbf{w}}$. The table provides no information on the real cost, $C(\hat{x} \mid \mathbf{w})$, of a nonoptimal design $\hat{x}$ evaluated with the true parameter values $\mathbf{w}$.

When evaluating the cost of using incorrect parameter values, three costs may be considered:

1. The minimal cost of the optimal design x^* correctly evaluated with the true parameter values, $C(x^* \mid \mathbf{w})$.
2. The cost of the nonoptimal design incorrectly evaluated, $C(\hat{x} \mid \hat{\mathbf{w}})$.
3. The cost of the nonoptimal design correctly evaluated, $C(\hat{x} \mid \mathbf{w})$.

Here $C(x^* \mid \mathbf{w})$ and $C(\hat{x} \mid \hat{\mathbf{w}})$ are calculated as specified in equations 3.85 and 3.89. The appropriate definition of the cost of the nonoptimal design correctly evaluated, $C(\hat{x} \mid \mathbf{w})$, is not always clear. One can assume that both the first and all subsequent expansions are of size $\hat{x}$. Alternatively, it may be more reasonable to assume that while the first expansion of size $\hat{x}$ is not optimal, subsequent expansions will be of an optimal size even though future errors undoubtedly will occur. The appropriate assumption may depend on one's problem. The assumption is made here that organizations would not continue to make the same mistake, although different errors should not be ruled out. The cost of the nonoptimal design evaluated with knowledge of the true parameter values is calculated as

$$C(\hat{x} \mid \mathbf{w}) = a(\hat{x})^b + C(x^* \mid \mathbf{w})e^{-r\hat{x}/g} \tag{3.90}$$

The difference between the two methods for calculating $C(\hat{x} \mid \mathbf{w})$ is small in this example.

Misspecification of a parameter may result in significant *loss of economic efficiency*, expressed here as a percentage of the minimum possible cost:

$$\text{LEE} = 100\left[\frac{C(\hat{x} \mid \mathbf{w}) - C(x^* \mid \mathbf{w})}{C(x^* \mid \mathbf{w})}\right] \tag{3.91}$$

The loss of economic efficiency LEE is the amount by which the actual costs of implementing the nonoptimal decision exceeds the costs that would result if the optimal design x^* were selected. Because x^* achieves the minimal value of $C(x \mid \mathbf{w})$, LEE is never less than zero. A large value of LEE indicates that a large misallocation of resources would result from planning using the incorrect values of the design parameters.

Table 3.8 expands upon Table 3.7 and reports for each $\hat{\mathbf{w}}$ the cost of the nonoptimal design incorrectly evaluated, $C(\hat{x} \mid \hat{\mathbf{w}})$; its cost correctly evaluated, $C(\hat{x} \mid \mathbf{w})$; and the loss of economic efficiency, LEE. An examination of the table indicates that only the errors in b result in more than a 1% loss of economic efficiency. Under- and overestimation of b by 25% results in losses of economic efficiency equal to only 3.1% and 2.3%. The reason for this is illustrated in Figure 3.7, which shows the total discounted cost $C(\hat{x} \mid \mathbf{w})$ as a function of x, the capacity of the first expansion. Large changes in the decision variable $\hat{x}$ result in only small changes in the total cost. This immunity from serious resource misallocation may account for the traditional disinterest in refined economic analysis in some branches of civil engineering.

Another important kind of error is denoted here as MMC, for *misrepresentation of minimal costs*. When planning, it is necessary (1) to estimate the costs of alternative plans, (2) to decide whether any plan should be undertaken, and, if so, (3) to select the desired plan. If estimated costs are in error, an approach for solving a problem may be selected that results in significantly higher costs than would be incurred by an alternative. An example of such a decision would be whether to supply water to a remote seashore village by

TABLE 3.8. Evaluation of the Consequences of Misspecification of Parameters Values

Case	*Parameter in Error*[a]	$\hat{x}$	$C(\hat{x}\mid\hat{\mathbf{w}})$[b]	$C(\hat{x}\mid\mathbf{w})$[c]	*Loss of Economic Efficiency, LEE* (%)	*Misrepresentation of Minimal Cost, MMC* (%)	*Error in Actual Cost, EAC* (%)
	Consequences of Overestimating Design Parameters by 25%						
1	None	6.7	7.755	7.755	—	—	—
2	$\hat{a} = 1.25$	6.7	9.694	7.755	0	25.	25.
3	$\hat{r} = 0.125$	5.4	6.633	7.775	0.26	−15.	−14.
4	$\hat{b} = 0.875$	2.6	10.078	7.932	2.3	30.	28.
5	$\hat{g} = 1.25$	8.4	9.066	7.784	0.37	17.	17.
	Consequences of Underestimating Design Parameters by 25%						
1	None	6.7	7.755	7.755	—	—	—
6	$\hat{a} = 0.75$	6.7	5.816	7.755	0	−25.	−25.
7	$\hat{r} = 0.075$	9.0	9.485	7.809	0.69	22.	22.
8	$\hat{b} = 0.525$	11.6	5.275	7.992	3.1	−32.	−35.
9	$\hat{g} = 0.75$	5.0	6.340	7.789	0.44	−18.	−19.

[a] True parameter values are $a = 1, g = 1, r = 0.10$, and $b = 0.70$; in each case, at most, one parameter is incorrectly specified.

[b] $C(\hat{x}\mid\hat{\mathbf{w}}) = \hat{a}(\hat{x})^{\hat{b}} + e^{-\hat{r}\hat{x}/\hat{g}}\, C(\hat{x}\mid\hat{\mathbf{w}})$; cost of nonoptimal design, incorrectly evaluated.

[c] $C(\hat{x}\mid\mathbf{w}) = a(\hat{x})^{b} + e^{-r\hat{x}/g}\, C(x^*\mid\mathbf{w})$; only first decision is nonoptimal; cost of nonoptimal design, correctly evaluated.

gravity pipeline from a remote reservoir, by pumping from groundwater, or by installation of a desalination unit.

Misrepresentation of minimal costs calculated as a percentage of the minimum possible cost is

$$\text{MMC} = 100\left[\frac{C(\hat{x}\mid\hat{\mathbf{w}}) - C(x^*\mid\mathbf{w})}{C(x^*\mid\mathbf{w})}\right] \tag{3.92}$$

The quantity MMC may be positive or negative. Table 3.8 indicates that errors in any of the parameters can result in large errors of this type. Whether or not such errors are important depends on whether or not plans or projects are undertaken which would or would not have been undertaken if their true costs were known.

A third type of error can also be important, particularly when some parties have contracted to perform a specific service for a given amount. Parameter errors result in incorrect estimates of the actual cost to construct facilities or to provide service. This *error in actual cost*, again as a percentage of the minimum possible cost, is measured by

$$\text{EAC} = 100\left[\frac{C(\hat{x}\mid\hat{\mathbf{w}}) - C(\hat{x}\mid\mathbf{w})}{C(x^*\mid\mathbf{w})}\right] \tag{3.93}$$

EAC differs from MMC in that EAC compares the incorrectly calculated cost $C(\hat{x}\mid\hat{\mathbf{w}})$ for the proposed design $\hat{x}$ with what it will actually cost, $C(\hat{x}\mid\mathbf{w})$,

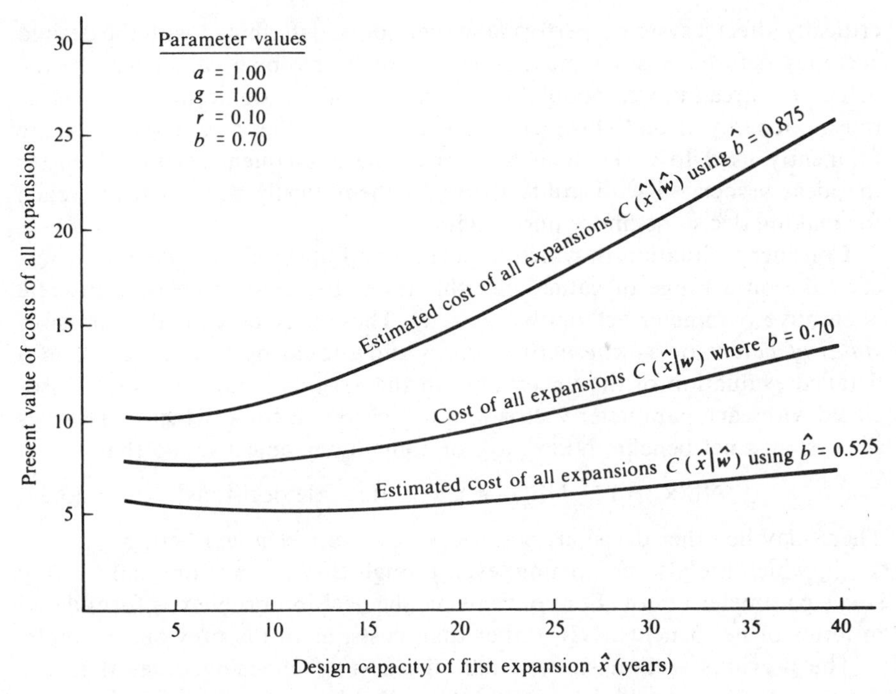

FIGURE 3.7. Comparison of the cost of various plans evaluated with correct w and incorrect $\hat{w}$ values of *b*.

and not with the cost for which the service could be provided, $C(x^* \mid \mathbf{w})$. As shown in Table 3.8, for this example, the error in actual costs EAC differs only slightly from the misrepresentation of minimal costs MMC. This is because $C(x^* \mid \mathbf{w})$ and $C(\hat{x} \mid \mathbf{w})$, the costs of the two designs correctly evaluated, are very close.

It is important to note that these measures of the effect of parameter error *ignore the sensitivity of the optimal decision* to the errors in the parameters. As can be seen from Tables 3.7 and 3.8, errors in *g*, *r*, and *b* result in significant changes in what would be the optimal decision. This example shows that one must not be misled into thinking that when the optimal decision is sensitive to parameter errors, economic benefits are as well. As Table 3.8 and Figure 3.7 show, *there can still be little loss of economic efficiency, even when parameter errors result in decisions very different from the optimal one.*

3.5.2 Incorporating Uncertainty into Decisions

Sensitivity analysis considers the possible costs of making other than the optimal decision. It does not address the question of what decision should be made when the future values of parameters and other variables that will critically affect a system's performance cannot be determined with the desired accuracy before decisions must be made. Such unknown values may be the values of streamflows, populations, agricultural or domestic water usage rates, or energy prices. This section presents approaches or criteria which are frequently used to make such decisions. The subsequent section discusses the ideas associated with utility theory, a theoretically consistent approach for making decisions under uncertainty.

Consider a situation in which important and uncertain model parameters can take on a range of values. Let this range be represented by n discrete alternative parameter vectors $\mathbf{w}_1, \ldots, \mathbf{w}_n$. These may be viewed as possible *states of nature* or as *alternative futures* and should be chosen based on a detailed evaluation of the uncertainty in the various parameters [37]. Associated with each parameter vector $\mathbf{w}_i$ is a decision vector $\mathbf{x}_i$ which maximizes the system's net benefits $\mathrm{NB}(\mathbf{x}_i \mid \mathbf{w}_i)$, or some other objective, so that

$$\mathrm{NB}(\mathbf{x}_i \mid \mathbf{w}_i) \geq \mathrm{NB}(\mathbf{x} \mid \mathbf{w}_i), \ \mathbf{x} \in \{\text{feasible decisions}\} \tag{3.94}$$

There may be other decisions besides $\{\mathbf{x}_1, \ldots, \mathbf{x}_n\}$, denoted here $\{\mathbf{x}_{n+1}, \ldots, \mathbf{x}_{n+m}\}$, which are also competing, even though they are not optimal for any single parameter vector. For convenience the decision problem is formulated in terms of net benefits, NB, rather than costs, as in the previous example.

The previous section showed that the cost of choosing other than the optimal decision could be measured by the loss of economic efficiency, $\mathrm{LEE}(\mathbf{x}_i \mid \mathbf{w}_j)$, sometimes called *regret*. The loss of economic efficiency or regret is the nonnegative amount by which the net benefits could have been increased by selection of the optimal decision vector $\mathbf{x}_j$ for parameter vector $\mathbf{w}_j$ rather than decision x_i:

$$\begin{aligned} \mathrm{LEE}(\mathbf{x}_i \mid \mathbf{w}_j) &= \max_k \,[\mathrm{NB}(\mathbf{x}_k \mid \mathbf{w}_j)] - \mathrm{NB}(\mathbf{x}_i \mid \mathbf{w}_j) \\ &= \mathrm{NB}(\mathbf{x}_j \mid \mathbf{w}_j) - \mathrm{NB}(\mathbf{x}_i \mid \mathbf{w}_j) \geq 0 \end{aligned} \tag{3.95}$$

Tables 3.9 and 3.10 contrast the net benefit matrix and the regret matrix for a hypothetical irrigation development. The three values of $\mathbf{w}$ correspond to possible levels of future demand in the region. Design vectors $\mathbf{x}_1$, $\mathbf{x}_2$, and $\mathbf{x}_3$ correspond to large, medium, and small levels of development where $\mathbf{w}_1$, $\mathbf{w}_2$, and $\mathbf{w}_3$ correspond to high, medium, and low levels of future demand. Design vector $\mathbf{x}_4$ corresponds to a staged development which foregoes economies of scale in construction to ensure that the facilities constructed can be utilized.

Decisions $\mathbf{x}_i$ for $i = 1, 2, 3$ are the optimal decision vectors for parameters $\mathbf{w}_i$. These decisions do not always do well if $\mathbf{w}$ takes on a value other than the assumed value. The design $\mathbf{x}_4$ is a compromise which is never the best performer, although it always does fairly well. This is illustrated by the regret matrix, which shows that the loss of economic efficiency associated with selection of $\mathbf{x}_4$ never exceeds 10.

TABLE 3.9. Net-Benefit Matrix $NB(\mathbf{x}_i|\mathbf{w}_j)$, Expected Net Benefits, and Smallest Possible Net Benefits Achievable with Each Decision for a Hypothetical Irrigation Project

		$NB(\mathbf{x}_i\|\mathbf{w}_j)$				
Parameter Sets:		$\mathbf{w}_1$	$\mathbf{w}_2$	$\mathbf{w}_3$	$E_{\mathbf{w}}[NB(\mathbf{x}_i\|\mathbf{w})]$[a]	$\min_j NB(\mathbf{x}_i\|\mathbf{w}_j)$
Decisions	$\mathbf{x}_1$	50	30	0	26.7	0
	$\mathbf{x}_2$	30	45	5	26.7	5
	$\mathbf{x}_3$	30	25	20	25.0	20
	$\mathbf{x}_4$	40	35	10	28.3	10

[a]Assuming that $\mathbf{w}_1$, $\mathbf{w}_2$, and $\mathbf{w}_3$ are equally likely.

TABLE 3.10. Regret Matrix $LEE(\mathbf{x}_i|\mathbf{w}_j)$ and Expected Loss of Economic Efficiency for Hypothetical Irrigation Project

		$LEE(\mathbf{x}_i\|\mathbf{w}_j)$			
Parameter Sets:		$\mathbf{w}_1$	$\mathbf{w}_2$	$\mathbf{w}_3$	$E_{\mathbf{w}}[LEE(\mathbf{x}_i\|\mathbf{w})]$[a]
Decisions	$\mathbf{x}_1$	0	15	20	11.7
	$\mathbf{x}_2$	20	0	15	11.7
	$\mathbf{x}_3$	20	20	0	13.3
	$\mathbf{x}_4$	10	10	10	10.0

[a]Assuming that $\mathbf{w}_1$, $\mathbf{w}_2$, and $\mathbf{w}_3$ are equally likely.

What plan should be chosen? This decision should depend on the attitudes and preferences toward benefits and risks held by the farmers, taxpayers, and others affected by the project. These value judgments cannot be replaced by mathematical calculations. The contribution of models and calculations to the decision-making process should be the information necessary to make informed decisions. However, sometimes values or preferences can be expressed in such a way that calculations can help identify the preferred decision.

To try and narrow down the number of projects that need to be considered in any analysis, the idea of *dominance* is generally employed. A decision vector $\mathbf{x}$ is dominated by another $\mathbf{x}'$ if for every $\mathbf{w}$

$$NB(\mathbf{x}\,|\,\mathbf{w}) \leq NB(\mathbf{x}'\,|\,\mathbf{w})$$

and for some $\mathbf{w}'$ (3.96)

$$NB(\mathbf{x}\,|\,\mathbf{w}') < NB(\mathbf{x}'\,|\,\mathbf{w}')$$

If $\mathbf{x}$ is dominated by $\mathbf{x}'$, then $\mathbf{x}$ can be dropped from the subsequent analysis because one would always do as well and sometimes better by selection of

$\mathbf{x}'$. None of the designs in Table 3.9 is dominated by another design, so none can be eliminated using this criterion. However, if $NB(\mathbf{x}_3 | \mathbf{w}_3)$ were decreased so that it was less than or equal to 10, then decision vector $\mathbf{x}_3$ would be dominated by decision vector $\mathbf{x}_4$.

Other criteria are useful for indicating which of a set of nondominated decisions may be preferred. The *max-min decision criterion* is one of the simplest and most pessimistic criteria that is used. The max-min decision is that decision which maximizes the smallest benefits that might be obtained and thus is the solution to

$$\max_{\mathbf{x}_i} [\min_{\mathbf{w}_j} NB(\mathbf{x}_i | \mathbf{w}_j)] \tag{3.97}$$

The rightmost column in Table 3.9 contains $\min_{\mathbf{w}_j} NB(\mathbf{x}_i | \mathbf{w}_j)$ for each decision. If individuals want to protect themselves from the worst possible outcome, they should select design $\mathbf{x}_3$, which achieves the maximum value in this column. Decision $\mathbf{x}_3$ ensures that benefits of at least 20 are obtained. The max-min decision may be appropriate when subsistence farmers choose what crops to plant; a very poor outcome may mean starvation. Thus one may be willing to forego a chance of large benefits (50 with $\mathbf{x}_1$ if $\mathbf{w} = \mathbf{w}_1$) to avoid the possibility of doing poorly (0 with $\mathbf{x}_1$ if $\mathbf{w} = \mathbf{w}_3$).

In most cases, people weigh not only the severity of the alternative outcomes, but also their likelihood. This requires assigning probabilities to each of the parameter sets. In general, there is no single best way of estimating probabilities of uncertain events. For an event that has taken place in the past, and for which past outcomes can be used to predict the likelihood of future outcomes (e.g., rainfall or streamflows), a probability distribution based on past observations may be derived. If these *objective probabilities* cannot be determined, one may proceed by first projecting the limits of the range of possible outcomes, and subdividing this range into subranges. Then, weights (probabilities) reflecting one's assessment of the relative likelihood of each subrange can be assigned. Alternatively, a reasonable probability distribution can describe the uncertain parameters' distribution based on subjective estimates of the distribution's mean, variance, or various quantiles. Estimates of probabilities or probability distributions should not be biased by considerations of risk preference or risk aversion. The estimation of discrete probabilities or of a continuous probability distribution should reflect what unbiased experts believe to be the actual likelihoods of possible outcomes.

For the simple example in Tables 3.9 and 3.10, the opinion of experts indicates that each of the parameter sets has an equal chance of occurrence. The tables show the resultant estimates of the average or *expected net benefits* and *expected loss of economic efficiency*. For each project, the expected loss of economic efficiency equals the average value of the maximum possible net benefits

$$E[\max_{\mathbf{w}} \text{NB}(\mathbf{x} \mid \mathbf{w})] = \frac{50 + 45 + 20}{3} = 38.33 \tag{3.98}$$

minus the expected net benefits of the project. Hence the ranking of the projects by decreasing expected benefits or increasing expected loss of economic efficiency yields the same ordering.

Design vector $\mathbf{x}_4$, by always doing well, achieves the maximum expected benefits with $\mathbf{x}_1$ and $\mathbf{x}_2$ not far behind, and $\mathbf{x}_3$ clearly last. If the expected value criteria is appropriate for selecting the design, the best decision vector is $\mathbf{x}_4$, which would never be selected if one knew the true parameter values.

A frequent trade-off in system design is between projects such as $\mathbf{x}_4$, which maximizes expected benefits, and $\mathbf{x}_3$, which maximizes the benefits of the worst outcome, but in so doing achieves the smallest expected benefits. Should one forego the certainty of receiving net benefits of 20 with project $\mathbf{x}_3$ for the larger expected benefits of $\mathbf{x}_4$ and the risk of receiving net benefits of only 10? The techniques and ideas in this section are only able to help the planner pose such a question. As is generally considered appropriate, the planner must make the value judgment. The next section presents utility theory, a methodology for describing a planner's attitudes toward risks and benefits in such a way as to be able to specify which of several projects is preferred.

3.5.3 Utility Theory

The expected benefit criterion and the max-min benefit criterion are imperfect. The max-min criterion overreacts to the risk of poor outcomes, while the expected benefits criterion ignores one's aversion to undesirable outcomes. The need for a methodology to incorporate risk preferences into project evaluations is illustrated by the problem of the *gambler's ruin*.

A gambler with D dollars to gamble has the opportunity to wager on a game of chance. If the gambler wins any play of the game, the gambler receives twice what is bet, and if the gambler loses, the bet is lost. The gambler knows in this instance that the probability of winning any play of the game is 60%. If the gambler wishes to maximize expected earnings, the gambler should bet all that he has on each play of the game, because the expected winnings when B dollars are bet is

$$0.40(0) + 0.60(\$2B) = 1.20(\$B) \tag{3.99}$$

which clearly increases with B (initially $0 \leq B \leq D$).

Following the strategy of betting all of one's earnings on each of n plays, when starting with D dollars, yields expected earnings of $(1.20)^n D$. Herein lies the gambler's ruin. The probability that the gambler has any earnings after n plays of the game is the probability of winning all n plays, or $(0.60)^n$. For n as small as 15, $(0.60)^n$ equals 0.05%. While the strategy of betting all of one's winnings at each play results in large expected winnings, the gambler is almost certain to lose everything. To most persons, the gambler's strategy is unreasonable. Who, beginning with $D = \$1$ and winning the first 14 games, would bet their entire earnings of $16,384 on the next game? The risk of losing everything is too great. When the consequences of failure are severe, expected values are generally not an adequate decision criterion. On the other hand, the max-min strategy would be never to bet anything. This strategy also seems unreasonable in view of the favorable odds.

A methodology exists to incorporate a "rational" decision maker's preference toward risk into the analysis, so that it is possible to determine which of several decisions is preferred. The fundamental concepts of this methodology, presented by von Neumann and Morgenstern [34], center on use of a *utility function.* Utility theory has not found wide application in water resource systems engineering for a number of reasons to be discussed at the end of this section. This occurs primarily because the theory requires, as it must, that the decision maker quantify his or her attitudes toward risk; this is extremely difficult, for it is very hard for anyone to determine and articulate how they value an investment whose return is a random variable. Still, it is very useful for the water resource systems planner to understand how the evaluation of plans yielding uncertain benefits could be performed. The systems planner should then be better able to conceptualize the issues raised by uncertain outcomes and to understand the practical implications of uncertainty.

Consider again the hypothetical irrigation project analyzed in the previous section. Which design a decision maker prefers can be determined by converting the outcomes of each design vector into equivalent plans (or lotteries) which yield the most preferred and least preferred outcomes with specified probabilities. In this instance, the best thing that can happen corresponds to net benefits of 50; the worst thing that can happen corresponds to net benefits of 0. To convert each decision into an equivalent plan, as far as the decision maker is concerned, one needs the decision maker's utiliy function $U(\text{NB})$. The utility $U(\text{NB})$ of net benefits NB will equal the probability p at which the decision maker is indifferent between a project that yields net benefits of NB for certain, and a project that yields net benefits of 50 with probability p and net benefits of 0 with probability $1 - p$.

Several characteristics of the utility function follow from its definition and the decision maker's preference for maximum net benefits and maximum reliability. The utility of net benefits of 50, denoted $U(50)$, must equal 1 because the decision maker will prefer net benefits of 50 for certain to any plan which yields net benefits of 50 (the best possible outcome) with a probability p less than 1. Likewise, the utility of net benefits of 0, denoted $U(0)$, must be 0.

$$U(50) = 1$$

and

$$U(0) = 0 \tag{3.100}$$

Moreover, the larger the net benefits a project yields for certain, the larger the probability of net benefits of 50 must be in an equivalent plan (yielding either net benefits of 50 or of 0) for a decision maker to be indifferent to which of the two plans is implemented. This implies that

$$\frac{dU(\text{NB})}{d(\text{NB})} > 0 \tag{3.101}$$

Assume that a decision maker's utility for net benefits of 35, 25, and 10, as assessed in an interview, have been found to be

$$\begin{aligned} U(35) &= 0.80 \\ U(25) &= 0.65 \\ U(10) &= 0.40 \end{aligned} \tag{3.102}$$

The decision maker is thus indifferent to a plan that yields net benefits of 25 for certain or to one that yields NB = 50 with probability 0.65 and NB = 0 with probability 1 − 0.65, or 0.35. While the first plan yields net benefits of 25 for certain, the expected benefits of the second plan are (0.65)(50), or 32.5. In the decision maker's view, the larger expected benefits of the second plan are just sufficient to compensate for the possibility of ending up with net benefits of only 0.

A continuous utility function was fit to these specified utilities and the values of the fitted function are listed in Table 3.11. This utility function can be used to transform the possible outcomes of each decision in Table 3.9 into an equivalent plan (as far as the decision maker is concerned) yielding net benefits of either 50 or 0. For example, decision x_1 yields net benefits of 50, 30, and 0 with probabilities $\frac{1}{3}$, $\frac{1}{3}$, and $\frac{1}{3}$. Now, the decision maker is

TABLE 3.11. Assessed Utilities and Fitted Utility Function for Decision Maker

NB	*Assessed Utility,* p^u	*Utility Function,* $0.10NB^{0.588}$
50	1.00	1.00
45		0.94
40		0.88
35	0.80	0.81
30		0.74
25	0.65	0.66
20		0.58
10	0.40	0.39
5		0.25
0	0.00	0.00

indifferent between net benefits of 30 for certain and (from Table 3.11) a 74%/26% chance of net benefits of 50 and 0, respectively. Hence, the decision maker also should be indifferent between a 33.3% chance of net benefits of 30, and a (0.74)(0.333) chance of NB = 50 and a (0.26)(0.333) chance of NB = 0. Replacing a 33.3% chance of net benefits of 30 by a 33.3% chance of the equivalent plan yielding net benefits of either 50 or 0, one finds that the decision maker is indifferent between design $\mathbf{x}_1$ and a design that yields net benefits of 50 with probability

$$(1)\tfrac{1}{3} + (0.74)\tfrac{1}{3} = 0.58 \tag{3.103}$$

and net benefits of 0 with probability

$$(0.26)\tfrac{1}{3} + (1)\tfrac{1}{3} = 0.42 \tag{3.104}$$

as shown in Figure 3.8.

Equation 3.103 corresponds to the expected value of the utility function $U(\text{NB})$ that results from implementing project $\mathbf{x}_1$. For all of the projects, the probability of net benefits of 50 in an equivalent plan yielding net benefits of either 50 or 0 is given by the expected utility

$$\begin{aligned} E\{U(\text{NB}[\mathbf{x}_i \mid \mathbf{w}])\} = \tfrac{1}{3}U(\text{NB}[\mathbf{x}_i \mid \mathbf{w}_1]) + \tfrac{1}{3}U(\text{NB}[\mathbf{x}_i \mid \mathbf{w}_2]) \\ + \tfrac{1}{3}U(\text{NB}[\mathbf{x}_i \mid \mathbf{w}_3]) \end{aligned} \tag{3.105}$$

where $\text{NB}[\mathbf{x}_i \mid \mathbf{w}_j]$ are the net benefits obtained with design $\mathbf{x}_i$ when the true system parameters are $\mathbf{w}_j$ for $j = 1, 2, 3$. (Each $\mathbf{w}_j$ occurs with probability $\frac{1}{3}$ in this example.)

While equation 3.105 gives the probability of net benefits of 50 (the best outcome), one minus that quantity gives the probability of net benefits of 0 (the worst outcome). Hence the preferred design vector achieves the largest expected utility, the probability of the best possible outcome. This is the *fundamental theorem of utility theory*: Given that a decision maker's preferences satisfy a number of axioms involving the substitutibility of one lottery for another and the consistency of his preferences, the action preferred by the decision maker has the largest expected utility.

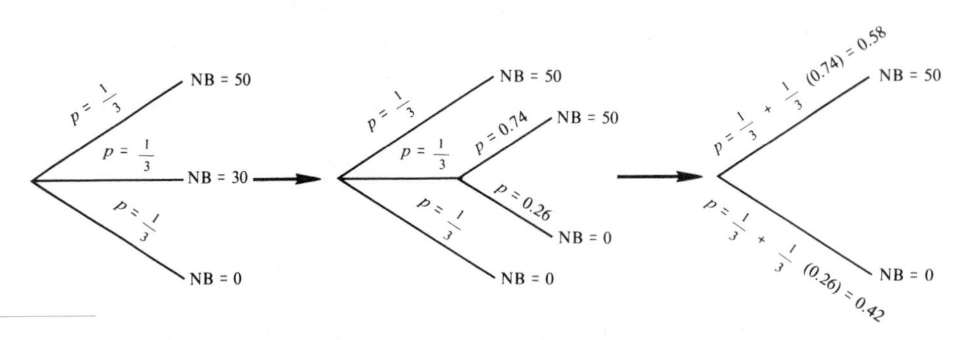

FIGURE 3.8. Replacement of NB = 30 by an equivalent lottery between NB = 50 and NB = 0.

TABLE 3.12. Evaluation of Four Designs in Table 3.9

Decision $\mathbf{x}_i$	$\underset{\mathbf{w}}{E}[\mathbf{x}_i \mid \mathbf{w}]$	$\underset{j}{min}\ NB(\mathbf{x}_i \mid \mathbf{w}_j)$	$\underset{\mathbf{w}}{E}\{U(NB[\mathbf{x}_i \mid \mathbf{w}])\}$
$\mathbf{x}_1$	26.7	0	0.58
$\mathbf{x}_2$	26.7	5	0.64
$\mathbf{x}_3$	25.0	20	0.66
$\mathbf{x}_4$	28.3	10	0.69

The estimated utilities of the designs in Table 3.9 are reported in Table 3.12. The expected utility criterion indicates that the decision maker prefers design vector $\mathbf{x}_4$ to $\mathbf{x}_3$, $\mathbf{x}_3$ to $\mathbf{x}_2$, and $\mathbf{x}_2$ to $\mathbf{x}_1$. This is a different ranking than provided by either the max-min or the expected value criteria. The poor performance of design vectors $\mathbf{x}_1$ and $\mathbf{x}_2$ when $\mathbf{w} = \mathbf{w}_3$ and the decision maker's distaste for such poor outcomes cause these to be least preferred by the decision maker even though they yield high expected benefits. Design vector $\mathbf{x}_4$ combines the highest expected value of net benefits and a satisfactory level of performance for all $\mathbf{w}$, so that overall the decision maker's preferences (as indicated in Table 3.11) imply that $\mathbf{x}_4$ is the preferred plan.

In actual use, utility analysis need not stop here. Changes can be made in the estimated probabilities, net benefits, or the utility function to see if they would change the ranking of the decisions. Alternatively, a utility function may be expanded to include many attributes of the possible outcomes, as explained by Keeney and Raiffa [18]. Multiattribute or multiobjective planning is discussed in more detail in Chapter 4. This section is only an introduction to the concepts of utility theory. The basic theory is covered elsewhere [18, 34] while both Benjamin and Cornell [1] and Keeney and Wood [17] present examples in water resources planning.

Advocates of the use of utility theory in actual planning situations point to several favorable properties of the methodology. It explicitly incorporates a planner's preferences toward risk, producing a ranking of the alternative plans. The sensitivity of this ranking to the planner's preferences can be explored. The use of utility functions makes the assessment of plans more formal and the criteria and judgments used more explicit and objective. The stated preferences then provide a focus and forum for debate.

Unfortunately, the use of utility functions also has important drawbacks for the decision maker. First, the decision maker must expend the effort necessary to understand the method and to carefully determine and quantify the values and trade-offs involved in a particular situation, if he or she can. The quantification of a decision maker's preferences is by nature a very difficult task and care must be exercised that the decision maker is not inconsistent when asked to assess probabilities for hypothetical and unrealistic lotteries. The decision maker's stated preferences then serve as a surrogate for the preferences of all those affected by the project. Decision makers may be uncomfortable or unable to describe the overall social value that should be attached to different attributes of a proposed project. By making their preferences explicit, decision makers also make themselves particularly vulnerable to criticism by those groups that do not see their interest served.

Problems also arise from the method's prescriptive nature. Based on the stated preferences, the utility function *determines* which decision is preferred. Most other criteria, such as expected net benefits, are only *descriptive* of the consequences of a decision. If a decision maker does not have an intuitive feeling or understanding of why the decision that maximizes his or her stated utility function should be selected, he or she is placed in an awkward position; the decision maker must either make a recommendation with which he or she is not comfortable and cannot defend, or must reject the utility analysis based on his or her stated preferences. As is appropriate, the latter generally occurs.

3.6 ANALYZING SYSTEMS WITH DYNAMIC UNCERTAINTY

The previous section addresses methods of planning when the values of particular system parameters are not known with certainty. In this section, the discussion is expanded to the analysis of situations where important components of the system are stochastic processes whose values change with time. A typical example is unregulated streamflows. The emphasis in this section is on how stochastic processes can be incorporated into water resource system models. The major questions concerning how one evaluates uncertain outcomes have just been addressed, and this section reflects only briefly on those themes.

This section starts with a discussion and example of stochastic simulation. This is certainly the most flexible and widely used tool for the analysis of complex water resources systems. The second section presents stochastic optimization models which are extensions of the deterministic techniques presented in Chapter 2. Both stochastic dynamic programming and the two types of stochastic linear programming presented in that section find use in Chapter 7.

3.6.1 Stochastic Simulation

Simulation was defined in Chapter 2 as the solution of a management model by trial and error. As with optimization models, simulation models may be either deterministic or stochastic. One of the most useful tools in water resource systems planning is stochastic simulation. The many optimization models discussed in this book are often of limited use for the detailed study of the operation of complex stochastic systems. Stochastic optimization models provide insight and guidance into how systems should be designed and how they can be operated, but seldom can they adequately deal with all the complexities that need to be considered. Stochastic simulation of complex systems on digital computers provides planners with a powerful tool for the evaluation of the probability distribution of performance indices of complex stochastic water resources systems.

When simulating any system, the modeler designs an experiment. Initial conditions must be specified: reservoirs can start full, empty, or at random representative conditions. The modeler also determines what data are to be collected on system performance and operation and how they are to be summarized. The length of time the simulation is to be run must be specified and, in the case of stochastic simulations, the number of runs to be made must also be determined. These considerations are discussed in more detail by Fishman [9] and in other books on simulation. The use of stochastic simulation and the analysis of the output of such models are introduced here primarily in the context of an example to illustrate what goes into a simulation model and how one can deal with the information that is generated.

Generating Random Variables. Included in any stochastic simulation model is some provision for the generation of sequences of random numbers that represent particular values of events such as rainfall, streamflows, or floods. To generate a sequence of values for a random variable, probability distributions for the variables must be specified. Historical data and an understanding of the physical processes are used to select appropriate distributions and to estimate their parameters (as discussed in Section 3.3).

Most computers have algorithms for generating random numbers uniformly distributed between zero and one; that is,

$$F_U(u) = u \qquad \text{for } 0 \leq u \leq 1$$

and

$$f_U(u) = \begin{cases} 1 & 0 \leq u \leq 1 \\ 0 & \text{otherwise} \end{cases} \tag{3.106}$$

These uniform random variables can then be transformed into random variables with any desired distribution. If $F_Q(q_t)$ is the distribution function of a random variable Q_t in period t, then Q_t can be generated as

$$Q_t = F_Q^{-1}[U_t] \tag{3.107}$$

Here U_t is the uniform random number used to generate Q_t. This is illustrated in Figure 3.9. Analytical expressions for the inverse of many distributions, such as the normal distribution, are not known, so that special algorithms are employed to efficiently generate deviates with these distributions [9]. The generation of stochastic processes is the subject of Chapter 6.

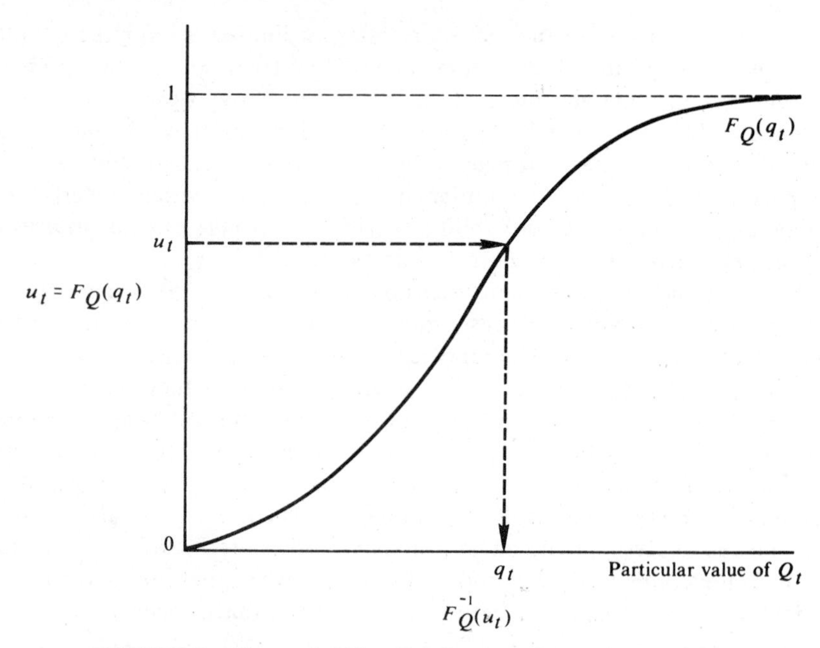

FIGURE 3.9. The probability distribution of a random variable can be inverted to produce values of the random variable.

River Basin Simulation. An example will demonstrate the use of stochastic simulation in the design and analysis of water resource systems. Assume that farmers in a particular valley have been plagued by frequent shortages of irrigation water. They currently draw water from an unregulated river to which they have water rights. A government agency has proposed construction of a moderate-size dam on the river upstream from points where the farmers withdraw water. The dam would be used to increase the quantity and reliability of irrigation water available to the farmers during the summer growing season.

After preliminary planning, a reservoir with an active capacity of 4×10^7 m^3 has been proposed for a natural dam site. It is anticipated that because of the increased reliability and availability of irrigation water, the quantity of water desired will grow from an initial level of 3×10^7 m^3/yr after construction of the dam to 4×10^7 m^3/yr within 6 years. After that, demand will grow more slowly, to 4.5×10^7 m^3/yr, the estimated maximum reliable yield. The projected demand for summer irrigation water is shown in Table 3.13.

A simulation study will evaluate how the system can be expected to perform over a 20-year planning period. Table 3.14 contains statistics that describe the hydrology at the dam site. The estimated moments are computed from the 45-year historic record.

TABLE 3.13. Projected Water Demand for Irrigation Water

Year	*Water Demand* ($\times 10^7$ m^3/yr)
1	3.0
2	3.2
3	3.4
4	3.6
5	3.8
6	4.0
7	4.1
8	4.2
9	4.3
10	4.3
11	4.4
12	4.4
13	4.4
14	4.4
15	4.5
16	4.5
17	4.5
18	4.5
19	4.5
20	4.5

TABLE 3.14. Characteristics of the River Flow

	Winter	*Summer*	*Annual*	
Mean flow	4.0	2.5	6.5	($\times 10^7$ m^3)
Standard deviation	1.5	1.0	2.3	($\times 10^7$ m^3)
Correlation of flows				
Winter with following summer		0.65		
Summer with following winter		0.60		

Using the techniques discussed in Chapter 6, a Thomas-Fiering model which produces lognormally distributed streamflows is used to generate 25 synthetic streamflow sequences. The statistical characteristics of the synthetic flows are those listed in Table 3.14. Use of only the 45-year historic flow sequence would not allow examination of the system's performance over the large range of streamflow sequences which could occur during the 20-year planning period. Jointly, the synthetic sequences should be a description of the range of inflows that the system might experience. A larger number of sequences could be generated.

The Simulation Model. The simulation model is composed primarily of continuity constraints and the proposed operating policy. The volume of water stored in the reservoir at the beginning of seasons 1 (winter) and 2 (summer) in year y are denoted by S_{1y} and S_{2y}. The reservoir's winter operating policy is to store as much of the winter's inflow Q_{1y} as possible. The winter release R_{1y} is determined by the rule

$$R_{1y} = \begin{cases} S_{1y} + Q_{1y} - K & \text{if } S_{1y} + Q_{1y} - R_{\min} > K \\ R_{\min} & \text{if } K \geq S_{1y} + Q_{1y} - R_{\min} \geq 0 \\ S_{1y} + Q_{1y} & \text{otherwise} \end{cases} \quad (3.108)$$

where K is the reservoir capacity of 4×10^7 m^3 and $R_{\min}$ is 0.50×10^7 m^3, the minimum release to be made if possible. The volume of water in storage at the beginning of the year's summer season is

$$S_{2y} = S_{1y} + Q_{1y} - R_{1y} \quad (3.109)$$

The summer release policy is to meet each year's projected demand or target release D_y, if possible, so that

$$R_{2y} = \begin{cases} S_{2y} + Q_{2y} - K & \text{if } S_{2y} + Q_{2y} - D_y > K \\ D_y & \text{if } K \geq S_{2y} + Q_{2y} - D_y \geq 0 \\ S_{2y} + Q_{2y} & \text{otherwise} \end{cases} \quad (3.110)$$

This operating policy is illustrated by Figure 3.10. The volume of water in storage at the beginning of the next winter season is

$$S_{1,y+1} = S_{2y} + Q_{2y} - R_{2y} \quad (3.111)$$

Simulation of the Basin. The question to be addressed by this simulation study is: How well will the reservoir meet the farmers' water requirements? The answering of this question can be viewed as having three steps. First, one must define the performance criteria or indices to be used to describe the system's performance. The appropriate indices will, of course, depend on the problem at hand and the specific concerns of the users and managers of a water resource system. In our reservoir-irrigation system, several indices will be used relating to the reliability with which target releases are met and the severity of any shortages.

The next step is to simulate the proposed system to evaluate the specified indices. For our reservoir-irrigation system, the reservoir's operation was simulated 25 times using the 25 synthetic streamflow sequences, each 20 years in length. Each of the 20 simulated years consisted of first a winter and then a summer season. At the beginning of the first winter season, the reservoir was taken to be empty ($S_{1y} = 0$ for $y = 1$) because construction would just have been completed. The target release or demand for water in each year is given in Table 3.13.

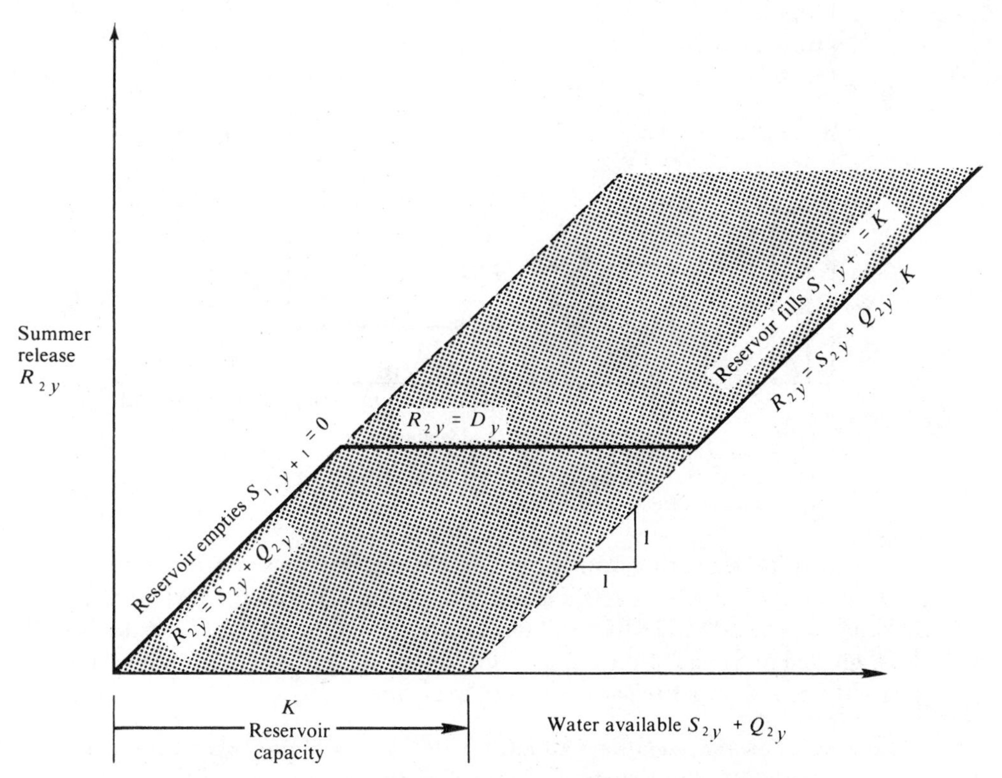

FIGURE 3.10. Summer reservoir operating policy. The shaded area denotes the region of reservoir operation.

After simulating the system, one must proceed to interpret the resulting information so as to gain an understanding of how the system might perform both with the proposed design and operating policy and with modifications in either the system's design or its operating policy. To see how this may be done, consider the operation of our example reservoir-irrigation system.

The reliability p_y of the target release in year y is the probability that the target release D_y is met or exceeded in that year:

$$p_y = \Pr[R_{2y} \geq D_y] \tag{3.112}$$

The system's reliability is a function of the target release D_y, the hydrology of the river, the size of the reservoir, and the operating policy of the system. In this example, the reliability also depends on the year in question. Figure 3.11 shows the total number of failures that occurred in each year of the 25 simulations. In 3 of the 25 simulations, the reservoir did not contain sufficient water after the initial winter season to meet the demand the first summer. After year 1, few failures occur in years 2 through 9 because of the low demand. Surprisingly few failures occur in years 10 and 13, when demand

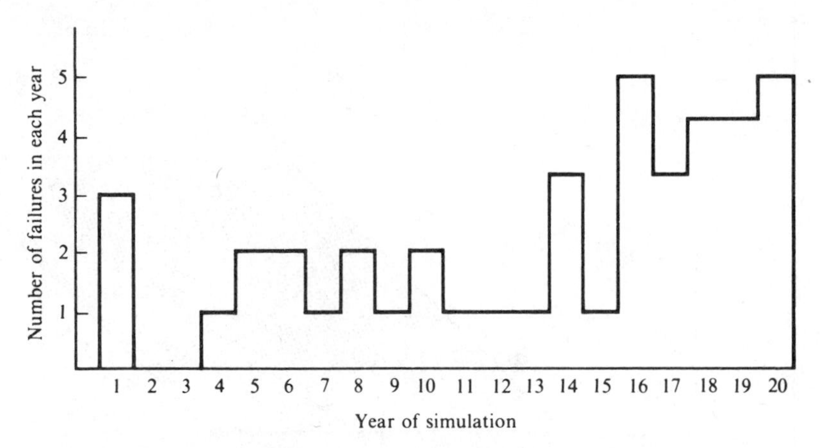

FIGURE 3.11. Number of failures in each year of 25 twenty-year simulations.

has reached its peak; this results because the reservoir was normally full at the beginning of this period as a result of lower demand in the earlier years. Starting in years 14 and after, failures occurred more frequently because of the high demand placed on the system. Thus one has a sense for how the reliability of the target releases changes over time.

Interpretation of Simulation Output. Table 3.15 contains several summary statistics of the 25 simulations. Column 2 of the table contains the average failure frequency in each simulation, which equals the number of years the target release was not met divided by 20, the number of years simulated. At the bottom of column 2 and the other columns are several statistics that summarize the 25 values of the different performance indices. The sample estimates of the mean and variance of each index are given as one way of summarizing the distribution of the observations. Another approach is specification of the sample median, the approximate interquartile range $x_{(6)} - x_{(20)}$, and/or the range $x_{(1)} - x_{(25)}$ of the observations, where $x_{(i)}$ is the ith largest observation. Either set of statistics could be used to describe the center and spread of each index's distribution.

Suppose that one is interested in the distribution of the system's failure frequency or, equivalently, the reliability with which the target can be met. Table 3.15 reports that the mean failure rate for the 25 simulations is 0.084, implying that the average reliability over the 20-year period is $1 - 0.084 = 0.916 \simeq 92\%$. The median failure rate is 0.05, implying a median reliability of 95%. These are both reasonable estimates of the center of the distribution of the failure frequency. Note that the actual failure frequency ranged from 0.00 (seven times) to 0.30. Thus the system's reliability ranged from 100% to as low as 70%, 75%, and 80% in runs 17, 8, and 11. Certainly, the farmers

TABLE 3.15. Results of 25 Simulations

Simulation Number, i	*Frequency of Failure to Meet Target*	*Frequency of Failure to Meet 80% of Target*	*Total Shortage TS* ($\times 10^7\ m^3$)	*Average Deficit, AD*
1	0.10	0.0	1.25	0.14
2	0.15	0.05	1.97	0.17
3	0.10	0.05	1.79	0.20
4	0.10	0.05	1.67	0.22
5	0.05	0.0	0.21	0.05
6	0.0	0.0	0.00	0.00
7	0.15	0.05	1.29	0.10
8	0.25	0.10	4.75	0.21
9	0.0	0.0	0.00	0.00
10	0.10	0.0	0.34	0.04
11	0.20	0.0	1.80	0.11
12	0.05	0.05	1.28	0.43
13	0.05	0.0	0.53	0.12
14	0.10	0.0	0.88	0.11
15	0.15	0.05	1.99	0.15
16	0.05	0.0	0.23	0.05
17	0.30	0.05	2.68	0.10
18	0.10	0.0	0.76	0.08
19	0.0	0.0	0.00	0.00
20	0.0	0.0	0.00	0.00
21	0.0	0.0	0.00	0.00
22	0.05	0.05	1.47	0.33
23	0.0	0.0	0.00	0.00
24	0.0	0.0	0.00	0.00
25	0.05	0.0	0.19	0.04
Mean, $\bar{x}$	0.084	0.020	1.00	0.106
Standard deviation of values, s_X	0.081	0.029	1.13	0.110
Median	0.05	0.00	0.76	0.10
Approximate interquartile range, $x_{(6)} - x_{(20)}$	0.0–0.15	0.0–0.05	0.0–1.79	0.0–0.17
Range, $x_{(1)} - x_{(25)}$	0.0–0.30	0.0–0.10	0.0–4.75	0.0–0.43

are interested not only in knowing the mean or median failure frequency but also the range of failure frequencies they are likely to experience.

If one knew the form of the distribution function of the failure frequency, one could use the mean and standard deviation of the observations to determine a confidence interval within which the observations would fall with some prespecified probability. For example, if the observations are normally distributed, there is a 90% probability that the index falls within the interval $\mu_X \pm 1.65\sigma_X$. Thus, if the simulated failure rates are normally distributed, there is about a 90% probability the actual failure rate is within the interval $\bar{x} \pm 1.65 s_X$. In our case this interval would be

$$[0.084 - 1.65(0.081), 0.084 + 1.65(0.081)] = [-0.050, 0.218]$$

Clearly, the failure rate cannot be less than zero, so that this interval makes little sense in our example.

A more reasonable approach to describing the distribution of a performance index whose probability distribution function is not known is to use the observations themselves. As discussed in Section 3.2.3, if the observations are of a continuous random variable, the interval $x_{(i)} - x_{(n+1-i)}$ provides a reasonable estimate of an interval within which the random variable falls with probability

$$P = \frac{n+1-i}{n+1} - \frac{i}{n+1} = \frac{n+1-2i}{n+1} \tag{3.113}$$

In our example, the range $x_{(1)} - x_{(25)}$ of the 25 observations is an estimate of an interval in which a continuous random variable falls with probability $(25 + 1 - 2)/(25 + 1) = 92\%$, while $x_{(6)} - x_{(20)}$ corresponds to probability $(25 + 1 - 2 \times 6)/(25 + 1) = 54\%$.

Table 3.15 reports that for the failure frequency, $x_{(1)} - x_{(25)}$ equals $0 - 0.30$, while $x_{(6)} - x_{(20)}$ equals $0 - 0.15$. Reflection on how the failure frequencies are calculated reminds us that the failure frequency can only take on the *discrete*, nonnegative values $\frac{0}{20}, \frac{1}{20}, \frac{2}{20}, \ldots, \frac{20}{20}$. Thus, the random variable X cannot be less than zero. Hence, if the lower endpoint of an interval is zero, as is the case here, then $0 - x_{(k)}$ is an estimate of an interval within which the random variable falls with a probability of at least $k/(n + 1)$. For k equal to 20 and 25, the corresponding probabilities are 77% and 96%.

Often, the analysis of a simulated system's performance centers on the average value of performance indices, such as the failure rate. It is important to know the accuracy with which the mean value of an index approximates the true mean. This is done by the construction of confidence intervals, discussed in greater detail in Appendix 3A. A confidence interval is an interval that will contain the unknown value of a parameter with a specified probability. Confidence intervals for a mean are constructed using the t statistic,

$$t = \frac{\bar{x} - \mu_X}{s_X/\sqrt{n}} \tag{3.114}$$

which for large n has approximately a standard normal distribution. Certainly, $n = 25$ is not very large, but the approximation to a normal distribution may be sufficiently good to obtain a rough estimate of how close the average frequency of failure $\bar{x}$ is likely to be to μ_X. A $100(1-2\alpha)\%$ confidence interval for μ_X is, approximately,

$$\bar{x} - t_\alpha \frac{s_X}{\sqrt{n}} \leq \mu_X \leq \bar{x} + t_\alpha \frac{s_X}{\sqrt{n}}$$

or

$$0.084 - t_\alpha\left(\frac{0.081}{\sqrt{25}}\right) \leq \mu_X \leq 0.084 + t_\alpha\left(\frac{0.081}{\sqrt{25}}\right) \tag{3.115}$$

If $\alpha = 0.05$, then $t_\alpha = 1.65$ and equation 3.115 becomes

$$0.057 \leq \mu_X \leq 0.11 \tag{3.116}$$

Hence, based on the simulation output, one can be about 90% sure that the true mean failure frequency lies between 5.7% and 11%. This corresponds to a reliability between 89% and 94%. By performing additional simulations to increase the size of n, the width of this confidence interval can be decreased. However, this increase in accuracy may be an illusion, because the uncertainty in the parameters of the streamflow model has not been incorporated into the analysis. Confidence intervals for medians can be constructed as described in Appendix 3A.

Failure frequency or system reliability are indices which describe only one dimension of the system's performance. Table 3.15 contains additional information on the system's performance related to the severity of shortages. Column 3 contains the frequency with which the shortage exceeded 20% of that year's demand. This occurred in approximately 2% of the years, or in 24% of the years in which a failure occurred. Taking another point of view, failures in excess of 20% of demand occurred in 9 out of 25, or 36%, of the simulation runs.

Columns 4 and 5 contain two other indices, which pertain to the severity of the failures. The total shortfall in column 4 is calculated as

$$\text{TS} = \sum_{y=1}^{20} [D_{2y} - R_{2y}]^+ \tag{3.117}$$

where

$$[Q]^+ = \begin{cases} Q & \text{if } Q > 0 \\ 0 & \text{otherwise} \end{cases}$$

The total shortfall equals the total amount by which the target release is not met in years in which shortages occur.

Related to the total shortfall is the average deficit. The deficit is defined as the shortfall in any year divided by the target release in that year. The average deficit is

$$\text{AD} = \frac{1}{m} \sum_{y=1}^{20} \frac{[D_{2y} - R_{2y}]^+}{D_{2y}} \tag{3.118}$$

where m is the number of failures (deficits) or nonzero terms in the sum.

Both the total shortfall and the average deficit measure the severity of shortages. The mean total shortfall $\overline{\text{TS}}$, equal to 1.00 for the 25 simulation runs, is a difficult number to interpret. While no shortage occurred in 7 runs, the total shortage was 4.7 in run 8, in which the shortfall in two different years exceeded 20% of the target. The median of the total shortage values, equal to 0.76, is an easier number to interpret in that one knows that half the time the total shortage was greater and half the time less than this value.

The mean average deficit $\overline{\text{AD}}$ is 0.106, or 11%. However, this mean includes an average deficit of zero in the 7 runs in which no shortages occurred. The average deficit in the 18 years in which shortages occurred is $(11\%)(\frac{25}{18}) = 15\%$. The average deficit in individual simulations in which shortages occurred ranges from 4% to 43%, with a median of 11.5%.

After examining the results reported in Table 3.15, the farmers might determine that the probability of a shortage exceeding 20% of a year's target release is higher than they would like. They can deal with more frequent minor shortages, not exceeding 20% of the target, with little economic hardship, particularly if they are warned at the beginning of the growing season that less than the targeted quantity of water will be delivered. Then they can curtail their planting or plant crops which require less water.

In an attempt to find out how better to meet the farmers' needs, the simulation program was rerun with the same streamflow sequences and a new operating policy in which only 80% of the growing season's target release is provided (if possible) if the reservoir is less than 80% full at the end of the previous winter season. This gives the farmers time to adjust their planting schedules and may increase the quantity of water stored in the reservoir to be used the following year if the drought persists.

As the simulation results with the new policy in Table 3.16 demonstrate, this new operating policy appears to have the expected effect on the system's operation. With the new policy, only 6 severe shortages in excess of 20% of demand occur in the 25 twenty-year simulations, as opposed to 10 such shortages with the original policy. In addition, these severe shortages are all less severe than the corresponding shortages that occur with the same streamflow sequence when the original policy is followed.

The decrease in the severity of shortages is obtained at a price. The overall failure frequency has increased from 8.4% to 14.2%. However, the latter figure is misleading because in 14 of the 25 simulations, a failure occurs in the first simulation year with the new policy, whereas only 3 occur with the original policy. Of course, these first-year failures occur because the reservoir starts empty at the beginning of the first winter and often does not fill that season. Ignoring these first-year failures, the failure rates with the two policies over the subsequent 19 years are 8.2% and 12.0%. Thus the frequency of failures in excess of 20% of demand is decreased from 2.0% to 1.2% by increasing the frequency of all failures after the first year from 8.2% to 12.0%. If the farmers are willing to put up with more frequent minor short-

TABLE 3.16. Results of 25 Simulations with Modified Operating Policy to Avoid Severe Shortages

Simulation Number, i	*Frequency of Failure to Meet Target*	*Frequency of Failure to Meet 80% of Target*	*Total Shortage TS* ($\times 10^7 m^3$)	*Average Deficit, AD*
1	0.10	0.0	1.80	0.20
2	0.30	0.0	4.70	0.20
3	0.25	0.0	3.90	0.20
4	0.20	0.05	3.46	0.21
5	0.10	0.0	1.48	0.20
6	0.05	0.0	0.60	0.20
7	0.20	0.0	3.30	0.20
8	0.25	0.10	5.45	0.26
9	0.05	0.0	0.60	0.20
10	0.20	0.0	3.24	0.20
11	0.25	0.0	3.88	0.20
12	0.10	0.05	1.92	0.31
13	0.10	0.0	1.50	0.20
14	0.15	0.0	2.52	0.20
15	0.25	0.05	3.76	0.18
16	0.10	0.0	1.80	0.20
17	0.30	0.0	5.10	0.20
18	0.15	0.0	2.40	0.20
19	0.0	0.0	0.0	0.0
20	0.05	0.0	0.76	0.20
21	0.10	0.0	1.80	0.20
22	0.10	0.05	2.37	0.26
23	0.05	0.0	0.90	0.20
24	0.05	0.0	0.90	0.20
25	0.10	0.0	1.50	0.20
Mean, $\bar{x}$	0.142	0.012	2.39	0.201
Standard deviation of values, s_X	0.087	0.026	1.50	0.050
Median	0.10	0.00	1.92	0.20
Approximate interquartile range, $x_{(6)} - x_{(20)}$	0.05–0.25	0.0–0.0	0.90–3.76	0.20–0.20
Range, $x_{(1)} - x_{(25)}$	0.0–0.30	0.0–0.10	0.0–5.45	0.0–0.31

ages, it appears they can reduce their risk of experiencing shortages of greater severity.

The preceding discussion has ignored the statistical issue of whether the differences between the indices obtained in the two simulation experiments are of sufficient statistical reliability to support the analysis. If care is not taken, observed changes in a performance index from one simulation experiment to another may be due to sampling fluctuations rather than to modifications of the water resource system's design or operating policy.

As an example, consider the change that occurred in the frequency of shortages. Let X_{1i} and X_{2i} be the simulated failure rates using the ith streamflow sequence with the original and modified operating policies. The random variables

$$Y_i = X_{1i} - X_{2i} \tag{3.119}$$

for i equal 1 through 25 are independent of each other if the streamflow sequences are generated independently, as they were.

One would like to confirm that the random variable Y tends to be negative more often than it is positive and hence that policy 2 indeed results in more failures overall. A direct test of this theory is provided by the sign test discussed in Appendix 3B. Of the 25 paired simulation runs, $y_i < 0$ in 21 cases and $y_i = 0$ in 4 cases. We can ignore the times when $y_i = 0$. Note that if $y_i < 0$ and $y_i > 0$ were equally likely, then the probability of observing $y_i < 0$ in all 21 cases when $y_i \neq 0$ is $2^{-21} \simeq 5 \times 10^{-7}$. This is exceptionally strong proof that the new policy has increased the failure frequency. A similar analysis of the frequency with which the release is less than 80% of the target can be made. In only 4 of the 25 simulation runs with the two policies do these failure frequencies differ. However, in all 4 cases where they differ, the new policy resulted in fewer severe failures. The probability of such a lopsided result, were it equally likely that either policy would result in a lower frequency of failures in excess of 20% of the target, is $2^{-4} = 0.0625$. This is fairly strong evidence that the new policy indeed decreases the frequency of severe failures.

Another approach to this problem is to ask if the difference between the average failure rates $\bar{x}_1$ and $\bar{x}_2$ is statistically significant; that is, can the difference between $\bar{x}_1$ and $\bar{x}_2$ be attributed to the fluctuations that occur in the average of any finite set of random variables? In this example the significance of the difference between the two means can be tested using the random variable Y_i defined in equation 3.119. The mean of the observed y_i's is

$$\bar{y} = \frac{1}{25} \sum_{i=1}^{25} (x_{1i} - x_{2i}) = \bar{x}_1 - \bar{x}_2 = 0.084 - 0.142 = -0.058 \tag{3.120}$$

and their standard deviation is

$$s_y^2 = \frac{1}{25} \sum_{i=1}^{25} (x_{1i} - x_{2i} - \bar{y})^2 = (0.0400)^2 \tag{3.121}$$

Now if the sample size n, equal to 25 here, is sufficiently large, then

$$t = \frac{\bar{y} - \mu_Y}{s_Y/\sqrt{n}} \tag{3.122}$$

has approximately a standard normal distribution. The closer the distribution of Y is to that of the normal distribution, the faster the convergence of the distribution of t is to the standard normal distribution with increasing n. If $X_{1i} - X_{2i}$ is normally distributed, which is not the case here, then each Y_i has a normal distribution and t in equation 3.122 has Student's t-distribution, discussed in Appendix 3A.

If $E[x_{1i}] = E[x_{2i}]$, then μ_Y equals zero and upon substituting the observed values of $\bar{y}$ and s_Y^2 into equation 3.122, one obtains

$$t = \frac{-0.0580}{0.0400/\sqrt{25}} = -7.25 \tag{3.123}$$

The probability of observing a value of t equal to -7.25 or smaller is less than 0.1% if n is sufficiently large that t is normally distributed. Hence it appears very improbable that μ_Y equals zero.

This example provides an illustration of the advantage of using the same streamflow sequences when simulating both policies. Suppose that different streamflow sequences were used in all the simulations. Then the expected value of Y would not change, but its variance would be given by

$$\begin{aligned} \text{Var}(Y) &= E[X_1 - X_2 - (\mu_1 - \mu_2)]^2 \\ &= E[(X_1 - \mu_1)^2] - 2E[(X_1 - \mu_1)(X_2 - \mu_2)] \\ &\quad E[(X_2 - \mu_2)^2] \\ &= \sigma_{X_1}^2 - 2\,\text{Cov}(X_1, X_2) + \sigma_{X_2}^2 \end{aligned} \tag{3.124}$$

where $\text{Cov}(X_1, X_2) = E[(X_1 - \mu_1)(X_2 - \mu_2)]$ and is the *covariance* of the two random variables. The covariance between X_1 and X_2 will be zero if they are independently distributed as they would be if different randomly generated streamflow sequences were used in each simulation. Estimating $\sigma_{X_1}^2$ and $\sigma_{X_2}^2$ by their sample estimates, an estimate of what the variance of Y would be if $\text{Cov}(X_1, X_2)$ were zero is

$$\hat{\sigma}_Y^2 = s_{X_1}^2 + s_{X_2}^2 = (0.081)^2 + (0.087)^2 = (0.119)^2 \tag{3.125}$$

The actual sample estimate s_Y equals 0.040; if independent streamflow sequences are used in all simulations, s_Y will take a value near 0.119 rather than 0.040 (equation 3.121). A standard deviation of 0.119 yields a value of the test statistic

$$t = \left.\frac{\bar{y} - \mu_Y}{0.119/\sqrt{25}}\right|_{\mu_Y = 0} = -2.44 \tag{3.126}$$

If t is normally distributed, the probability of observing a value less than -2.44 is about 0.8%. This illustrates that use of the same streamflow sequences in the simulation of both policies allows one to better distinguish the differences in the policies' performance. By using the same streamflow sequences, or other random inputs, one can construct a simulation experiment in which variations in performance caused by different random inputs are confused as little as possible with the differences in performance caused by changes in the system's design or operating policy.

3.6.2 Stochastic Optimization

Given the uncertainities inherent in the prediction of economic, hydrologic, technologic, and other factors affecting the performance of water resources systems, deterministic planning models are often inadequate for effective preliminary plan formulation and evaluation. Hence many of the optimization techniques discussed in Chapter 2 must be extended to incorporate mathematical descriptions of various random processes. In this section the subject of optimization under uncertainty is discussed. More detailed stochastic models for river basin planning are presented in Chapter 7.

Recall that the general form of an optimization problem is

$$\text{maximize } f(\mathbf{X})$$

subject to

$$g_i(\mathbf{X}) = b_i \qquad i = 1, 2, \ldots, m \tag{3.127}$$

in which the vector **X** contains the decision variables whose values define a particular plan and operating policy. Uncertainty may arise in the objective function $f(\cdot)$, one or more constraint functions $g_i(\cdot)$, or right-hand-side values b_i. Because of this uncertainty, it may not be possible to specify the value of all decision variables until events have unfolded and the options available have been determined.

Consider first uncertainty that affects only the problem's objective function $f(\mathbf{X})$ and is not related to the constraints $g_i(\mathbf{X}) = b_i$. Such uncertainty arises from imprecise knowledge of the value of the future benefits and costs resulting from alternative decisions. Such uncertainty can often be handled by substitution of the expected value $E[\text{NB}(\mathbf{X})]$ for the uncertain net benefits function $\text{NB}(\mathbf{X})$. Use of the expected value of the objective is satisfactory if the alternatives are not too extreme so that the expected value can be substituted for the expected utility of a decision.

Uncertainty in the constraints $g_i(\mathbf{X}) = B_i$ can be handled in different ways. If the uncertainty is small, it may be satisfactory to use

$$E[g_i(\mathbf{X})] = E[B_i] \tag{3.128}$$

Many deterministic problems may be thought of as examples of equation 3.128, because few functions and quantities are known precisely. However, the substitution of expected values for random quantities is unacceptable when large variation in some quantities will result in significant violations of the constraints.

When only the right-hand side B_i of one or more inequality constraints is random, chance constraints can be written that define the probability P_i that the constraint can fail. Thus instead of specifying that

$$g_i(\mathbf{X}) \leq E[B_i] \tag{3.129}$$

for those inequality constraints in which B_i is random, a *chance constraint*

$$\Pr[g_i(\mathbf{X}) \leq B_i] \geq 1 - P_i \tag{3.130}$$

can be defined, indicating that the constraint can be violated no more than $100P_i\%$ of the time.

As an example, suppose that B_i is the total quantity of water to be available, and this quantity is uncertain. Further assume that the function $g_i(\mathbf{X})$ is the demand for water. If this demand should be met at least 95% of the time, equation 3.130 becomes

$$\Pr[g_i(\mathbf{X}) \leq B_i] \geq 0.95 = 1 - 0.05 \tag{3.131}$$

indicating that 95% of the time demand $g_i(\mathbf{X})$ should be less than the supply B_i. This means that a 5% probability of a water shortage is acceptable.

The probabilities P_i have to be given prior to model solution. It is rarely obvious what they should be. The resultant model may also be faulted for ignoring what happens during the $100P_i\%$ of the time that the constraint is violated.

The major advantage of these types of chance constraints is that they can be converted into deterministic equivalents given knowledge of the distribution function $F_i(b_i)$. By definition,

$$\Pr[B_i \leq b_i] = F_i(b_i) \tag{3.132}$$

which implies that

$$\Pr[B_i \leq F_i^{-1}(P_i)] = P_i \tag{3.133}$$

as illustrated in Figure 3.12. Hence there is a $100P_i\%$ chance that the value of B_i will be less than or equal to the quantity $F_i^{-1}(P_i)$, which is denoted $b_i^{(P_i)}$. Thus, for a fixed decision $\mathbf{X}$, if

$$g_i(\mathbf{X}) \leq b_i^{(P_i)} \tag{3.134}$$

then the constraint $g_i(\mathbf{X}) \leq B_i$ will be violated at most the $100P_i\%$ of the time that B_i is less than or equal to $b_i^{(P_i)}$. For continuous random variables, equation 3.134 is a deterministic constraint which is equivalent to the chance constraint in equation 3.130.

An alternative to the use of chance constraints is to include in the optimization model an explicit description of the uncertainty in the constraints $g_i(\mathbf{X}) = B_i$ and to allow the decision variables, when possible, to depend on the values of various random variables. This allows the constraints to be met for different values of the random variables and allows the operating policy to exploit the availability of any extra resources. Uncertainty in the

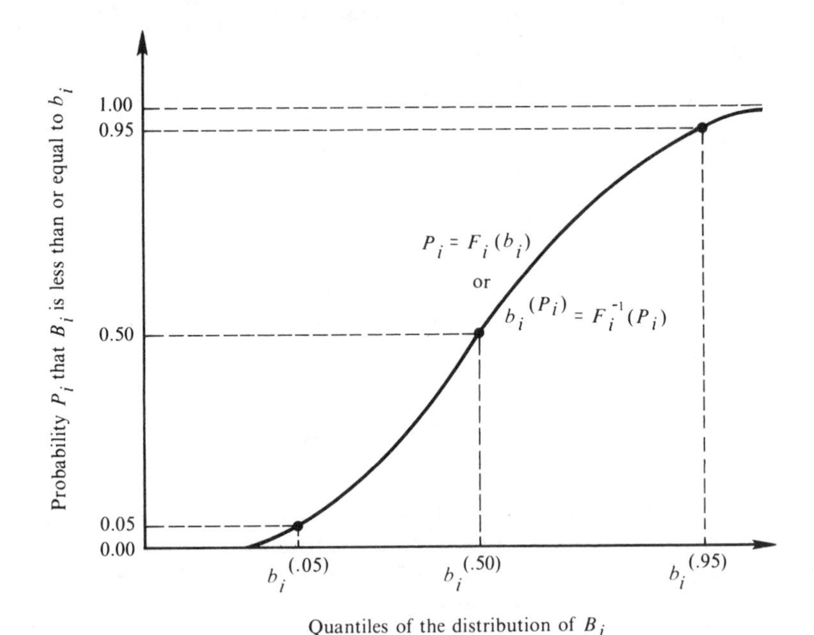

FIGURE 3.12. The cumulative distribution function $F_i(b_i)$ relates probabilities P_i with values $b_i^{(P_i)}$.

constraints $g_i(\mathbf{X}) = B_i$ can be modeled in different ways, depending on the optimization technique. Stochastic dynamic programming can be used to study very complicated situations, as long as few state variables are required. Stochastic linear programming techniques require more restrictive assumptions concerning the structure of the uncertainty but can solve many problems very efficiently.

Stochastic Dynamic Programming. The generalization of deterministic dynamic programming to the stochastic case is straightforward. Consider a system that can be in any one of m discrete states, $s_1, \ldots, s_m$. If $\text{NB}_t(s_i, s_j, k)$ are the net benefits during period t when the system starts in state s_i and ends in state s_j when decision k is made, then the recursive equation of deterministic backward-moving dynamic programming is

$$f_t(s_i) = \max_k \left[\text{NB}(s_i, s_j, k) + f_{t+1}(s_j)\right] \tag{3.135}$$

where $f_t(s_i)$ are the maximum net benefits obtainable from time period t onward starting in state s_i in period t. In the deterministic case, the subsequent state s_j is a deterministic function of k and of the initial state s_i.

Often, the next state that will be occupied is not a deterministic function of the current state and decision. It may depend on uncertain events such as rainfall, streamflow, population growth, or political decisions. To model such a situation, let $p^t_{ij}(k)$ equal the probability that the state in period $t + 1$ is s_j, given that the state in period t is s_i and decision k is made.

$$p^t_{ij}(k) = \Pr[s^{t+1} = s_j \mid s^t = s_i \text{ and decision } k] \tag{3.136}$$

These are the transition probabilities of a time-dependent decision-dependent Markov chain. The policy that maximizes the expected net benefits is found by solution of

$$f_t(s_i) = \max_k \{\sum_{j=1}^{m} p^t_{ij}(k) \cdot [\text{NB}_t(s_i, s_j, k) + f_{t+1}(s_j)]\} \tag{3.137}$$

These equations are the basic recurrence relation of stochastic dynamic programming in Markov chains. To include the transition uncertainty, it is necessary to calculate, at each step, the expected net benefits resulting from each decision. Except for this extra step, when the underlying uncertainty can be represented by a Markov chain, stochastic dynamic programming is equivalent to backward-moving deterministic dynamic programming. Examples of stochastic dynamic programming may be found in Chapter 7.

Stochastic Linear Programming. Solution of stochastic problems using linear programming results either in a large increase in the number of variables and constraints over the deterministic case to account for alternative possible scenarios or in the use of chance constraints. These two approaches are illustrated by a simple example.

Two-Stage Linear Programming. Assume that a water manager is charged with diverting the water from an unregulated stream during the dry season to three users: a municipality, an industrial concern, and the agricultural sector. The industrial concern and the agricultural sector are expanding and want to know how much water they can expect. If insufficient water is available, they will curtail their expansion plans.

The water manager can formulate the problem as follows to maximize the expected value of economic activity in the region. Let T_i for $i = 1$, 2, and 3 be the quantity of water that is promised to each user i. If this water is delivered, the resulting net benefits to the local economy per unit of water allocated are estimated to be NB_i. However, if promised water is not delivered, either water must be obtained from alternative and more expensive sources, or demand must be curtailed by rationing of residential users, by reduced production and recycling by the industrial concern, or by not irrigating as planned in the agricultural sector. This results in a reduction of net benefits to user i of C_i per unit of water not delivered ($C_i > NB_i$).

Letting Q be a random variable equal to the total water available, the water manager's problem can be formulated as

$$\text{maximize } (\sum_{i=1}^{3} NB_i T_i) - E[\sum_{i=1}^{3} C_i D_{iQ}] \tag{3.138}$$

subject to

$$Q \geq T_1 + T_2 + T_3 - [D_{1Q} + D_{2Q} + D_{3Q}]$$
$$T_i^{\max} \geq T_i \geq D_{iQ} \geq 0 \quad \text{for all } i$$

where the T_i's are fixed allocations or delivery targets and the D_{iQ}'s are the amount by which target T_i is not met when the seasonal flow is Q. All T_i's and D_{iQ}'s are unknown. Q is a random variable.

To solve this problem with linear programming, the distribution of Q must be approximated by a discrete distribution. Let Q take values q_j with probability p_j for $j = 1, 2, \ldots, n$. The reformulation of the foregoing problem is:

$$\text{maximize} \sum_{i=1}^{3} NB_i \cdot T_i - \sum_{i=1}^{3} \sum_{j=1}^{n} p_j C_i D_{ij} \tag{3.139}$$

subject to

$$q_j \geq T_1 + T_2 + T_3 - D_{1j} - D_{2j} - D_{3j} \quad \text{for all } j$$
$$T_i^{\max} \geq T_i \geq D_{ij} \geq 0 \quad \text{for all } i \text{ and } j$$

For each value of j it is necessary to repeat the basic constraint set which relates the resources available q_j to the targets T_i and deficits D_{ij}. This can cause stochastic linear programs to become quite large. This formulation of the problem is called a *two-stage* linear program because the targets T_1, T_2, and T_3 are set at the first stage before the inflows are known, while the deficits D_{ij} are set at the second stage when the flow is known and the targets are fixed. Additional stages can be added to describe how a problem unfolds over time. However, for more than two or three stages, the linear program often becomes too big to justify this approach, unless the number of possible combinations in later stages can be restricted.

Table 3.17 presents a very simple example of the application of this technique. Seasonal flows of 4, 10, and 17 units occur with probabilities 0.20, 0.60, and 0.20. The optimal solution is for the water manager to promise all three users the water they requested for maximum development $T_i^{\max}$. However, in the case of insufficient water, first the allotment to farmers should be decreased and then if necessary the allotment to industry. This general conclusion is obtained by examination of the shadow prices associated with the various constraints.

Linear Programming with Chance Constraints. Adding uncertainty to a simple allocation problem as done in the previous section greatly increases the size of the otherwise simple linear program. This increase can be avoided if the problem is reformulated. Suppose that the water manager does not want to estimate the shortage-cost coefficients C_i, but instead wants to specify the reliability with which the targets are met. In particular, suppose that he or she decides to

$$\text{maximize} \sum_{i=1}^{3} NB_i T_i \tag{3.140}$$

TABLE 3.17. Two-Stage Linear Programming Example

DEVELOPMENT AND ECONOMIC DATA

Use	*Use Index, i*	T_i^{max}	NB_i	C_i
Municipal	1	2	100	250
Industrial	2	3	50	75
Agricultural	3	5	30	60

DESCRIPTION OF STREAMFLOW DISTRIBUTION

Flow Index, j	q_j	p_j
1	4	0.20
2	10	0.60
3	17	0.20

OPTIMAL SOLUTION

Use Index, i	*Target, T_i*	SHORTAGE, D_{ij}			ALLOCATION, A_{ij}		
		$j=1$	$j=2$	$j=3$	$j=1$	$j=2$	$j=3$
1	2	0	0	0	2	2	2
2	3	1	0	0	2	3	3
3	5	5	0	0	0	5	5

Value of objective $= 100(2) + 50(3) + 30(5) - 0.2[75(1) + 60(5)] = 425$

subject to

$$\Pr[Q \leq T_1 + T_2 + T_3] \leq P$$
$$T_i^{max} \geq T_i \geq 0$$

Noting that $\Pr[Q \leq T_1 + T_2 + T_3]$ is equivalent to $F_Q[T_1 + T_2 + T_3]$, where F_Q is the distribution function of Q, the last constraint becomes

$$F_Q(T_1 + T_2 + T_3) \leq P$$

or

$$T_1 + T_2 + T_3 \leq F_Q^{-1}(P) = q^{(P)} \tag{3.141}$$

where Q is less than or equal to $q^{(P)}$ with probability P. Thus, if the total water allocation is less than or equal to $q^{(P)}$, a deficit will occur no more than 100P% of the time. Equation 3.141 is a chance constraint. The water manager's problem now consists of one linear constraint plus upper bound and nonnegativity restrictions:

$$\text{maximize} \sum_{i=1}^{3} NB_i T_i \tag{3.142}$$

subject to

$$T_1 + T_2 + T_3 \leq q^{(P)}$$
$$T_i^{max} \geq T_i \geq 0$$

In the solution to the two-stage linear programming formulation of this problem, the targets were violated 20% of the time. For $P = 0.20$, $q^{(P)} \simeq 6$ and the solution to the chance-constraint model in equation 3.144 is $(T_1^*, T_2^*, T_3^*) = (2, 3, 1)$, a different answer than is obtained by the two-stage linear programming formulation of the problem.

The two-stage formulation of the problem took account of the costs of failure and the benefits of large targets and found that it was advantageous to plan for full development, $T_i = T_i^{\max}$ for $i = 1, 2, 3$, even though frequent failures in the agricultural target would result. The chance-constraint formulation of the problem restricts the total allocation to be less than $q^{(P)} \simeq 6$ regardless of the economic consequences. Thus the full development proposal is no longer considered a feasible alternative because $T_1^{\max} + T_2^{\max} + T_3^{\max}$ equals 10.

The chance-constraint approach has the desirable property that the linear programming formulation of a stochastic problem is no more complex than the deterministic problem in which demands and available resources are known. It does this by requiring only the specification of $q^{(\beta)}$ (the resource available with probability $1 - \beta$) and/or $d^{(1-\alpha)}$ (the demand exceeded with probability α), and then by not considering the consequences when the demand actually exceeds $d^{(1-\alpha)}$ or when the supply is less than $q^{(\beta)}$. The optimization of α or β needs to be a part of the problem, and this can only be done by evaluating the costs and consequences of failures. Still, it may not be possible to select values of α and β which yield the optimal solution identified by more detailed models.

Another difficulty with these chance-constraint models is their pessimism, particularly when they contain several chance constraints. Planning targets are set as if demands are high $d^{(1-\alpha)}$ and resources are few $q^{(\beta)}$. It is unlikely that both of these situations will occur simultaneously, and it is often unreasonable for one to plan as if they would.

3.7 CONCLUDING REMARKS

This chapter has brought together some important ideas associated with uncertainty in the planning process. The first four sections provided an introduction to the basic tools and ideas used to describe uncertainty. These included probability theory and alternative approaches and methods of fitting probability distributions to data. The ideas of a stochastic process and a time series were introduced and serve as a basis for the stochastic streamflow and river basin models in Chapters 6 and 7.

The last two sections of the chapter described how one can assess and model the stochastic performance of water resource systems. Determination of the economic importance of misspecification of parameters used in planning models was addressed at length because of its frequent if not universal occurrence. The related topic of how to choose between projects yielding uncertain benefits was addressed and the use of utility functions discussed. Uncertainty in population estimates, water usage rates, and other parameters, as well as the variability in streamflows, rainfall, and other natural phenomena, have a major impact on the performance of water resource system projects in a physical, ecological, and economic sense. Although one can model uncertainty and explain its consequences by arrays of indicators, one generally cannot eliminate uncertainty or reduce its impact to the point of unimportance. Major sources of uncertainty will always be present.

Although the water resource systems planner can describe the consequences of uncertainty, it is unclear as to whether or not society can determine exactly what to do with that information. While multiattribute utility theory could allow us to determine which of a set of projects society prefers, it requires that one first determine society's willingness to trade off increased expected benefits for decreased risk. This is an impossible task, for who can really know and articulate what a pluralistic society's preferences are? Still, if the water resource systems analysts do their job well, then dominated solutions or those which perform poorly or those which have unacceptably large risks can be identified and eliminated from consideration. If public officials cannot discriminate between the marginal differences among the remaining plans, then society may be indifferent to which plan is implemented. This, of course, does not mean that particular groups of individuals will not be benefited more by one or another plan and hence will express intense preferences.

APPENDIX 3A

CONFIDENCE INTERVALS FOR QUANTILES AND MOMENTS

Selected sample quantiles and the sample mean and variance are often used as summary statistics to describe the distribution of an observed random variable. People report the mean and variance of such natural phenomena as streamflow, rainfall, evaporation and runoff rates. Similarly, in stochastic simulation studies discussed in Section 3.6.1, selected sample quantiles and the mean and variance of performance statistics such as net benefits or power production may be reported. In all cases, it is advisable to assess the accuracy with which the sample quantiles or the mean and variance can be expected to approximate the true quantiles or the mean and variance of the phenomena.

The precision of these estimates can be expressed by their standard error or in terms of *confidence intervals*. For integers l and u, a confidence interval

$[X_{(l)}, X_{(u)}]$ is an interval whose endpoints depend on the ordered observed sample values $\{X_{(1)}, \ldots, X_{(n)}\}$. It has the property that the probability that the interval contains a specified population parameter, such as the quantile x_p, is at least a prespecified value. Thus a $100(1-\alpha)\%$ confidence interval for the quantile x_P, would be such that

$$\Pr[X_{(l)} \leq x_P \leq X_{(u)}] \geq 1 - \alpha \tag{3A.1}$$

where $100(1-\alpha)\%$ is the *level of confidence*. Note that x_p is unknown but fixed, while the endpoints of the interval are considered random.

Consider construction of a confidence interval for the population quantile x_p using only the ordered observations $X_{(1)} \leq \ldots \leq X_{(n)}$. For any two integers $r < s$, logic insures that

$$\begin{aligned} &\Pr[X_{(r)} \leq x_P] \\ &\quad = \Pr[\{X_{(r)} \leq x_p \text{ and } X_{(s)} < x_p\} \text{ or } \{X_{(r)} \leq x_p \leq X_{(s)}\}] \\ &\quad = \Pr[X_{(s)} < x_p] + \Pr[X_{(r)} \leq x_p \leq X_{(s)}] \end{aligned} \tag{3A.2}$$

so that

$$\Pr[X_{(r)} \leq x_p \leq X_{(s)}] = \Pr[X_{(r)} \leq x_p] - \Pr[X_{(s)} < x_p] \tag{3A.3}$$

For continuous random variables

$$\Pr[X \leq x_p] = F_X(x_P) = p \tag{3A.4}$$

Then the probability that x_p is contained in the confidence interval $[X_{(r)}, X_{(s)}]$ is exactly

$$\Pr[X_{(r)} \leq x_p \leq X_{(s)}] = \sum_{k=r}^{s-1} \frac{n!}{k!(n-k)!} p^k (1-p)^{n-k} \tag{3A.5}$$

Values of r and s with $s = n + 1 - r$ for which $X_{(r)}$ and $X_{(s)}$ provide a 90% confidence interval for the population median are given in Table 3A.1.

Confidence intervals for population quantiles can be constructed without knowledge of the true distribution of the observations. Construction of confidence intervals for the population mean and variance in general requires that one assume that the observations are drawn from a statistically convenient distribution, usually the normal distribution. If X has a normal distribution, then the precision with which the sample mean $\bar{x}$ approximates μ_X can be assessed with the t statistic

$$t = \frac{\bar{x} - \mu_X}{\sqrt{s_X^2/n}} \tag{3A.6}$$

If the X_i's are independent and normally distributed with the same mean and variance, then t has *Student's t-distribution* with $n - 1$ degrees of freedom. Knowing the distribution of t, one can construct a confidence interval for μ_X. Rearranging equation 3A.6 yields

$$\bar{x} - \frac{ts_X}{\sqrt{n}} \leq \mu_X \leq \bar{x} + \frac{ts_X}{\sqrt{n}} \tag{3A.7}$$

TABLE 3A.1. At Least 90% Confidence Intervals for Median $\Pr[X_{(r)} \leq x_{0.50} \leq X_{(n+1-r)}] = 1 - \alpha \geq 0.90$

n	r	*Actual Value of* α
7	1	0.016
8	2	0.070
9	2	0.039
10	2	0.021
11	3	0.065
12	3	0.039
13	4	0.092
14	4	0.057
15	4	0.035
16	5	0.077
17	5	0.049
18	6	0.096
19	6	0.064
20	6	0.041
21	7	0.078
22	7	0.017
23	8	0.093
24	8	0.064
25	8	0.043
30	11	0.099
35	13	0.090
40	15	0.082
45	17	0.074
50	19	0.066
60	24	0.093

If a value t_α is selected for t such that

$$\Pr\left[\left|\frac{\bar{x} - \mu_X}{\sqrt{s_X^2/n}}\right| \leq t_\alpha\right] = 1 - 2\alpha \tag{3A.8}$$

for fixed probability α, then intervals such as those in equation 3A.7 will contain the true mean $100(1 - 2\alpha)\%$ of the time.

Values of t_α depend on α and the number of degrees of freedom equal to $n - 1$, one less than the sample size. Values of t_α are given in Table 3A.2 More complete tables appear in most books on statistics.

If X is not normally distributed, the exact distribution of t in equation 3A.6 generally is not known. However, the central limit theorem proves that if X has a distribution with finite mean and variance, then for large n the distribution of t approaches that of the normal distribution with mean zero and variance 1. For infinite "degrees of freedom," the distribution of Stu-

TABLE 3A.2. Percentage Points of Student's t-Distribution $\Pr[|t| \leq t_\alpha] = 1 - 2\alpha$

	VALUES OF 2α			
Degrees of Freedom	*0.20*	*0.10*	*0.05*	*0.01*
5	1.476	2.015	2.571	4.032
10	1.372	1.812	2.228	3.169
15	1.341	1.753	2.131	2.947
20	1.325	1.725	2.086	2.845
25	1.316	1.708	2.060	2.787
30	1.310	1.697	2.042	2.750
40	1.303	1.684	2.021	2.704
60	1.296	1.671	2.000	2.660
120	1.289	1.658	1.980	2.617
∞	1.282	1.645	1.960	2.576

dent's t is identical to that of the normal distribution. Hence, if X_t has finite mean and variance and one has a large number of observations, an approximate confidence interval for $\bar{x}$ can be calculated by assuming that t has a standard-normal distribution.

APPENDIX 3B

USEFUL STATISTICAL TESTS

In simulation studies, the validation of quantitative models, and the analysis of experimental results and field data, it is necessary to determine the statistical significance of various results. In a simulation study, the analyst will want to determine if a change in an operating policy or a variation in the system's parameters affected the system's performance or if observed variations in performance may be due to the stochastic variation in the system's operation. The problems of drawing inferences about the real underlying behavior or properties of a system based on a set of observations lie in the domain of statistics. This section presents several statistical tests which are useful for addressing questions and issues that arise in water resource systems engineering. Emphasis is placed on *distribution-free* or *nonparametric* methods. These tests do not rely on unrealistic assumptions about the distribution of the observations. They are often overlooked and not taught in introductory statistics courses. In the situations encountered in water resources systems engineering, distribution-free tests are often as powerful, if not more so, than standard tests which assume that the observations are normally distributed. These and other tests are discussed in more detail in Bickel and Doksun [2], Bradley [7], Mosteller and Rourke [27], and other books that discuss nonparametric statistics.

This appendix discusses the use of ranks in statistical tests and illustrates the basic ideas of hypothesis testing. The first section discusses the problem of determining if a stochastic streamflow model provides an adequate representation of the true distribution of flows. The second section presents the two-sample Wilcoxon test for determining if it is reasonable to conclude that two independent sets of observations were drawn from the same distribution. The third section addresses the same issue when the observations occur in pairs, a situation that arises in well-designed simulation experiments. Statistical tests related to the value of particular population quantiles or the mean value of a random variable can also be constructed using the confidence intervals discussed in the previous section. Table 3B.1 provides a guide to appropriate tests for a number of statistical problems. Some of these tests are discussed here while the others are discussed in the references.

TABLE 3B.1. Appropriate Statistical Tests for Different Questions and Circumstances

Question	*Information Available*	*Appropriate Test*	*Section of Book*
Does $F_X = F_Y$ or is X stochastically smaller than Y?	Independent samples $\{x_i\}, \{y_j\}$	(1) Wilcoxon two-sample test	Appendix 3B.2
		(2) Student's t test if observations normal	Sec. 3.6.1 and Appendix 3A
Does $F_X = F_Y$ or is X stochastically smaller than Y?	Paired observations $\{(x_i, y_i)\}$	(1) Sign test applied to $x_i - y_i$	Sec. 3.6.1 and Appendix 3B.3
		(2) Wilcoxon signed rank test	Appendix 3B.3
		(3) Student's t if observations normal	Sec. 3.6.1 and Appendix 3A
Is F_X a particular distribution G?	Single sample $\{x_i\}$	(1) Kolmogorov–Smirnov test for general G	Sec. 3.3.1
		(2) Shapiro–Wilk and correlation tests if G is normal	Sec. 3.3.1
Are X and Y independent or do they increase and decrease together?	Paired observations	(1) Spearman rank correlation	Not discussed see [2, 7, 21]
		(2) Pearson correlation coefficient if observations normal	Not discussed see [1, 2, 13, 39]

3B.1 Introduction to Statistical Ideas

As an introduction to statistical concepts and the way statistical questions are posed, consider the task of validating a stochastic streamflow model developed for a reservoir design study. One should attempt to determine if the important characteristics of droughts produced by the streamflow model are consistent with the historic streamflow record. This question can be asked formally. Let S_H be the active storage capacity of the smallest reservior which receiving the historical streamflows as input can release at least a given volume of water Y (the firm yield) in every period. Let S_i be the required storage capacities for the reservoir that one would calculate using each of a set of synthetic streamflow sequences, each having the same length as the historical streamflow record. Consider now two possible situations, which are called *hypotheses*. First, the *null hypothesis* is that the calculated capacity based on the historical flows is one observation of a random variable whose distribution is the same as the distribution of the capacities S_i required with the synthetic flows.

$$H_0: S_H \text{ and } \{S_i\} \text{ have a common cumulative distribution function } F_S$$

One wants to determine if it is reasonable to accept the null hypothesis and hence conclude that the streamflow model produces flow sequences which require capacities S_i consistent with that required with the historical record. A frequent failing of streamflow models is that they fail to adequately reproduce the persistence of actual flows and the historical capacity requirements S_H tend to be larger than those S_i calculated with synthetic flow sequences. In statistical terminology, the random variable S_i is said to be *stochastically smaller* than S_H when

$$F_{S_i}(y) \geq F_{S_H}(y) \qquad \text{for all } y \tag{3B.1}$$

This means that for any s, the probability that S_i is less than or equal to each s is greater than or equal to the probability that S_H is less than or equal to s. Hence S_i tends to be smaller. Our *alternative hypothesis*, which is what might be true if the null hypothesis were false, is

$$H_a: S_i \text{ is stochastically smaller than } S_H$$

Now the true distribution of the S's under either hypothesis is not known and it is preferable not to complicate the situation by introducing a guess as to what their distribution might be. All that is really known is how the observed value s_H compares to the s_i's or, to be precise, how many of the s_i's are less than s_H. This is often summarized by the *rank* R_H of s_H in the combined sample $\{s_H, s_1, \ldots, s_n\}$.

$$R_H = k + 1 \text{ if exactly } k \text{ of the } s_i\text{'s are less than } s_H$$

If the null hypothesis is true and the calculated historical and synthetic capacity requirements are independent, the combined sample is a set of $m + 1$ independent identically distributed random variables drawn from the same distribution. Then there would be nothing special about the rank of s_H, and R_H would have an equal probability of taking each of the integer values 1, 2, 3, . . . , $m + 1$.[1] Hence

$$\Pr[R_H = k] = \frac{1}{m+1} \qquad k = 1, \ldots, m + 1 \tag{3B.2}$$

This is all that is needed to construct a test of whether it is "reasonable" to accept or believe the null hypothesis or if it should be rejected in favor of the alternative. For our alternative hypothesis, a value of R_H equal to 1, or any integer less than $m/2$, is evidence that would appear to refute the alternative. However, if R_H equals $m + 1$, m, $m - 1$, or some other large integer, one might find the null hypothesis unreasonable. In fact, the larger R_H, the more one is inclined to reject H_0 and to accept the alternative hypothesis H_a that S_i is stochastically smaller than S_H.

The exact value of R_H at or above which one chooses to reject the null hypothesis in favor of the alternative depends on how much evidence one demands before one will consider H_0 unreasonable. Because R_H under H_0 has an equal probability of taking on each of the values $1, \ldots, m + 1$, the probability α of observing a particular value of R_H or a value more likely under H_a is

$$\alpha = \Pr[R_H \geq k] = \frac{m - k + 2}{m + 1} \tag{3B.3}$$

Typically, statisticians choose values of α, the *confidence level* of the test, equal to 5% or 10%. Hence they will incorrectly reject the null hypothesis 5% or 10% of the time when it is in fact true.

For our problem, if one generated 49 synthetic flow sequences and hence 49 values of S_i, the probability of R_H equaling 46 or more is $(49 - 46 + 2)/(49 + 1) = 0.1$, or 10%. Thus values of R_H equal to 46 or more is often considered sufficient evidence to reject the reasonableness of the null hypothesis.

Of course, one need not specify the value of α before examining the available data. Upon calculating R_H, one can compute $(m - R_H + 2)/(m + 1)$ and then decide how to proceed. If the calculated value of R_H suggests that the model is inadequate but one judges the evidence insufficient to make a definite decision, then more observations S_i may be generated in an attempt to resolve the issue.

At some point, one must decide whether to accept the null hypothesis or the alternative. In this idealized situation where only one hypothesis or the other can be true, the analyst can make one of two possible errors: (1) H_a can

[1] In reality S_H is not independent of S_i if the parameters of the streamflow model were estimated using the historical streamflow record. Then the exact distribution of the rank of S_H is not known, although one would expect that extreme values of R_H would become less likely.

True state of nature

Decision made	H_0 is true	H_a is true
H_0 accepted	Correct decision	Probability of this error is β
H_a accepted	Probability of this error is α	Correct decision

FIGURE 3B.1. Possible errors when deciding between hypotheses.

be accepted when H_0 is true, or (2) H_0 can be accepted when H_a is true. These two types of errors are illustrated in Figure 3B.1. The probability of the first type of error is specified by the analysts when α is selected. The probability of the second type of error β frequently cannot be determined because of the vagueness of the alternative hypothesis, as is the case in our example. Still one can see that among efficient tests, there is a clear trade-off between the error probabilities α and β. For our example, consider the test where one rejects H_0 if $R_H \geq r$ corresponding to $\alpha(r) = (m - r + 2)/(m + 1)$. Another test might reject H_0 whenever $R_H \geq r + 1$ corresponding to $\alpha(r + 1) = (m - r + 1)/(m + 1)$. The first test has a larger α-error probability but a smaller β-error probability: in the first test the alternative hypothesis has a smaller probability of being rejected when it is in fact true.

$$\begin{aligned} \beta(r + 1) &= \Pr[R_H < r + 1 \mid H_a \text{ true}] \\ &= \Pr[R_H = r \mid H_a \text{ true}] + \Pr[R_H < r \mid H_a \text{ true}] \qquad (3B.4) \\ &= \Pr[R_H = r \mid H_a \text{ true}] + \beta(r) \geq \beta(r) \end{aligned}$$

In practice, the appropriate values of α and β and the appropriate choice for the two hypotheses is an important decision. In general, the null hypothesis is accepted unless sufficient evidence is available to support the alternative hypothesis. Thus one places the burden of proof on the alternative hypothesis. In questions of safety, it is of course appropriate to assume that systems are unsafe until proven otherwise. Generally, one must consider the issues involved and the cost or consequences of making either type of error when designing a statistical test and selecting the appropriate value of α.

3B.2 Two-Sample Wilcoxon Test

In the previous example, a comparison was made of the storage capacity s_H required with the historical flows and those $\{s_1, \ldots, s_n\}$ required with sets of synthetic flows of the same length. A similar problem arises when one has two sets of independent observations $\{x_1, \ldots, x_n\}$ and $\{y_1, \ldots, y_m\}$ of two random variables X and Y and wants to test the hypothesis:

H_0: X and Y have a common cumulative distribution function

versus

H_a: X is stochastically smaller than Y

For example, x_i could be the annual flood flows at a gaging station before a major modification of a basin and y_j the observed flows after the modification. To illustrate the test, consider a situation that arises in Section 3.6.1. The amount of water promised but not supplied by a reservoir in 10 replicate simulations of the planning period with each of two different operating policies is given in Table 3B.2.

TABLE 3B.2. Total Shortage

Policy 1: x_i	*Policy 2:* y_j
1.25	1.80
1.97	4.70
1.79	3.90
1.67	3.46
0.21	1.48
0.00	0.60
1.29	3.30
4.75	5.45
0.00	0.60
0.34	3.24

It is necessary to demonstrate that, in fact, the total shortage X experienced with policy 1 is stochastically smaller than the total shortage Y experienced with policy 2. Complication of the analysis by a guess of what the common distribution of X and Y might be if the null hypothesis were true is unnecessary. Instead, one can use the ranks of the observed x_i's in the combined sample $\{x_1, \ldots, x_{10}, y_1, \ldots, y_{10}\}$ to construct an efficient statistical test.

Here the rank of each x_i, denoted R_i, will range from 1 if x_i is the smallest of all the observations to $n + m = 20$ if x_i is the largest of all the observations. When ties occur so that a unique rank cannot be assigned to each observation, the average of the ranks corresponding to the tied observations should be used as the rank of each of the tied observations. This is illustrated in Table 3B.3. Two observed values of X had value 0; hence they are each assigned rank $(1 + 2)/2 = 1.5$, because they correspond to the two smallest values.

TABLE 3B.3. Determination of Ranks of Observation for Wilcoxon Two-Sample Test

	Rank	*Ordered x's*	*Ordered y's*
	1.5	0.00	
	1.5	0.00	
	3	0.21	
	4	0.34	
	5.5		0.60
	5.5		0.60
	7	1.25	
	8	1.29	
	9		1.48
	10	1.67	
	11	1.79	
	12		1.80
	13	1.97	
	14		3.24
	15		3.30
	16		3.46
	17		3.90
	18		4.70
	19	4.75	
	20		5.45
Sum of ranks	$W =$	78	132

An efficient test of the null hypothesis for the given alternative is provided by the *Wilcoxon statistic* W, which is the sum of the ranks of the x_i's:

$$W = \sum_{i=1}^{n} R_i \tag{3B.5}$$

If, indeed, X is stochastically smaller than Y, then one expects W to be smaller than would be the case if the X's and Y's were drawn from the same distribution. For n and m greater than 9, regardless of the distribution of X and Y, W is essentially normally distributed with mean and variance

$$\begin{aligned} E[W] &= \frac{n(n+1)}{2} + \tfrac{1}{2}nm \\ \text{Var}(W) &= \tfrac{1}{12}nm(n+m+1) \end{aligned} \tag{3B.6}$$

so that

$$Z = \frac{W - E[W]}{\sqrt{\text{Var}(W)}} \tag{3B.7}$$

has a standard normal distribution. For smaller values of n and m, tables are available [2, 7, 27]. Using Table 3B.3, for our example

$$W = 1.5 + 1.5 + 3 + \ldots + 13 + 19 = 78 \tag{3B.8}$$

where

$$E[W] = 105$$
$$\text{Var}(W) = (13.2)^2 \tag{3B.9}$$

so that

$$z = \frac{78 - 105}{13.2} = -2.05 \tag{3B.10}$$

Examination of standard normal tables reveals that the probability a standard-normal variate is less than -2.05 is about 2%. Hence one can reject the null hypothesis at the 2% confidence level.

3B.3 Sign and *t*-Test for Paired Observations

When validating models and in well-designed field studies and simulation experiments, one frequently obtains pairs of observations (X_i, Y_i). Here X_i and Y_i may correspond to an observation and a model's prediction. In stochastic simulation experiments X_i and Y_i may be a system's performance with different operating policies or different parameter sets for each i, where all other factors affecting the systems performance are the same. This is actually the case for the total water shortage reported in Table 3B.2; for each i, the total shortages (x_i, y_i) were obtained by using the same synthetic streamflow sequence in both simulations. The difference between each x_i and y_i is due solely to having changed the operating policy. Hence $\{x_1, \ldots, x_{10}\}$ and $\{y_1, \ldots, y_{10}\}$ are not independent samples as was assumed in the previous section.

When one has paired observations (X_i, Y_i), the sign test provides a simple and frequently adequate test of the null hypothesis that X and Y have a common distribution against the alternative that X is stochastically smaller than Y. Consider the differences $Z_i = X_i - Y_i$. If the null hypothesis is true, then

$$\Pr[Z_i > 0 \mid Z_i \neq 0] = \Pr[Z_i < 0 \mid Z_i \neq 0] = \tfrac{1}{2} \tag{3B.11}$$

The probabilities are made conditional on $Z_i \neq 0$ to emphasize that pairs for which $x_i = y_i$ can be disregarded when performing this test.

If the alternative hypothesis is true, then

$$\Pr[Z_i > 0 \mid Z_i \neq 0] < \tfrac{1}{2} < \Pr[Z_i < 0 \mid Z_i \neq 0] \tag{3B.12}$$

A reasonable test statistic is S, the number of the Z_i that are greater than zero. If the alternative hypothesis is true, one expects to observe a smaller value of S than one would observe if the null hypothesis were true. If there are m z_i's not equal to zero, then under the null hypothesis, S has a binomial distribution and

$$\Pr[S \leq s] = \sum_{k=0}^{s} \frac{m!}{k!(m-k)!}(\tfrac{1}{2})^m \tag{3B.13}$$

For the example in the previous section, $z_i < 0$ for all i and thus $S = 0$.

For $m = 10$, the probability that $S \leq 0$ is $(0.5)^{10} < 0.001$. This demonstrates that it is very unlikely that the X's and Y's are observations drawn from the same distribution.

Another test is often used with paired data. If the differences $Z_i = X_i - Y_i$ have a normal distribution, then one could consider the hypothesis that

$$H_0\colon \mu_X = \mu_Y \text{ and thus } \mu_Z = 0$$

versus

$$H_a\colon \mu_X < \mu_Y \text{ and thus } \mu_Z < 0$$

As discussed in the previous section, for normally distributed Z,

$$t = \frac{\bar{z} - \mu_Z}{s_Z/\sqrt{n}} \tag{3B.14}$$

has Student's t-distribution with $n - 1$ degrees of freedom when there are n observations used to compute $\bar{z}$ and s_Z^2. This implies that $\sqrt{n}\,\bar{z}/s_Z$ has Student's t-distribution under the null hypothesis that $\mu_Z = 0$. If $\mu_Z < 0$, one expects $\sqrt{n}\,\bar{z}/s_Z$ to be smaller than it would otherwise be under H_0, thus one should reject H_0 if $\sqrt{n}\,\bar{z}/s_Z$ is too small. For the total shortage example,

$$\begin{aligned} \bar{z} &= -1.53 \\ s_Z &= 0.91 \end{aligned} \tag{3B.15}$$

and

$$\sqrt{n}\,\frac{\bar{z}}{s_Z} = \sqrt{10}\,\frac{-1.53}{0.91} = -5.32$$

The probability that t is less than -5.32 is less than 0.01%. The t-test appears to be more powerful than the sign test, and it often is. However, one should remember that the t-test is based on the often invalid assumption that Z is normally distributed. Hence the significance of $t = -5.32$ is really not known unless the distribution of Z is specified. When one has paired observations and the sign test has insufficient power for one's purpose, the Wilcoxon signed rank test discussed in Mosteller and Rourke [27] and Bickel and Doksum [2] should be used if the assumption of normality is questionable.

APPENDIX 3C

PROPERTIES OF TIME-SERIES STATISTICS

The statistics most frequently used to describe the distribution of a stationary stochastic process are the sample mean, variance, and various autocorrleations. Several properties of the sample mean and variance were presented in Section 3.2.3 and Appendix 3A for the case when the observations are independent. Statistical dependence among the observations, as is frequently the case in time series, can have a marked affect on the distribution of these statistics. This appendix reviews the sampling properties of these statistics when the observations are a realization of a stochastic process.

Let $\{x_t\}$ be the observed values of a stationary stochastic process. The sample mean

$$\bar{x} = \frac{1}{n}\sum_{t=1}^{n} x_t \tag{3C.1}$$

is an unbiased estimate of the mean of the process μ_X, because

$$E[\bar{X}] = \frac{1}{n}\sum_{t=1}^{n} E[X_t] = \mu_X \tag{3C.2}$$

However, correlation among the X_t's, so that $\rho_X(k) \neq 0$ for $k \neq 0$, affects the variance of $\bar{X}$.

$$\begin{aligned} \text{Var}(\bar{X}) &= E[(\bar{X} - \mu_X)^2] = \frac{1}{n^2} E\left\{\sum_{t=1}^{n}\sum_{s=1}^{n}(X_t - \mu_X)(X_s - \mu_X)\right\} \\ &= \frac{\sigma_X^2}{n}\left\{1 + 2\sum_{k=1}^{n-1}\left(1 - \frac{k}{n}\right)\rho_X(k)\right\} \end{aligned} \tag{3C.3}$$

The variance of $\bar{X}$, equal to σ_X^2/n for independent observations, is inflated by the factor within the brackets. For $\rho_X(k) \geq 0$, as is often the case, this factor is a nondecreasing function of n, so that the variance of $\bar{X}$ is inflated by a factor whose importance does not decrease with increasing sample size.

A common model of stochastic series has

$$\rho_X(k) = [\rho_X(1)]^k = \rho^k \tag{3C.4}$$

This correlation structure arises from the autoregressive Markov model discussed at length in Chapter 6. For this correlation structure

$$\text{Var}(\bar{X}) = \frac{\sigma_X^2}{n}\left\{1 + \frac{2\rho}{n}\frac{[n(1 - \rho) - (1 - \rho^n)]}{(1 - \rho)^2}\right\} \tag{3C.5}$$

Substitution of the sample estimates for σ_X^2 and $\rho_X(1)$ in the equation above often yields a more realistic estimate of the variance of $\bar{X}$ than does the estimate v_X^2/n if the correlation structure $\rho_X(k) = \rho^k$ is reasonable; otherwise, equation 3C.3 may be employed. Table 3C.1 illustrates the affect of correlation among the X's on the standard error of their mean.

TABLE 3C.1. Standard Error of $\bar{X}$ When $\sigma_X = 0.25$ and $\rho_X(k) = \rho^k$

	Correlation of Observations		
Sample Size	$\rho = 0$	0.3	0.6
$n =$ 25	0.050	0.067	0.096
50	0.035	0.048	0.069
100	0.025	0.034	0.050

The properties of the estimate of the variance of X,

$$v_X^2 = \frac{1}{n} \sum_{t=1}^{n} (x_t - \bar{x})^2 \tag{3C.6}$$

are also affected by correlation among the X_t's. The expected values of v_X^2 becomes

$$E[v_X^2] = \sigma_X^2 \left\{ 1 - \frac{1}{n} - \frac{2}{n} \sum_{k=1}^{n-1} \left(1 - \frac{k}{n}\right) \rho_X(k) \right\} \tag{3C.7}$$

The bias in v_X^2 depends on terms involving $\rho_X(1)$ through $\rho_X(n-1)$. Fortunately, the bias in v_X^2 decreases with n and is generally unimportant when compared to its variance. Correlation among the x_t's also affects the variance of v_X^2. Assuming that X has a normal distribution (here the variance of v_X^2 depends on the fourth moment of X), the variance of v_X^2 for large n is approximately

$$\text{Var}(v_X^2) \simeq 2\frac{\sigma_X^4}{n} \left\{ 1 + 2 \sum_{k=1}^{\infty} \rho_X^2(k) \right\} \tag{3C.8}$$

where for $\rho_X(k) = \rho^k$, equation 3C.8 becomes

$$\text{Var}(v_X^2) \simeq 2\frac{\sigma_X^4}{n} \left(\frac{1 + \rho^2}{1 - \rho^2} \right) \tag{3C.9}$$

Like the variance of $\bar{x}$, the variance of v_X^2 is inflated by a factor whose importance does not decrease with n. This is illustrated by Table 3C.2, which gives the coefficient of variation of v_X^2 as a function of n and ρ when the observations have a normal distribution and $\rho_X(k) = \rho^k$.

A fundamental problem of time-series analyses is the estimation or description of the relationship between the random variable at different times. The statistics used to describe this relationship are the autocorrelations. Several estimates of the autocorrelations have been suggested; a simple and satisfactory estimate recommended by Jenkins and Watts [15] is

$$\hat{\rho}_X(k) = r_k = \frac{\sum_{t=1}^{n-k} (x_t - \bar{x})(x_{t+k} - \bar{x})}{\sum_{t=1}^{n} (x_t - \bar{x})^2} \tag{3C.10}$$

TABLE 3C.2. Coefficient of Variation of v_X^2 When Observations Have a Normal Distribution and $\rho_X(k) = \rho^k$

	Correlation of Observations		
Sample Size	*$\rho = 0$*	*0.3*	*0.6*
$n =$ 25	0.28	0.31	0.41
50	0.20	0.22	0.29
100	0.14	0.15	0.21

Here r_k is the ratio of two sums where the numerator contains $n - k$ terms and the denominator contains n terms. The estimate r_k is biased, but unbiased estimates frequently have larger mean square errors [15]. A comparison of the bias and variance of r_1 is provided by the case when the X_t's are independent normal variates. Then [19]

$$E[r_1] = -\frac{1}{n} \tag{3C.11}$$

and

$$\text{Var}(r_1) = \frac{(n-2)^2}{n^2(n-1)} \simeq \frac{1}{n}$$

For $n = 25$, the expected value of r_1 is -0.04 rather than the true value of zero; its standard deviation is 0.188. This results in a mean square error of

$$(E[r_1])^2 + \text{Var}(r_1) = 0.0016 + 0.0353 = 0.0369 \tag{3C.12}$$

Clearly, the variance of r_1 is the dominant term.

For X_t's that are not independent, exact expressions for the variance of r_k generally are not available. However, for normally distributed X_t and large n [19],

$$\begin{aligned} \text{Var}(r_k) \simeq \frac{1}{n} \sum_{l=-\infty}^{+\infty} [\rho_X^2(l) + \rho_X(l+k)\rho_X(l-k) \\ - 4\rho_X(k)\rho_X(l)\rho_X(k-l) + 2\rho_X^2(k)\rho_X^2(l)] \end{aligned} \tag{3C.13}$$

If $\rho_X(k)$ is essentially zero for $k > q$, then the simpler expression [6]

$$\text{Var}(r_k) \simeq \frac{1}{n}\left[1 + 2\sum_{l=1}^{q} \rho_X^2(l)\right] \tag{3C.14}$$

is valid for r_k corresponding to $k > q$; thus for large n, $\text{Var}(r_k) \geq 1/n$ and values of r_k will frequently be outside the range of $\pm 1.65/\sqrt{n}$, even though $\rho_X(k)$ may be zero.

If $\rho_X(k) = \rho^k$, equation 3C.13 reduces to

$$\text{Var}(r_k) \simeq \frac{1}{n}\left[\frac{(1+\rho^2)(1-\rho^{2k})}{1-\rho^2} - 2k\rho^{2k}\right] \tag{3C.15}$$

In particular for r_1, this gives

$$\text{Var}(r_1) \simeq \frac{1}{n}(1-\rho^2) \tag{3C.16}$$

Approximate values of the standard deviation of r_1 for different values of n and ρ are given in Table 3C.3.

The estimates of r_k and r_{k+l} are highly correlated for small l; this causes plots of r_k versus k to exhibit slowly varying cycles when the true values of $\rho_X(k)$ may be zero. This increases the difficulty of interpreting the sample autocorrelations.

TABLE 3C.3. Approximate Standard Deviation of r_1 When Observations Have a Normal Distribution and $\rho_X(k) = \rho^k$

Sample Size	*Correlation of Observations* $\rho = 0$	*0.3*	*0.6*
$n =$ 25	0.20	0.19	0.16
50	0.14	0.13	0.11
100	0.10	0.095	0.080

EXERCISES

3-1. Give an example of a water resources planning study with which you have some familiarity. Make a list of the basic information used in the study and of the methods and models used to transform that information into decisions, recommendations, and conclusions.

(a) Indicate the major sources of uncertainty and possible error in the basic information and in the transformation of that information into decisions, recommendations, and conclusions.

(b) In systems studies, sources of error and uncertainty are sometimes grouped into three categories:

1. Uncertainty due to the natural variability of rainfall, temperature, and streamflows which affect a system's operation.
2. Uncertainty due to errors made in estimation of the models' parameters with a limited amount of data.
3. Uncertainty or errors introduced into the analysis because conceptual and/or mathematical models do not reflect the true nature of the relationships being described.

Indicate, if applicable, into which category each of the sources of error or uncertainty you have identified falls.

3-2. The following matrix displays the joint probabilities of different weather conditions and of different recreation benefit levels obtained from use of a reservoir in a state park:

	POSSIBLE RECREATION BENEFIT LEVELS		
Weather	RB_1	RB_2	RB_3
Wet	0.10	0.20	0.10
Dry	0.10	0.30	0.20

(a) Compute the probabilities of recreation levels RB_1, RB_2, RB_3 and of dry and wet weather.
(b) Compute the conditional probabilities $P(\text{wet} \mid RB_1)$, $P(RB_3 \mid \text{dry})$, and $P(RB_2 \mid \text{wet})$.

3-3. In flood protection planning, the 100-year flood, which is an estimate of the quantile $x_{0.99}$, is often used as the design flow. Assuming that the floods in different years are independently distributed:
(a) Show that the probability of at least one 100-year flood in a 5-year period is 0.049.
(b) What is the probability of at least one 100-year flood in a 100-year period?
(c) If floods at 1000 different sites occur independently, what is the probability of at least one 100-year flood at some site in any single year?

3-4. The price to be charged for water by an irrigation district has yet to be determined. Currently it appears as if there is a 60% probability that the price will be \$10 per unit of water and a 40% probability that the price will be \$5 per unit. The demand for water is uncertain. The estimated probabilities of different demands given alternative prices are as follows:

	QUANTITY DEMANDED GIVEN PRICE				
Price	*30*	*55*	*80*	*100*	*120*
\$ 5	0.00	0.15	0.30	0.35	0.20
\$10	0.20	0.30	0.40	0.10	0.00

(a) What is the most likely value of future revenue from water sales?
(b) What are the mean and variance of future water sales?
(c) What is the median value and interquantile range of future water sales?
(d) What price will maximize the expected revenue from the sale of water?

3-5. Plot the following data on possible recreation losses and irrigated agricultural yields. Show that use of the expected storage level or expected allocation underestimates the expected value of the convex function describing reservoir losses while it overestimates the expected value of the concave function describing crop yield. A concave function $f(x)$ has the property that $f(x) \leq f(x_0) + f'(x_0)(x - x_0)$ for any x_0; prove that use of $f(E[X])$ will always overestimate the expected value of a concave function $f(X)$ when X is a random variable.

Irrigation Water Allocation	*Crop Yield/Hectare*	*Probability of Allocation*
10	6.5	0.20
20	10.0	0.30
30	12.0	0.30
40	11.0	0.20

Summer Storage Level	*Decrease in Recreation Benefits*	*Probability of Storage Level*
200	5	0.10
250	2	0.20
300	0	0.40
350	1	0.20
400	4	0.10

3-6. Complications can be added to the economic evaluation of a project by uncertainty concerning the useful life of the project. For example, the time at which the useful life of a reservoir will end due to silting is never known with certainty when the reservoir is being planned. If the discount rate is high and the life is relatively long, the uncertainty may not be very important. However, if the life of a reservoir, or of a wastewater treatment facility, or of any other such project, is relatively short, the practice of using the expected life to calculate present costs or benefits may be misleading.

In this problem, assume that a project resulting in \$1000 of net benefits at the end of each year is expected to last between 10 and 30 years, each year within the range of 11 to 30 being equally likely. Given a discount rate of 10%:

(a) Compute the present value of net benefits NB_0, assuming a 20-year project lifetime.

(b) Compare this with the expected present net benefit $E[NB_0]$ taking account of the uncertainty in the project's lifetime.

(c) Compute the probability that the actual present net benefit is at least \$1000 less than NB_0, the benefit estimate based on the 20-year life.

(d) What is the chance of getting \$1000 more than the original estimate NB_0?

3-7. A continuous random variable that could describe the proportion of fish or other animals in different large samples which have some distinctive feature is the beta distribution whose density is ($\alpha > 0$, $\beta > 0$):

$$f_X(x) = \begin{cases} cx^{\alpha-1}(1-x)^{\beta-1} & 0 \le x \le 1 \\ 0 & \text{otherwise} \end{cases}$$

(a) Directly calculate the value of c and the mean and variance of X for $\alpha = \beta = 2$.

(b) In general, $c = \Gamma(\alpha + \beta)/\Gamma(\alpha)\Gamma(\beta)$, where $\Gamma(\alpha)$ is the gamma function equal to $(\alpha - 1)!$ for integer α. Using this information, derive the general expressions for the mean and variance of X given in the text. To obtain a formula which gives the values of the integrals of interest, note that the expression for c must be such that the integral over (0, 1) of the density function is unity for any α and β.

3-8. The joint probability density of rainfall at two places on rainy days could be described by

$$f_{XY}(x, y) = \begin{cases} \dfrac{2}{(x+y+1)^3} & x, y \ge 0 \\ 0 & \text{otherwise} \end{cases}$$

Calculate and graph:

(a) $F_{XY}(x, y)$, the joint distribution function of X and Y.

(b) $F_Y(y)$, the marginal cumulative distribution function of Y, and $f_Y(y)$, the density function of Y.

(c) $f_{Y|X}(y|x)$, the conditional density function of Y given that $X = x$, and $F_{Y|X}(y|x)$, the conditional cumulative distribution function of Y given that $X = x$ (the cumulative distribution function is obtained by integrating the density function).

Show that

$$F_{Y|X}(y|x = 0) > F_Y(y) \qquad \text{for } y > 0$$

Find a value of x_0 and y_0 for which

$$F_{Y|X}(y_0|x_0) < F_Y(y_0)$$

3-9. Let X and Y be two continuous independent random variables. Prove that

$$E[g(X)h(Y)] = E[g(X)]E[h(Y)]$$

for any two real-valued functions g and h. Then show that $\text{Cov}(X, Y) = 0$ if X and Y are independent.

3-10. A frequent problem is that observations (X, Y) are taken on such quantities as flow and concentration and then a derived quantity $g(X, Y)$ such as mass flux is calculated. Given that one has estimates of the standard deviations of the observations X and Y and their correlation, an estimate of the standard deviation of $g(X, Y)$ is needed. Using a second-order Taylor series expansion for $g(X, Y)$ about μ_X and μ_Y, obtain an expression for the mean of $g(X, Y)$ as a function of its partial derivatives and of the means, variances, and covariance of X and Y.

Using a first-order approximation of $g(X, Y)$, obtain an estimate of the variance of $g(X, Y)$ as a function of its partial derivatives and the moments of X and Y. Note, the covariance of X and Y equals

$$E[(X - \mu_X)(Y - \mu_Y)] = \sigma_{XY}^2$$

3-11. A study of the behavior of water waves impinging upon and reflecting off a breakwater located on a sloping beach was conducted in a small tank. The height (crest-to-trough) of the waves was measured a short distance from the wave generator and at several points along the beach different distances from the breakwater were measured and their mean and standard error recorded.

Location	*Mean Wave Height (cm)*	*Standard Error of Mean (cm)*
Near wave generator	3.32	0.06
1.9 cm from breakwater	4.42	0.09
5.0 cm from breakwater	2.59	0.09
10.0 cm from breakwater	3.26	0.06

At which points were the wave heights significantly different from the height near the wave generator assuming that errors were independent?

Of interest to the experimenter is the ratio of the wave heights near the breakwater to the initial wave height near the wave generator in the deep water. Using the results in Exercise 3.10, estimate the standard error of this ratio at the three points assuming that errors made in measuring the height of waves at the three points and near the wave generator are independent. At which points does the ratio appear to be significantly different from 1.00?

Using the results of Exercise 3.10, show that the ratio of the mean wave heights is probably a biased estimate of the actual ratio. Does this bias appear to be important?

3-12. Derive Kirby's bound, equation 3.43, on the estimate $\hat{\gamma}_X$ of the coefficient of skewness by computing the sample estimate of the skewness of the most skewed sample it would be possible to observe. Derive also the upper bound $\sqrt{n-1}$ for the estimate of the population coefficient of variation

$$\frac{v_X}{\bar{x}}$$

when all the observations must be nonnegative.

3-13. The errors in the predictions of water quality models are sometimes described by the double exponential distribution whose density is

$$f(x) = \frac{\alpha}{2} \exp(-\alpha|x - \beta|) \qquad -\infty < x < +\infty$$

What are the maximum likelihood estimates of α and β. Note that

$$\frac{d}{d\beta}|x-\beta| = \begin{cases} -1 & x > \beta \\ +1 & x < \beta \end{cases}$$

Is there always a unique solution for β?

3-14. Derive the equations that one would need to solve to obtain maximum likelihood estimates of the two parameters α and β of the gamma distribution. Note an analytic expression for $d\,\Gamma(\alpha)/d\alpha$ is not available so that a closed form expression for the maximum likelihood estimate of α is not available. What is the maximum likelihood estimate of β as a function of the maximum likelihood estimates of α?

3-15. The log-Pearson Type III distribution is often used to model flood flows. If X has a log-Pearson Type III distribution then

$$Y = \ell\text{n}(X) - m$$

has a two-parameter gamma distribution where e^m is the lower bound of X if $\beta > 0$ and e^m is the upper bound of X if $\beta < 0$. The density function of Y can be written

$$f_Y(y)\,dy = \frac{(\beta y)^{\alpha-1}}{\Gamma(\alpha)} \exp(-\beta y)\,d(\beta y) \qquad \text{for } 0 < \beta y < +\infty$$

Calculate the mean and variance of X in terms of α, β and m. Note that

$$E[X^r] = E[(\exp(Y+m))^r] = \exp(rm)\,E[\exp(rY)]$$

To evaluate the required integrals remember that the constant terms in the definition of $f_Y(y)$ ensure that the integral of this density function over the range of βy must be unity for any values of α and β so long as $\alpha > 0$ and $\beta y > 0$. For what values of r and β does the mean of X fail to exist? How do the values of m, α and β affect the shape and scale of the distribution of X?

3-16. When plotting observations to compare the empirical and fitted distributions of streamflows, or other variables, it is necessary to assign a cumulative probability to each observation. These are called *plotting positions*. As noted in the text, for the ith largest observations $X_{(i)}$,

$$E[F_X(X_{(i)})] = \frac{i}{n+1}$$

Thus the *Weibull plotting position* $i/(n+1)$ is one logical choice. Other commonly used plotting positions are the *Hazen plotting position* $(i - 1/2)/n$ and plotting position $(i - 3/8)/(n + 1/4)$. The plotting position $(i - 3/8)/(n + 1/4)$ is a reasonable choice because its use provides a good approximation to the expected value of $X_{(i)}$. In particular, for standard normal random variables

$$E[X_{(i)}] \cong \Phi^{-1}\left(\frac{i - 3/8}{n + 1/4}\right)$$

where $\Phi(\cdot)$ is the cumulative distribution function of a standard normal random variable. While much debate centers on the appropriate plotting position to use to estimate $p_i = F_X(X_{(i)})$, often people fail to realize how imprecise all such estimates must be. Noting that

$$\text{Var}(p_i) = \frac{i(n - i + 1)}{(n + 1)^2(n + 2)}$$

contrast the difference between the estimates $\hat{p}_i$ of p_i provided by these three plotting positions and the standard deviation of p_i. Provide a numerical example. What do you conclude?

3-17. The following data represent a sequence of annual flood flows, the maximum flow rate observed each year, for the Sebou River at the Azib Soltane gaging station in Morocco.

Date	*Maximum Discharge* (m^3/s)	*Date*	*Maximum Discharge* (m^3/s)
26–03–33	445	13–03–54	750
11–12–33	1410	27–02–55	603
17–11–34	475	08–04–56	880
13–03–36	978	03–01–57	485
18–12–36	461	15–12–58	812
15–12–37	362	23–12–59	1420
08–04–39	530	16–01–60	4090
04–02–40	350	26–01–61	376
21–02–41	1100	24–03–62	904
25–02–42	980	07–01–63	4120
20–12–42	575	21–12–63	1740
29–02–44	694	02–03–65	973
21–12–44	612	23–02–66	378
24–12–45	540	11–10–66	827
15–05–47	381	01–04–68	626
11–05–48	334	28–02–69	3170
11–05–49	670	13–01–70	2790
01–01–50	769	04–04–71	1130
30–12–50	1570	18–01–72	437
26–01–52	512	16–02–73	312
20–01–53	613		

(a) Construct a histogram of the Sebou flood flow data to see what the flow distribution looks like.

(b) Calculate the mean, variance, and sample skew. Based on Table 3.2, does the sample skew appear to be significantly different from zero?

(c) Fit a normal distribution to the data and use the Kolmogorov–Smirnov test to determine if the fit is adequate. Draw a quantile–

quantile plot of the fitted quantiles $F^{-1}[(i - 3/8)/(n + 1/4)]$ versus the observed quantiles $x_{(i)}$ and include on the graph the Kolmogorov–Smirnov bounds on each $x_{(i)}$, as shown in Figures 3.2 and 3.3.

(d) Repeat part (c) using a two-parameter lognormal distribution.

(e) Repeat part (c) using a three-parameter lognormal distribution. The Kolmogorov–Smirnov test is now approximate if applied to $\log_e[X_{(i)} - \tau]$, where τ is calculated using equation 3.60 or some other method of your choice.

(f) Repeat part (c) for two- and three-parameter versions of the gamma distribution. Again, the Kolmogorov–Smirnov test is approximate.

(g) A powerful test of normality is provided by the correlation test. As described by Filliben [8], one should approximate $p_i = F_X(X_{(i)})$ by its median value which is nearly

$$\hat{p}_i = \begin{cases} 1 - (0.5)^{1/n} & i = 1 \\ (i - 0.3175)/(n + 0.365) & i = 2, \ldots, n - 1 \\ (0.5)^{1/n} & i = n \end{cases}$$

Then one obtains a test for normality by calculation of the correlation r between the ordered observations $X_{(i)}$ and m_i the median value of the ith largest observation in a sample of n standard normal random variables so that

$$m_i = \Phi^{-1}(\hat{p}_i)$$

where $\Phi(x)$ is the cumulative distribution function of the standard normal distribution. The value of r is then

$$r = \frac{\sum_{i=1}^{n} (x_{(i)} - \bar{x})(m_i - \bar{m})^2}{\sqrt{\sum_{i=1}^{n} (x_{(i)} - \bar{x})^2 \sum_{j=1}^{n} (m_j - \bar{m})^2}}$$

Some significance levels for the value of r are [8]

	SIGNIFICANCE LEVEL		
n	1%	5%	10%
10	0.876	0.917	0.934
20	0.925	0.950	0.960
30	0.947	0.964	0.970
40	0.958	0.972	0.977
50	0.965	0.977	0.981
60	0.970	0.980	0.983

The probability of observing a value of r less than the given value, were the observations actually drawn from a normal distribution, equals the specified probability. Use this test to determine whether a normal or two-parameter lognormal distribution provides an adequate model for these flows.

3-18. A small community is considering the immediate expansion of its wastewater treatment facilities so that the expanded facility can meet the current deficit of 0.25 MGD and the anticipated growth in demand over the next 25 years. Future growth is expected to result in the need of an additional 0.75 MGD. The expected demand for capacity as a function of time is

$$\text{Demand} = 0.25\text{ MGD} + G(1 - e^{-0.23t})$$

where t is the time in years and $G = 0.75$ MGD. The initial capital costs and maintenance and operating costs related to capital are \$1.2 $\times 10^6 C^{0.70}$ where C is the plant capacity (MGD). Calculate the loss of economic efficiency LEE that would result if a designer incorrectly assigned G a value of 0.563 or 0.938 ($\pm$ 25%) when determining the required capacity of the treatment plant. [Note: When evaluating the true cost of a nonoptimal design which provides insufficient capacity to meet demand over a 25-year period, include the cost of building a second treatment plant; use an interest rate of 7% per year to calculate the present value of any subsequent expansions.] In this problem, how important is an error in G compared to an error in the elasticity of costs equal to 0.70? One MGD, a million gallons per day, is equivalent to 0.0438 m^3/s.

3-19. A municipal water utility is planning the expansion of their water acquisition system over the next 50 years. The demand for water is expected to grow and is given by

$$D = 10t(1 - 0.006t)$$

where t is time in years. It is expected that two pipelines will be installed along an acquired right-of-way to bring water to the city from a distant reservoir. One pipe will be installed immediately and then a second pipe when the demand just equals the capacity of the first pipe. The present value of installing a pipe of capacity C in year t is

$$\text{PV} = (\alpha + \beta C^{\gamma})e^{-rt}$$

where

$$\alpha = 29.5$$
$$\beta = 5.2$$
$$\gamma = 0.5$$
$$r = 0.07/\text{yr}$$

Using a 50-year planning horizon, what is the capacity of the first pipe which minimizes the total present value of the construction of the two pipelines? When is the second pipe built? If a $\pm 25\%$ error is made in estimating γ or r, what are the losses of economic efficiency (LEE) and the misrepresentations of minimal costs (MMC)? When finding the optimal decision with each set of parameters, find the time of the second expansion to the nearest year; a computer program that finds the total present value of costs as a function of the time of the second expansion t for $t = 1, \ldots, 50$ would be helpful. (*Note:* A second pipe need not be built.)

3-20. A national planning agency for a small country must decide how to develop the water resources of a region. Three development plans have been proposed, which are denoted d_1, d_2, and d_3. Their respective costs are $200f$, $100f$, and $100f$ where f stands for 1 million farths, the national currency. The national benefits which are derived from the chosen development plan depend, in part, on the international market for the goods and agricultural commodities that would be produced. Consider three possible international market outcomes, m_1, m_2, and m_3. The national benefits if development plan 1 is selected would be, respectively, 400, 290, and 250. The national benefits from selection of plan 2 would be 350, 160, and 120, while the benefits from selection of plan 3 would be 250, 200, and 160.
   (a) Is any plan inferior or dominated?
   (b) If one felt that probabilities could not be assigned to m_1, m_2, and m_3 but wished to avoid poor outcomes, what would be an appropriate decision criterion, and why? Which decision would be selected using this criterion?
   (c) If $\Pr[m_1] = 0.50$ and $\Pr[m_2] = \Pr[m_3] = 0.25$, how would each of the expected net benefits and expected regret criteria rank the decisions?

3-21. Show that a utility function need not correspond to a probability. In particular, prove that if a and b are constants ($b > 0$), and $U(\text{NB})$ is a utility function (the expected value of U produces the correct preference ranking of projects), then the utility function $W(\text{NB})$ equal to $a + bU(\text{NB})$ is an equivalent utility function (it produces the same ranking of any set of projects).

3-22. As shown by Exercise 3-21, the scale of the utility function is not unique. Hence it makes no sense in the hypothetical irrigation project discussed in Sections 3.5.2 and 3.5.3 to say that project x_4 is worth 0.03 "utility units" more than x_3. As a measure of the amount by which x_4 is preferred to x_3, determine the amount m so that the decision maker is indifferent between design x_4 and a design that yields net benefits $\text{NB}(x_3 \mid w_i) + m$ for each w_i. The amount m can be said to be the amount by which design x_4 is preferred to x_3.

3-23. A commonly used idea in applications of utility theory is that of a *certainty equivalent.* If X corresponds to the random variable that equals the net benefits to be derived from a project, then the certainty equivalent is that quantity x_{ce} such that

$$U(x_{ce}) = E[U(X)]$$

where $U(\cdot)$ is the decision maker's utility function; thus the decision maker is indifferent between having x_{ce} for certain and having the project yielding uncertain benefits X.

Assuming that the standard deviation of X is fairly small, so that $U(\cdot)$ can be replaced by the first three terms of its Taylor series, obtain an approximate expression for x_{ce} as a function of the first and second moments of X and the derivatives of $U(\cdot)$. Explain why your result is reasonable. (Will x_{ce} increase if σ_X increases?) [*Hint:* To avoid dealing with the inverse of $U(\cdot)$, replace $U(x_{ce})$ by $U(\mu_X + r) \simeq U(\mu_X) + U'(\mu_X)r$ and solve for r, where $r = x_{ce} - \mu_X$ is often called the risk premium for the risky but actuarially neutral investment $X - \mu_X$.]

3-24. Show that if one has a choice between two water management plans yielding benefits X and Y, where X is stochastically smaller than Y, then for any reasonable utility function, plan Y is preferred to plan X.

3-25. A reservoir system was simulated for 100 years and the average annual benefits and their variance was found to be

$$\bar{B} = 4.93$$

$$s_B^2 = 3.23$$

The correlation of annual benefits was also calculated and is:

k	r_k
0	1.000
1	0.389
2	0.250
3	0.062
4	0.079
5	0.041

(a) Assuming that $\rho(l) = 0$ for $l > k$, compute (using equation 3.74) the standard error of the calculated average benefits for $k = 0$, 1, 2, 3, 4, and 5. Also calculate the standard error of the calculated benefits, assuming that annual benefits may be thought of as a stochastic process with a correlation structure $\rho_B(k) = [\rho_B(1)]^k$. What is the effect of the correlation among the observed benefits on the standard error of their average?

(b) At the 90% and 95% levels, which of the r_k are significantly different from zero, assuming that $\rho_B(l) = 0$ for $l \geq k$?

3-26. Replicated reservoir simulations using two operating policies produced the following results:

	Replicate	BENEFITS	
		Policy 1	*Policy 2*
	1	6.27	4.20
	2	3.95	2.58
	3	4.49	3.87
	4	5.10	5.70
	5	5.31	4.02
	6	7.15	6.75
	7	6.90	4.21
	8	6.03	4.13
	9	6.35	3.68
	10	6.95	7.45
	11	7.96	6.86
Mean, $\bar{X}_i$		6.042	
Standard deviation of values, s_{X_i}		1.217	

(a) Construct 90% confidence limits for each of the two means $\bar{X}_i$.

(b) With what confidence can you state that policy 1 produces higher benefits than policy 2 using the sign test and using the t-test?

(c) If the corresponding replicates with each policy were independent, estimate with what confidence one could have concluded that policy 1 produces higher benefits with the t-test.

3-27. Assume that annual streamflows at a gaging site have been grouped into three categories or states. State 1 is 5 to 15 m³/s, state 2 is 15 to 25 m³/s, and state 3 is 25 to 35 m³/s, and these groupings contain all the flows on record. The following transition probabilities have been computed from the record:

P_{ij}	$j = 1$	$j = 2$	$j = 3$
$i = 1$	0.5	0.3	0.2
$i = 2$	0.3	0.3	0.4
$i = 3$	0.1	0.5	0.4

(a) If the flow for the current year is between 15 and 25 m^3/s, what is the probability that the annual flow 2 years from now will be in the range 25 to 35 m^3/s?

(b) What is the probability of a dry, an average, and a wet year many years from now?

3-28. A Markov chain model for the streamflows in two different seasons has the following transition probabilities

STREAMFLOW IN SEASON 1	STREAMFLOW FOLLOWING SEASON 2		
	0–3 m^3/s	3–6 m^3/s	$\geq$6 m^3/s
0–10 m^3/s	0.25	0.50	0.25
$\geq$10 m^3/s	0.05	0.55	0.40

STREAMFLOW IN SEASON 2	STREAMFLOW FOLLOWING SEASON 1	
	0–10 m^3/s	$\geq$10 m^3/s
0–3 m^3/s	0.70	0.30
3–6 m^3/s	0.50	0.50
$\geq$6 m^3/s	0.40	0.60

Calculate the steady-state probabilities of flows in each interval in each season.

3-29. Assume that there exist two possible discrete flows Q_{it} into a small reservoir in each of two periods t each year having probabilities P_{it}. Find an optimal steady-state operating policy (release as a function of initial reservoir volume and *current periods inflow*) for the reservoir. Limit the storage volumes to integer values that vary from 3 to 5. Take as your objective the minimization of the expected sum of squared deviations from a storage volume target of 4 and a release target of 2 in each period t. (Assume only integer values of all state and decision variables and that each period's inflow is known at the beginning of the period.)

Period, t	FLOWS, Q_{it}		PROBABILITIES, P_{it}	
	$i = 1$	$i = 2$	$i = 1$	$i = 2$
1	1	2	0.17	0.83
2	3	4	0.29	0.71

3-30. Assume that the streamflow Q at a particular site has cumulative distribution function $F_Q(q) = q/(1+q)$ for $q \geq 0$. The withdrawal x at that location must satisfy a chance constraint of the form $\Pr[x \leq Q] \geq \alpha$, or equivalently $\Pr[x \geq Q] \leq 1 - \alpha$. Write the deterministic equivalent of the following chance constraints:

$$\Pr[x \leq Q] \geq 0.90 \qquad \Pr[x \geq Q] \leq 0.80$$
$$\Pr[x \leq Q] \leq 0.95 \qquad \Pr[x \leq Q] \geq 0.10$$
$$\Pr[x \geq Q] \geq 0.75$$

3-31. Why is the allocation problem in Section 3.6.2 not amendable to solution by dynamic programming?

3-32. Assume that a potential water user can withdraw water from an unregulated stream, and that the probability distribution function $F_Q(q)$ of the available streamflow Q is known. Calculate the withdrawal target T that will maximize the expected net benefits from the water's use given the two short-run benefit functions specified below.

(a) The benefits from streamflow Q when the target was T are

$$B(Q \mid T) = \begin{cases} B_0 + \beta T + \gamma(Q - T) & Q \geq T \\ B_0 + \beta T + \delta(Q - T) & Q < T \end{cases}$$

where $\delta > \beta > \gamma$. In this case, the optimal target T^* can be expressed as a function of p^*, the probability that the random streamflow Q will be less than or equal to T. Prove that p^* equals $(\beta - \gamma)/(\delta - \gamma)$.

(b) The benefits from streamflow Q when the target was T are

$$B(Q \mid T) = B_0 + \beta T - \delta(Q - T)^2$$

3-33. If a random variable is discrete, what effect does this have on the specified confidence of a confidence interval for the median or any other quantile? Give an example.

3-34. (a) Use the Wilcoxon test for unpaired samples to test the hypothesis that the distribution of the total shortage TS in Table 3.15 is stochastically less than the total shortage TS reported in Table 3.16. Use only the data from the second 10 simulations reported in the table. Use the fact that observations are paired (i.e., simulation j for $11 \leq j \leq 20$ in both tables were obtained with the same streamflow sequence) to perform the analysis with the sign test.

(b) Use the sign test to demonstrate that the average deficit with policy 1 (Table 3.15) is stochastically smaller than with policy 2 (Table 3.16); use all 25 simulations.

3-35. The accompanying table provides an example of the use of nonparametric statistics for examining the adequacy of synthetic streamflow generators. Appendix 3B considered comparison of the storage capacity required to deliver a specified yield or demand with the historical and with synthetic flow series. Here the maximum yield that can be supplied with a given size reservoir is considered. The following table gives the rank of the maximum yield obtainable with the historic flows among the set consisting of the historic yield and the maximum yield achievable with 1000 synthetic sequences. The synthetic flows were generated by 12 monthly ARMA models fit to the normalized monthly flow sequences of 25 North American rivers. These models are discussed in Chapter 6.

(a) Plot the histogram of the ranks for reservoir storage sizes $S/\mu_Q =$ 0.85, 1.35, and 2.00. (*Hint:* Use the intervals 0–100, 101–200, 201–300, etc.) Do the ranks look uniformly distributed?

Rank of the Maximum Historic Yield among 1000 Synthetic Yields

River	NORMALIZED ACTIVE STORAGE, S/μ_Q			
Number	*0.35*	*0.85*	*1.35*	*2.00*
1	47	136	128	235
2	296	207	183	156
3	402	146	120	84
4	367	273	141	191
5	453	442	413	502
6	76	92	56	54
7	413	365	273	279
8	274	191	86	51
9	362	121	50	29
10	240	190	188	141
11	266	66	60	118
12	35	433	562	738
13	47	145	647	379
14	570	452	380	359
15	286	392	424	421
16	43	232	112	97
17	22	102	173	266
18	271	172	260	456
19	295	162	272	291
20	307	444	532	410
21	7	624	418	332
22	618	811	801	679
23	1	78	608	778
24	263	902	878	737
25	82	127	758	910

Source: A. I. McLeod and K. W. Hipel, Critical Drought Revisited, paper presented at International Symposium on Risk and Reliability in Water Resources, Waterloo, Ont., June 26–28, 1978.

(b) Do you think that this streamflow generation model produces streamflows which are consistant with the historical flows when one uses as a criterion the maximum possible yield? Construct a statistical test to support your conclusion and show that it does support your conclusion. (*Idea:* You might want to consider if it is equally likely that the rank of the historical yield is 500 and below or 501 and above. You could then use the binomial distribution to determine the significance of the results.)

(c) Use the Kolmogorov–Smirnov test to check if the distribution of the ranks of the yields obtainable with storage $S/\mu_Q = 1.35$ is significantly different from uniform $F_U(u) = u$ for $0 \leq u \leq 1$. How important do you feel this result is?

3-36. Appendix 3C dismisses the bias in v_X^2 for correlated X's as unimportant compared to its variance.

(a) Calculate the approximate bias in v_X^2 for the cases corresponding to Table 3C.2 and determine if this assertion is justified.

(b) By numerically evaluating the bias and variance of v_X^2 when $n = 25$, determine if the same result holds if $\rho_X(k) = 0.5(0.9)^k$, which is the autocorrelation function of an ARMA (1, 1) process sometimes used to describe annual streamflow series.

REFERENCES

1. Benjamin, J. R., and C. A. Cornell, *Probability, Statistics and Decisions for Civil Engineers*, McGraw-Hill Book Company, New York, 1970.
2. Bickel, P. J., and K. A. Doksum, *Mathematical Statistics: Basic Ideas and Selected Topics*, Holden-Day, Inc., San Francisco, 1977.
3. Binkley, C., An Analysis of Design Life Standards for Interceptor Sewers, *Water Resources Research*, Vol. 14, No. 2, 1978, pp. 365–369.
4. Bobee, B., The Log Pearson Type 3 Distribution and Its Applications in Hydrology, *Water Resources Research*, Vol. 11, No. 5, 1975, pp. 681–689.
5. Bobee, B., and R. Robitaille, The Use of the Pearson Type 3 and Log Pearson Type 3 Distribution Revisited, *Water Resources Research*, Vol. 13, No. 2, 1977, pp. 427–443.
6. Box, G. E. P., and G. M. Jenkins, *Time Series Analysis Forecasting and Control*, Holden-Day, Inc., San Francisco, 1976.
7. Bradley, J. V., *Distribution-free Statistical Tests*, Prentice-Hall, Inc., Englewood Cliffs, N.Y., 1968.
8. Filliben, J. J., The Probability Plot Correlation Test for Normality, *Technometrics*, Vol. 17, No. 1, 1975, pp. 111–117.
9. Fishman, G. S., *Concepts and Methods in Discrete Event Digital Simulation*, John Wiley & Sons, Inc., New York, 1973.

10. GIESBRECHT, F., and O. KEMPTHORNE, Maximum Likelihood Estimation in the Three-Parameter Log Normal Distribution, *Journal of the Royal Statistical Society B.*, Vol. 38, No. 3, 1976, pp. 257–264.

11. GREENWOOD, J. A., and D. DURAND, Aids for Fitting the Gamma Distribution by Maximum Likelihood, *Technometrics*, Vol. 2, No. 1, 1960, pp. 55–65.

12. GUMBEL, E. J., *Statistics of Extremes*, Columbia University Press, New York, 1958.

13. HAAN, C. T., *Statistical Methods in Hydrology*, Iowa State University Press, Ames, Iowa, 1977.

14. HAHN, G. J., and S. S. SHAPIRO, *Statistical Models in Engineering*, John Wiley & Sons, Inc., New York, 1967.

15. JENKINS, G. M., and D. G. WATTS, *Spectral Analysis and Its Application*, Holden-Day, Inc., San Francisco, 1968.

16. KACZMAREK, Z., *Statistical Methods in Hydrology and Meteorology*, published for the U.S. Geological Survey by the Foreign Scientific Publications Department of the National Center for Scientific, Technical and Economic Information, Warsaw, Poland, 1977.

17. KEENEY, R. L., and E. F. WOOD, An Illustrative Example of the Use of Multiattribute Utility Theory for Water Resources Planning, *Water Resources Research*, Vol. 13, No. 4, 1977, pp. 705–712.

18. KEENEY, R. L., and H. RAIFFA, *Decisions with Multiple Objectives*, John Wiley & Sons, Inc., New York, 1976.

19. KENDALL, M. G., and A. STUART, *The Advanced Theory of Statistics*, Vol. 3, Hafner Publishing Company, Inc., New York, 1966.

20. KIRBY, W., Computer Oriented Wilson–Hilferty Transformation That Preserves the First 3 Moments and Lower Bound of the Pearson Type 3 Distribution, *Water Resources Research*, Vol. 8, No. 5, 1972, pp. 1251–1254.

21. KIRBY, W., Algebraic Boundness of Sample Statistics, *Water Resources Research*, Vol. 10, No. 2, 1974 pp. 220–222.

22. KITE, G. W., *Frequency and Risk Analysis in Hydrology*, Water Resources Publications, Fort Collins, Colo., 1977.

23. KOENIG, L., Optimal Fail-Safe Process Design, *J. Water Pollution Control Federation*, Vol. 45, No. 4, 1973, pp. 647–654.

24. MANNE, A. S., Capacity Expansion and Probabilistic Growth, *Econometrica*, Vol. 29, 1961, pp. 632–641.

25. MATALAS, N. C., and J. R. WALLIS, Eureka! It Fits a Pearson Type 3 Distribution, *Water Resources Research*, Vol. 9, No. 3, 1973, pp. 281–289.

26. Meta Systems, Inc., *Systems Analysis in Water Resources Planning*, Water Information Center, Inc., Port Washington, N.Y., 1975, pp. 180–202.

27. MOSTELLER, F., and R. E. K. ROURKE, *Sturdy Statistics: Nonparametrics and Order Statistics*, Addison-Wesley Publishing Co., Inc., Reading, Mass., 1973.

28. SHAPIRO, S. S., and M. B. WILK, An Analysis of Variance Test for Normality (Complete Samples), *Biometrika*, Vol. 52, 1965, pp. 591–610.

29. STEDINGER, J. R., Fitting Log Normal Distributions to Hydrologic Data, *Water Resources Research*, Vol. 16, No. 3, 1980.

30. STEPHENS, M., E.D.F. Statistics for Goodness of Fit, *Journal of the American Statistical Association*, Vol. 69, 1974, pp. 730–737.
31. THOM, H. C. S., A Note on the Gamma Distribution, *Monthly Weather Review*, Vol. 86, No. 4, 1958, pp. 117–122.
32. THOMAS, H. A., Sensitivity Analysis of Design Parameters of Water Resource Systems, Discussion Paper, Environmental Systems Program, Harvard University, Cambridge, Mass., April 1971.
33. U.S. Water Resources Council, Hydrology Committee, *Guidelines for Determining Flood Flow Frequencies*, Bulletin 17, Washington, D.C., 1976.
34. VON NEUMANN, J., and O. MORGENSTERN, *Theory of Games and Economic Behavior*, Princeton University Press, Princeton, N.J., 1974.
35. WALLIS, J. R., N. C. MATALAS, and J. R. SLACK, Just a Moment!, *Water Resources Research*, Vol. 10, No. 2, 1974, pp. 211–219.
36. WALLIS, J. R., N. C. MATALAS, and J. R. SLACK, *Just a Moment! Appendix*, National Technical Information Service, PB-231 816, Springfield, Va., 1974.
37. WHITFORD, P. W., Residential Water Demand Forecasting, *Water Resources Research*, Vol. 8, No. 4, 1972, pp. 829–839.
38. WILK, M. B., and R. GNANADESIKAN, Probability Plotting Methods for the Analysis of Data, *Biometrika*, Vol. 55, No. 1, 1968, pp. 1–17.
39. YEVJEVICH, V. M., *Probability and Statistics in Hydrology*, Water Resources Publications, Fort Collins, Colo., 1972.
40. YEVJEVICH, V. M., *Stochastic Processes in Hydrology*, Water Resources Publications, Fort Collins, Colo., 1972.

CHAPTER 4

Water Resources Planning Objectives

4.1 INTRODUCTION

Identifying relevant planning objectives, and defining the relative importance of each of these objectives, is one of the most difficult aspects of water resource systems planning. There are always many individuals and groups affected by any project and involved in the decision-making process. Each individual often has more than one objective. In addition, the relative importance of each objective may change during the planning process. Furthermore, some objectives are difficult to quantify, and those that can be quantified are frequently expressed in incomparable units.

Consider, for example, the economic and environmental interests involved with the development of coastal wetlands or the damming of a free-flowing river. There is certainly no one best plan that satisfies all the conflicting economic and environmental objectives. Hence, those responsible for plan formulation and analysis must identify plans that represent reasonable trade-offs among conflicting goals and interests.

Until relatively recently, economic objectives appeared to dominate water resources planning methodology in the United States. The Flood Control Act of 1936 defined feasible federal flood control projects as those for which "the benefits, to whomsoever they may accrue, are in excess of the estimated costs." This law is often considered the beginning of the tradition of benefit–cost analysis in water resources planning. A progression of federal regulations, agency reports, and academic studies eventually formalized the *contribution to national income* as the primary objective in water resource systems planning. This implied that the main objective should be the maximization of aggregated net monetary benefits to all parties affected by a water resources project.

In the 1970s, new directions in water resources planning became apparent. The difficulties of associating monetary benefits with all water resources planning objectives had long been a concern of economists working in water resources [9, 20]. With the heightened interest in environmental quality and social welfare came the formal recognition that national income alone did not adequately reflect the public interest. This was a belated admission that a respectable benefit–cost ratio was often viewed as a constraint, and once satisfied other objectives dominated. Benefit–cost data were generated and used often only to demonstrate economic respectability when, in fact, other nonmonetary objectives were actually being considered.

In the United States a set of federal water resources planning procedures, the Water Resources Council's *Principles and Standards* [27], were adopted by presidential order in 1973 and were revised in 1979. The *Principles and Standards* established two equally important objectives for federal water resources projects:

1. *National economic development:* "Enhance national economic development by increasing the value of the Nation's output of goods and services and improving national economic efficiency."
2. *Environmental quality:* "Enhance the quality of the environment by the management, conservation, preservation, creation, restoration, or improvement of the quality of certain natural and cultural resources and ecological systems."

These trends in U.S. water resources planning can be seen in the planning methods used in other developed and developing nations. When planning, any government agency should consider the trade-offs between such objectives as economic growth, regional development, regional autonomy, resource development, environmental quality, employment, population control, agricultural self-sufficiency, foreign trade, national security, energy dependence, and/or public health. Each of these objectives, to a greater or lesser extent, is affected by the reliability of water quantity and quality in time and space.

This chapter is divided into two sections. The first section outlines the basic principles of benefit–cost analysis. Economic objectives are still important, and an understanding of these principles is required for an appreciation of multiobjective water resources planning. Moreover, benefit–cost analysis is directly applicable to any national economic development objective. The second section of the chapter outlines some of the concepts and methods used in plan formulation and plan selection in multiobjective planning.

4.2 ECONOMIC BENEFIT–COST OBJECTIVES

The rationale for benefit–cost analysis is based on two fundamental economic concepts: scarcity and substitution. The first of these implies that the supplies of natural, man-made, and human resources are limited, and that these resources should be used efficiently. The concept of substitution indicates that individuals, social groups, and institutions are generally willing to trade off a certain amount of one objective for more of another. As applied to water resources, maximization of the net-benefit objective requires the efficient and reliable allocation, over both time and space, of water (in its two dimensions: quantity and quality) to its many uses, including hydropower, recreation, water supply, flood control, navigation, irrigation, cooling, waste disposal, and waste assimilation. The following example illustrates the maximization of the benefits obtained from operation of a multiple-purpose reservoir.

Consider a reservoir under construction by a public agency. Although the reservoir is designed primarily for irrigation, the newly formed lake will also be used for recreation (boating and swimming). Irrigation and recreation are not very compatible, however, since the latter requires high reservoir elevations throughout the summer recreation season, just when the irrigation demand would normally cause a drop in the reservoir storage level. Thus the project has two conflicting purposes: provision of irrigation water and improvement of recreation opportunities.

Some basic principles of benefit–cost analysis can be illustrated with this example. Let X be the quantity of irrigation water to be delivered to farmers each year and Y the number of visitor-days of recreation use on the reservoir. Possible levels of irrigation and recreation are shown in Figure 4.1. The solid line in Figure 4.1, termed the *production-possibility frontier*, is the boundary of the feasible combinations of X and Y [13].

Any combination of X and Y within the shaded area can be obtained by operation of the reservoir (i.e., by regulating the amount of water released for irrigation and other uses). One is really interested in obtaining as much of both X and Y as possible. Thus attention generally focuses on the production-possibility frontier which comprises those combinations of X and Y that are *technologically efficient* in the sense that more of either X or Y cannot be obtained without production of less of the other. The shape and location of the production-possibility frontier is determined by the quantity of avail-

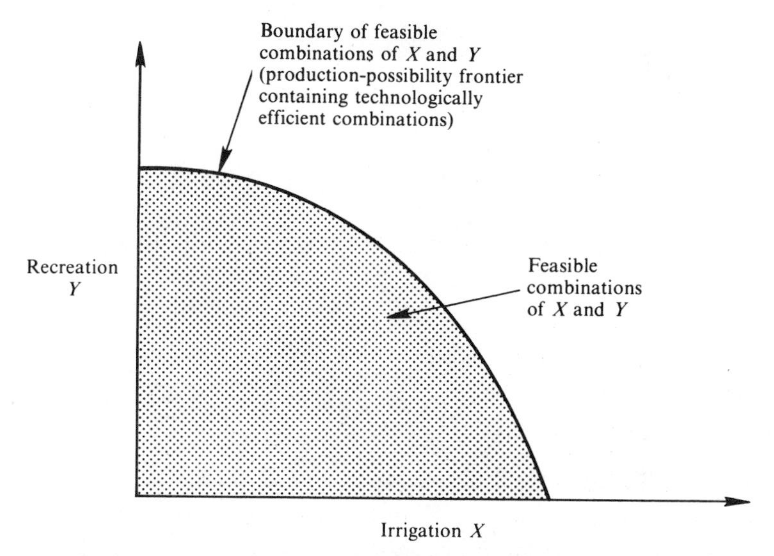

FIGURE 4.1. A two-purpose planning problem.

able resources (water, reservoir storage capacity, recreation facilities, etc.) and their ability to generate both X and Y.

To continue the example, assume that this two-purpose reservoir is owned by a private entrepreneur in a competitive environment (i.e., there are a number of competing irrigation water suppliers and reservoir recreation sites). Let the unit market prices for irrigation water and recreation opportunities be p_x and p_y, respectively. Also assume that the entrepreneur's costs are fixed and independent of X and Y. Then, his total income is

$$I = p_x X + p_y Y \tag{4.1}$$

Values of X and Y that result in fixed income levels $I_1 < I_2 < I_3$ are plotted in Figure 4.2. The value of X and Y that maximizes the entrepreneur's income is indicated by the points on the production-possibility frontier yielding an income of I_3. Income greater than I_3 is impossible.

Returning now to the original case with public reservoir ownership and assuming that competitive conditions prevail, the prices p_x and p_y reflect the *value* of the irrigation water and recreation opportunities to the users. The aggregated value of the project is indicated by the user's *willingness to pay* for the irrigation and recreation outputs. In this case, this willingness to pay is $p_x X + p_y Y$, which would be equivalent to the entrepreneur's income. Private operation of the reservoir to maximize income or government operation to maximize user benefits both should, under competitive conditions, produce the same result.

When applied to water resources planning, benefit–cost analysis presumes a similarity between decision making in the private and public sectors. It

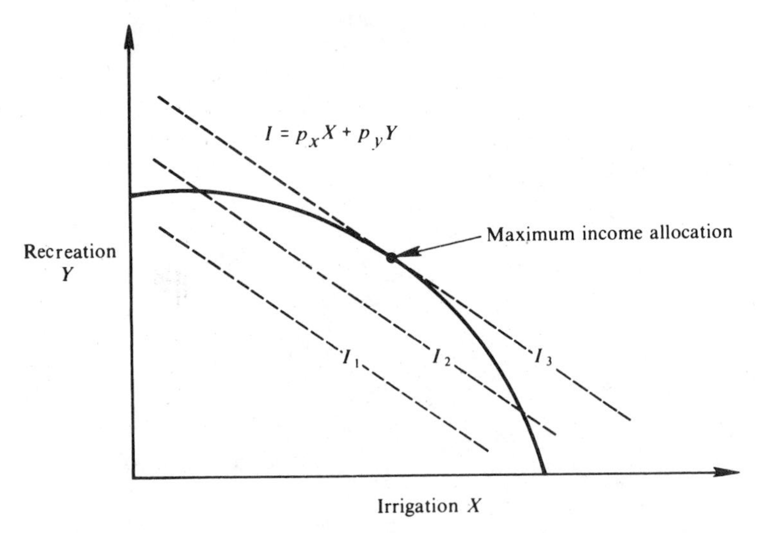

FIGURE 4.2. Two-purpose project and various income levels.

assumes that the aggregation of the monetary values to the users of the products of water resources projects are reasonable surrogates for their social value.

4.2.1 Benefit and Cost Estimation

In benefit–cost analysis, one may need to estimate the monetary value of irrigation water, shoreline property, land inundated by a lake, lake recreation, fishing opportunities, scenic vistas, hydropower production, navigation, or the loss of a wild river [16]. The situations in which benefits and costs may need to be estimated are sometimes grouped into four categories, reflecting the ease with which prices can be determined [14]. These situations are:

1. Situations in which market prices exist and are an accurate reflection of marginal social values (i.e., marginal willingness to pay for all individuals). This situation often occurs in the presence of competitive market conditions. An example would be agricultural commodities that are not subsidized or that do not have supported prices.
2. Situations in which market prices exist but for various reasons do not reflect marginal social values. Examples include price-supported agricultural crops, labor that would otherwise be unemployed, or inputs whose production generates pollution, the social cost of which is not included in its price.
3. Situations in which market prices are essentially nonexistent, but for which it is possible to infer or determine what users or consumers would pay if a market existed. An example is outdoor recreation.

4. Situations in which no real or simulated marketlike process is easily conceived. This category may be relatively rare. Although scenic amenities and historic sites are often considered appropriate examples, both are sometimes privately owned and managed to generate income.

For the first three categories, benefits and costs can be measured as the *aggregate net willingness to pay* of those affected by the project. Assume, for example, that a set of water resources planning alternatives or projects $\mathbf{X}_1, \mathbf{X}_2, \ldots$, has been defined. Let $B(\mathbf{X}_j)$ equal the amount the beneficiaries of the plan $\mathbf{X}_j$ are willing to pay rather than forego the project. This represents the aggregate value of the project to the beneficiaries. Let $D(\mathbf{X}_j)$ equal the amount the nonbeneficiaries of plan $\mathbf{X}_j$ are willing to pay to prevent it from being implemented. This includes the social value of the resources that will be unavailable to society if project or plan $\mathbf{X}_j$ is implemented. The aggregate net willingness to pay $W(\mathbf{X}_j)$ for plan $\mathbf{X}_j$ is equal to the difference between $B(\mathbf{X}_j)$ and $D(\mathbf{X}_j)$,

$$W(\mathbf{X}_j) = B(\mathbf{X}_j) - D(\mathbf{X}_j) \tag{4.2}$$

Plans $\mathbf{X}_j$ can be ranked according to the aggregate net willingness to pay $W(\mathbf{X}_j)$. If, for example, $W(\mathbf{X}_j) > W(\mathbf{X}_i)$, it is inferred that plan $\mathbf{X}_j$ is preferable or superior to plan $\mathbf{X}_i$.

One rationale for the willingness-to-pay criterion is that if $B(\mathbf{X}_j) > D(\mathbf{X}_j)$, the beneficiaries could compensate the losers and everyone would benefit from the project. However, there is, in general, no means for this compensation actually to be paid. That is, while many resources required for the project, such as land, labor, and machinery, will be bought and paid for, those people who lose favorite scenic sites or the opportunity to use a wild river, who must breathe dirty air, or who suffer a decrease in the value of their property because of the project are seldom compensated.

This compensation criterion also ignores the resultant income redistribution, which should be considered during the plan selection process. The compensation criterion implies that the marginal social value of income to all affected parties is the same. If a project's benefits accrue primarily to affluent individuals and the costs are borne by lower-income groups, $B(\mathbf{X}_j)$ may be larger than $D(\mathbf{X}_j)$ simply because the beneficiaries can pay more than the nonbeneficiaries.

In addition to these and other conceptual difficulties related to the willingness-to-pay criterion, practical measurement problems also exist. Many of the products of water resources plans are *public* or *collective goods*. This means that they are essentially indivisible, and once provided for any one individual it is very difficult not to provide them to others. Collective goods often also have the property that consumption by one person does not prohibit or infringe upon its consumption or use by others. Community flood protection is an example of a public good. Once protection is provided for one individual, it is simultaneously provided for many others. As a result, it is not in an individual's self-interest to volunteer to help pay for the project by contributing an amount equal to his or her actual benefits if others are willing to pay for the project. However, if others are going to pay for the project, individuals may exaggerate their own benefits to ensure that the project is undertaken.

Determining what benefits should be attributed to a project is not always simple and the required accounting can become rather involved. In a benefit–cost analysis, economic conditions should first be projected for a base case in which no project is implemented and the benefits and costs are estimated for that scenario. Then the benefits and costs for each project are measured as the incremental economic impacts that occur in the economy over these baseline conditions. The appropriate method for benefit and cost estimation depends on whether or not the market prices reflecting true social values are available or if such prices can be constructed.

Market Prices Equal Social Values. Consider the estimation of irrigation benefits in the irrigation–recreation example discussed in the previous section. Let x_1 be the quantity of irrigation water supplied by the project each year. If the prevailing market price p_0 before the project's construction reflects the marginal social value of water and if the project's operation will have little effect on the price, the value of the water is just p_0x_1. However, it often happens that large water projects have a major impact on the prices of the commodities they supply. In such a case, the value of water from our example irrigation district would be neither p_0x_1 or p_1x_1, where p_0 and p_1 are, respectively, prevailing prices before and after project construction. Rather, the total value of the resource to the users is the total amount they would be willing to pay for the resource delivered.

Let $D(p)$ be the amount of water the consumers would want to buy at a price p. For any price, consumers will continue to buy the water until the value of another unit of water is less than or equal to the price. $D(p)$ defines what is called the demand function. As illustrated by Figure 4.3, the lower the price, the more water that will be demanded.

To determine how much farmers would be willing to pay for the water x_1 to be supplied by the project, one can think of selling the water in small amounts Δ. If one first sells an amount Δ, it can be sold at a price $D^{-1}(\Delta)$ producing a revenue $\Delta D^{-1}(\Delta)$ if the farmers are not told that more water will be sold later at the lower price. Next, one can sell an additional amount Δ at a price $D^{-1}(2\Delta)$. The farmers will buy this additional water because the value to them of an additional Δ units of water is greater than $\Delta D^{-1}(2\Delta)$ or else they would not have been willing to buy 2Δ units of water at a price of

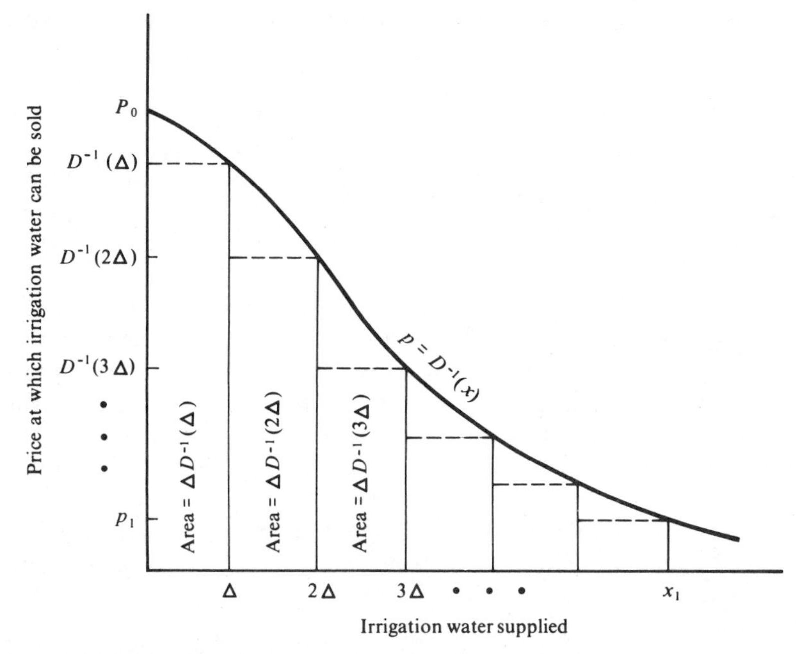

FIGURE 4.3. Willingness to pay is given by the area under the demand curve.

$D^{-1}(2\Delta)$. Continuing in this manner, as illustrated in Figure 4.3, the farmers would be willing to pay for an amount of water $n\Delta = x_1$ at least

$$\sum_{j=1}^{n} \Delta D^{-1}(j\Delta) \tag{4.3}$$

As Δ becomes very small, one extracts from the farmers all they would be willing to pay, so that

$$\text{willingness to pay} = \int_0^{x_1} D^{-1}(x)\, dx \tag{4.4}$$

which is just the area under the demand curve between $x = 0$ and $x = x_1$. Consumer's willingness to pay for a product is an important concept in welfare economics.

Market Prices Not Equal to Social Values. It frequently occurs that market prices do not truly reflect the social value of the various inputs to or goods and services supplied by a water resources project. In such cases it is necessary for the planner to estimate the appropriate value of the quantities in question. There are several procedures that can be used depending on the situation. A rather simple technique which can reach absurd conclusions if incorrectly applied is to estimate the benefits of a service supplied by a project as the cost of supplying the service by the least expensive alternative method. Thus the benefits from hydroelectricity generation could be estimated as the cost of generating that electricity by the least-cost alternative method using geothermal, coal-fired, or nuclear energy sources. Clearly, this approach to benefit estimation is only valid if in the absence of the project's adoption, the service in question would in fact be supplied by the least-cost alternative method. The pitfalls associated with this method of benefit and cost estimation should be avoidable if one clearly identifies reasonable with- and without-project economic scenarios.

In other situations, simulated or imagined markets can be used to derive the demand function for a good or service and to estimate the value of the amount of that quantity generated or consumed by the project. The following hypothetical example, based on a similar example in Knetsch [19], illustrates how this technique can be used to estimate the value of outdoor recreation.

A unique reservoir recreation area is to be developed which will serve two population centers. Center A has a population of 10,000 and the more distant center B has a population of 30,000. From questionnaires it is estimated that 20,000 visits per year will be made from center A at an average round-trip travel cost of $1. Similarly 30,000 visits per year will come from center B, at an average round-trip cost of $2.

The benefits derived from the proposed recreation area can be estimated from an imputed demand curve. First, as illustrated in Figure 4.4, a graph of travel cost as a function of the average number of visits per capita can be constructed. Two points are available: two visits per capita (from center A) at a cost of $1/visit, and one visit per capita (from center B) at a cost of $2/visit. This travel cost data is extrapolated to the ordinate and abscissa.

Even assuming that there are no plans to charge admission at the site, if users respond to an entrance fee as they respond to travel costs, it is possible to estimate what the user response might be if an entrance fee was charged. This information will provide a demand curve for recreation at the site. Consider first a $1 admission price to be added to the travel cost for recreation. The total cost to users from population center A would then be $2 per visit. From Figure 4.4 at $2 per visit, one visit per capita is made; hence 10,000 visits per year can be expected from center A. The resulting cost to users from center B is $3 per visit, and hence from Figure 4.4 no visits would be made. Therefore, one point on the demand curve (10,000 visits at $1) is obtained. Similarly, it can be inferred that at a $2 admission price, there will be no visits from either city. A final point, corresponding to a zero admission price (no added costs) is just the estimated site attendance (50,000 visits). The resulting demand curve is shown in Figure 4.5. Recreation willingness to pay benefits are equal to the area under the demand curve or $35,000 ($0.70 per visitor-day).

Obviously, this example is illustrative only. Additional factors, such as

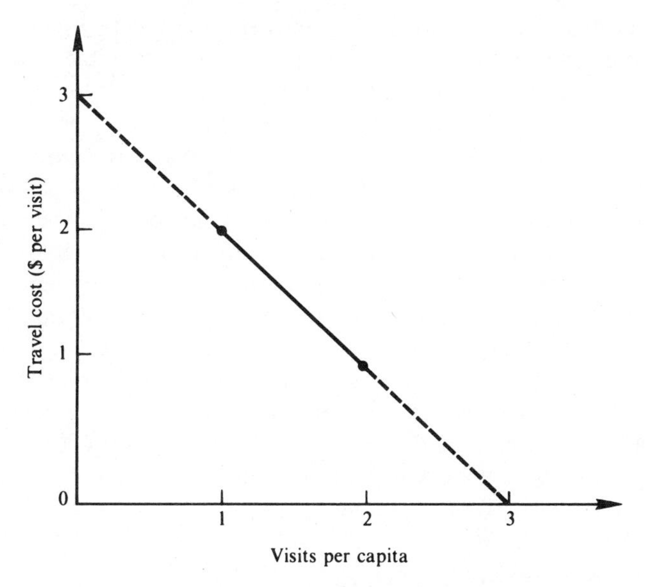

FIGURE 4.4. Estimated relationship between travel cost and visits per capita.

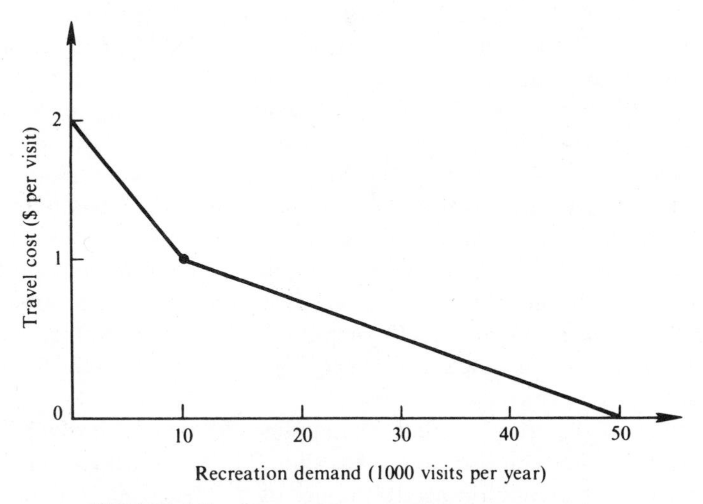

FIGURE 4.5. Demand function for recreation site.

the cost of travel time (which differs for each population center) and the availability of alternative sites, must be included in a more detailed analysis [19].

No Market Processes. In the absence of any marketlike process (real or simulated) it is extremely difficult to associate specific monetary values with benefits. The benefits associated with aesthetics and with many aspects of environmental quality have long been considered "intangible" (i.e., difficult to quantify in monetary or perhaps in any other terms). Although attempts have been made to estimate water quality benefits, the results have had limited success. Clearly, in the United States, the Water Resources Council's *Principles and Standards* does not encourage assignment of monetary benefits to intangible objectives. The Water Pollution Control Act Amendments of 1972 and the Clean Air Act of 1970 have set environmental goals which are to be met in a cost-effective manner. These goals are set by legislative and administrative processes which forego the explicit determination of benefits.

To a certain extent, environmental and aesthetic objectives, if quantified, can be incorporated into the multiobjective modeling techniques discussed in Section 4.3. In this sense, the values of these objectives may be decided in the political decision-making process. However, this falls short of the explicit assignment of monetary benefits that is often possible for the first three categories of benefits.

4.2.2 A Note Concerning Costs

To be consistent in the estimation of net benefits from water resources projects, cost estimates should reflect *opportunity costs*, the value of resources in their most productive alternative uses. This principle is much easier stated than implemented, and as a result true opportunity costs are seldom included in a benefit–cost analysis. For example, if land must be purchased for a reservoir project, is the purchase price (which would typically be used in a benefit–cost analysis) the land's opportunity cost? Suppose that the land is currently a wild area and its alternative use is as a wildlife preserve and camping area. The land's purchase price may be low, but this is unlikely to reflect its true value to society. Furthermore, assume that individuals who would otherwise be unemployed are hired to construct the reservoir. The opportunity cost for such labor is the marginal value of leisure forgone, since there is no alternative productive use. Yet, labor wages have to be included in the budget for the project.

The results of rigorous benefit–cost analyses seldom dictate which of competing water resources projects and plans should be implemented. This occurs in part because of the multiobjective nature of the decisions; one must consider environmental impacts, income redistribution effects, and other local, regional, and national goals. Other important considerations are the financial, technical, and political feasibilities of alternative plans. Particularly important when a plan is undertaken by government agencies are the relative political and legal clout of those who want and those who oppose any plan. Still, a plan's economic efficiency is an important measure of a plan's value to society and an indicator of whether it should be considered at all.

4.2.3 Long- and Short-Run Benefit Functions

When planning water resource systems, it is convenient to think of two types of benefit functions: long-run and short-run. In *long-run planning*, the capacities of proposed facilities and the target allocation of water to alternative uses are assumed to be unknown decision variables that can all be varied to achieve the most efficient use of available resources. In *short-run planning*, the capacity of proposed facilities and the quantity of water promised or targeted to alternative uses has been fixed by prior decisions and one must allocate water and other resources as best one can given those capacities and planned allocations.

Thus long-run benefits are those benefits one could obtain if capacities, targets, and other strategic decision variables could be assigned their most efficient values. Of course, one seldom has sufficient information to determine what the best values of capacities and targets will be, nor can one vary fixed capacities and targets from one period to the next. Hence one fixes capacities and targets at values that maximize expected net benefits or at levels that ensure that targets can be met with a reasonable reliability. Short-run benefits are the benefits one actually obtains by optimizing the operation of a system to best use the available resources after capacities and target allocations have been fixed by prior decisions. If the resources available in the short run are those that were expected when long-run decisions were made, then estimated long-run benefits and the actual short-run benefits can be the same.

The distinction between long-run and short-run benefits can be illustrated by considering a potential water user (e.g., an irrigation district, an industry, or a hydroelectric plant) that may be established at a particular site near a water body. Assume that the long-run net benefits associated with various allocations of water to that use can be estimated. These long-run net benefits are those that will be obtained if the actual allocation Q equals the target allocation T. This long-run net benefit function can be denoted as $B(T)$. Next assume that for various fixed values of the target T the actual net benefits derived from various allocations Q can be estimated. These short-run benefit functions $b(Q \mid T)$ are dependent on the target T and the actual allocation Q. The relationship between the long-run net benefits $B(T)$ and the short-run net benefits $b(Q \mid T)$ for a particular target T is illustrated in Figure 4.6.

The long-run function $B(T)$ in Figure 4.6 reflects the benefits users receive when they have adjusted their plans in anticipation of receiving an allocation equal to the target T and actually receive it. The short-run benefits function specifies the benefits users actually receive when a particular allocation is less (e.g., Q_1) or more (e.g., Q_2) than their anticipated allocation T and they cannot completely adjust their plans to the resulting deficit or surplus. The short-run loss of any actual allocation Q equals the long-run benefit of the target allocation T minus the short-run benefit of the actual allocation Q

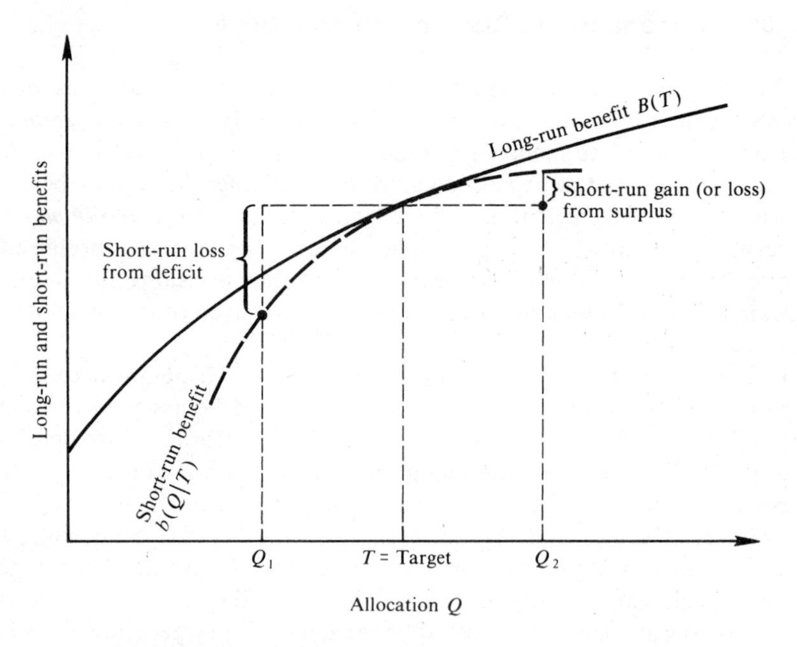

FIGURE 4.6. Long-run and short-run benefit functions.

[i.e., $B(T) - b(Q|T) = L(Q|T)$]. Clearly, when the actual allocation equals the target allocation, the short-run loss is zero.

The short-run benefit function depends on the value of the target allocation. If the short-run losses associated with any deficit allocation $(T - Q_1)$ or surplus allocation $(Q_2 - T)$ are relatively constant over a reasonable range of targets, it may not be necessary to define the loss as a function of the target T, but only of the deficit D or surplus (excess) E. Both the deficit D or excess E can be defined by the constraint

$$Q = T - D + E \tag{4.5}$$

The costs of building many components of multipurpose projects are not easily expressed as functions of the targets associated with each use (e.g., reservoir capacity costs). These capacity costs can usually be stated as a function of the project's capacity K and must include both the amortized capital costs as well as the annual operation, maintenance, and repair costs.

Assuming that the benefit and loss functions reflect annual benefits and losses, the annual net benefits from all projects is the sum of each project's long-run benefits $B_j(T_j)$ less short-run losses $L_j(Q_j|T_j)$ and capacity costs $C_j(K_j)$.

$$\text{NB} = \sum_j [B_j(T_j) - L_j(Q_j|T_j) - C_j(K_j)] \tag{4.6}$$

The monetary net benefits accrued by alternative groups of water users can also be determined so that the income redistribution effects of a project can be evaluated.

The formulation of benefit functions as either short- or long-run is, of course, a simplification of reality. In reality, planning takes place on many time scales and for each time scale one could construct a benefit function. Consider the planning problems of farmers. In the very long run, they decide whether or not to own farms in a particular area. On a shorter time scale, farmers allocate their resources to different activities such as beef cattle, dairy cattle, dry farming, and/or irrigated agriculture. Different activities, of course, require capital investments in farm machinery, tanks, pipes, and pumps, some of which cannot easily be transferred to another use. At least on an annual basis, the farmers reappraise these resource allocations in light of the projected market price of the generated commodities and the availability and cost of water, energy, and other required inputs; the farmers can then make marginal adjustments in the size of herds and the land devoted to different activities and the specific crops grown within the bounds allowed by available capital. Within any growing season, some changes can be made in response to changes in prices and the actual availability of water. If the farmers frequently find that insufficient water is available in the short run to meet cattle and crop requirements, then they will reassess and perhaps change their long-run plans by shifting to less water-intensive activities, seeking additional water from other sources (such as deep wells), or selling their farms and engaging in other activities. For the purposes of modeling, this planning hierarchy can generally be described by two levels, denoted as long- and short-run. However, the appropriate decisions included in each category will depend on the time scale of a model.

4.3 Multiobjective Models

As previously mentioned, national or regional income maximization is only one of many possible and important planning objectives. Others that are not so easily expressed in monetary terms include income redistribution, environmental quality, social well-being, national security, self-sufficiency, regional growth and stability, and preservation of natural areas. Insofar as nonmonetary objectives can be quantified, they too can be explicitly incorporated into the objective function and/or the constraint set of a model for defining and evaluating water resource management alternatives.

Multiple-objective analyses do not yield single optimal solutions, but are more useful at identifying the trade-offs among conflicting noncommensurable objectives. The selection of the *best compromise* solution is a political decision. Multiobjective analyses should assist those responsible for making these political decisions by illustrating the range of possible decisions and the impacts of the alternative and competing plans.

The irrigation–recreation example presented earlier in this chapter illustrates some basic concepts in multiobjective planning. As indicated in Figure 4.1, one of the functions of multiobjective planning is the formulation of plans which are technologically efficient in that they lie on the production-possibility frontier. Plans which are inside this frontier are inferior in the sense that it is always possible to identify alternatives which will increase one or more objectives without decreasing the others. Although the formulation of efficient plans is seldom a trivial matter, it is conceptually straightforward. The selection of one of these efficient plans as the best compromise is quite another matter. Social welfare functions which could provide a basis for selection are impossible to construct, and the reduction of multiple objectives to a single criterion (as in Figure 4.2) is seldom possible.

When the various objectives of a water resources planning project cannot be combined into a single scalar objective function, a single objective management model may no longer be appropriate. Rather, a *vector optimization* formulation of the problem may be more useful [5]. Assuming that each objective $Z_j(\mathbf{X})$ is to be maximized, the model can be written

$$\text{maximize } [Z_1(\mathbf{X}), Z_2(\mathbf{X}), \ldots, Z_p(\mathbf{X})] \tag{4.7}$$

subject to

$$g_i(\mathbf{X}) = b_i \qquad i = 1, 2, \ldots, m$$

In this model $\mathbf{X}$ is the vector of decision variables, $\mathbf{X} = [X_1, X_2, \ldots, X_n]$, and $Z_j(\mathbf{X}), j = 1, \ldots, p$, are the objective functions. The objective in equation 4.7 is a vector consisting of p separate objectives. The constraints impose technical feasibility.

The vector optimization model is a concise way of formulating a multiobjective problem. In reality, a vector can be maximized only if comparisons can be made between its various components and thereby reduce the problem to a scalar one. Thus the multiobjective planning problem given in equation 4.7 cannot, in general, be solved without additional information. Multiobjective planning involves three steps: *objective quantification*, *formulation of alternatives*, and *plan selection*. Each step involves a distinct set of issues and is discussed separately below.

4.3.1 Objective Quantification

Quantification of an objective is the adoption of some quantitative (numerical) scale which provides an indicator for how well the objective would be achieved. For example, one of the objectives of a watershed conservation program might be protection or preservation of wildlife. In order to rank how the several plans meet this objective, a numerical criterion is needed, such as acres of preserved habitat or populations of key wildlife species.

Quantification does not require that all objectives be described in comparable units. The same project could have a flood control objective quantified as the height of the protected flood stage and a regional development objective quantified as increased local income. Quantification does not require that monetary costs and benefits be assigned to all objectives.

Once objectives are quantified, the steps of plan formulation and selection may proceed. Modeling techniques which are appropriate for plan formulation and for plan selection are discussed in the next two sections.

4.3.2 Plan Formulation

The goal of multiobjective plan formulation is the generation of a set of technologically efficient and acceptable plans. Two common approaches for plan formulation are the *weighting* and *constraint* methods. Both methods require numerous solutions to a single-objective management model to generate points on the objective functions' production-possibility frontier.

The weighting approach involves assigning a relative weight to each objective to convert the objective vector (equation 4.7) to a scalar which is the weighted sum of the separate objective functions. The multiobjective model becomes

$$\text{maximize } Z = w_1 Z_1(\mathbf{X}) + w_2 Z_2(\mathbf{X}) + \dots + w_p Z_p(\mathbf{X}) \tag{4.8}$$

subject to

$$g_i(\mathbf{X}) = b_i \qquad \forall i$$

where the nonnegative weights $w_1, w_2, \dots, w_p$ are specified constants. They are varied systematically, and the model is solved for each set of values to generate a set of technically efficient (or noninferior) plans.

The foremost attribute of the weighting approach is that the trade-offs or marginal rate of substitution of one objective for another at each identified point on the objective functions' production-possibility frontier is explicitly specified by the relative weights. The marginal rate of substitution between any two objectives Z_j and Z_k is

$$-\left.\frac{dZ_j}{dZ_k}\right|_{Z=\text{constant}} = \frac{w_k}{w_j} \tag{4.9}$$

when each of the objectives is continuously differentiable at the point in question.

These relative weights can be varied over reasonable ranges to generate a wide range of plans which reflect different priorities. Alternatively, specific values of the weights can be selected to reflect preconceived ideas of the relative weight which should be placed on each objective. It is clear that the prior selection of weights requires value judgments. For many projects within developing countries, these weights are often estimated by the agencies financing the projects. The weights specified by these agencies can, and often do, differ from those implied by national or regional policy. But regardless of who does it, the estimation of appropriate weights requires a study of the impacts on the economy, society, and development priorities involved. Additional information on the selection and use of weights, especially for developing regions, can be found elsewhere [6, 21, 29].

The principal disadvantage of the weighting approach is that it cannot generate the complete set of efficient plans unless the efficiency frontier is strictly convex, as it is in Figure 4.1. If the frontier, or any portion of it, is concave, only the endpoints of the concave region will be identified as illustrated in Figure 4.7.

The constraint method for multiobjective planning can always produce the entire set of efficient plans. In its general form, the constraint model is

$$\text{maximize } Z_j(\mathbf{X}) \tag{4.10}$$
$$g_i(\mathbf{X}) = b_i \qquad \forall i$$
$$Z_k(\mathbf{X}) \geq L_k \qquad \forall k \neq j$$

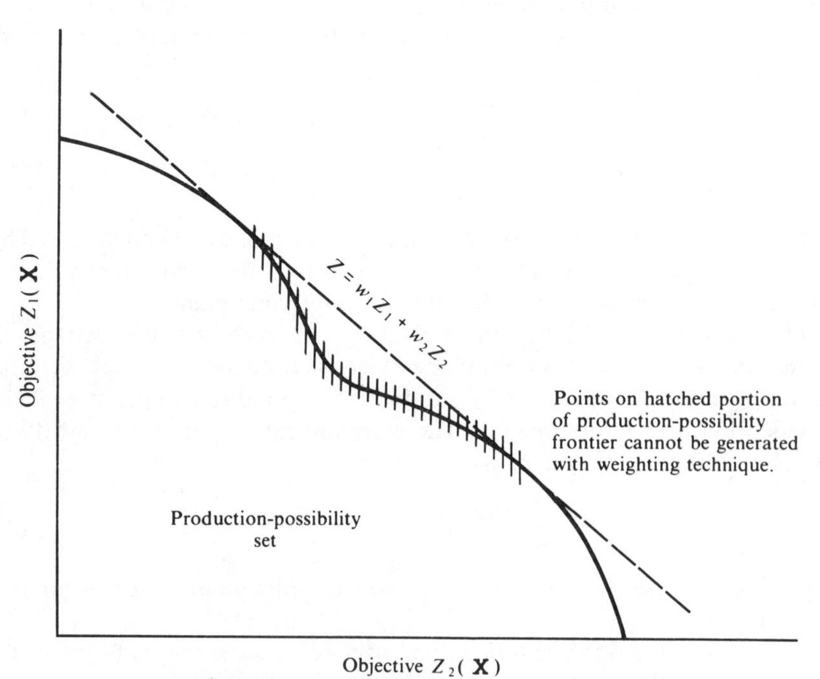

FIGURE 4.7. Example where weighting method cannot generate all noninferior solutions.

In this model, one objective $Z_j(\mathbf{X})$ is maximized subject to lower limits L_k on the other objectives. The solution of the model, corresponding to any set of feasible lower limits L_k, produces an efficient alternative if the constraint on each $Z_k(\mathbf{X})$, $k \neq j$, is binding. The constraint model is particularly attractive when it can be solved using linear programming. Then the sensitivity analysis feature of most linear programming computer packages will rapidly generate alternative solutions for different values of the L_k's. Also note that the dual variables associated with the right-hand-side values L_k are the marginal rates of substitution or rate of change of $Z_j(\mathbf{X})$ per unit change in L_k [or $Z_k(\mathbf{X})$, if binding].

Clearly, the problem facing the analyst is that the socially optimal values of the weights w_k or the lower limits L_k are unknown. They are even unknown to the decision makers until they have had an opportunity to examine the details and impacts of each resulting plan. Decision makers cannot be expected to know what they want until they know what they can get. However, several sets of weights or lower limits can be selected and the model solved for each set. Each feasible solution is a point on the efficiency frontier which may serve as a point of departure for discussion and further search.

The weighting and constraint methods are among many techniques available for generating efficient or noninferior solutions [5]. The use of generating techniques alone assumes that once all the noninferior alternatives have been identified, the decision makers will be able to select the best compromise alternative from among the many efficient alternatives without further aid from analysts. In some situations this has worked. Undoubtedly, there will be planning activities in the future where the use of generation techniques alone will continue to be of value. However, in many other planning situations, they alone will not be sufficient. Often, the number of feasible noninferior alternatives is simply too large, and decision makers will not have the time or patience to examine and evaluate each alternative plan. Decision makers may also need help in identifying which alternatives they prefer. If they are willing to work with the analyst, the analyst can help them identify what alternatives they prefer without generating all the noninferior plans.

There are a number of multiobjective modeling methods that require information pertaining to the preference of one or more decision makers. Some of these methods are useful after a number of alternative plans have been identified; other methods require preference information during the plan generation process. Both types of multiobjective planning methods will be discussed in the next section.

4.3.3 Plan Selection

By definition, the plan selection phase of multiobjective planning involves those who are responsible for approving or disapproving alternative plans. A number of iterative and interactive schemes have been proposed and used that require the participation of these decision makers as well as the planners or analysts. Some of these are extensions to the multiobjective planning methods just discussed. Each method differs in the types of information asked of the decision makers; hence the best method for a particular situation will depend not only on the method itself but also on the decision makers, the decision-making process, and the responsibilities accepted by the analyst and the decision makers.

If multiple plans have already been generated, and must be compared and ranked based on the values of various objectives derived from each plan, there are a number of methods available for assisting in this plan selection process. Three such techniques, called dominance, satisficing, and lexicography, are relatively simple, but limited. They involve the comparison of alternative decision vectors $\mathbf{X}_k$ based on the values of $Z_j(\mathbf{X}_k)$ of each separate objective j [12].

A plan $\mathbf{X}$ *dominates* all others if it results in an equal or superior value for all objectives, and at least one objective value is strictly superior to those of each other plan. In symbols, assuming that all objectives j are to be maximized, alternative $\mathbf{X}_{k^*}$ dominates if

$$Z_j(\mathbf{X}_{k^*}) \geq Z_j(\mathbf{X}_k) \qquad \text{for all objectives } j \text{ and plans } k \tag{4.11}$$

and for each $k \neq k^*$ there is an l such that

$$Z_l(\mathbf{X}_{k^*}) > Z_l(\mathbf{X}_k) \tag{4.12}$$

It is infrequent that one plan dominates all others. However, it usually occurs that some plans are inferior or are dominated by others. In particular, if there exists two plans k and h such that $Z_j(\mathbf{X}_k) \geq Z_j(\mathbf{X}_h)$ for all j and for some l, $Z_l(\mathbf{X}_k) > Z_j(\mathbf{X}_h)$, then plan $\mathbf{X}_k$ dominates plan $\mathbf{X}_h$ and plan $\mathbf{X}_h$ can be dropped from further consideration. This assumes that the modeled objectives $Z_1(\mathbf{X}), \ldots, Z_p(\mathbf{X})$ truly incorporate all the relevant planning considerations, so that there can be no reason to retain plan $\mathbf{X}_h$ in the analysis. Note that a dominance analysis only requires that a decision maker specify the objectives that are to be maximized or minimized. It does not require the assessment of the relative importance of each objective. Noninferior, efficient, or nondominated solutions are often called *Pareto optimal* because they satisfy the conditions proposed by the economist Pareto, namely that to increase the value of any one objective, one should have to accept less of another objective.

One method of further reducing the number of alternatives is called *satisficing*. It requires that the decision maker specify a minimum acceptable value for each objective. Those alternatives that do not meet these minimum values are eliminated from consideration. Those that remain can again be screened if the decision maker increases the minimal acceptable values of one or more objectives. When used in an iterative fashion, the number of noninferior alternatives can be reduced down to a single best compromise or a set of plans between which the decision makers are essentially indifferent. Of course, sometimes the decision maker will be unwilling or unable to sufficiently narrow down the set of available noninferior plans with the iterative application of satisficing. Then it may be necessary to examine the possible trade-offs among the competing alternatives.

Another simple approach is *lexicography*. To use this approach, the decision maker must rank the objectives by priority. Then, from among the noninferior plans that satisfy minimum levels of each objective, the plans can be ranked by the values of each objective in order of priority. Thus the preferred plan will achieve the largest value of the highest-priority objective. If there is more than one such plan, then among this set the preferred plan achieves the highest value of the next priority objective, and so on. This method is useful only if such a ranking of the objectives is possible. Often the relative values of the objectives of each alternative plan are of more importance to the decision makers.

Other more involved methods are available to aid in plan selection. One method of selecting the best compromise uses indifference analysis. This technique is best applied when there exists only one decision maker.

To illustrate the possible application of indifference analysis to the plan selection phase of water resources planning, consider a simple situation in which there are only two alternative plans (**A** and **B**) and two planning objectives (1 and 2) being considered. Let Z_1^A and Z_2^A be the levels of the two respective objectives for plan **A** and let Z_1^B and Z_2^B be the two objective levels for plan **B**. Indifference analysis first requires the selection of an arbitrary value for one of the objectives, say Z_2^* for objective 2. Next, the decision maker is asked to specify a value of objective 1, say Z_1, such that he or she is indifferent between the hypothetical plan that would have as its objective values (Z_1, Z_2^*) and plan **A** that has as its objective values (Z_1^A, Z_2^A). In other words, Z_1 must be determined such that (Z_1, Z_2^*) is as desirable to the decision maker as (Z_1^A, Z_2^A).

$$(Z_1, Z_2^*) \simeq (Z_1^A, Z_2^A) \tag{4.13}$$

Next the decision maker must identify another value of the first objective, say Z_1', such that he or she would be indifferent between a hypothetical plan (Z_1', Z_2^*) and the objective values (Z_1^B, Z_2^B) associated with plan **B**.

$$(Z_1', Z_2^*) \simeq (Z_1^B, Z_2^B) \tag{4.14}$$

These comparisons yield two hypothetical plans, one for each actual plan. These hypothetical plans differ only in the value of objective 1 and hence they are easily compared. If Z_1 is larger than Z_1', then the first hypothetical plan yielding Z_1 is preferred to the second hypothetical plan yielding Z_1'. Since the two hypothetical plans are equivalent to plans **A** and **B**, one can conclude that plan **A** is preferred to plan **B**. Conversely, if Z'_1 is larger than Z_1, then plan **B** is preferred to plan **A**.

This process can be extended to a larger number of objectives and plans, all of which may be ranked by a common objective. For example, assume that there are three objectives Z^i_1, Z^i_2, and Z^i_3 and n alternative plans i. A reference value Z^*_3 for objective Z_3 can be chosen and a value $\tilde{Z}^i_1$ estimated for each alternative plan i such that the decision maker will be indifferent between $(\tilde{Z}^i_1, Z^i_2, Z^*_3)$ and (Z^i_1, Z^i_2, Z^i_3). Assuming that each objective is to be maximized, and $Z^*_3 < Z^i_3$, then $\tilde{Z}^i_1$ will no doubt be greater than Z^i_1; conversely, if $Z^*_3 > Z^i_3$, then $\tilde{Z}^i_1 < Z^i_1$.

Next, a reference value Z^*_2 can be selected together with values $\hat{Z}^i_1$ such that the decision maker is indifferent between $(\hat{Z}^i_1, Z^*_2, Z^*_3)$ and $(\tilde{Z}^i_1, Z^i_2, Z^*_3)$. Hence for all plans i, the decision maker is indifferent between two hypothetical plans and the actual one.

$$(\hat{Z}^i_1, Z^*_2, Z^*_3) \sim (\tilde{Z}^i_1, Z^i_2, Z^*_3) \sim (Z^i_1, Z^i_2, Z^i_3) \tag{4.15}$$

Note that each plan comparison is based on only two pairs of objective values.

Having done this for all n plans, there are now n hypothetical plans that differ only in the value of $\hat{Z}^i_1$. It can be concluded that objective values $(\hat{Z}^i_1, Z^*_2, Z^*_3)$ are preferred to objective values $(\hat{Z}^j_1, Z^*_2, Z^*_3)$ only if $\hat{Z}^i_1 > \hat{Z}^j_1$. Hence, from equation 4.15, plan alternative i is preferred to plan alternative j if $\hat{Z}^i_1 > \hat{Z}^j_1$. Proceeding in this fashion, it is clear that all n plans can be ranked according to the decision maker's preferences for various values of only a single objective.

Indifference analysis can be extended to include utility theory [17], introduced in Section 3.5. Keeney and Wood applied such an approach to the development of the Tisza River in Hungary [18].

Each of these plan selection techniques requires the prior identification of alternative plans. Some selection techniques combine plan generation and selection. These methods require (and are aided by) the participation of the decision makers.

Perhaps one of the simplest of such iterative and interactive plan generation and selection methods is the goal attainment method. This modeling approach combines some of the advantages of both the weighting and constraint plan generation methods discussed in the previous section. The decision maker specifies a set of goals or targets T_k for each objective and, if possible, a weight w_k that reflects the relative importance of meeting that goal compared to meeting other goals. If the decision maker is unable to specify these weights, the analyst must select them and then later change them on the basis of the decision makers' reactions to the generated plans.

The goal attainment method identifies the plans that minimize the maximum weighted deviations of the objectives from specified targets. The problem to be solved is

$$\text{minimize } D \tag{4.16}$$

subject to

$$g_i(\mathbf{X}) = b_i \qquad i = 1, 2, \ldots, m$$

$$w_k[T_k - Z_k(\mathbf{X})] \leq D \qquad k = 1, \ldots, p$$

This problem is always feasible and can generate all possible efficient or noninferior plans. By interactively adjusting the weights and targets, analysts and decision makers can identify alternative solutions which achieve reasonable levels of all objectives [4].

Other techniques involving decision makers and analysts for plan generation as well as plan selection include compromise programming (CP), goal programming (GP), the step method (STEM), and the surrogate worth tradeoff method (SWT). These techniques have all seen some limited application in water resources system planning and are representative of many plan generation-selection methods [5].

Compromise programming [8, 30] is based on a geometric notion of best. The objective in compromise programming is to find that feasible solution that minimizes the "distance" from an ideal, but infeasible, solution. For weights w_j and ideal values Z_j^*, where $Z_j^* \geq Z_j(\mathbf{X})$ for all $\mathbf{X}$, the objective is to

$$\text{minimize} \left\{ \sum_{j=1}^{p} w_j^{\alpha}[Z_j^* - Z_j(\mathbf{X})]^{\alpha} \right\}^{1/\alpha} \tag{4.17}$$

The weights w_j are specified by the decision maker and the exponent α is varied from 1 to ∞ to obtain a range of solutions. Duckstein [8] suggests that if the objectives are in noncommensurable units, each objective function $Z_j(\mathbf{X})$ should be scaled so that its value is confined to a given range, such as [0, 1]. This can be done by dividing each deviation $Z_j^* - Z_j(\mathbf{X})$ by the difference between Z_j^* and the minimum value of objective j.

For $\alpha = 1$, the objective 4.17 becomes a linear weighting

$$\text{maximize} \sum_{j=1}^{p} w_j Z_j(\mathbf{X}) \tag{4.18}$$

which will yield an extreme point in objective space that maximizes the weighted sum independent of the ideal values Z_j^*.

When $\alpha = 2$, the solution of objective 4.17 or equivalently the objective

$$\text{minimize} \left\{ \sum_{j=1}^{p} w_j^2[Z_j^* - Z_j(\mathbf{X})]^2 \right\}^{1/2} \tag{4.19}$$

will be the noninferior feasible solution which is the closest (in terms of a weighted geometric distance) to the ideal solution $\{Z_j^*\}$. If $\alpha = \infty$, the largest weighted deviation determines the preferred solution. Hence equation 4.17 with $\alpha = \infty$ is equivalent to minimizing the maximum weighted deviation, such as was done with the goal attainment method, equations 4.16. Thus compromise programming generates, for various $\alpha \geq 1$, a subset of all noninferior alternative plans. This subset may or may not contain the alternative plan considered best by the decision makers.

Goal programming [3] is perhaps the most widely used multiobjective plan generation and selection method. It is similar to compromise programming with $\alpha = 1$. The ideal values Z_j^* are replaced by goals or targets T_j and are selected, along with the weights w_j, by the decision maker. The objective is to

$$\text{minimize} \sum_{j=1}^{p} w_j |T_j - Z_j(\mathbf{X})| \tag{4.20}$$

It is not always easy for the decision maker to identify what weights w_j are most appropriate; hence a range of weights is sometimes used to identify a range of plans. Unless $T_j \geq Z_j^*$, some plans may be inferior.

The step method [2, 22] objective is similar to the compromise programming objective when $\alpha = \infty$. The step method (denoted STEM) is an iterative method, requiring preference information from the decision maker at each iteration. This information affects the values of the weights w_j and identifies constraints on various objective values. The decision makers are never asked for values of the weights w_j. Rather, the weights are calculated based on the relative range of values each objective can assume, and on whether or not the decision maker has indicated satisfaction regarding a particular objective value obtained from a previous solution. If the decision maker is satisfied with, say, an objective $Z_j(\mathbf{X})$, then that value less what he or she is willing to give up to get more of other objective values is specified as a lower bound on $Z_j(\mathbf{X})$. Then w_j is set to 0, and the weights of all remaining objectives are recalculated. The problem is again solved. The process is repeated until some best compromise plan is identified.

This STEM approach guides the decision maker and analyst among noninferior alternatives toward the plan or solution the decision maker considers best without requiring an exhaustive generation of all noninferior alternatives. However, the decision maker must be willing to indicate how much of some objective value can be given up in order to obtain some unknown amount of other objective values. This is not as easy as indicating how much more is desired of any or all unsatisfactory objectives.

The *surrogate worth trade-off method* [10, 11] is based on a decision maker's assessment of trade-offs between objective k, the numeraire, and every other objective, assuming that the remaining objectives are fixed. The trade-off between objective l and objective k is evaluated by solution of

$$\text{maximize } Z_k(\mathbf{X}) \tag{4.21}$$

subject to

$$Z_l(\mathbf{X}) \geq L_l \tag{4.22}$$

$$Z_j(\mathbf{X}) = L_j \qquad \forall\, j \neq k, l \tag{4.23}$$

The original p objective problem becomes a series of two objective problems. The value of L_l or equivalently $Z_l(\mathbf{X})$ is varied to trace out a trade-off between Z_k and Z_l. This trade off $\partial Z_k / \partial Z_l$ depends on the particular values of all other L_j in equation 4.23. The decision maker is then asked to state his or her preferences for those trade-offs in terms of a surrogate worth function of ordinal values ranging from -10 to 10. These values indicate the relative worth to the decision maker of the marginal change (gain) in the value of objective Z_k per unit change (loss) in the value of objective Z_l. Similarly, trade-offs and associated surrogate worth functions are obtained for all objectives $l \neq k$. Of course, the surrogate worth of each trade-off between an objective $l(\neq k)$ and objective k depends on the values assigned to the other $p - 2$ objectives. Thus each surrogate worth function $\text{SWF}_{l,k}$ is a function of the values L_j for $j \neq l$ or k. The surrogate worth trade-off method finds the best compromise solution by identifing those objective values $L_1^*, L_2^*, \ldots, L_p^*$ at which the surrogate worth of any small trade-off is zero.

This has been a very brief introduction to some of the approaches available for plan generation and selection. Details on these and other potentially useful techniques can be found in many books, some of which are devoted solely to this subject of multiobjective planning [4, 5, 11, 17, 24, 30].

4.4 CONCLUDING REMARKS

Many theoretical and practical approaches have been proposed in the literature for identifying and quantifying objectives and for considering multiobjectives in water resources planning. The discussion and techniques presented in this chapter serve merely as an introduction to this subject. Identifying and quantifying appropriate objectives and incorporating them into multiobjective planning models can be a tedious and frustrating process. Of course, there are many water resources planning problems that have single well-defined goals and hence do not require multiobjective modeling. For those more typical applications that dictate a multiobjective approach, the analyst or planner may only have to proceed through the plan formulation step and the generation of efficient plans. This may be sufficient if the public and the decision makers are then able to make a good choice from among those alternatives.

Water resource systems analysts face a more difficult challenge when they are required to recommend a single planning alternative for adoption from among many noninferior plans. This requires an involvement in the plan selection process and requires a balancing of the goals and values of the various individuals and groups concerned with the project. There is virtually no way in which the selection step can be a normative procedure; there can be no standard set of criteria or methods which result in identification of the preferred project. At best, an iterative procedure in which the analyst works with decision makers may result in some reasonable articulation of preferences and an eventual convergence on a plan that is politically as well as technically, socially, financially, and institutionally feasible.

To decision makers, many of these approaches for objective quantification and multiobjective planning may seem theoretical or academic. Political decision makers are often reluctant to learn complicated quantative policy analysis techniques or to spend time answering seemingly irrelevant questions that might lead eventually to a "compromise" plan. Reluctance to engage in quantification of trade-offs among alternative plans may stem from the support political leaders desire from distinct and often conflicting interest groups. In such situations it is obviously to their advantage not to be too explicit in quantifying political values. They might prefer that the "analyst" make these trade-offs. Planners, engineers, or analysts are often very willing to make these trade-offs because these pertain to subject areas in which they consider themselves expert.

Through the further development and use of analytical multiobjective identification and planning techniques, analysts can begin to interact with decision makers and can enlighten any who would argue that water resources policy evaluation and analyses should not be political. Through such interaction, analysts can learn more about the process of decision making, what information is most useful to that process, and how it can best be presented. Knowledge of these facts in a particular planning situation might dictate substantially the approach to objective identification and quantification and to plan formulation and selection.

EXERCISES

4-1. Distinguish between multiple purposes and multiple objectives and give some examples of complementary and conflicting purposes and objectives of water resources projects.

4-2. Assume that the farmers' demand for water q is a linear function $a - bp$ of the price p, where $a, b > 0$. Calculate the farmers' willingness to pay for a quantity of water q. If the cost of delivering a quantity of water q is cq, $c > 0$, how much water should a public agency supply to maximize willingness to pay minus total cost? If the local water district is owned and operated by a private firm whose objective is to maximize profit, how much water would they supply and how much would they earn? The farmers' *consumer surplus* is their willingness to pay minus what they must pay for the resource. Compare the farmers' consumer surplus in the two cases. Do the farmers lose more than the private firm gains by moving from the social optimum to the point that maximizes the firm's profit? Illustrate these relationships with a graph showing the demand curve and the unit cost c of water. Which areas on the graph represent the firm's profits and the farmers' consumer surplus?

4-3. Under what condition may the constraint method for generating non-inferior solutions generate inferior solutions?

4-4. A reservoir is planned for irrigation and low flow augmentation for water quality control. A storage volume of 6×10^6 m^3 will be available for these two conflicting uses each year. The maximum irrigation demand (capacity) is 4×10^6 m^3. Let X_1 be the allocation of water to irrigation and X_2 the allocation for downstream flow augmentation. Assume that there are two objectives, expressed as

$$Z_1 = 4X_1 - X_2$$
$$Z_2 = -2X_1 + 6X_2$$

(a) Write the multiobjective planning model using a weighting approach and a constraint approach.

(b) Define the efficiency frontier. This requires a plot of the feasible combinations of X_1 and X_2.

(c) Assuming that various values are assigned to the weight W_1 where W_2 is constant and equal to 1, verify the following solutions to the weighting model.

W_1	X_1	X_2	Z_1	Z_2
>6	4	0	16	-8
6	4	0 to 2	16 to 14	-8 to 4
<6 to >1.6	4	2	14	4
1.6	4 to 0	2 to 6	14 to -6	4 to 36
<1.6	0	6	-6	36

4-5. Show how the following benefit, loss, and cost functions can be included in a linear optimization problem for finding the annual storage volume target T^S, annual release target T^R and the actual reservoir releases R_t in each within-year period t, and the reservoir capacity K. The objective is to maximize annual net benefits from the construction and operation of the reservoir. Assume that the inflows are known in each of 12 within-year periods t. Note that the loss function associated with reservoir recreation is independent of the value of T^S, unlike the loss function associated with reservoir releases. Structure the complete linear programming model. Define all variables used that are not defined below. Let $\delta_t T^R$ be the known fraction of the unknown annual release target T^R that is the release target in period t.

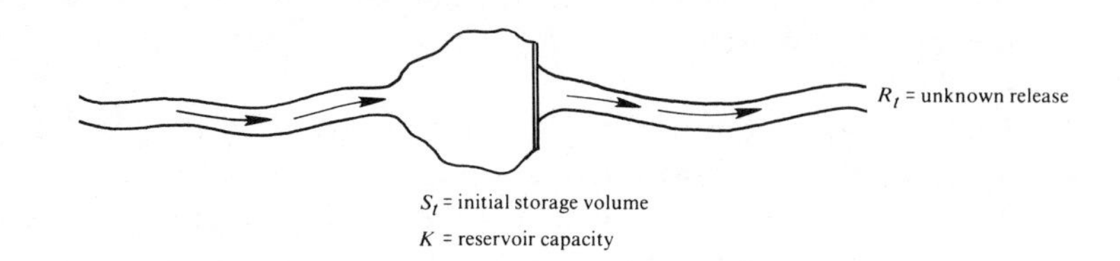

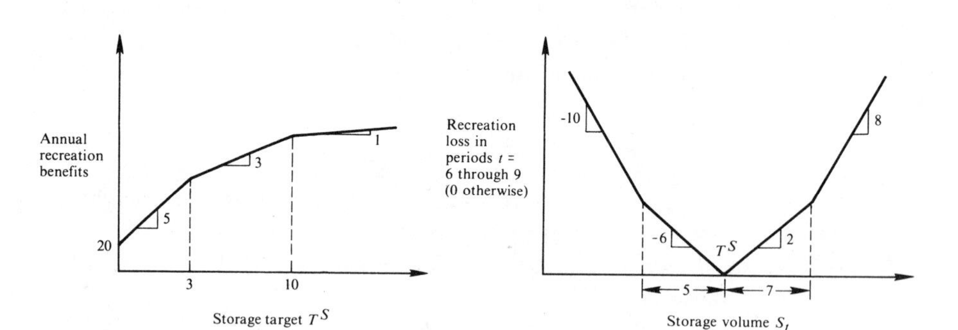

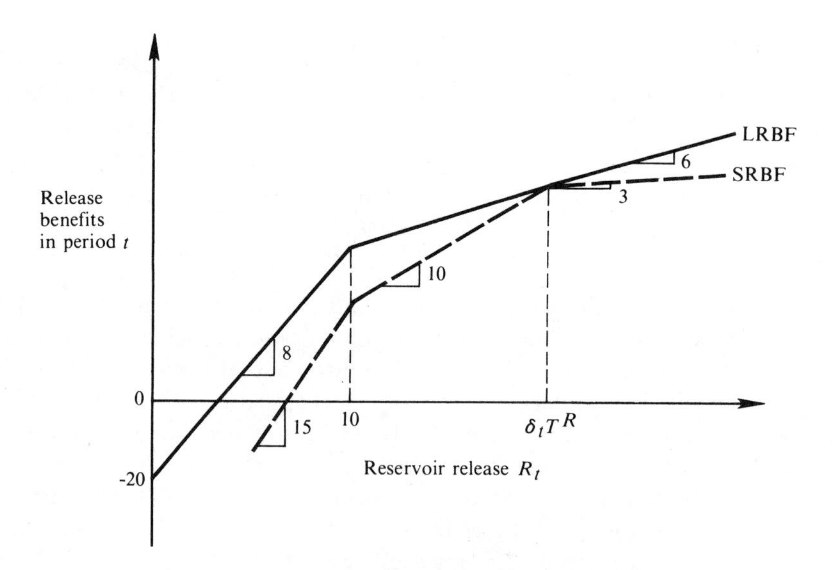

Note: Each unit of deficit within the reservoir release range of 0 to 10 has a loss of 15; otherwise, the loss is 10 per unit deficit.

EXERCISE 4-5.

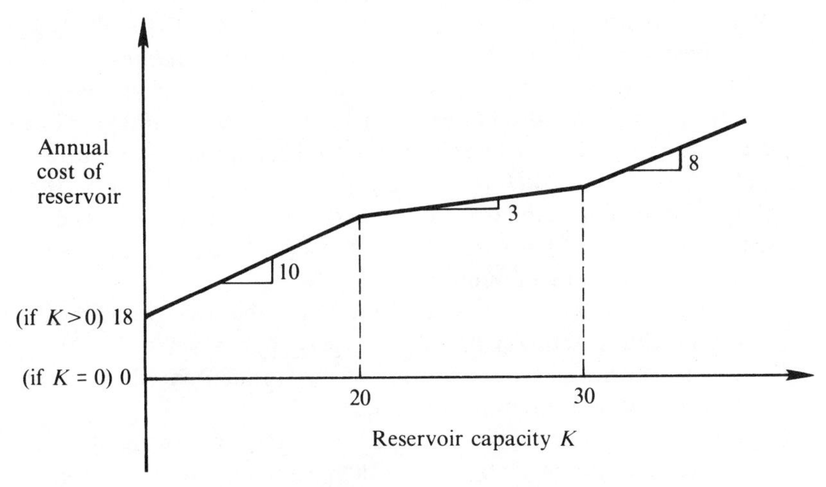

EXERCISE 4-5. (Cont.)

4-6. For the river basin shown, potential reservoirs exist at sites $i = 1, 2$, and 4 and a diversion can be constructed between sites $i = 1$ and 2. The cost of each reservoir is $C_i(K_i)$, where K_i is the storage volume capacity. The cost of the diversion canal is $C(Q)$, where Q is the canal capacity. The cost in period t of diverting a flow Q_t from site $i = 1$ or 2 to site $j = 2$ or 1 is $C_{ij}(Q_t)$. The two users, $i = 3$ and 5, have known target allocations T_{it} for each period t. Develop a model that could define the reservoir capacities required to meet various percentages of the specified targets T_{3t} and T_{5t} at a minimum cost. Assume that the return flow at use $i = 3$ is 40% of the allocation to that use, and that the unregulated flows Q_{it} at each site i are known.

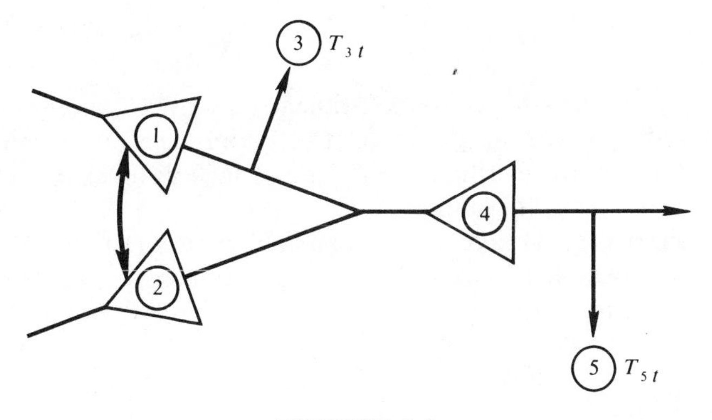

EXERCISE 4-6.

4-7. Suppose that there exist two polluters, A and B, who can provide additional treatment, X_A and X_B, at a cost of $C_A(X_A)$ and $C_B(X_B)$, respectively. Let W_A and W_B be the waste produced at sites A and B, and $W_A(1 - X_A)$ and $W_B(1 - X_B)$ be the resulting waste discharges at sites A and B. These discharges must be no greater than the effluent standards E_A^{max} and E_B^{max}. The resulting pollution concentration $a_{Aj}(W_A(1 - X_A)) + a_{Bj}(W_B(1 - X_B)) + q_j$ at various sites j must not exceed the stream standards S_j^{max}. Assume that total cost and cost inequity [i.e., $C_A(X_A) + C_B(X_B)$ and $|C_A(X_A) - C_B(X_B)|$] are management objectives to be minimized.

(a) Discuss how you would model this multiobjective problem using the weighting and constraint (or target) approaches.

(b) Discuss how you would use the model to identify efficient, noninferior (Pareto-optimal) solutions.

(c) Effluent standards at sites A and B and ambient stream standards at sites j could be replaced by other planning objectives (e.g., the minimization of waste discharged into the stream). What would these objectives be, and how could they be included in the multiobjective model?

4-8. (a) What condition must apply if the goal attainment method defined by equations 4.16 is to produce only noninferior alternatives for each assumed target T_k and weight w_k?

(b) Convert the goal programming objective 4.20 to a form suitable for solution by linear programming.

4-9. Water quality objectives are sometimes difficult to quantify. Various attempts have been made to include the many aspects of water quality in single water quality indices. One such index was proposed by Dinius (Social Accounting System for Evaluating Water Resources, *Water Resources Research*, Vol. 8, No. 5, 1972, pp. 1159–1177). Water quality, Q, measured in percent is given by

$$Q = \frac{w_1 Q_1 + w_2 Q_2 + \ldots + w_n Q_n}{w_1 + w_2 + \ldots + w_n}$$

where Q_i is the ith quality constituent (dissolved oxygen, chlorides, etc.) and w_i is the weight or relative importance of the ith quality constituent. Write a critique on the use of such an index in multiobjective water resources planning.

4-10. Let objective $Z_1(\mathbf{X}) = 5X_1 - 2X_2$ and objective $Z_2(\mathbf{X}) = -X_1 + 4X_2$. Both are to be maximized. Assume that the constraints on variables X_1 and X_2 are:

1. $-X_1 + X_2 \leq 3$
2. $X_1 \leq 6$

3. $X_1 + X_2 \leq 8$
4. $X_2 \leq 4$
5. $X_1, X_2 \geq 0$

(a) Graph the Pareto-optimal or noninferior solutions in decision space.
(b) Graph the efficient combinations of Z_1 and Z_2 in objective space.
(c) Reformulate the problem to illustrate the weighting method for defining all efficient solutions of part (a) and illustrate this method in decision and objective space.
(d) Reformulate the problem to illustrate the constraint method of defining all efficient solutions of part (a) and illustrate this method in decision and objective space.
(e) Solve for the compromise set of solutions using compromise programming (equations 4.17 through 4.19) with all weights equal to 1 and α equal to 1, 2, and ∞.

4-11. Illustrate the procedure for selecting among three plans, each having three objectives, using indifference analysis. Let Z_j^i represent the value of objective i for plan j. The values of each objective for each plan are given below. Assume that each objective is to be maximized. Assume that an identical indifference function for all trade-offs between pairs of objectives, namely one that implies you are willing to give up twice as many units of your higher (larger) objective value to gain one unit of your lower (smaller) objective value. [For example, you would be indifferent to two plans having as their three objective values (30, 5, 10) and (20, 5, 15).] Rank these three plans in order of preference.

Plan j \ Objective i	1	2	3
	Z_j^1	Z_j^2	Z_j^3
1	5	25	15
2	10	20	10
3	15	10	15

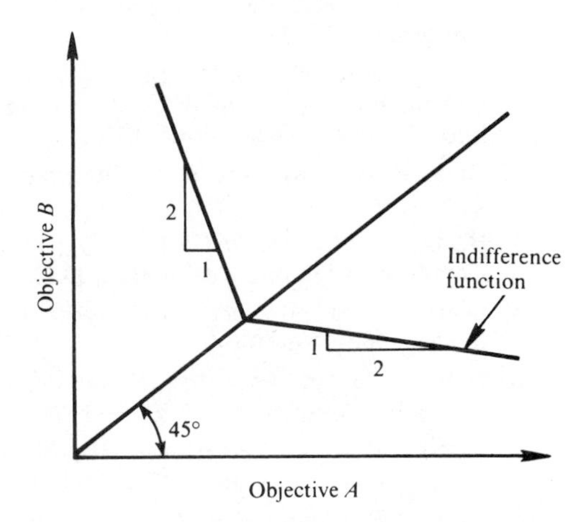

EXERCISE 4-11.

REFERENCES

1. ARROW, K. J., *Social Choice and Individual Values*, 2nd ed., John Wiley & Sons, Inc., New York, 1963.
2. BENAYOUN, R., et al., Linear Programming with Multiple Objective Functions: Step Method, *Mathematical Programming*, Vol. 1, No. 3, 1971.
3. CHARNES, A., and W. COOPER, *Management Models and Industrial Applications of Linear Programming*, Vol. 1, John Wiley & Sons, Inc., New York, 1961.
4. COHON, J. L., and D. H. MARKS, A Review and Evaluation of Multi-objective Programming Techniques, *Water Resources Research*, Vol. 11, No. 2, 1975, pp. 208–220.
5. COHON, J. L., *Multiobjective Programming and Planning*, Academic Press, Inc., New York, 1978.
6. DASGUPTA, P., A. SEN, and S. MARGLIN, *Guidelines for Project Evaluation*, UNIDO, United Nations, New York, 1972.
7. DERIDDER, N. A., and A. EREZ, *Optimum Use of Water Resources*, International Institute for Land Reclamation and Improvement, Wegeningen, Netherlands, 1977.
8. DUCKSTEIN, L., Imbedding Uncertainties into Multi-Objective Decision Models in Water Resources, in *Reliability in Water Resources Management*, E. A. McBean, K. W. Hipel, and T. E. Unny (eds.), Water Resources Publications, Littleton, Colo., 1979.
9. ECKSTEIN, O., *Water Resource Development: The Economics of Project Evaluation*, Harvard University Press, Cambridge, Mass., 1961.
10. HAIMES, Y. Y., and W. A. HALL, Multiobjectives in Water Resource Systems Analysis: The Surrogate Worth Trade Off Method, *Water Resources Research*, Vol. 10, No. 4, 1974, pp. 615–624.
11. HAIMES, Y. Y., W. A. HALL, and H. FREEDMAN, *Multiobjective Optimization in Water Resources Systems: The Surrogate Worth Trade Off Method*, Elsevier, Amsterdam, 1975.
12. HAITH, D. A., and D. P. LOUCKS, Multi-objective Water Resources Planning, in *Systems Approach to Water Management*, A. K. Biswas (ed.), McGraw-Hill Book Company, New York, 1976.
13. HENDERSON, J. M., and R. E. QUANDT, *Microeconomic Theory*, McGraw-Hill Book Company New York, 1958.
14. HOWE, C. W., *Benefit-Cost Analysis for Water System Planning*, American Geophysical Union, Washington, D.C., 1971.
15. JAMES, A. (ed.), *Mathematical Models in Water Pollution Control*, John Wiley & Sons, Inc., New York, 1978.
16. JAMES, L. D., and R. R. LEE, *Economics of Water Resources Planning*, McGraw-Hill Book Company, New York, 1971.
17. KEENEY, R. L., and H. RAIFFA, *Decisions with Multiple Objectives: Preferences and Value Tradeoffs*, John Wiley & Sons, Inc., New York, 1976.
18. KEENEY, R. L., and E. F. WOOD, An Illustrative Example of the Use of Multiattribute Utility Theory for Water Resource Planning, *Water Resources Research*, Vol. 13, No. 4, 1977, pp. 705–712.

19. KNETSCH, J. L., *Outdoor Recreation and Water Resources Planning*, American Geophysical Union, Washington, D.C., 1974.

20. KRUTILLA, J. V., and O. ECKSTEIN, *Multiple Purpose River Development: Studies in Applied Economic Analysis*, Johns Hopkins Press, Baltimore, Md., 1958.

21. LITTLE, I. M. D., and J. A. MIRRLEES, *Project Analysis and Planning for Developing Countries*, Basic Books, Inc., Publishers, New York, 1974.

22. LOUCKS, D. P., An Application of Interactive Multiobjective Water Resources Planning, *Interfaces*, Vol. 8, No. 1, 1977.

23. MAASS, A., M. M. HUFSCHMIDT, R. DORFMAN, H. A. THOMAS, JR., S. A. MARGLIN, and G. M. FAIR, *Design of Water Resource Systems*, Harvard University Press, Cambridge, Mass., 1962.

24. MAJOR, D. C., *Multiobjective Water Resource Planning*, Water Resources Monograph 4, American Geophysical Union, Washington, D.C., 1977.

25. MARGLIN, S. A., *Public Investment Criteria*, The MIT Press, Cambridge, Mass., 1967.

26. MCBEAN, E. A., K. W. HIPEL, and T. E. UNNY (eds.), *Reliability in Water Resources Management*, Water Resources Publications, Littleton, Colo., 1979.

27. Principles and Standards for Planning Water and Related Land Resources, Water Resources Council, *Federal Register*, Vol. 38, No. 174, part III, 1973. (Revision: Vol. 44, No. 242, Part X, 1979.)

28. SHEN, H. W. (ed.), *Stochastic Approaches to Water Resources*, Vols. 1 and 2, H. W. Shen, Fort Collins, Colo., 1976.

29. SQUIRE, L., and H. G. VAN DER TAK, *Economic Analysis of Projects*, Johns Hopkins University Press, Baltimore, Md., March 1975.

30. ZELENY, M., Compromise Planning, in *Multiple Criteria Decision Making*, J. Cochrane and M. Zeleny (eds.), University of South Carolina Press, Columbia, S.C., 1973, pp. 262–301.

PART III

Managing Surface-Water Quantity

The methods for analyzing water resource systems problems introduced and reviewed in the previous three chapters are now applied to multipurpose surface water planning and management problems. Chapter 5 introduces the typical model structure required to evaluate and schedule various river basin development alternatives. These alternatives may include municipal and industrial water supplies, flood control, reservoir recreation, hydropower, and irrigation. In this introductory chapter, hydrologic uncertainty is ignored. It is explicitly considered in chapters 6 and 7. Chapter 6 describes a wide range of models which can be used to generate synthetic streamflow sequences and other phenomena useful in stochastic simulation studies. Chapter 7 discusses several river basin operation and planning models which explicitly incorporate the stochastic nature of streamflows. Chapter 8 focuses on irrigation planning and operation models and the resources in addition to water that should be considered. It provides an example of how many of the benefit functions required for the economic evaluation of potential water uses can be derived for inclusion in multipurpose river basin planning models.

CHAPTER 5

Deterministic River Basin Modeling

5.1 INTRODUCTION

Deterministic models for river basin system planning do not explicitly consider uncertainty in hydrologic variables or model parameters. As such, deterministic models provide a limited representation of planning and management problems. Yet for preliminary analyses of alternative plans prior to more detailed stochastic optimization or simulation study, deterministic models using selected values of uncertain inputs, parameters, and variables can be useful. For example, Major and Lenton [16] report on the application of deterministic screening and dynamic sequencing models in an analysis of development alternatives for the Rio Colorado River in Argentina. There are numerous other examples in the literature. Whether or not deterministic models will be of value in a particular situation, river basin models are more easily explained and understood if at first uncertainty is ignored.

This chapter begins with a discussion of methods for estimating the unregulated streamflows at various sites of interest throughout a basin. Next, several methods for estimating reservoir storage requirements for water supplies are reviewed and compared. Following this, the model components associated with withdrawals and diversions, reservoir storage both for water supply and flood control, hydroelectric power generation, and flood control structures at potential damage sites in a river basin are introduced. These components are then combined into a multiple-purpose multiobjective planning model for a hypothetical river basin. The chapter concludes with an introduction to some dynamic models for assisting in the scheduling and sequencing over time of multiple projects within a river basin. The modeling procedures outlined in this chapter serve as the basis for the development of the stochastic optimization models for river basin planning introduced in Chapter 7.

5.2 STREAMFLOW ESTIMATION

5.2.1 Time Periods

When analyzing and evaluating various water management plans designed to distribute the natural unregulated flows over time and space, it is usually sufficient to consider monthly or seasonal flows, as opposed to daily or weekly flows. The shortest time period considered in such analyses is usually no less than the time it takes water to travel from the upper end of a river basin to the lower end of the basin. Except for problems involving floods that occur in shorter periods, this assumption will hold throughout this chapter. When flood routing in short hourly or daily time periods is not required, flows can be defined by simple mass-balance or continuity equations.

The actual length of each within-year period defined in a model may vary from period to period. The appropriate number and length of each modelled period will depend, in part, on the hydrologic variation (i.e., the variation in natural flows), the possible need for greater detail during certain seasons of a year (e.g., when modeling the variation in lake water levels or volumes in the summer recreation season), and the amount of computer time and capacity available. Each period that is defined adds to the number of equations and variables in the model and to its solution cost.

Another important factor to consider in making a decision as to the number and duration of periods to include in any model is the purpose for which the model is to be used. Some analyses are concerned only with identifying desirable designs and operating policies of various engineering projects for managing water resources at some fixed time (year) in the future. These static analyses are not concerned with investment project scheduling or sequencing. On the other hand, dynamic capacity expansion models must include equations to describe the system's operation in different years as well as different periods within each year. These dynamic multiperiod multiyear models include the change, if any, in economic, environmental and other objectives and parameters over time. As a result, dynamic expansion models generally contain many more periods than do static models, even though static models have more periods per year.

In the static and dynamic models developed and discussed in this chapter, the flow Q_t^s (L^3T^{-1}) at site s in period t refers either to the mean flow, a critical period flow, some historical flow, or else a particular flow exceeded with some specified probability.

Consider, for example, the simple river basin illustrated in Figure 5.1. The stream flows from north to south, and the streamflows have been recorded over a number of years at gage sites 1 and 9. Assume that each modelled period t is of sufficient length to permit a mass balance of the streamflows at any site in the basin. Knowledge of the flows Q_t^s in each period t at gage sites 1 and 9 permits the calculation of the incremental flow $Q_t^9 - Q_t^1$ between those gage sites in each period. The next task is to compute the streamflows at any site in the river basin, and then to develop the mass-balance equations that may describe the changes in these flows resulting from various management alternatives.

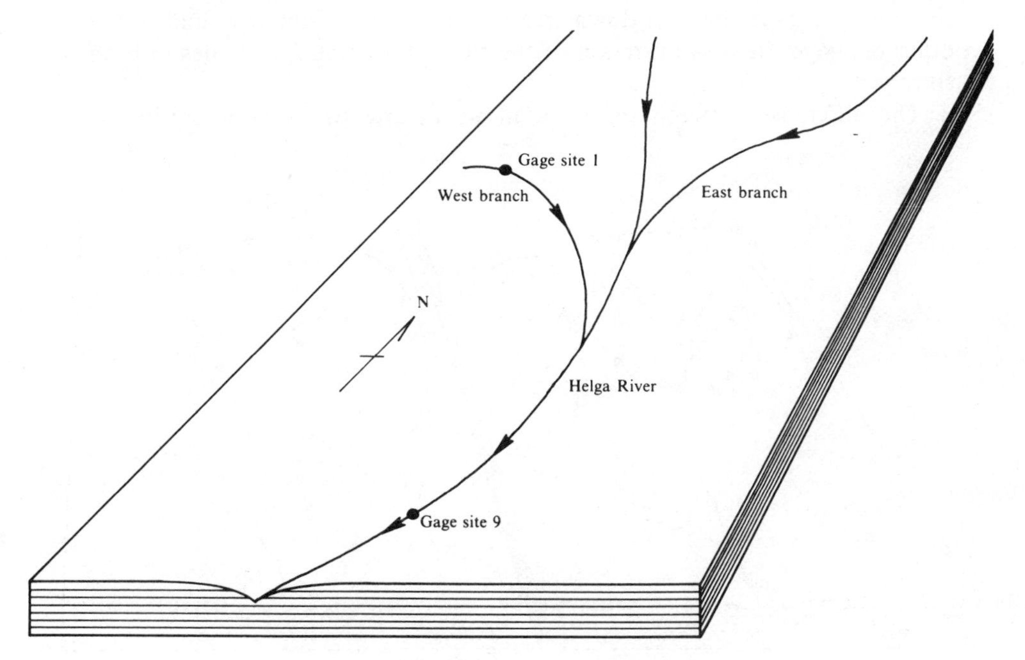

FIGURE 5.1. Helga River valley.

5.2.2 Estimating Unregulated Streamflows

It would be ideal if there existed a gage, with a long record of measured streamflows, at each potential withdrawal or reservoir storage site in the basin. The quantity of water that can be withdrawn or stored at any specific site is, in part, a function of the amount of water available at that site. In almost all situations, the streamflows, at each site and for each period of interest, must be estimated based on the measured flows at one or more nearby gage sites.

The method used to estimate flows will depend on the characteristics of the watershed of the river basin [21]. In humid regions where watersheds are generally homogeneous throughout the basin, the spatial distribution of monthly or seasonal rainfall does not significantly vary from one part of the river basin to another. In these situations, estimated flows Q_t^s at any site s can be based on the watershed areas A^s above those sites, and the streamflow $Q_t^{s'}$, and watershed area $A^{s'}$ above the nearest or most representative gage site s'.

$$Q_t^s = Q_t^{s'}\left(\frac{A^s}{A^{s'}}\right) \tag{5.1}$$

This approach for streamflow estimation is illustrated in Figure 5.2. In one situation the gage site s' is downstream of the site s of interest, and in the other the gage site s' is upstream of the site s. Equation 5.1 applies in both situations.

The difference between the streamflows at any two sites is called the

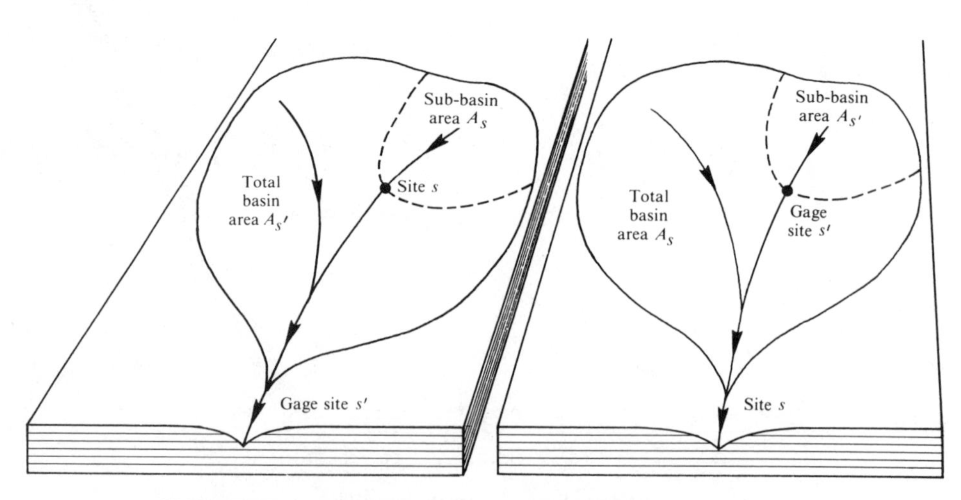

FIGURE 5.2. Two situations in which watershed areas may be used for streamflow estimation with equation 5.1.

incremental flow. Using equation 5.1 to estimate streamflows will result in positive incremental flows. In arid regions, incremental flows may be negative, indicating a net loss in flow in the downstream direction. This might be the case when a stream originates in a wet area and flows into a region that receives little rainfall. In such arid areas there may be no runoff into the stream but considerable evaporation and infiltration into the ground along the stream channel.

For stream channels where there are no known sites in which the stream abruptly enters the ground, but rather where there exist relatively uniform conditions affecting water loss, the average streamflow for a particular period t at site s will be a function of the nearest or most representative gage flow $Q_t^{s'}$ and of the length of channel $L_{s,s'}$ between the site s and the gage site s'. Assuming that the length L is positive in the downstream direction, one method for estimating flows in arid areas is through the use of an exponential decay function.

$$Q_t^s = Q_t^{s'} \cdot 10^{-\beta_t L_{s,s'}} \tag{5.2}$$

The parameter β_t is dependent on the particular watershed as well as on the period t. As illustrated in Figure 5.3, β_t can be estimated from a knowledge of the average streamflows and channel lengths at various gaging sites along the channel in the arid portion of the watershed. When using this procedure for estimating β_t, each period should be of sufficient length to average out short-term random changes in the soil moisture content within the period. Otherwise, there may be excessive variations in the distribution of data points used to estimate these parameters.

Equation 5.1 applies only to homogenous watersheds in which the total runoff volume increases in the downstream direction. Equation 5.2 applies only to portions of homogenous watersheds in which there is little or

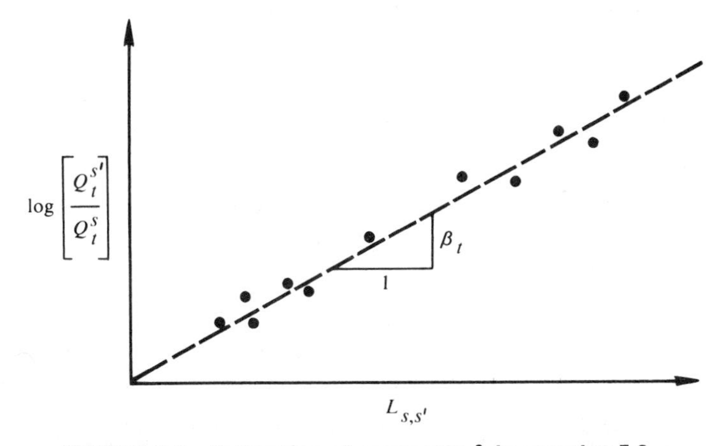

FIGURE 5.3. Estimation of parameter β_t in equation 5.2.

no runoff. In many cases neither equation may apply. This is often the case for watersheds characterized by significant elevation changes and consequently varying rainfall and runoff distributions. In such situations relationships between mean annual flow, drainage areas and elevations can be derived. This permits an estimate of the mean annual flow Q^s at any ungaged site s. Multiplying the estimated mean annual flow Q^s at site s by the ratio of each period's flow to the annual flow, yields an estimate of each period's mean flow Q_t^s at the site.

The selection of the appropriate gage site to use to estimate the streamflow Q_t^s at a particular site can be a matter of judgment. The best gage site need not necessarily be the nearest gage to site s, but rather the most similar with respect to important hydrologic variables. When using more than one gage site (such as the gages at site 1 and 9 in the Helga River valley in Figure 5.1) as the basis for estimating the streamflows at other sites, care must be taken so as not to create a negative incremental flow when positive increments are expected. In these situations, a weighted average of the flows at two (or more) gage sites may provide more consistent estimates of the flows at ungaged sites. For example, for any site s between gage sites 1 and 9 in the Helga River valley where equation 5.1 might apply, the streamflow may be estimated as

$$Q_t^s = w\left[Q_t^1\left(\frac{A^s}{A^1}\right)\right] + (1 - w)\left[Q_t^9\left(\frac{A^s}{A^9}\right)\right] \tag{5.3}$$

where the weight w would be between 0 and 1, depending on the relative distance between site s and gage sites 1 and 9. In equation 5.3 the closer site s is to a gage site, the greater the weight w that would normally be associated with that site.

In general for n gage sites u one could write

$$Q_t^s = \sum_{u=1}^{n} w_u\left[Q_t^u\left(\frac{A^s}{A^u}\right)\right] \tag{5.4}$$

where

$$\sum_{u=1}^{n} w_u = 1.$$

5.3 ESTIMATING RESERVOIR STORAGE REQUIREMENTS FOR WATER SUPPLY

The primary purpose of a reservoir is to provide a means of regulating surface water flows. The capacity of a reservoir, together with its operating policy, determine the extent to which streamflows can be stored for later release.

The use of reservoirs for temporarily storing streamflows often results in a net loss of total streamflow due to evaporation and seepage. While these losses may not be desired, the benefits derived from regulation of water supplies, from flood water storage, from hydroelectric power, and from any recreational activities at the reservoir site may offset the hydrologic losses and the costs of reservoir construction and operation. If not, the reservoir should not be built.

Reservoir storage capacity can be divided among three major uses: (1) the active storage used for streamflow regulation and for water supply; (2) the dead storage required for sediment collection, recreational development or hydropower production; and (3) the flood storage capacity reserved to reduce potential downstream flood damage. Often these components of reservoir capacity can be modeled separately and then added together to determine total reservoir capacity. This section will be limited to a discussion of several methods for estimating active storage requirements.

5.3.1 Mass Diagram Analyses

Perhaps one of the earliest methods used to calculate the active storage capacity required to meet a specified reservoir release R_t (L^3T^{-1}) in a sequence of periods t, was developed by Rippl in 1883 [9, 20]. His *mass diagram analysis* is still used today by many water resource planners. It involves finding the maximum positive cumulative difference between a sequence of specified reservoir releases R_t and known historical or simulated inflows Q_t.

Let d_t represent the positive or negative difference $R_t - Q_t$ between the release R_t and inflow Q_t in period t. The maximum positive cumulative difference d_t^* between releases and inflows during an interval of time beginning in period t and extending up to period T is

$$d_t^* = \underset{t \le j \le T}{\text{maximum}} \left(\sum_{\tau=t}^{j} d_\tau\right) \tag{5.5}$$

The required active storage capacity K_a is the maximum of the maximum cumulative differences d_t^*:

$$K_a = \underset{1 \le t \le T}{\text{maximum}} (d_t^*) \tag{5.6}$$

Combining equations 5.5 and 5.6, the active storage requirement is

$$K_a = \underset{1 \le i \le j \le T}{\text{maximum}} \left[\sum_{t=i}^{j} (R_t - Q_t)\right] \tag{5.7}$$

This procedure works only if, over the period of record, the sum of reservoir releases does not exceed the sum of reservoir inflows. If the critical sequence of periods which might determine the required reservoir capacity occurs at the end of the period of record, it is usually assumed that the observed flow sequence will repeat.

Equation 5.7 is the analytical equivalent of a variety of graphical procedures similar to the one originally proposed by Rippl for finding the active storage requirements. Two of these graphical procedures are illustrated in Figure 5.4. Rippl's original approach, shown in Figure 5.4a, involves plotting the cumulative inflow $\sum_{\tau=1}^{t} Q_\tau$, versus time t. Assuming a constant reservoir release R_t in each period t, a line with slope R_t is placed so that it is tangent to the cumulative inflow curve. To the right of these points of tangency the release R_t exceeds the inflow Q_t. The maximum vertical distance between the cumulative inflow curve and the release line of slope R_t equals the maximum water deficit, and hence the required active storage capacity. Clearly, if the annual average of the releases R_t is greater than the mean inflow, a reservoir of any capacity will not be able to meet the demand.

This graphical approach is easily performed if the releases R_t are the same in each period t. If the releases vary, it may be easier to plot the cumulative

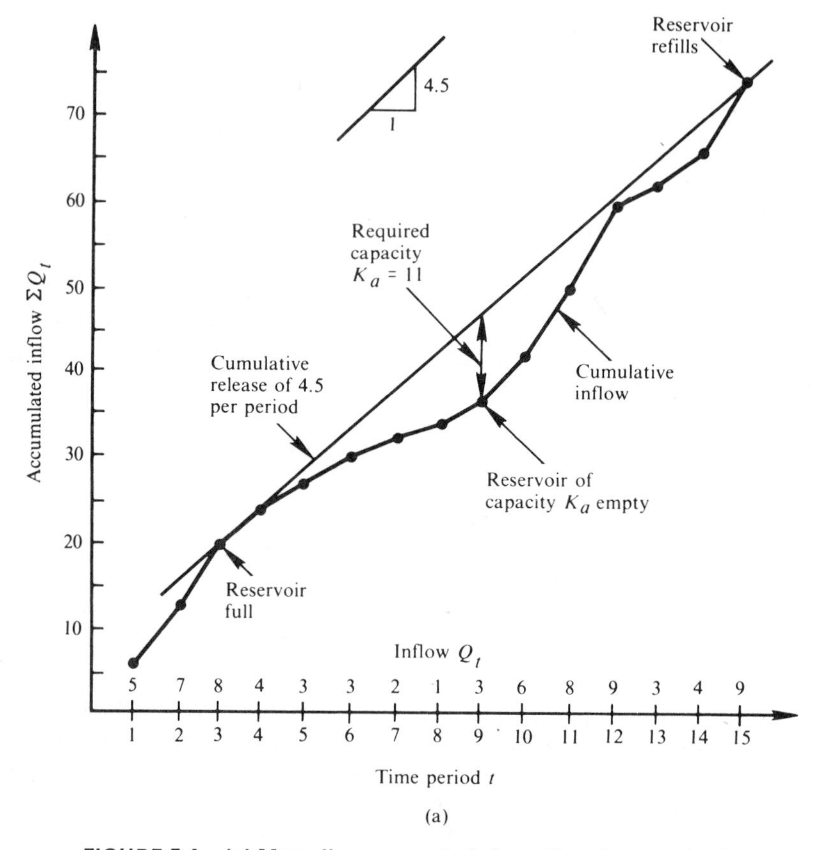

FIGURE 5.4. (a) Mass diagram analysis for estimating required active storage capacities for given inflows and releases.

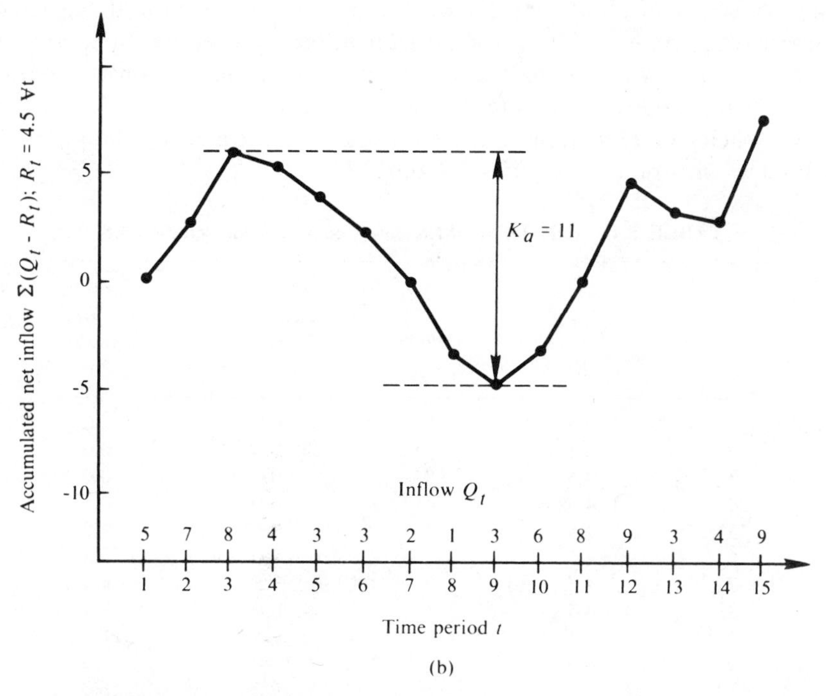

FIGURE 5.4. (b) Alternative mass diagram analysis for estimating required active storage capacities.

difference between Q_t and R_t, as shown in Figure 5.4b. In this case the maximum vertical distance between the highest peak of the cumulative difference curve and lowest valley to the right of this peak represents the required active storage capacity. Both Figures 5.4a and 5.4b assume a repetitive 15-period inflow sequence as shown and constant releases of 4.5 in each period.

5.3.2 Sequent Peak Analyses

The mass diagram method, although in use today, is cumbersome compared to a modification called the *sequent peak procedure* [22]. Let K_t be the storage capacity required at the beginning of period t, R_t be the required release in period t, and Q_t be the inflow. Setting K_0 equal to 0, the procedure involves calculating K_t using equation 5.8 consecutively for up to twice the total length of record. This assumes that the record repeats to take care of the case when the critical sequence of flows occurs at the end of the streamflow record.

$$K_t = \begin{cases} R_t - Q_t + K_{t-1} & \text{if positive} \\ 0 & \text{otherwise} \end{cases} \tag{5.8}$$

The maximum of all K_t is the required storage capacity for the specified releases $\{R_t\}$. Table 5.1 illustrates the sequent peak procedure for computing the active capacity K_a = maximum $\{K_t\}$ required to achieve a constant release $R_t = 4.5$ given the series of streamflows listed in Figure 5.4.

A capacity of 11 units is required, as is demonstrated by the graphical techniques illustrated in Figures 5.2 and 5.3.

TABLE 5.1. Calculation of Storage Capacity from Equation 5.8

Period t	*Release* [R_t	−	*Inflow* Q_t	+	*Previous Required Capacity* K_{t-1}][a]	=	*Current Required Capacity* K_t
1	4.5		5		0.0		0.0
2	4.5		7		0.0		0.0
3	4.5		8		0.0		0.0
4	4.5		4		0.0		0.5
5	4.5		3		0.5		2.0
6	4.5		3		2.0		3.5
7	4.5		2		3.5		6.0
8	4.5		1		6.0		9.5
9	4.5		3		9.5		11.0[b]
10	4.5		6		11.0		9.5
11	4.5		8		9.5		6.0
12	4.5		9		6.0		1.5
13	4.5		3		1.5		3.0
14	4.5		4		3.0		3.5
15[c]	4.5		9		3.5		0

[a] K_t equals the larger of zero and the quantity in brackets.
[b] Maximum required active storage = $K_a = 11$.
[c] The second cycle is identical to the first, since $K_{15} = K_0 = 0.0$

5.3.3 Optimization Analyses

While the mass diagram and sequent peak procedures are relatively simple, the latter even with changing releases R_t, they are not readily adaptable to reservoirs where evaporation losses and/or lake level regulation are important considerations, or to problems involving more than one reservoir. Mathematical programming (optimization) methods provide this capability. These optimization methods are based on mass-balance equations for routing flows through each reservoir. The mass-balance or continuity equations explicitly define storage volumes at the beginning of each period t. A knowledge of reservoir storage volumes is required to compute evaporation and seepage losses as well as to analyze reservoir-based recreation and hydro-electric power development alternatives.

Let S_t be the storage volume (L^3) in the reservoir at the beginning of period t. Continuity or conservation of flow requires that the initial storage volume S_t plus any inflow Q_t less the release R_t and evaporation and seepage losses L_t must equal the final storage volume in that period. This final storage volume is equal to the initial reservoir storage volume in the next period S_{t+1}.

$$S_t + Q_t - R_t - L_t = S_{t+1} \tag{5.9}$$

Evaporation and seepage losses in each period t are a function of the storage volumes in that period. Required for the estimation of evaporation losses is the reservoir storage volume/surface area function, illustrated in Figure 5.5. Also required is the average evaporation rate e_t for each period. Multiplying the average surface area (L^2) times the loss rate e_t (L) yields the volume (L^3) of evaporation loss in the period.

Ignoring seepage, the evaporation loss L_t in each period t may be approximated, as shown in Figure 5.5, by

$$L_t = A_a e_t\left(\frac{S_t + S_{t+1}}{2}\right) + A_0 e_t \tag{5.10}$$

Let

$$a_t = 0.5 A_a e_t \tag{5.11}$$

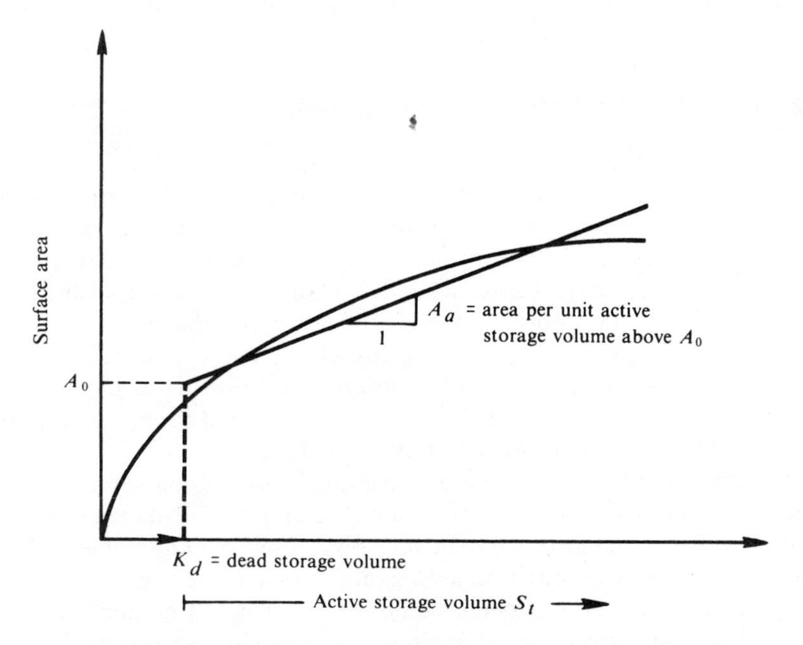

FIGURE 5.5. Storage area relationship and approximation of surface area per unit active storage volume.

Then combining equations 5.9, 5.10, and 5.11 and rearranging terms so that all unknown variables are on the left-hand side of the equation and all known variables are on the right-hand side yields

$$(1 + a_t)S_{t+1} - (1 - a_t)S_t = Q_t - R_t - A_0 e_t \tag{5.12}$$

Seepage losses, where significant, can be incorporated into equation 5.12 in a similar manner.

The initial storage volumes obviously cannot exceed the active reservoir storage capacity K_a:

$$S_t \leq K_a \tag{5.13}$$

The minimum active storage capacity required for a given set of releases R_t may be found by minimizing the active capacity K_a subject to equations 5.12 and 5.13 for all periods t. Setting the unknown initial storage S_1 in period $t = 1$ equal to the unknown final storage S_{T+1} in period T, the last period of record, results in a steady-state solution that is not biased by arbitrarily assumed initial or final storage volumes.

Obviously, if the active storage capacity is zero, then there is no need for dead storage capacity K_d and hence no fixed evaporation loss $A_0 e_t$. Stated another way, if the solution of any model indicates an active storage requirement just due to the fixed evaporation loss, then the correct solution is one in which the active storage capacity and fixed evaporation loss are both zero.

5.4 FLOOD CONTROL ALTERNATIVES

Two types of structural alternatives exist for flood control, namely upstream flood storage capacity in reservoirs that will reduce downstream peak flood flows or channel protection and/or flood-proofing works that will contain peak flood flows and reduce damage. This section introduces methods of modeling both of these alternatives for inclusion in either a benefit–cost analysis or in simply a cost-effectiveness analysis. The latter analyses apply to situations in which a significant portion of the flood control benefits cannot be expressed in monetary terms and the aim is to provide a specified level of flood protection at minimum cost.

The discussion will focus first on the estimation of flood storage capacity in a single reservoir upstream of a single potential flood damage site. This analysis will then be expanded to include several reservoir and potential flood damage sites. Finally, downstream channel capacity improvements will be included. Each of the modeling methods discussed will be appropriate for inclusion in multipurpose river basin planning models, as will be illustrated later in the chapter.

5.4.1 Reservoir Storage for Flood Control

In addition to the active storage capacities in a reservoir, some capacity may be allocated for the temporary storage of flood flows during certain periods in the year. Flood flows usually occur over time intervals lasting from a few hours up to a few days or weeks. Computational limitations discourage these relatively short duration flows from being explicitly defined in the form of continuity constraints similar to equation 5.9 or 5.12. If they were, flood routing equations would also have to be included in the model; a simple mass balance would not be sufficient. Nevertheless, the reservoir flood storage capacities that will provide any prespecified reduction in the magnitude or probability of various downstream flood peaks can be considered and included in monthly or seasonal river basin planning models. Flood routing procedures, together with a knowledge of the flood control operating policy and channel storage characteristics between a reservoir site and a downstream potential damage site, can be used to predict the influence on the downstream flood peaks of various discrete flood storage capacities in an upstream reservoir.

The probability or likelihood of a flood peak of a given magnitude is often described by its return period. The return period T of a flood is the expected number of years before the occurrence of a flood of equal or greater magnitude. The probability that a T-year flood will be exceeded in any given year is $1/T$. If PQ is the random annual peak flood flow and PQ_T is a particular peak flood flow having a return period of T years, then by definition the probability of PQ equaling or exceeding PQ_T is

$$\Pr[\text{PQ} \geq \text{PQ}_T] = \frac{1}{T} \tag{5.14}$$

The exceedence probability distribution just defined is simply 1 minus the distribution function $F_{\text{PQ}}(\cdot)$. Hence for continuous distributions

$$\Pr[\text{PQ} \geq \text{PQ}_T] = 1 - F_{\text{PQ}}(\text{PQ}_T) \tag{5.15}$$

or

$$F_{\text{PQ}}(\text{PQ}_T) = 1 - \frac{1}{T} \tag{5.16}$$

The expected annual flood damage at a potential flood damage site can be estimated from a knowledge of the exceedence probability distribution of peak flood flows and the resulting damage. The peak flow PQ_T at any potential damage site resulting from a flood of return period T will be a function $f_T(\cdot)$ of the upstream reserovir flood storage capacity K_f and the reservoir operating policy.

$$\text{PQ}_T = f_T(K_f) \tag{5.17}$$

Assuming a known operating policy for flood flow releases, the function $f_T(\cdot)$ can be defined by routing a series of floods through the upstream reservoir, having various flood storage capacities K_f and a known operating policy, to the downstream potential damage site. This will require a flood routing simulation model [14, 23].

Associated with any peak flow PQ_T at a potential damage site is a flood stage. This, together with the flow, will determine the extent of flood damage FD_T. Various flood damage functions $g(\cdot)$ can be derived from field surveys at the potential damage site [12].

$$FD_T = g(PQ_T) \tag{5.18}$$

The probability that flood damage of FD_T will be exceeded is precisely the same as the probability that the peak flow PQ_T that causes the damage will be exceeded. Letting FD be a random flood damage variable,

$$\Pr[FD \geq FD_T] = \frac{1}{T} \tag{5.19}$$

The expected annual flood damage $E[FD]$ can be shown to equal the area under the exceedence probability distribution.

$$E[FD] = \int_0^\infty \Pr[FD \geq FD_T]\, dFD_T \tag{5.20}$$

The expected annual flood damage can also be derived graphically. Figure 5.6a illustrates the information that can result from routing of flood flows, corresponding to three different return periods T, through a reservoir with various flood storage capacities. These are the functions f_T that are needed to eventually compute expected flood damages, as shown in Figure 5.6b.

The relationships between flood stage and damage, and flood stage and peak flow, defined in quadrants (1) and (2) of Figure 5.6b, must be known. The information in quadrant (3) is derived from information similar to that shown in Figure 5.6a and a knowledge of the exceedence probabilities of each peak flow. Then the probability of exceeding a specified level of flood damage in quadrant (4) can be derived by the graphical procedure illustrated. The areas under these exceedence probability distributions in quadrant (4) are the expected flood damages associated with, in this case, three particular flood storage capacities. This provides three points on the function shown in Figure 5.6c. This function is simply the difference between the expected flood damage with no upstream flood storage capacity and the expected flood damage given various upstream flood storage capacities. The eventual trade-off is between the expected flood reduction benefits as defined by Figure 5.6c and the costs of upstream reservoir flood storage capacity.

Total reservoir capacity K can be calculated as the sum of dead storage capacity K_d, active storage capacity K_a and flood storage capacity K_f. However, since the required active storage capacity varies throughout the year and flood storage may not be required in all seasons, a less conservative estimate of the total capacity K will be the sum of dead storage plus the maximum

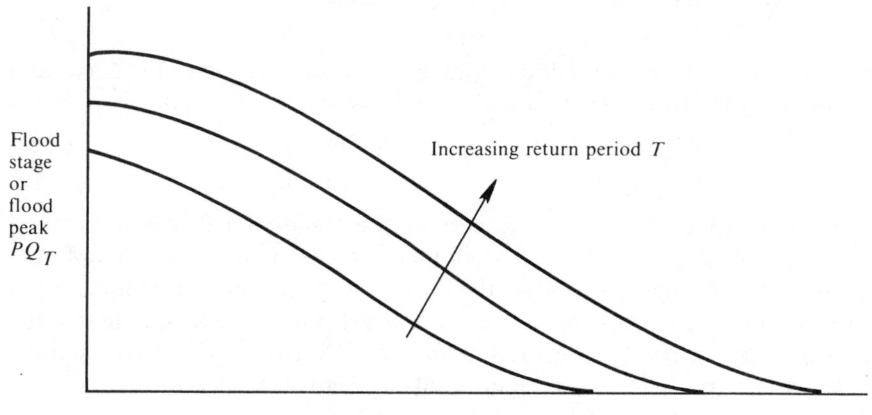

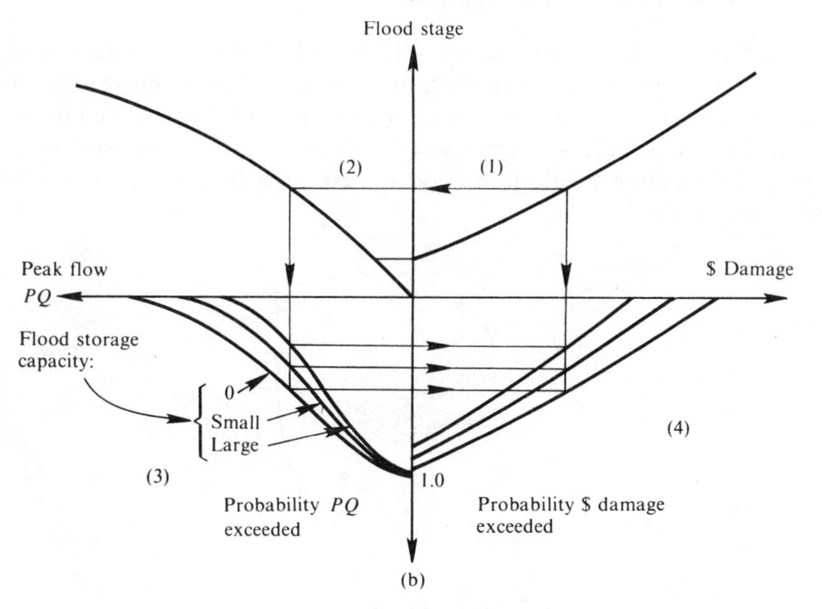

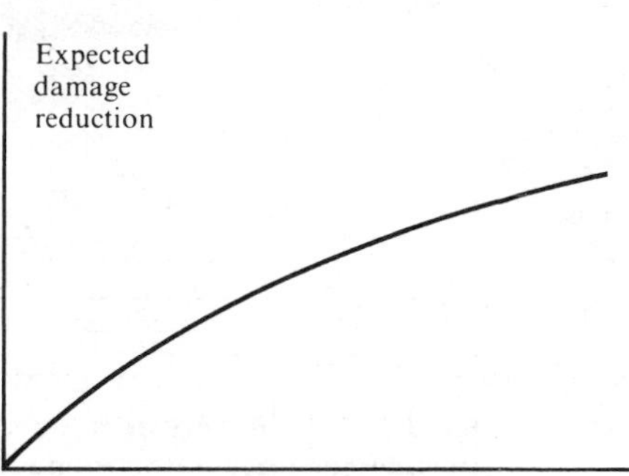

FIGURE 5.6. (a) Results of flood simulation studies to determine effect of upstream flood storage capacity on downstream flood peak. (b) Calculation of probability of flood damage exceedence as a function of reservoir flood storage capacity. (c) Expected damage reduction as a function of flood storage capacity derived from Fig. 5.6(b) in quadrant four.

required active storage and flood storage in the flood season or the maximum required active storage in nonflood seasons, whichever is greater.

$$K \geq K_d + S_t + K_f \qquad \forall t \text{ in flood season} \tag{5.21}$$

$$K \geq K_d + S_t \qquad \forall t \text{ not in flood season} \tag{5.21a}$$

This definition of total capacity permits the trade-off between active and flood storage capacity when and if there can be a trade-off in different seasons. In the equations above, the dead storage capacity is a known variable. It is included in the capacity equations 5.21 and 5.21a assuming that the active storage capacity is greater than zero. Clearly, if the active storage capacity is zero, there would be no need for dead storage.

5.4.2 Equivalent Flood Storage Capacities for Multireservoir Systems

So far this discussion of flood control has applied only to a single upstream reservoir site and a single downstream potential flood damage site. Now consider several reservoir and potential damage sites. Figure 5.7 illustrates such a planning problem and will serve to demonstrate an approximate procedure of estimating the desired flood storage capacities, if any, in multiple reservoir systems.

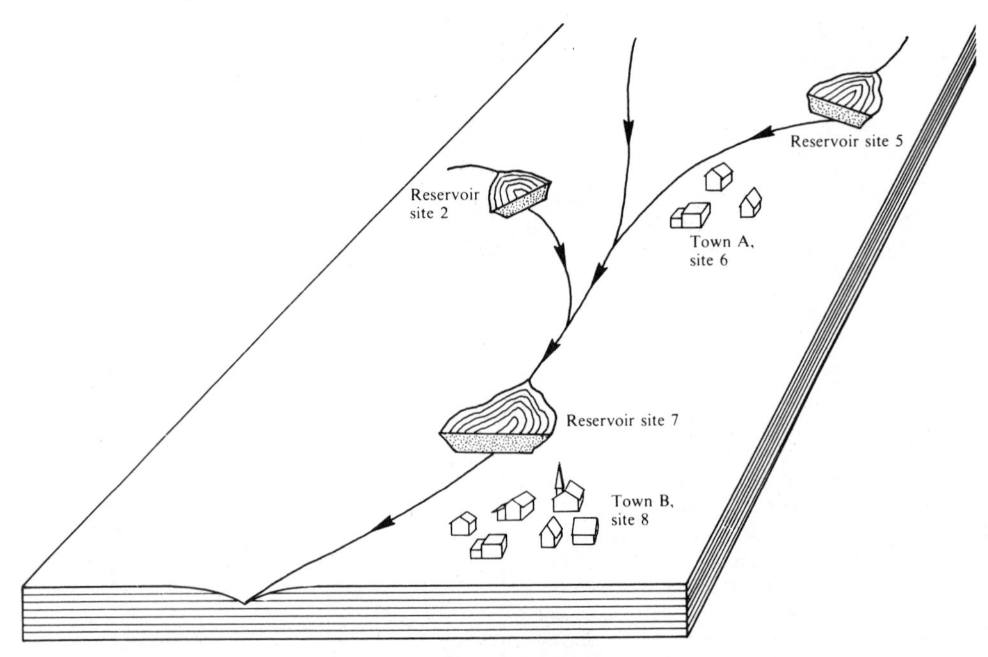

FIGURE 5.7. Reservoir flood storage alternatives and potential damage sites in the Helga River valley.

As illustrated in Figure 5.7, reservoir capacity at site 5 can be used to reduce the flood peaks at town A, and all three upstream reservoirs can be used to reduce the flood peaks at town B. The analysis of the flood peak reduction at town A that might result from storage capacity in reservoir 5 is similar to that just discussed. Of particular interest in this section is the allocation of flood storage capacities in all three upstream reservoirs for peak flow reduction at town B.

For the preliminary estimation of flood storage capacities in multiple-reservoir systems, a single design flood can be selected. Using this design flood, the system may be simulated to determine the effect that any combination of flood storage capacities has on reducing the flood peak at town B. The difficulty is in obtaining a simple expression for this effect as a function of all three upstream reservoir capacities. Hence it is useful to introduce the concept of an equivalent flood storage capacity at a fictitious or actual reservoir site just upstream of the potential damage site. In this example, the equivalent flood storage capacity EK_f^7 at reservoir site 7 will be that capacity which reduces the flood peak at town B the same amount as the actual flood storage capacities K_f^s in the three reservoirs at sites $s = 2$, 5, and 7.

To obtain the equivalent reservoir capacity at reservoir site 7 that reduces the downstream flood peak at town B (site 8) by the same amount as a given flood storage capacity at either upstream reservoir sites 2 or 5, the design flood must be routed through various storage capacities at site 2, assuming no capacity at site 5, and through various storage capacities at site 5, assuming no capacity at site 2. Comparing these simulations with those for various flood storage capacities only at site 7 will result in the functional relationships illustrated in Figure 5.8.

Figure 5.8a is a plot of the design peak (T-year) flood flow reduction QS_T^8 at site 8 associated with various flood storage capacities K_f^s at upstream reservoir sites $s = 2$ or 5, assuming that one or the other of the upstream reservoirs might be built, but not both. These peak flow reduction functions are denoted as $F_{s,8}(K_f^s)$. Figure 5.8b is a plot of the peak flood flow reduction at site 8, associated with various flood storage capacities at reservoir site 7 only, denoted as $F_{7,8}(K_f^7)$. These peak reduction functions $F_{s,8}(K_f^s)$ can be derived from the simulations represented in Figure 5.6a for a particular design flood of return period T.

Figure 5.8c merely translates the axes of that quadrant using a straight line. Thus Figure 5.8d is a plot of the flood storage capacities K_f^7 at reservoir site 7 that are equivalent to (reduces the T-year flood peak at site 8 the same amounts as do) various flood storage capacities K_f^s at upstream reservoir sites $s = 2$ or 5. The graphical procedure illustrated in Figure 5.8 for deriving these equivalent capacities at reservoir site 7 is simply a method of first equating the downstream peak reduction function $F_{7,8}(K_f^7)$ with each of the upstream functions $F_{s,8}(K_f^s)$ and then solving for K_f^7. Hence to find equivalent

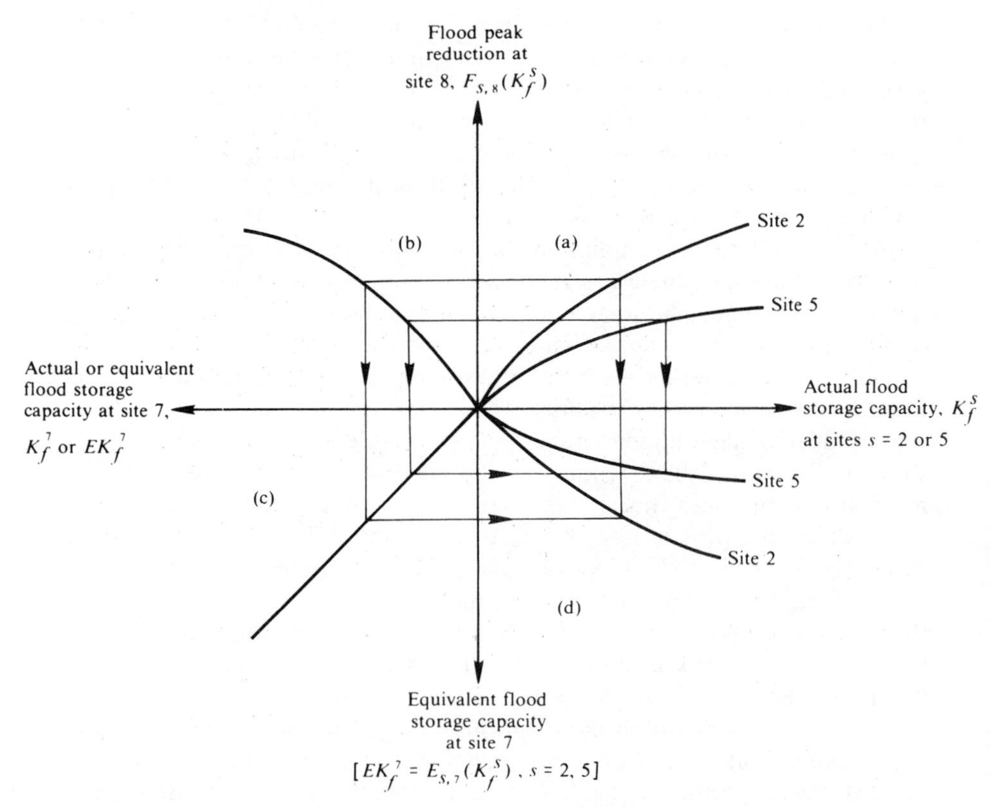

FIGURE 5.8. Derivation of equivalent flood storage capacity functions for a *T*-year flood in the Helga River valley.

flood storage capacities at site 7 for those at sites $s = 2$ and 5, one sets

$$F_{7,8}(K_f^7) = F_{s,8}(K_f^s) \tag{5.22}$$

to obtain

$$K_f^7 = F_{7,8}^{-1}[F_{s,8}(K_f^s)] = E_{s,7}(K_f^s) \tag{5.22a}$$

The function $E_{s,7}(K_f^s)$ defines the flood storage capacity at site 7 that is equivalent to the flood storage capacity at an upstream site s. It is based on the peak flood flow reduction at site 8 of a design T-year flood.

A first estimate of the total equivalent flood storage capacity EK_f^7 at reservoir site 7 is the sum of the individual equivalent capacities and the actual flood storage capacity, if any, at site 7.

$$EK_f^7 = E_{2,7}(K_f^2) + E_{5,7}(K_f^5) + K_f^7 \tag{5.23}$$

Clearly, the effectiveness of the reservoirs in reducing the flood peak at Town B (site 8) will depend on the operating policy of each flood control reservoir and the timing of releases from the upstream reservoirs. It is unlikely that the linear relationship defined by equation 5.23 will be a perfect estimate of the total equivalent capacity at reservoir site 7. But there is no simple way to determine what will be a good simple estimate without knowing the flood storage capacity K_f^s at sites $s = 2$, 5, and 7. Thus after a first solution is obtained, based on equation 5.23, the actual equivalent capacity at site 7 should be evaluated given the computed values of each capacity K_f^s. Let the ratio of the estimated equivalent capacity EK_f^7 to the actual equivalent capacity be α. This parameter should be used to scale the right-hand side of equation 5.23 when an improved new solution is obtained.

$$EK_f^7 = \alpha[E_{2,7}(K_f^2) + E_{5,7}(K_f^5) + K_f^7] \qquad (5.23a)$$

This process can be repeated until the change in α from one solution to the next is negligible.

The flood storage capacities K_f^s may be conservative estimates of the capacities needed to reduce flood peaks downstream if active storage capacity or dead storage capacity is included in the total reservoir capacity. Existing lakes will tend to increase channel widths and some flood storage may be provided by unused active reservoir storage capacity at the time of a flood. Both of these conditions result in reduced flood peaks downstream, even without additional flood storage capacity at the reservoir sites [15].

5.4.3 Flood Protection at Potential Damage Sites

In the previous sections on flood damage reduction, only reservoir flood storage capacities were considered. In this section the possibility of levees or dikes and other channel capacity or flood-proofing improvements at a downstream potential damage site will be included in the analysis. The approach will provide a means of estimating combinations of flood control storage capacity in upstream reservoirs and downstream channel capacity improvements that together will provide a prespecified level of flood protection at the downstream site. Nonstructural flood damage reduction measures will not be considered here, although certainly in any comprehensive floodplain planning study both structural and nonstructural alternatives should be evaluated [13].

The unregulated natural peak flow of a particular design flood at a potential flood damage site can be reduced by upstream reservoir flood storage capacity or it can be contained within the channel at the potential damage site by levees and other channel-capacity improvements. Let QN_T^s be the unregulated natural peak flow in the flood season having a return period of T years. Assume that this peak flood flow is the design flood for which protection is desired. To protect site s from this design peak flow, a portion QS_T^s of the peak flow at site s may be reduced by upstream flood storage capacity. The remainder of the peak flow QR^s must be contained within the channel. Hence at damage site s one requires that

$$QN_T^s \leq QS_T^s + QR^s \tag{5.24}$$

The reservoir flood storage capacity upstream of site s required to reduce the peak flow of any particular flood can be obtained by simulation, as illustrated in Figure 5.9. The required flood storage capacity K_f will be a function $K_{fT}^s(QS_T^s)$ of the required peak flow reduction QS_T^s. The functions $K_{fT}(QS_T^s)$ are dependent on the relative locations of the reservoir and the downstream potential damage site s, on the characteristics and length of the channel between the reservoir and downstream site s, and on the magnitude of the peak flood flow.

An objective function for evaluating these two structural flood control measures should include the cost of reservoir capacity and the cost of channel capacity improvements required to contain a flood flow of QR^s at site s. For a single-purpose, single-damage-site, single-reservoir flood control problem, the minimum total cost required to protect site s from a design flood peak of QN_T^s may be obtained by solving the model:

$$\text{minimize } \text{Cost}_K(K_f) + \text{Cost}_R(QR^s) \tag{5.25}$$

subject to

$$QN_T^s \leq QS_T^s + QR^s \tag{5.24}$$

$$K_f \geq K_{fT}^s(QS_T^s) \tag{5.26}$$

Again, equation 5.24 in the planning model assumes that a decision will be

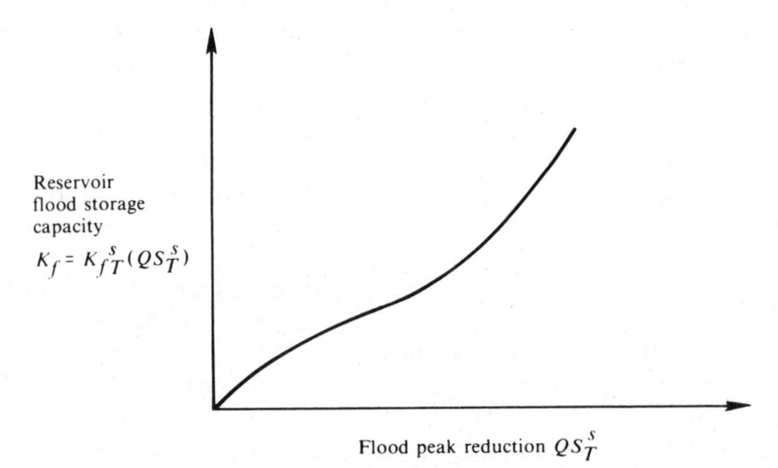

FIGURE 5.9. Functional relationship between upstream flood storage capacity and downstream flood peak reduction.

made to provide protection from a design flood QN_T^s of return period T; it is only a question of how to provide the required protection.

It is possible to extend this cost-effective approach to a benefit–cost analysis that considers more than a single design flood protection level at the potential damage site s. To do this, estimates must be made of the expected annual benefits BF_T^s that would be achieved at site s if it were protected from floods of various return periods T. (Such procedures were discussed in Section 5.4.1.) Next a 0, 1 integer variable X_T^s can be defined that will identify the most beneficial of all the discrete design flood protection levels considered. The variable X_T^s must be 0 for all but the most beneficial flood protection level, and for this level T^* the value of $X_{T^*}^s$ will be 1. Assuming that it is possible that no flood protection may be beneficial, the conditions on X_T^s are

$$\sum_T X_T^s \leq 1 \tag{5.27}$$

and

$$X_T^s = \text{integer} \qquad \forall T$$

This integer variable X_T^s can be multiplied by the estimated annual expected flood damage reduction benefits BF_T^s and then these terms $BF_T^s X_T^s$ can be included in the objective function 5.25. Furthermore, equation 5.24 must be modified and written for each return period T:

$$QN_T^s X_T^s \leq QS_T^s + QR^s \qquad \forall T \tag{5.24a}$$

Hence, the single-site single-purpose flood protection model can be written as follows:

$$\text{maximize} \sum_T (BF_T^s X_T^s) - \text{Cost}_R(QR^s) - \text{Cost}_K(K_f) \tag{5.25a}$$

subject to

$$QN_T^s X_T^s \leq QS_T^s + QR^s \qquad \forall T \tag{5.24a}$$

$$K_f \geq K_{fT}(QS_T^s) \qquad \forall T \tag{5.26a}$$

$$\sum_T X_T^s \leq 1 \tag{5.27}$$

$$X_T^s = \text{integer} \qquad \forall T$$

This model is a mixed-integer programming model. It can be made piecewise linear for solution by linear mixed-integer programming methods discussed in Chapter 2.

For multiple-damage-site problems, these constraints can be repeated for each site using the appropriate equivalent capacities EK_f in place of actual capacities K_f and assuming that the channel improvements at upstream sites will not significantly alter the flood peaks at downstream sites. If upstream channel improvements do significantly alter the downstream flood peaks, this preliminary screening approach (that considers both reservoir storage and channel improvements for flood damage reduction) may not be very effective.

5.5 HYDROELECTRIC POWER PRODUCTION

The production of hydroelectric energy during any period at any particular reservoir site is dependent on the installed plant capacity; the flow through the turbines; the average productive storage head; the number of hours in the period; the plant factor (to be defined shortly); and a constant for converting the product of flow, head, and plant efficiency to kilowatt-hours of electrical energy. The kilowatt-hours of energy KWH_t produced in period t are proportional to the product of the plant efficiency ϵ, the productive storage head H_t, and the flow q_t through the turbines.

A cubic meter of water, weighing 10^3 kg, falling a distance of 1 meter, acquires 9.81×10^3 joules (newton-meters) of kinetic energy. The energy generated in 1 second equals the watts (joules per second) of power produced. Hence an average flow of $\hat{q}_t$ cubic meters per second falling a height of H_t meters in period t yields $9.81 \times 10^3 \hat{q}_t H_t$ watts or $9.81 \hat{q}_t H_t$ kilowatts of power. Multiplying by the number of hours in period t yields the kilowatt-hours of energy produced from an average flow rate of $\hat{q}_t$. The total kilowatt-hours of energy KWH_t produced in period t assuming 100% efficiency in conversion of potential to electrical energy is

$$\text{KWH}_t = \frac{9.81 \hat{q}_t H_t \text{ (seconds in period } t)}{3.6 \times 10^3} \tag{5.28}$$

Since the *total flow* q_t in period t, in units of 10^6 m^3, equals the average *flow rate* $\hat{q}_t$ m^3/s times the number of seconds in the period divided by 10^6, the total kilowatt-hours of energy produced in period t given a plant efficiency of ϵ equals

$$\text{KWH}_t = 2730 q_t H_t \epsilon \tag{5.28a}$$

If the flow is measured in units of acre-feet and the head in feet, the same derivation procedure leads to

$$\text{KWH}_t = 1.024 q_t H_t \epsilon \tag{5.28b}$$

If one is to use a linear programming algorithm for solution of a model with hydroelectric power production capabilities, then the nonlinear relationships in equation 5.28a or 5.28b involving the product of head and flow must be replaced by a linear approximation. If the average heads H_t^0 and flows q_t^0 can be estimated for each period t, then these fixed constants can be used to obtain a linear approximation of the flow-head product term,

$$\begin{aligned} q_t H_t &\simeq q_t^0 H_t^0 + q_t^0 (H_t - H_t^0) + (q_t - q_t^0) H_t^0 \\ &= q_t^0 H_t + H_t^0 q_t - q_t^0 H_t^0 \end{aligned} \tag{5.29}$$

The model may need to be solved several times in order to identify reasonably accurate average flow and head estimates so that $q_t \simeq q_t^0$ and $H_t \simeq H_t^0$. Once

q_t^0 and H_t^0 are reasonably accurate, the linear approximation (equation 5.29) is valid.

The amount of electrical energy produced is limited by the installed kilowatts of plant capacity P as well as on the plant factor p_t. The plant factor is a measure of hydroelectric power plant use and is usually dictated by the characteristics of the power system supply and demand. It is defined as the average load on the plant for the period divided by the installed plant capacity. The plant factor accounts for the variability in the flow rate during each period t, and this variability must be specified by those responsible for energy production and distribution. It may or may not vary for different periods t. The total energy produced cannot exceed the product of the plant factor p_t, the number of hours in the period h_t, and the plant capacity P, measured in kilowatts.

$$\text{KWH}_t \leq p_t h_t P \tag{5.30}$$

5.6 WITHDRAWALS AND DIVERSIONS

Major demands for the withdrawal of water include those resulting from domestic or municipal uses, industrial uses (including cooling water), and agricultural uses such as irrigation. These uses require the withdrawal of water from a river system or other water body. The water withdrawn may be only partially consumed, and that which is not consumed may be returned to the river system, perhaps at a different site, at a later time period or with a different quality.

Water can also be allocated to instream uses that alter the distribution of flows in time and space. Such uses include (1) reservoir storage, possibly for recreational use as well as for water supply; (2) flow augmentation, possibly for water quality control or for navigation; and (3) hydroelectric power production. The instream uses may complement or compete with each other or with various municipal, industrial, and agricultural demands. One purpose of developing management models of river basin systems is to help derive policies that will best serve these multiple uses.

The allocated flow q_t^s to a particular use at site s in period t must be no greater than the total flow available Q_t^s at that site and in that period.

$$q_t^s \leq Q_t^s \tag{5.31}$$

The quantity of water that any particular user expects to receive in each particular period, is termed the *target allocation*. Given an annual (known or unknown) target allocation T^s at site s, some (usually known) fraction f_t^s of that annual target allocation will be expected in period t. If the actual allocation q_t^s is less than the target allocation $f_t^s T^s$, there will be a deficit D_t^s. If the allocation is greater than the target allocation, there will be an excess E_t^s. Hence

$$q_t^s = f_t^s T^s - D_t^s + E_t^s \tag{5.32}$$

where either D_t^s or E_t^s or both are zero. Whether or not any deficit or excess allocation should be allowed at site s depends on the quantity of water available and the losses associated with deficit or excess allocations to that site. At sites where the benefits derived in each period are independent of the allocations in other periods, the losses associated with deficits and the losses or benefits associated with excesses can be defined in each period t. For example, the benefits derived from the allocation of q_t^s for hydropower production in period t in some cases will be essentially independent of previous allocations.

For any use in which the benefits derived from a sequence of allocations are not independent, such as at irrigation sites, the measure of benefits may be based on the annual (or growing season) target allocations T^s if the particular distribution of within-year (or within-growing-season) allocations $f_t^s T^s$ is specified and assured by constraints, such as

$$q_t^s \geq f_t^s T^s \qquad \text{for all relevant } t \tag{5.33}$$

If, for any reason, an allocation q_t^s must be zero, then clearly from equation 5.33 the annual or growing season target allocation T^s would be zero.

The average flow Q_t^{s+} just downstream of the withdrawal site will equal the original flow Q_t^s at that site, less the quantity withdrawn q_t^s plus any fractions $\rho_{\tau,t}^s$ of the amount withdrawn in the current and previous periods τ that is returned to the stream during the current period t. Assuming that all the return flows enter just downstream of the point of withdrawal,

$$Q_t^{s+} = Q_t^s - q_t^s + \sum_{\tau<t}^{t} \rho_{\tau,t}^s q_\tau^s \tag{5.34}$$

Water stored in reservoirs can often be used to augment downstream flows for instream uses such as recreation, navigation, and water quality control. During natural low-flow periods in the summer season, it is not only the increased volume but also the lower temperature of the augmented flows that may provide the only means of maintaining certain species of fish and other aquatic life. Dilution of wastewater or runoff from nonpoint sources is another potential benefit from flow augmentation. These and other factors related to water quality management are discussed in greater detail in Chapter 9.

The benefits derived from navigation on a potentially navigable portion of a river system can usually be expressed as a function of the stage or depth Z_t of water in various periods t. Assuming known streamflow-stage relationships $Z_t^s(Q_t^s)$ at various sites s in the river, a possible constraint might require at least a minimum acceptable depth $Z_t^{\min}$ for all sites s.

$$Z_t^s(Q_t^s) \geq Z_t^{\min} \qquad \forall t, s \tag{5.35}$$

Stage-flow relationships $Z_t^s(Q_t^s)$ are concave functions of the streamflow Q_t^s and can be made piecewise linear for inclusion in linear optimization models.

Either the navigation benefits expressed as functions of the minimum design water depth $Z_t^{\min}$ can be included in the objective function of the river basin planning model, or some minimum acceptable depth $Z_t^{\min}$ may be specified. If the latter approach is used, it is equivalent to specifying some minimum acceptable flow $Q_{t,s}^{\min}$ at each site s in each period t. A more detailed analysis of navigation benefits and operating policies may be necessary, especially when lockages are required. If investments in locks are necessary, their fixed costs as a function of the design depth of the channel, together with any expected operating costs at each lock site s, should be included in the planning model.

5.7 MODEL SYNTHESIS

At this point, each of the model components already discussed are combined into a deterministic, static, multipurpose, multisite, multiobjective river basin planning model for the Helga River valley, shown with various water uses and water management alternatives in Figure 5.10.

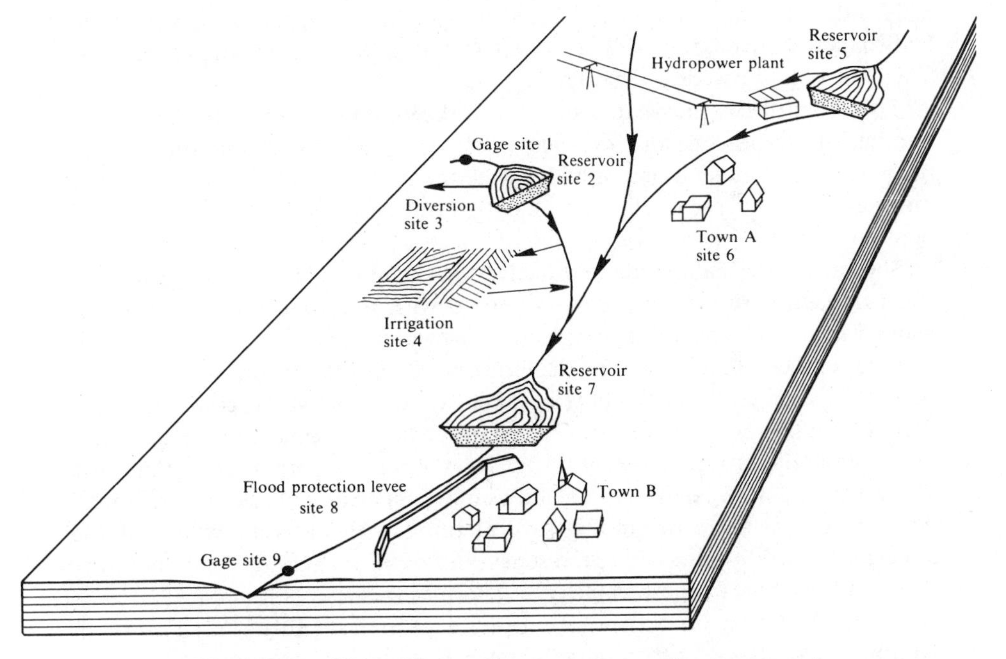

FIGURE 5.10. Helga River valley water use and management alternatives.

As illustrated in Figure 5.10, the Helga River valley has two streamflow gages (at sites 1 and 9), two towns (at sites 6 and 8), three potential reservoirs (at sites 2, 5, and 7), a diversion (to site 3) out of the basin (from site 2), a potential irrigation area (at site 4), and a potential hydropower plant (at reservoir site 5). The water targets for the two towns are known, but the target allocations to all other uses are to be determined along with the actual allocations to all uses, the reservoir capacity requirements, and any channel improvement for flood protection at Town B (site 8).

5.7.1 Number and Duration of Time Periods

Before beginning the development of the river basin model, one must decide the number of time periods t to include in the model and the length of each period. If a reservoir system is to contain storage for distribution of water between years, called over-year storage, then a large number of periods encompassing several years of operation must be included in the model to allow for the evaluation of the possible benefits of storing excess water in wet years for release in dry years.

Many reservoir systems completely fill almost every year, and in such cases one is only concerned with the within-year operation of the system. This is the problem addressed here, although this model can be easily modified to include overyear storage as well as within-year if and when necessary. This will be discussed in Chapter 7.

To model the within-year operation of the system, a year is broken into a number of periods of equal or unequal length. The number and duration of the periods will depend on the hydrology, the particular objectives, and computer capacity, as previously discussed. In this case 12 monthly periods will be used.

Several choices can be made for selecting the unregulated flows Q_t^s at each site s in each period t. If one is concerned with the maximum quantities of water for agriculture, for hydroelectric energy, and for stream quality control that can be obtained during droughts or years with unusually low natural flows, the selected flows Q_t^s may be taken to be the flows observed in a particular drought or critical year. On the other hand, if one is concerned with the average operation of the system in an average or normal year, the flows Q_t^s may be taken to be the average flows in each period t at each site s. It may be desirable to solve the model several times with different values of the selected flows Q_t^s to determine the sensitivity of various design and operating decision variables to these values.

In the Helga River valley, the flood season lasts from April through October, periods $t = 4$ through 10. Town B is a potential flood damage site which is to be protected from a flood up to the magnitude expected once every 200 years. Since this is to be a static analysis of a single year sometime in the future, the reservoir storage volumes at the beginning and end of the year will be undetermined, but equal to one another.

5.7.2 Objective Functions

One objective of water resources development in the Helga River valley will be to increase the net benefits derived from the water resources and to achieve an improved distribution of those benefits within the region. Let the net benefits that each interest group i receives from a project at each site s be denoted as NB_i^s. The net benefits NB_i^s that each group i receives from each site s are defined by the benefits they derive, if any, from the target allocation T^s to each water use less any losses and costs. Losses may result from actual allocations that do not meet the target allocation. Costs will include what must be paid for each water use or project. Hence for all groups i and sites s, where appropriate,

$$\text{NB}_i^s = B_i^s(T^s) - \sum_t [L_{it}^s(D_t^s) - G_{it}^s(E_t^s)] - C_i^s(K^s) \tag{5.36}$$

where

$B_i^s(T^s)$ = net benefits to group i from an annual target allocation of T^s at site s

$L_{it}^s(D_t^s)$ = losses to group i from an allocation deficit of D_t^s at site s in period t

$G_{it}^s(E_t^s)$ = gain (or loss, if negative) to group i from an excess allocation E_t^s at site s in period t

$C_i^s(K^s)$ = annual cost to group i of the capacity K^s (or P^s) of the water resources project at site s (includes capital and OMR costs)

NB_i^s = total net benefit (which may be positive or negative) to group i, derived from site s

At the reservoir and hydropower site 5, the costs of both the reservoir storage and hydropower plant capacities will be included in equation 5.36, as will be the benefits and losses associated with the storage of water and the production of hydroelectric power. The cost of hydropower plant capacity will also be a function of the maximum head (as illustrated in Figure 5.11). This cost function can be estimated for an assumed maximum head and later adjusted based on the computed maximum head.

$$\begin{aligned}\text{NB}_i^5 = {} & B_i^{S5}(T^{S5}) - \sum_t [L_{it}^{S5}(D_t^{S5}) - G_{it}^{S5}(E_t^{S5})] - C_i^{S5}(K^5) \\ & + B_i^{p5}(T^{p5}) - \sum_t [L_{it}^{p5}(D^{p5}) - G_{it}^{p5}(E_t^{p5})] - C_i^{p5}(P^5)\end{aligned} \tag{5.37}$$

where the superscript S represents the storage reservoir and the superscript p represents the hydropower plant. The annual recreation storage volume target is T^{S5} and the annual firm energy production target is T^{p5} kWh. The reservoir capacity is denoted as K^5 and the hydropower plant capacity is P^5.

The level of development at each project site s will influence the distribution of net benefits among the various interest groups. Hence, as discussed in Chapter 4, alternative distributions of net benefits can be obtained with the use of weights w_i associated with each interest group i. The overall objective will be to maximize the total weighted net benefits:

$$\text{maximize} \sum_i w_i \sum_s \text{NB}_i^s \tag{5.38}$$

The net benefits will be computed from solving the model, but the weights must be estimated, and varied, in an effort to define the relative trade-offs between one interest group and another. These different solutions can then be presented as efficient alternatives to those responsible for planning and managing the water resources in the basin. The alternatives are efficient in the sense that in order to increase one interest group's net benefits, it will be necessary to decrease the net benefits of other interest group(s). This multi-objective approach identifies the trade-offs between the various interest groups. It is left to the political decision-making process to identify an acceptable compromise.

5.7.3 Model Constraints

The constraints, in addition to equations 5.36 and 5.37, associated with each site are listed below. Recall that for this static analysis, for period $t = 12$, period $t + 1 = 1$. The unknown variables will be included in the left-hand side of each constraint and the known variables on the right-hand side.

Site 2. Reservoir and diversion. For each monthly period t in the year:

$$\begin{aligned}
&(1 + a_t^2)S_{t+1}^2 - (1 - a_t^2)S_t^2 + R_t^2 + R_t^3 \\
&\begin{bmatrix}\text{final active}\\ \text{storage volume,}\\ \text{adjusted for}\\ \text{losses in}\\ \text{period } t\end{bmatrix} - \begin{bmatrix}\text{initial}\\ \text{storage}\\ \text{volume,}\\ \text{adjusted}\\ \text{for losses}\end{bmatrix} + \begin{bmatrix}\text{reservoir}\\ \text{release,}\\ \text{excluding}\\ \text{diversion}\end{bmatrix} + \begin{bmatrix}\text{reservoir}\\ \text{release to}\\ \text{diversion}\\ \text{at site 3}\end{bmatrix} \\
&= Q_t^2 - L_0^2 \\
&= \begin{bmatrix}\text{unregulated}\\ \text{inflow}\end{bmatrix} - \begin{bmatrix}\text{fixed}\\ \text{volume}\\ \text{loss}\\ \text{(possibly 0)}\end{bmatrix}
\end{aligned} \tag{5.39}$$

$$K^2 \quad - \quad S_t^2 \quad - \quad K_f^2 \quad \geq \quad K_d^2 \tag{5.40}$$

$$\begin{bmatrix}\text{total}\\ \text{capacity}\end{bmatrix} - \begin{bmatrix}\text{initial}\\ \text{active}\\ \text{storage}\\ \text{volume}\end{bmatrix} - \begin{bmatrix}\text{flood storage}\\ \text{capacity for}\\ t = 4 \text{ to } 11;\\ 0 \text{ otherwise}\end{bmatrix} \geq \begin{bmatrix}\text{dead}\\ \text{storage}\\ \text{capacity}\\ (0 \text{ if all } S_t = 0)\end{bmatrix}$$

$$S_t^2 \quad - \quad f_t^2 T^2 \quad + \quad D_t^2 \quad - \quad E_t^2 \quad = 0 \tag{5.41}$$

$$\begin{bmatrix}\text{initial}\\ \text{storage}\end{bmatrix} - \begin{bmatrix}\text{storage}\\ \text{target}\end{bmatrix} + \begin{bmatrix}\text{deficit}\\ \text{storage}\end{bmatrix} - \begin{bmatrix}\text{excess}\\ \text{storage}\end{bmatrix} = 0$$

Site 3. For each period t:

$$R_t^3 \quad - \quad f_t^3 T^3 \quad + \quad D_t^3 \quad = 0 \tag{5.42}$$

$$\begin{bmatrix}\text{diversion}\end{bmatrix} - \begin{bmatrix}\text{diversion}\\ \text{target}\end{bmatrix} + \begin{bmatrix}\text{deficit}\end{bmatrix} = 0$$

Site 4. Irrigation area. For each period t:

$$q_t^4 \quad - \quad R_t^2 \quad \leq \quad Q_t^4 \quad - \quad Q_t^2 \tag{5.43}$$

$$\begin{bmatrix}\text{allocation}\\ \text{to irrigation}\\ \text{area}\end{bmatrix} - \begin{bmatrix}\text{reservoir}\\ \text{release}\\ \text{at site 2}\end{bmatrix} \leq \begin{bmatrix}\text{unregulated}\\ \text{flow at}\\ \text{site 4}\end{bmatrix} - \begin{bmatrix}\text{unregulated}\\ \text{flow at}\\ \text{site 2}\end{bmatrix}$$

$$q_t^4 \quad - \quad f_t^4 T^4 \quad \geq 0 \tag{5.44}$$

$$\begin{bmatrix}\text{allocation}\\ \text{to irrigation}\\ \text{area}\end{bmatrix} - \begin{bmatrix}\text{target}\\ \text{allocation}\\ \text{in period}\end{bmatrix} \geq 0$$

Note that at this site the annual costs of the irrigation area can be expressed as a function of the total target allocation, and hence $B_i^4(T^4)$ are the net benefits derived from a target allocation of T^4 units of water.

Site 5. Reservoir and hydropower. For each period t:

$$(1 + a_t^5)S_{t+1}^5 - (1 - a_t^5)S_t^5 + \quad R_t^5 \quad + \quad q_t^5$$

$$\begin{bmatrix}\text{final}\\ \text{storage}\\ \text{volume}\\ \text{adjusted}\\ \text{for losses}\\ \text{in period } t\end{bmatrix} - \begin{bmatrix}\text{initial}\\ \text{storage}\\ \text{volume}\\ \text{adjusted}\\ \text{for losses}\\ \text{in period } t\end{bmatrix} + \begin{bmatrix}\text{release}\\ \text{to}\\ \text{stream}\end{bmatrix} + \begin{bmatrix}\text{release}\\ \text{to}\\ \text{hydropower}\\ \text{plant}\end{bmatrix} \tag{5.45}$$

$$= \quad Q_t^5 \quad - \quad L_0^5$$

$$= \begin{bmatrix}\text{unregulated}\\ \text{inflow}\end{bmatrix} - \begin{bmatrix}\text{fixed}\\ \text{volume}\\ \text{losses}\\ (0 \text{ if all } S_t = 0)\end{bmatrix}$$

$$K^5 - S_t^5 - K_f^5 \geq K_d^5 \tag{5.46}$$

$$\begin{bmatrix}\text{total}\\ \text{capacity}\end{bmatrix} - \begin{bmatrix}\text{initial}\\ \text{storage}\\ \text{volume}\end{bmatrix} - \begin{bmatrix}\text{flood}\\ \text{storage}\\ \text{capacity}\\ \text{for periods}\\ t = 4 \text{ to } 11;\\ 0 \text{ otherwise}\end{bmatrix} \geq \begin{bmatrix}\text{dead}\\ \text{storage}\\ \text{capacity}\\ (0 \text{ if all } S_t = 0)\end{bmatrix}$$

$$S_t^5 - f_t^{S5} T^{S5} + D_t^{S5} - E_t^{S5} = 0 \tag{5.47}$$

$$\begin{bmatrix}\text{initial}\\ \text{storage}\\ \text{volume}\end{bmatrix} - \begin{bmatrix}\text{target}\\ \text{storage}\\ \text{volume}\end{bmatrix} + \begin{bmatrix}\text{deficit}\\ \text{storage}\end{bmatrix} - \begin{bmatrix}\text{excess}\\ \text{storage}\end{bmatrix} = 0$$

$$\text{KWH}_t^5 - (k) \cdot (q_t^5) \cdot H^5(K_d^5, S_t^5, S_{t+1}^5) \cdot \epsilon = 0 \tag{5.48}$$

$$\begin{bmatrix}\text{kilowatt}\\ \text{hours of}\\ \text{energy}\end{bmatrix} - \begin{bmatrix}\text{unit}\\ \text{conversion}\\ \text{constant}\end{bmatrix} \cdot \begin{bmatrix}\text{flow}\\ \text{through}\\ \text{turbines}\end{bmatrix} \cdot \begin{bmatrix}\text{average}\\ \text{productive}\\ \text{storage}\\ \text{head}\end{bmatrix} \cdot \begin{bmatrix}\text{hydropower}\\ \text{efficiency}\end{bmatrix} = 0$$

$$\text{KWH}_t^5 - (P^5) \cdot (h_t) \cdot (p_t^5) \leq 0 \tag{5.49}$$

$$\begin{bmatrix}\text{kilowatt-}\\ \text{hours of}\\ \text{energy}\end{bmatrix} - \begin{bmatrix}\text{plant}\\ \text{capacity,}\\ \text{kW}\\ \text{(unknown)}\end{bmatrix} \cdot \begin{bmatrix}\text{hours in}\\ \text{period } t\\ \text{(known)}\end{bmatrix} \cdot \begin{bmatrix}\text{plant}\\ \text{factor}\\ \text{(known)}\end{bmatrix} \leq 0$$

$$\text{KWH}_t^5 - f_t^{p5} T^{p5} + D_t^{p5} - E_t^{p5} = 0 \tag{5.50}$$

$$\begin{bmatrix}\text{kilowatt-}\\ \text{hours of}\\ \text{energy}\end{bmatrix} - \begin{bmatrix}\text{firm}\\ \text{energy}\\ \text{target}\end{bmatrix} + \begin{bmatrix}\text{deficit}\\ \text{energy}\\ \text{production}\end{bmatrix} - \begin{bmatrix}\text{excess}\\ \text{energy}\\ \text{production}\end{bmatrix} = 0$$

Site 6. Town A. For each period t:

$$q_t^6 - R_t^5 - q_t^5 \leq Q_t^6 - Q_t^5 \tag{5.51}$$

$$\begin{bmatrix}\text{allocation}\\ \text{to town A}\end{bmatrix} - \begin{bmatrix}\text{reservoir}\\ \text{release to}\\ \text{stream at}\\ \text{site 5}\end{bmatrix} - \begin{bmatrix}\text{reservoir}\\ \text{release to}\\ \text{hydropower}\\ \text{plant at}\\ \text{site 5}\end{bmatrix} \leq \begin{bmatrix}\text{unregulated}\\ \text{flow at}\\ \text{site 6}\end{bmatrix} - \begin{bmatrix}\text{unregulated}\\ \text{flow at}\\ \text{site 5}\end{bmatrix}$$

$$q_t^6 + D_t^6 - E_t^6 = f_t^6 T^6 \tag{5.52}$$

$$\begin{bmatrix}\text{allocation}\\ \text{to town}\end{bmatrix} + \begin{bmatrix}\text{deficit}\\ \text{allocation}\end{bmatrix} - \begin{bmatrix}\text{excess}\\ \text{allocation}\end{bmatrix} = \begin{bmatrix}\text{target}\\ \text{allocation}\end{bmatrix}$$

Site 7. Reservoir. For each period t:

$$(1 + a_t^7)S_{t+1}^7 - (1 - a_t^7)S_t^7 - I_t^7 + R_t^7 = L_0^7 \quad (5.53)$$

[final storage volume adjusted for losses in period t] − [initial storage volume adjusted for losses in period t] − [total inflow into reservoir] + [reservoir release] = [fixed volume losses (0 if all $S_t = 0$)]

$$I_t^7 - R_t^2 - (R_t^5 + q_t^5) + \delta_t^4 q_t^4 + \delta_t^6 q_t^6 = Q_t^7 - Q_t^2 - Q_t^5 \quad (5.54)$$

[inflow into reservoir at site 7] − [reservoir release at site 2] − [reservoir release at site 5] + [fraction of allocation to irrigation area not returned to stream] + [fraction of allocation to town A not returned to stream]

= [unregulated flow at site 7] − [unregulated flow at site 2] − [unregulated flow at site 5]

$$K^7 - S_t^7 - K_f^7 \geq K_d^7 \quad (5.55)$$

[total capacity] − [initial storage volume] − [flood storage capacity for periods $t = 4$ to 11; 0 otherwise] ≥ [dead storage capacity (0 if all $S_t = 0$)]

$$S_t^7 - f_t^7 T^7 + D_t^7 - E_t^7 = 0 \quad (5.56)$$

[initial storage volume] − [target storage volume] + [deficit storage] − [excess storage] = 0

For the design flood having a return period of 200 years,

$$EK_f^7 - \alpha(E_{2,7}(K_f^2) + E_{5,7}(K_f^5) + K_f^7) = 0 \quad (5.57)$$

[equivalent flood storage capacity at site 7] − [proportion of sum of equivalent capacities at site 7 of actual flood storage capacities K_f^2 and K_f^5 and actual flood storage capacity K_f^7] = 0

Site 8. Town B. For each period t:

$$q_t^8 - R_t^7 \leq Q_t^8 - Q_t^7 \tag{5.58}$$

$$\begin{bmatrix}\text{allocation}\\ \text{to town}\end{bmatrix} - \begin{bmatrix}\text{reservoir}\\ \text{release}\end{bmatrix} \leq \begin{bmatrix}\text{unregulated}\\ \text{flow at}\\ \text{site 8}\end{bmatrix} - \begin{bmatrix}\text{unregulated}\\ \text{flow at}\\ \text{site 7}\end{bmatrix}$$

$$q_t^8 + D_t^8 - E_t^8 = f_t^8 T^8 \tag{5.59}$$

$$\begin{bmatrix}\text{allocation}\\ \text{to town}\end{bmatrix} + \begin{bmatrix}\text{deficit}\\ \text{allocation}\end{bmatrix} - \begin{bmatrix}\text{excess}\\ \text{allocation}\end{bmatrix} = \begin{bmatrix}\text{target}\\ \text{allocation}\end{bmatrix}$$

For the design flood:

$$QS_{200}^8 + QR^8 = QN_{200}^8 \tag{5.60}$$

$$\begin{bmatrix}\text{peak flood flow at}\\ \text{site 8 reduced by}\\ \text{reservoir flood}\\ \text{storage capacity}\\ \text{upstream}\end{bmatrix} + \begin{bmatrix}\text{peak flood flow}\\ \text{contained in}\\ \text{channel at site}\\ \text{8}\end{bmatrix} = \begin{bmatrix}\text{peak flood flow}\\ \text{at site 8}\end{bmatrix}$$

$$EK_f^7 - K_{f,200}^8(QS_{200}^8) = 0 \tag{5.61}$$

$$\begin{bmatrix}\text{equivalent reservoir}\\ \text{flood storage}\\ \text{capacity at site 7}\end{bmatrix} - \begin{bmatrix}\text{function of}\\ \text{flood peak}\\ \text{reduction at site 8}\end{bmatrix} = 0$$

Note that the channel capacity QR^8 at site 8 is the relevant capacity term K^8 in the net-benefit equation 5.36 for site $s = 8$.

5.7.4 Model Solution

This completes the river basin model. Many of these constraint equations can now be combined, if desired, to reduce the number of variables and equations. As written above, there are 243 constraints, not including equations 5.36 and 5.37 and those required to define the nonlinear (but separable) functions in a piecewise linear form for solution by linear programming methods. Models of this size and complexity, even though this is in fact a rather simple model, are usually solved by linear programming algorithms simply because the algorithms are inexpensive to use and readily available. If linear programming is used, the nonlinear nonseparable hydroelectric energy equations 5.48 and the equivalent flood storage capacity equation 5.57 require some trial-and-error adjustments of assumed heads and/or flows and the α term, respectively. This requires several solutions of the model for a single set of assumed weights w_i. The benefit and cost functions in the objective function, equations 5.36 and 5.37, have the general forms indicated in Figure 5.11 and also must be approximated by concave or convex piecewise-linear functions, as appropriate.

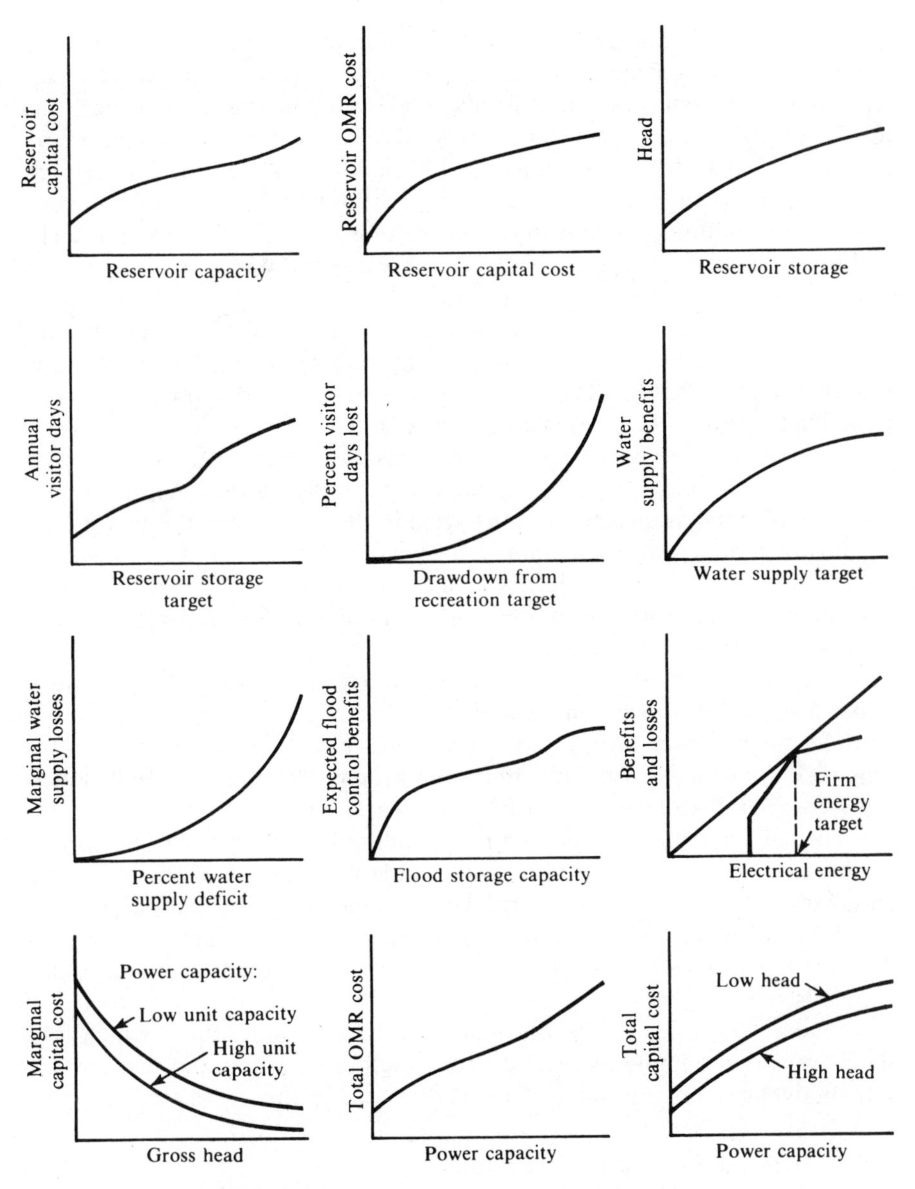

FIGURE 5.11. Typical functional relationships.

5.8 PLANNING THE EXPANSION OF WATER RESOURCES SYSTEMS

The river basin models discussed thus far in this chapter deal with static planning situations. Project capacities, targets, and operating policies take on fixed values and one examines "steady-state" solutions for a particular year in the future. These "snapshots" only allow for fluctuations caused by hydrologic variability. Unfortunately, the nonhydrologic world is seldom static. Demands and targets change in response to population growth, investment in agriculture and industry, and shifting priorities for water use. In addition, financial resources available for water resources investment are limited and may vary from year to year.

Dynamic planning models can aid those responsible for the long-run development and expansion of water resources systems. Although static models can identify good targets and system configurations for a particular period in the future [11], they are not well adapted to long-run planning for a staged expansion of water resource systems over a 25-, 50-, or 100-year period. In such situations, the optimal system configurations identified by a sequence of static models for various years in the future may not be the best to construct in stages given limited financial resources for water resource investments. In addition, projects which the static models indicate should be built in the earlier years may not be present in the solutions for later years or may be present at a different scale.

Whereas static models consider how a water resources system operates under a single set of economic conditions, dynamic expansion models must consider the sequence of economic conditions that may exist over the planning period. For this reason, dynamic expansion models are potentially much more complex than are their static counterparts. However, to keep the size and cost of dynamic models within the limitations of most studies, these models are generally restricted to very simple descriptions of the economic and hydrologic variables of concern. All the models considered in this section use deterministic hydrology and are constrained either to stay within predetermined investment budget constraints or to meet predetermined future demands.

Dynamic expansion models fall into two basic classes: those which rely on integer programming to select the construction sequence, and those which rely on dynamic programming. Each is discussed in turn.

5.8.1 Integer Programming Models

There are several issues that dynamic expansion models can address. Consider a situation in which n fixed-scale projects may be built during the planning period. The *scheduling problem* is to determine which of the projects to build and in what order. Solution of this problem generally requires resolution of the *timing problem*, which is: When should each project be built? In the integer programming formulation of this problem, the following information is assumed to be available:

Y = set of years in which projects may be built

I = set of possible project sites $\{1, 2, 3, \ldots, n\}$

NB_{sy} = present value of total net benefits if project at site $s \in I$ is built in year $y \in Y$

C_s = cost of constructing project at site s

$\$_y$ = funds available for investment in year y

K_s = capacity of project at site s

The integer decision variables are

$$X_{sy} = \begin{cases} 1 & \text{if project at site } s \text{ is built in year } y \\ 0 & \text{otherwise} \end{cases}$$

The problem of scheduling the projects so as to maximize the present value of total net benefits may be formulated as

$$\text{maximize} \sum_{y \in Y} \sum_{s \in I} \text{NB}_{sy} X_{sy} \tag{5.62}$$

such that total expenditures do not exceed the funds available up to any time $y' \in Y$:

$$\sum_{y \leq y'} \sum_{s \in I} C_s X_{sy} \leq \sum_{y \leq y'} \$_y \tag{5.63}$$

and where each project can be built at most once,

$$\sum_{y \in Y} X_{sy} \leq 1 \qquad \text{for all } s \in I \tag{5.64}$$

The drawback to use of this model in river basin planning is that the net benefits NB_{sy} associated with each project should depend on which other projects are built and when. It is generally necessary to include a static river basin model for each year $y \in Y$ within the dynamic expansion model. The objective function then includes the present value of the benefits generated by the entire system and the present value of the costs associated with project construction and operation. The static river basin models contain continuity constraints at each potential reservoir site s for reservoir storage S_{ty}^s and releases R_{ty}^s for each period t of each year $y \in Y$,

$$S_{t+1,y}^s = S_{ty}^s + Q_t^s - R_{ty}^s + \sum_{s' \in I_s} (R_{ty}^{s'} - q_t^{s'}) \tag{5.65}$$

where Q_t^s is the inflow at site s in period t in every planning year y and I_s is the set of all reservoir sites whose releases flow directly into reservoir s and all sites in between at which net withdrawals $q_t^{s'}$ are made. Additional variables and constraints may be needed to specify what diversions are made and to ensure that instream flow requirements are met. Capacity constraints are required to ensure that the storage S^s_{ty} never exceeds the reservoir capacity K_s which is built in year y,

$$S^s_{ty'} \leq K_s(\sum_{y \leq y'} X_{sy}) \qquad \text{for all } t \text{ and } y' \tag{5.66}$$

These maximum reservoir capacity variables are assumed known, determined perhaps from static planning models. Thus for each planning year the constraints of an entire static river basin model are used to determine the net benefits derived from the system.

Additional constraints and variables can be added to this dynamic model to enforce requirements that some projects precede others or that if one project is built another is infeasible.

Another issue that capacity expansion models can address is the *sizing problem*. Frequently, the scale or capacity of each reservoir, pipeline, pumping station, or irrigation project is variable and needs to be determined by the dynamic expansion model concurrently with the resolution of the scheduling and timing problems. To solve the sizing problem, the costs and capacities in the scheduling model become variables. One simple case is when the cost of a project is approximately a linear function of capacity over a range of reasonable values. Given fixed costs a_s and variable unit costs b_s at each site s,

$$C_s(K_{sy}) = a_s + b_s K_{sy} \qquad \text{for } 0 \leq K^{\min}_s \leq K_{sy} \leq K^{\max}_s \qquad \forall y, s \tag{5.67}$$

The model's budget constraint then becomes

$$\sum_{y \leq y'} \sum_{s \in I} (a_s X_{sy} + b_s K_{sy}) \leq \sum_{y \leq y'} \$_y \qquad \forall y' \tag{5.68}$$

with the additional constraints on each capacity variable,

$$K^{\min}_s X_{sy} \leq K_{sy} \leq K^{\max}_s X_{sy} \qquad \text{for } y \in Y \tag{5.69}$$

to ensure that $K_{sy} = 0$ if $X_{sy} = 0$ and $K^{\min}_s \leq K_{sy} \leq K^{\max}_s$ when $X_{sy} = 1$. It is also necessary to change the capacity constraints given by equation 5.66 to

$$S^s_{ty} \leq \sum_{y \leq y'} K_{sy} \qquad \text{for all } y' \in Y \tag{5.70}$$

If the linear approximation to the cost function is inadequate, then more complex expressions for $C_s(K_{sy})$ need to be substituted into equation 5.68 using the piecewise-linear approximation techniques described in Chapter 2.

If cost functions over the relevant range of capacities can be expressed as in equation 5.67, the integer programming formulation of the sizing and scheduling problem is essentially no more difficult than the scheduling problem. However, if the cost function for some project is not convex, so that one or more additional integer variables must be introduced, such variables must be added for each planning year. Reservoir capital cost in Figure 5.11 is such a function. These approximation techniques are discussed in detail in Chapter 2.

These dynamic investment models can be very expensive to solve. For a 50-year planning period with years 0, 5, 10, 15, 20, . . . , 50 serving as planning years, there are 11 planning periods. If each of 10 projects can be built in each year, the integer programming model has $10 \times 11 = 110$ integer variables. An additional 11 integer variables must be added for each additional linear segment needed to approximate a nonconvex cost function. O'Laoghaire and Himmelblau [19] present a branch-and-bound integer programming solution procedure which employs several tricks to try and make the procedure as efficient as possible. The need for inexpensive models that can consider a large number of projects with a fine time resolution has motivated the development of dynamic programming formulations of the scheduling and sizing problems.

5.8.2 Dynamic Programming Models

Although integer programming models are exceedingly flexible, they can often be prohibitively expensive to solve. Dynamic programming algorithms are much more efficient but require that the user invest the time required to construct and debug the required program. This section presents a simple approach for the optimal sequencing and timing of alternative interdependent fixed-scale projects that are to meet a growing demand [4]. Of concern in this section is how to meet a growing demand at minimum cost. Extensions of the model to include project sizing are also discussed.

The demand to be met by the system is assumed to be a continuous increasing function $D(\tau)$ of time τ, where the current demand $D(0)$ is being met with existing capability. As before, let I represent the total set of projects. Let X be a partial set of those projects. The largest demand that can be met by a set of projects X is given by a function $Z(X)$. Because existing capacity is assumed to be sufficient to meet demand, the capacity of no additional projects must be at least $D(0)$, the demand at time zero. The value of $Z(X)$ may be estimated by a deterministic river basin model similar to that discussed earlier in this chapter. It is assumed that each additional project s increases the system's ability to meet the demand. In other words,

$$Z(X \text{ and } s) \geq Z(X) \qquad \text{for } s \notin X \tag{5.71}$$

The cost of adding a project s when projects X have been built is denoted $C_s(X)$. This formulation allows for the costs of modifying existing projects when a new project is constructed. This annual cost $C_s(X)$ includes all maintenance and replacement costs necessary to maintain the project s until the end of the planning period.

Now if all projects in the set X have been built and project $s \notin X$ is to be added, the discounting of construction cost ensures that it will not be optimal to build project s at time τ as long as the available supply $Z(X)$ exceeds the demand $D(\tau)$.

Ignoring the time required for construction, the optimal time to build project s is when $Z(X)$ just equals $D(\tau)$, or

$$\tau = D^{-1}[Z(X)] = \tau(X) \tag{5.72}$$

The function $\tau(X)$ denotes the latest time when the capacity supplied by the set X is just adequate to meet demand. Letting ϕ stand for the empty set, the dynamic programming recursion relationship necessary to determine the cost effective construction schedule is

$$C(\phi) = 0 \tag{5.73}$$

$$C(X) = \min_{s \in X} \{C(X - s) + C_s(X - s)e^{-r\tau(X-s)}\} \qquad X \neq \phi \tag{5.74}$$

where $X - s$ is the set of projects left in X after eliminating s and where $C(X)$ is the minimum discounted cost of building the projects in X so as to meet demand from time 0 to $\tau(X)$. If X contains k projects, then $C(X)$ in equation 5.74 is the minimum cost of first completing the $(k - 1)$ projects $X - s$ and then constructing s in year $\tau(X - s)$ when the projects $X - s$ are just able to meet the demand $D(\tau)$.

Solution of the scheduling problem with this algorithm is best illustrated by a simple example. Consider a situation in which demand grows linearly:

$$D(\tau) = 4\tau \text{ for } \tau \in [0, 25 \text{ years}] \tag{5.75}$$

Hence the time when the supply $Z(X)$ will equal the demand $D(\tau)$ is

$$\tau(X) = \frac{Z(X)}{4} \tag{5.76}$$

Assume that construction costs $C_s(X)$ are independent of X and that available supplies add, so that

$$Z_s(X) = \sum_{s \in X} Z(\{s\}) \tag{5.77}$$

The project cost and supply data are given in Table 5.2. Using these data and

TABLE 5.2. Cost Data for Simple Capacity Expansion Example

Project, s	*Cost of Construction, C_s*	*Supply Resulting from Project s, $Z(\{s\})$*
1	20	10
2	35	20
3	40	30
4	50	40

equation 5.74, the minimum-cost solution will be found for a continuous discount rate of $r = 0.0488$ corresponding to $1 + 0.05 = \exp(r)$.

The first step in the solution of equation 5.74 is to find $C(X)$ for X equal to all possible single projects: {1}, {2}, {3}, {4}. In these cases, $C(X)$ is simply the cost 20, 35, 40, or 50 of building the respective project at time $\tau = 0$ as given in Table 5.2 and as listed in Table 5.3.

TABLE 5.3. Dynamic Programming Solution to Supply Capacity Expansion Problem

Set of Projects, X	*Minimum Discounted Cost, C(X)*	*Optimal Sequence*	*Available Supply, Z(X)*	*Time Supply Equals Demand, τ(X)*
{1}	20	1	10	2.5
{2}	35	2	20	5.0
{3}	40	3	30	7.5
{4}	50	4	40	10
{1, 2}	50.7	2, 1	30	7.5
{1, 3}	53.9	3, 1	40	10
{1, 4}	62.3	4, 1	50	12.5
{2, 3}	64.3	3, 2	50	12.5
{2, 4}	71.5	4, 2	60	15
{3, 4}	74.7	3, 4	20	17.5
{1, 2, 3}	75.2	3, 2, 1	60	15
{1, 2, 4}	81.1	4, 2, 1	70	17.5
{1, 3, 4}	83.2	3, 4, 1	80	20
{2, 4, 3}	89.6	3, 4, 2	90	22.5
{1, 2, 3, 4}	96.3	3, 4, 2, 1	100	25

The next step is to find $C(X)$ when X corresponds to all pairs of two projects {1, 2}, {1, 3}, {1, 4}, {2, 3}, {2, 4}, and {3, 4}. For example, using equation 5.74, $C(\{1, 2\})$ is computed as

$$C(\{1, 2\}) = \min \begin{cases} C(\{1\}) + C_2 e^{-r\tau(\{1\})} = 50.98 \\ C(\{2\}) + C_1 e^{-r\tau(\{2\})} = 50.67 \end{cases} \tag{5.78}$$

The top line of equation 5.78 corresponds to building project 1 in year 0 and then building project 2 after 2.5 years when 1's capacity of 10 units is used up. The bottom line corresponds to building project 2 in year 0, then project 1 in year 5 when 2's capacity is exhausted. Although both sequences meet the demand until year 7.5, the second construction sequence has a lower discounted cost.

A similar set of calculations is required when X is allowed to take all possible combinations of three projects. For example, when $X = \{1, 2, 3\}$,

$$C(\{1, 2, 3\}) = \min \begin{cases} C(\{1, 2\}) + C_3 e^{-r\tau(\{1,2\})} = 78.44 \\ C(\{1, 3\}) + C_2 e^{-r\tau(\{1,3\})} = 75.39 \\ C(\{2, 3\}) + C_1 e^{-r\tau(\{2,3\})} = 75.17 \end{cases} \tag{5.79}$$

If the first three projects are to be constructed, then of the schedules that satisfy demand from time 0 to time $\tau(\{1, 2, 3\})$, building projects 2 and 3 in some order followed by project 1 results in the minimum discounted cost. Examination of the line in Table 5.3 corresponding to $X = \{2, 3\}$ reveals that if these two projects are built first, project 3 should be constructed first followed by 2. Hence the optimal sequence is 3, 2, 1.

In the final step, the minimum cost and optimal sequence for building all four projects is determined. If the planning horizon is 25 years, all projects must be built if the projected demand is realized. In this case the optimal sequence is 3, 4, 2, 1.

Table 5.3 contains additional useful information about the cost and optimal sequencing of many competing sets of projects. Suppose that the planning horizon was 20 rather than 25 years. Then three sets of projects are feasible and the sequence 3, 4, 1 produces the minimal cost. Of the three feasible sets of projects for a time horizon of 20 years or more, the optimal sequences all start with project 3, followed by 4. Hence one can be quite confident that these projects represent a good initial construction schedule with respect to this objective and given this schedule for the future demand.

The extra information provided by Table 5.3 concerning alternative sets of projects is very valuable. It allows the analyst to identify sets of projects which do well with respect to the particular objective and which may do better with respect to other environmental, social, or economic objectives than the identified "optimal" solution. This function of the table is particularly important when one recalls that the purpose of systems analysis is to identify good plans that decision makers can consider taking into account various constraints and objectives which pertain to the decision, and the uncertainty of the future [17].

Many modifications to this algorithm can be made to improve its efficiency and to allow for projects to be constructed at alternative scales [5, 18]. In addition, the ability to delay the construction of projects when demand exceeds supply by importing water (or exporting sewage) at some cost can be included [7, 8].

As formulated in this chapter, the dynamic programming algorithm efficiently calculates the minimum discounted cost of constructing every possible subset of the n possible projects. There are 2^n sets that must be considered unless some of the sets can be excluded for one reason or another (e.g., because some sets can supply more water than is ever demanded). Ignoring such possibilities, if there are 10 projects, there are $2^{10} = 1024$ sets to consider, certainly a manageable number. However, the computational requirements increase rapidly if sizing must be considered. If n projects can be built at two different scales in addition to not being built at all, then there are 3^n combinations to evaluate; for $n = 10$, $3^n = 59{,}049$. Clearly, the need to determine scale as well as the optimal schedule can make the problem much harder. In such cases it may be advantageous to use a dynamic programming formulation which does not guarantee identification of an optimal solution but should produce a good one [6]; such an approach is presented by Becker and Yeh [1, 2].

5.9 CONCLUDING REMARKS

This chapter on deterministic river basin planning models introduces the art of river basin modeling. Purposely ignored during the development of these models were the uncertainties associated with hydrologic and economic parameters and data. As discussed in Chapter 3, these uncertainties may have a substantial effect on model solution and the decision taken. An analysis of the Delaware River basin [11] in the northeastern United States indicated that it was precisely the variability of monthly streamflows, and the need for over-year storage capacity, that made it economically desirable to develop additional reservoir storage in the basin. No such additional storage was needed to regulate average monthly streamflows. In this basin, as in others [16], a deterministic analysis using average flows did not prove to be very helpful in identifying reasonable development and management alternatives.

Another limitation of deterministic models is the deterministic character of their reservoir operating and streamflow allocation policies. Reservoir storage volumes and streamflows in each period of the year will rarely, if ever, exactly equal the value indicated by the deterministic model solution. Still, much work has gone into the development of deterministic multi-reservoir operating models to help those responsible for system operation. (See, for example, the models proposed by Hall et al. [10], Becker and Yeh [3], and Yeh et al. [24].)

The limitations of deterministic planning models have resulted in the development of methods to account for many of the effects of hydrologic uncertainty in river basin planning. The following two chapters will focus on some of these methods.

Whether or not one decides to develop a deterministic planning model, or a modification of it that includes some aspects of hydrologic or economic uncertainty, it is important to emphasize once again that these optimization models are for the preliminary screening of design and operating policy alternatives. The solution of these screening models, and any associated sensitivity analyses, should be evaluated using more precise simulation modeling methods. Screening models should be developed and used in a way that ensures the elimination of clearly inferior alternative solutions, not in a way that attempts to define a single optimal solution. Preliminary screening of river basin systems, especially given multiple objectives, is a challenge to accomplish in an efficient and effective manner. The static deterministic models discussed in this chapter serve as an introduction to that art and also to the equations most often used in planning models of multiple purpose water resource systems.

EXERCISES

5-1. Using the following information pertaining to the drainage area and discharge in the Han River in South Korea, verify equation 5.1 for predicting the natural unregulated flow at any site in the river, by plotting average flow as a function of catchment area. What does the slope of the line equal?

Gage Point	*Catchment Area* (km^2)	*Average Flow* ($10^6\ m^3/yr$)
First bridge of the Han River	25,047	17,860
Pal Dang dam	23,713	16,916
So Yang dam	2,703	1,856
Chung Ju dam	6,648	4,428
Yo Ju dam	10,319	7,300
Hong Chun dam	1,473	1,094
Dal Chun dam	1,348	1,058
Kan Yun dam	1,180	926
Im Jae dam	461	316

5-2. In watersheds characterized by significant elevation changes, one can often develop reasonable predictive equations for average annual runoff per hectare as a function of elevation. Describe how one would use such a function to estimate the natural average annual flow at any gage in a watershed which is marked by large elevation changes and little loss of water from stream channels due to evaporation or seepage.

5-3. Compute the storage-yield function for a single reservoir system by the mass diagram and modified sequent peak methods given the following sequence of annual inflows: (7, 3, 5, 1, 2, 5, 6, 3, 4). Next assume that each year has two distinct hydrologic seasons, one wet and the other dry, and that 80% of the annual inflow occurs in season $t = 1$ and 80% of the yield is desired in season $t = 2$. Using the modified sequent peak method, show the increase in storage capacity required for the same annual yield resulting from within-year redistribution requirements.

5-4. Write two different linear programming models for estimating the maximum constant reservoir release or yield Y given a fixed reservoir capacity K, and for estimating the minimum reservoir capacity K required for a fixed yield Y. Assume that there are T time periods of historical flows available. How could these models be used to define a storage capacity-yield function indicating the yield Y available from a given capacity K?

5-5. (a) Construct an optimization model for estimating the least-cost combination of active storage capacities, K_1 and K_2, of two reservoirs located on a single stream, used to produce a constant flow or yield downstream of the two reservoirs. Assume that the cost functions $C_s(K_s)$ at each reservoir site s are known and there is no dead storage and no evaporation. (Do not linearize the cost functions; leave them in their functional form.) Assume that 10 years of monthly unregulated flows are available at each site s.

(b) Describe the two-reservoir operating policy that you would incorporate into a simulation model to check the solution obtained from the optimization model.

5-6. Given the information in the accompanying tables, compute the reservoir capacity that maximizes the net expected flood damage reduction benefits less the annual cost of reservoir capacity.

Reservoir Capacity	FLOOD STAGE FOR FLOOD OF RETURN PERIOD T					*Annual Capacity Cost*
	$T = 1$	$T = 2$	$T = 5$	$T = 10$	$T = 100$	
0	30	105	150	165	180	10[a]
5	30	80	110	120	130	25
10	30	55	70	75	80	30
15	30	40	45	48	50	40
20	30	35	38	39	40	70

[a] 10 is fixed cost if capacity > 0; otherwise, it is 0.

Flood Stage	*Cost of Flood Damage*
30	0
50	10
70	20
90	30
110	40
130	50
150	90
180	150

5-7. Develop a deterministic, static, within-year model for evaluating the development alternatives in the river basin shown in the accompanying figure. Assume that there are $t = 1, 2, 3, \ldots, n$ within-year periods and that the objective is to maximize the total annual net benefits in the basin. The solution of the model should define the reservoir capacities (active capacity + flood storage capacity), the annual allocation targets, the levee capacity required to protect site 4 from a T-year flood, and the within-year period allocations of water to the uses at sites 3 and 7. Clearly define all variables and functions used, and indicate how the model would be solved to obtain the maximum-net-benefit solution.

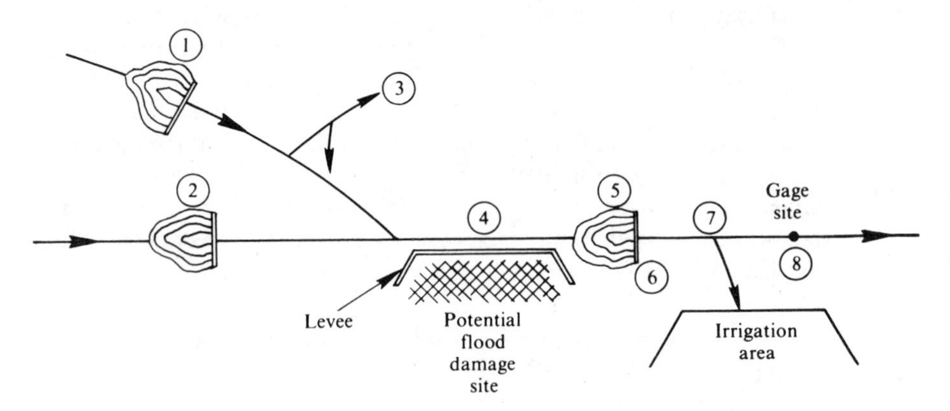

EXERCISE 5.7.

Site	*Fraction of Gage Flow*	
1	0.1	Potential reservoir for water supply, flood control
2	0.3	Potential reservoir for water supply, flood control
3	0.15	Diversion to a use, 60% of allocation returned to river
4	—	Existing development, possible flood protection from levee
5	0.6	Potential reservoir for water supply, recreation
6	—	Hydropower; plant factor = 0.30
7	0.9	Potential diversion to an irrigation district
8	1.0	Gage site

5-8. List the potential difficulties involved when attempting to structure models for defining:
(a) Water allocation policies for irrigation during the growing season.
(b) Energy production and capacity of hydroelectric plants.
(c) Dead storage volume requirements in reservoirs.
(d) Active storage volume requirements in reservoirs.
(e) Flood storage capacities in reservoirs.
(f) Channel improvements for flood damage reduction.
(g) Evaporation and seepage losses from reservoirs.
(h) Water flow or storage targets using long-run benefit and short-run loss functions.

5-9. Assume that demand for water supply capacity is expected to grow as $t(60 - t)$, for t in years. Determine the minimum present value of construction cost of some subset of the water supply options described below so as to always have sufficient capacity to meet demand over the next 30 years. Assume that the water supply network currently has no excess capacity so that some project must be built immediately. In this problem, assume project capacities are independent and thus can be summed. Use a discount factor equal to $\exp(-0.07\,t)$. Before you start, what is your best guess at the optimal solution?

Project Number	*Construction Cost*	*Capacity*
1	100	200
2	115	250
3	190	450
4	270	700

5-10. Briefly outline some of the factors that would apply to the design and operation of a canal to be used for navigation. Assume that water for operating the canal is available in reservoirs and that locks will be necessary at various sites along the canal. In your discussion, identify possible objectives, decision variables, the functional relationships between decision variables and objectives, and constraints.

5-11. (a) Construct a flow diagram for a simulation model designed to define a storage-yield function for a single reservoir given known inflows in each month t for n years. Indicate how you would obtain a steady-state solution not influenced by an arbitrary initial storage volume in the reservoir at the beginning of the first period. Assume that evaporation rates (mm per month) and the storage volume/surface area functions are known.

(b) Write a flow diagram for a simulation model to be used to estimate the reliability of any specific reservoir capacity, K, used to satisfy a series of known release demands, r_t, downstream given unknown future inflows, i_t. You need not discuss how to generate possible future sequences of streamflows, only how to use them to solve this problem.

5-12. (a) Referring to Figure 5.7, develop an optimization model for finding the cost-effective combination of flood storage capacities at reservoir sites 2, 5, and 7, and channel improvements at sites 6 and 8 that will protect sites 6 and 8 from a prespecified design flood of return period T. Define all variables and functions used in the model.

(b) How could this model be modified to consider a number of design floods T and the benefits from protecting sites 6 and 8 from those design floods? Let BF_T^s be the annual expected flood protection benefits at sites $s = 6$ and 8 for various floods having return periods of T.

(c) How could the model be further modified to include water supply requirements of A_t^s to be withdrawn at sites $s = 6$ and 8 in each month t? Assume known natural flows Q_t^s at each site s in the basin in each month t.

(d) How could the model be enlarged to include recreation benefits or losses at site 7? Let T^7 be the unknown storage volume target and D_t^7 be the difference between the storage volume S_t^7 and the target T^7 if $S_t^7 - T^7 > 0$, and E_t^7 be the difference if $T^7 - S_t^7 > 0$. Assume that the annual recreation benefits $B(T^7)$ are a function of the target storage volume T^7 and the losses $L^D(D_t^7)$ and $L^E(E_t^7)$ are associated with the deficit D_t^7 and excess E_t^7 storage volumes.

5-13. Given the hydrologic and economic data listed below, develop and solve a linear programming model for estimating the reservoir capacity K, the flood storage capacity K_f, and the recreation storage volume target T that maximizes the annual expected flood control benefits, $B_f(K_f)$, plus the the annual recreation benefits, $B(T)$, less all losses $L^D(D_t)$ and $L^E(E_t)$ associated with deficits D_t or excesses E_t in the periods of the recreation season, minus the annual cost $C(K)$ of the reservoir capacity K. Assume that the reservoir must also provide a constant release or yield of $Y = 30$ in each period t. The flood season begins at the beginning of period 3 and lasts through period 6. The recreation season begins at the beginning of period 4 and lasts through period 7.

Period t	1	2	3	4	5	6	7	8	9
Inflows to reservoir	50	30	20	80	60	20	40	10	70

$$B_f(K_f) = \begin{cases} 12K_f & \text{if } K_f \leq 5 \\ 60 + 8(K_f - 5) & \text{if } 5 \leq K_f \leq 15 \\ 140 + 4(K_f - 15) & \text{if } K_f \geq 15 \end{cases}$$

$$C(K) = \begin{cases} 0 & \text{if } K = 0 \\ 45 + 10K & \text{if } 0 < K \leq 5 \\ 95 + 6(K - 5) & \text{if } 5 \leq K \leq 20 \\ 185 + 10(K - 20) & \text{if } 20 \leq K \leq 40 \\ 385 + 15(K - 40) & \text{if } K \geq 40 \end{cases}$$

$B(T) = 9T$ where T is a particular unknown value of reservoir storage

$L^D(D_t) = 4D_t$ where D_t is $T - S_t$ if $T \geq S_t$

$L^E(E_t) = 2E_t$ where E_t is $S_t - T$ if $S_t \geq T$

5-14. The optimal operation of multiple reservoir systems for hydropower production presents a very nonlinear and often difficult problem.

(a) Use dynamic programming to determine the operating policy that maximizes the total annual hydropower production of a two reservoir system, one downstream of the other. The releases R_1 from the upstream reservoir plus unregulated incremental flow $(Q_{2t} - Q_{1t})$ constitute the inflow to the downstream dam. The flows Q_1 into the upstream dam in each of four seasons along with the incremental flows $(Q_{2t} - Q_{1t})$ and constraints on reservoir releases are given in the accompanying two tables:

Upstream Dam

Season t	*Inflow,* Q_{1t}	*Minimum Release*	*Maximum Release Through Turbines*
1	60	20	90
2	40	30	90
3	80	20	90
4	120	20	90

Downstream Dam

Season t	*Incremental Flow,* $(Q_{2t} - Q_{1t})$	*Minimum Release*	*Maximum Release Through Turbines*
1	50	30	140
2	30	40	140
3	60	30	140
4	90	30	140

Note that there is a limit on the quantity of water that can be released through the turbines for energy generation in any season due to the limited capacity of the power plant and the desire to produce hydropower during periods of peak demand.

Additional information that affects the operation of the two reservoirs are limitations on the fluctuations in the pool levels (head) and the storage-head relationships:

Data	*Upstream Dam*	*Downstream Dam*
Maximum head, H^{max}	70 m	90 m
Minimum head, H^{min}	30 m	60 m
Maximum storage volume, S^{max}	150×10^6 m^3	400×10^6 m^3
Storage-head relationship	$H = H^{max}(S/S^{max})^{0.64}$	$H = H^{max}(S/S^{max})^{0.62}$

In solving the problem, discretize the storage levels in units of 10×10^6 m^3. Do a preliminary analysis to determine how large a variation in storage might occur at each reservoir. Assume that the conversion of potential energy equal to the product R_iH_i to electric energy is 70% efficient independent of R_i and H_i. In calculating the energy produced in any season t at reserovir i, use the average head during the season

$$\bar{H}_i = \tfrac{1}{2}[H_i(t) + H_i(t + 1)]$$

Report your operating policy and the amount of energy generated per year. Find another feasible policy and show that it generates less energy than the optimal policy.

(b) Use linear programming to solve for the optimal operating policy by approximating the product term $R_i\bar{H}_i$ by a linear expression as described in this chapter (equation 5.29).

REFERENCES

1. Becker, L., and W. W-G. Yeh, Optimal Timing, Sequencing, and Sizing of Multiple Reservoir Surface Water Supply Facilities, *Water Resources Research*, Vol. 10, No. 1, 1974, pp. 57–62.
2. Becker, L., and W. W-G. Yeh, Timing and Sizing of Complex Water Resource

Systems, *Journal of the Hydraulics Division, ASCE,* Vol. 100, No. HY10, 1974, pp. 1457–1470.

3. Becker, L., and W. W-G. Yeh, Optimization of Real Time Operation of a Multiple Reservoir System, *Water Resources Research,* Vol. 10, No. 6, 1974, pp. 1107–1112.
4. Erlenkotter, D., Sequencing of Interdependent Hydroelectric Projects, *Water Resources Research,* Vol. 9, No. 1, 1973, pp. 21–27.
5. Erlenkotter, D., Sequencing Expansion Projects, *Operations Research,* Vol. 21, No. 2, 1973, pp. 542–553.
6. Erlenkotter D., Comment on "Optimal Timing, Sequencing, and Sizing of Multiple Reservoir Surface Water Supply Facilities" by L. Becker and W. W-G. Yeh, *Water Resources Research,* Vol. 11 No. 2, 1975, pp. 380–381.
7. Erlenkotter, D., Coordinating Scale and Sequencing Decisions for Water Resources Projects, *Economic Modeling for Water Policy Evaluation,* North-Holland/TIMS Studies in the Management Sciences, Vol. 3, 1976, pp. 97–112.
8. Erlenkotter D. and J. S. Rogers, Sequencing Competitive Expansion Projects, *Operations Research,* Vol. 25, No. 6, 1977, pp. 937–951.
9. Fair, G. M., J. C. Geyer, and D. A. Okun, *Water and Wastewater Engineering,* Vol. 1, *Water Supply and Wastewater Removal,* John Wiley & Sons, Inc., New York, 1966.
10. Hall, W. A., G. W. Tauxe, and W. W-G. Yeh, An Alternative Procedure for Optimization of Operations for Planning with Multiple River, Multiple Purpose Systems, *Water Resources Research,* Vol. 5, No. 6, 1969, pp. 1367–1372.
11. Jacoby, H. D., and D. P. Loucks, Combined Use of Optimization and Simulation Models in River Basin Planning, *Water Resources Research,* Vol. 8, No. 6, 1972, pp. 1401–1414.
12. James, L. D., and R. R. Lee, *Economics of Water Resources Planning,* McGraw-Hill Book Company, New York, 1971.
13. Johnson, W. K., *Physical and Economic Feasibility of Non-Structural Flood Plain Management Measures,* Hydrologic Engineering Center, U.S. Army Corps of Engineers, Davis, Calif., May 1977.
14. Lawler, E. A., Flood Routing, in *Handbook of Applied Hydrology,* V. T. Chow (ed.), McGraw-Hill Book Company, New York, 1964.
15. Loucks, D. P., Surface Water Quantity Management, Chap. 5 in *Systems Approach to Water Management,* A. K. Biswas (ed.), McGraw-Hill Book Company, New York, 1976.
16. Major, D. C., and R. L. Lenton, *Applied Water Resources Systems Planning,* Prentice-Hall, Inc., Englewood Cliffs, N.J., 1979.
17. Morin, T. L., Optimal Sequencing of Capacity Expansion Projects, *Journal of the Hydraulics Division, ASCE,* Vol. 99, No. HY9, 1973, pp. 1605–1622.
18. Morin, T. L., and A. M. O. Esogbue, A Useful Theorem in the Dynamic Programming Solution of Sequencing and Scheduling Problems Occurring in Capital Expenditure Planning, *Water Resources Research,* Vol. 10, No. 1, 1974, pp. 49–50.
19. O'Laoghaire, D. T., and D. M. Himmelblau, *Optimal Expansion of a Water Resources System,* Academic Press, Inc., New York, 1974.

20. RIPPL, W., The Capacity of Storage Reservoirs for Water Supply, *Proceedings of the Institute of Civil Engineers* (Brit.), Vol. 71, 1883.

21. THOMAS, D. M., and M. A. BENSON, *Generalization of Streamflow Characteristics from Drainage-Basin Characteristics*, U.S. Geological Survey Water-Supply Paper 1975, 1970.

22. THOMAS, H. A., JR., and R. P. BURDEN, *Operations Research in Water Quality Management*, Harvard Water Resources Group, Cambridge, Mass., 1963, pp. 1–17.

23. VIESSMAN, W., JR., J. W. KNAPP, G. L. LEWIS, and T. E. HARBAUGH, *Introduction to Hydrology*, 2nd ed., Dun-Donnelley Publishing Corp., New York, 1977.

24. YEH, W. W-G., L. BECKER, and W-S. CHU, Real-Time Hourly Reservoir Operation, *Journal of the Water Resources Planning and Management Division, ASCE,* Vol. 105, No. WR2, 1979, pp. 187–203.

CHAPTER 6

Synthetic Streamflow Generation

6.1 INTRODUCTION

This chapter is concerned primarily with the techniques for the generation of streamflows and other stochastic processes used in simulation studies of water resource systems. Although the models and techniques discussed here can, with some modification, be used to generate any number of quantities used in simulation studies, the discussion will be directed toward the generation of streamflows. This is done because of the historic development and frequent use of these models in that context. Generated streamflows have been called *synthetic* or *operational* to distinguish them from historic observations. The field has been called operational hydrology, and more recently, stochastic hydrologic modeling.

River basin simulation studies use many sets of streamflow, rainfall, evaporation, and/or temperature sequences to evaluate the statistical properties of the performance of alternative water resources systems. For this purpose, synthetic flows and other generated quantities should resemble those sequences which are likely to be experienced during the planning period. Figure 6.1 illustrates how synthetic streamflow, rainfall, and other stochastic sequences are used in conjunction with projections of future demands and

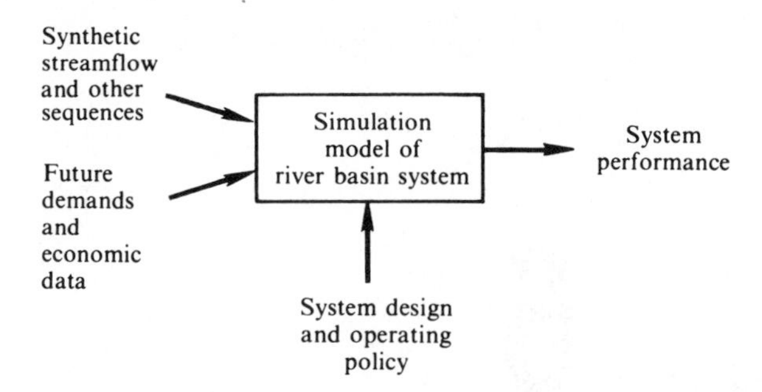

FIGURE 6.1. Structure of a simulation study, indicating the transformation of a synthetic streamflow sequence, future demands, and a system design and operating policy into system performance.

other economic data to determine how different system designs and operating policies might perform. Use of only the historic flow or rainfall record in water resource studies does not allow for the testing of alternative designs and policies against the range of sequences that are likely to occur in the future. By testing designs and policies against a range of sequences that could occur, the variability and range of possible future performance is better understood and better system designs and policies can be selected.

This is particularly true for water resource systems having large amounts of over-year storage so that use of the historic hydrologic records in system simulation yields only one time history of how the system would operate from year to year. In water resource systems having relatively little storage so that reservoirs and/or groundwater aquifers refill almost every year, synthetic hydrologic sequences may not be needed if historic sequences of a reasonable length are available. In this second case, a 25-year historic record provides 25 descriptions of the possible within-year operation of the system. This is sufficient for many studies.

Two basic techniques are used for streamflow generation. If the streamflow population can be described by a stationary stochastic process, a process whose parameters do not change over time, and if a long historic streamflow record exists, then a statistical streamflow model may be fit to the historic flows. This statistical model can then generate synthetic sequences that reproduce selected characteristics of the historic flows. The next few sections discuss a number of these models.

The assumption of stationarity is not always plausible, particularly in river basins that have experienced marked changes in runoff characteristics due to changes in land cover, land use, climate, or the utilization of groundwater during the period of flow record. Similarly, if the physical characteristics of a basin will change substantially in the future, the historic streamflow record may not provide reliable estimates of the distribution of future unregulated flows. In the absence of the stationarity of streamflows or a representative historic record, an alternative scheme is to assume that precipitation is a stationary stochastic process and to route either historic or synthetic precipitation sequences through an appropriate rainfall–runoff model of the river basin. Streamflow generation from precipitation data is discussed briefly in a later section of this chapter.

6.2 STATISTICAL STREAMFLOW GENERATION MODELS

The first step in the construction of a statistical streamflow generating model is to extract from the historic streamflow record the fundamental information about the joint distribution of flows at different sites and at different times. A streamflow model should ideally reproduce what is judged to be the fundamental characteristics of the joint distribution of the flows. The specification of what characteristics are fundamental is of primary importance.

One may want to model as closely as possible the true marginal distribution of seasonal flows and/or the marginal distribution of annual flows. These describe both how much water may be available at different times and also how variable is the supply. Also, modeling the joint distribution of flows at a single site in different months, seasons, and years may be required. The persistance of high flows and of low flows, often described by their correlation, affects the reliability with which a reservoir of a given size can provide a specified yield [13]. For multicomponent reservoir systems, reproduction of the joint distribution of flows at different sites and at different times will also be important.

Sometimes, a streamflow model is said to statistically resemble the historic flows if the streamflow model produces flows with the same mean, variance, skew coefficient, autocorrelations, and/or cross correlations as were observed in the historic series (for example, Ref. 38). Here the emphasis is placed on producing flows that have the same moments as the historic flows. This definition of statistical resemblance is attractive because it is very operational and requires that an analyst need only find a model that can reproduce the observed statistics. The drawback of this approach is that it shifts the modeling emphasis away from trying to find a good model of marginal distributions of the observed flows and their joint distribution over time and over space, given the available data, to just reproducing arbitrarily selected statistics. Defining statistical resemblance in terms of moments may also be faulted for specifying that the parameters of the fitted model be determined using the observed sample moments, or their unbiased counterparts. As discussed in Section 3.3, other parameter estimation techniques are often more efficient.

For any particular river basin study, one must determine what streamflow characteristics need to be modeled. The decision should depend on what characteristics are important to the operation of the system being studied, the data available, and how much time can be spared to build and test a stochastic streamflow model. If time permits, it is good practice to see if the simulation results are in fact sensitive to the streamflow generation model and its parameters by using an alternative model and parameter values. If the model's results are sensitive to changes, then, as always, one must exercise judgment in selecting the appropriate model and parameter values to use.

This chapter presents a range of statistical models for the generation of synthetic streamflow records. The necessary sophistication of a streamflow model depends on the intended use of the synthetic flows. The next section presents the simple autoregressive Markov model for generating annual flow sequences. This model alone is too simple for many practical studies but is useful for illustrating the fundamentals of the more complex models that follow. Therefore, considerable time is spent exploring the properties of this model.

Subsequent sections discuss how flows with any marginal distribution can be produced and present models for generating sequences of flows that can reproduce the persistence of historic flow sequences. Other sections present models to generate concurrent flows at several sites and to generate seasonal or monthly flows while preserving the characteristics of annual flows. For those wishing to study synthetic streamflow models in greater depth, there are numerous references in the text to more advanced material. Matalas and Wallis [38] and Lawrence and Kottegoda [26] provide more advanced surveys, and Yevjevich [60] provides a more complete introduction to the field.

6.3 A SIMPLE AUTOREGRESSIVE MODEL

A simple model of annual streamflows is the autoregressive Markov model. The historic annual flows q_y are thought of as a particular value of a stationary stochastic process Q_y. Here as in Chapter 3, random variables are denoted by capital letters and the values they assume are denoted by the corresponding lowercase letter.

The generation of annual streamflows would be a simple matter if annual flows were independently distributed. In general, this is not the case and a generating model should reproduce the relationship between flows in different years. A common and reasonable assumption is that annual flows are the result of a Markov process.

Assume also that annual streamflows are normally distributed. In some areas, the distribution of annual flows is in fact nearly normal. Streamflow models that produce nonnormal streamflows are discussed in the next section.

The joint normal density function of two streamflows Q_y and Q_w in years y and w having mean μ, variance σ^2, and correlation between flows ρ is

$$f_{Q_yQ_w}(q_y, q_w) = \frac{1}{2\pi\sigma^2(1-\rho^2)^{0.5}} \cdot \exp\left[\frac{(q_y-\mu)^2 - 2\rho(q_y-\mu)(q_w-\mu) + (q_w-\mu)^2}{2\sigma^2(1-\rho^2)}\right] \tag{6.1}$$

The joint normal distribution for two random variables with the same mean and variance depends only on their mean μ, variance σ^2, and correlation ρ (or covariance $\rho\sigma^2$).

The sequential generation of synthetic streamflows requires the conditional distribution of the flow in one year given the value of the flows in previous years. However, if the streamflows are a first-order (lag 1) Markov process, then the dependence of the distribution of the flow in year $y + 1$ on flows in previous years is summarized by the value of the flow in year y. In addition, if the annual streamflows have a multivariate normal distribution, then the conditional distribution of Q_{y+1} is normal with mean and variance

$$\begin{aligned} E[Q_{y+1} \mid Q_y = q_y] &= \mu + \rho(q_y - \mu) \\ \text{Var}(Q_{y+1} \mid Q_y = q_y) &= \sigma^2(1-\rho^2) \end{aligned} \tag{6.2}$$

where q_y is the value of Q_y in year y. This relationship is illustrated in Figure 6.2. Notice that the larger the absolute value of the correlation ρ between the flows, the smaller the conditional variance of Q_{y+1} which does not depend at all on the value q_y.

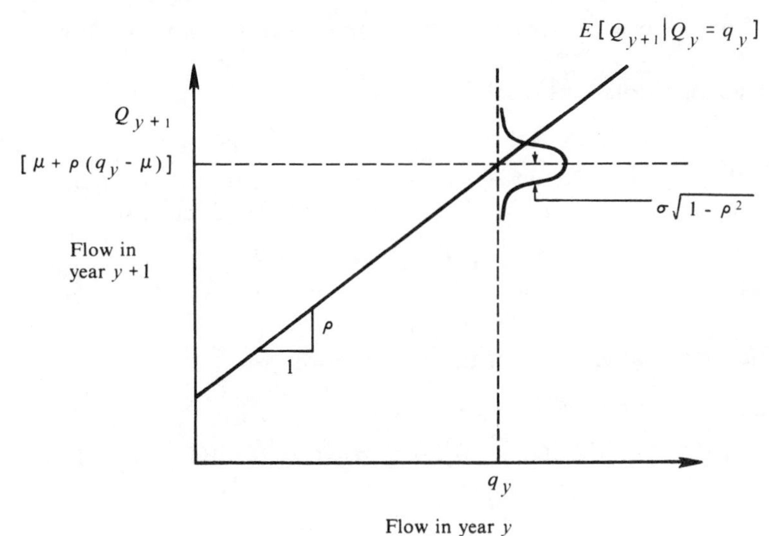

FIGURE 6.2. Conditional distribution of Q_{y+1} given $Q_y = q_y$ for two normal random variables.

Synthetic normally distributed streamflows which have mean μ, variance σ^2, and year-to-year correlation ρ are produced by the model

$$Q_{y+1} = \mu + \rho(Q_y - \mu) + V_y \sigma \sqrt{1 - \rho^2} \tag{6.3}$$

where V_y is a *standard normal* random variable, meaning that its mean $E[V_y] = 0$ and variance $E[V_y^2] = 1$. In addition, each V_y is independent of past flows Q_w where $w \leq y$, and V_y is independent of V_w for $w \neq y$. These restrictions imply that

$$E[V_w V_y] = 0 \qquad w \neq y \tag{6.4}$$

and

$$E[(Q_w - \mu)V_y] = 0 \qquad w \leq y \tag{6.5}$$

Clearly, Q_{y+1} will have a normal distribution if both Q_y and V_y are normally distributed because sums of independent normally distributed random variables are normally distributed. It is a straightforward procedure to show that this model produces streamflows with the specified moments.

The conditional mean of Q_{y+1} given that Q_y equals q_y is

$$E[Q_{y+1} \mid q_y] = E[\mu + \rho(q_y - \mu) + V_y \sigma \sqrt{1 - \rho^2}] = \mu + \rho(q_y - \mu) \tag{6.6}$$

using the fact that $E[V_y] = 0$. Using equation 6.6, the conditional variance of Q_{y+1} is

$$\begin{aligned} \mathrm{Var}(Q_{y+1} \mid q_y) &= E[\{Q_{y+1} - E[Q_{y+1} \mid q_y]\}^2 \mid q_y] = E[\{\mu + \rho(q_y - \mu) \\ &\quad + V_y \sigma \sqrt{1 - \rho^2} - [\mu + \rho(q_y - \mu)]\}^2] \\ &= E[V_y \sigma \sqrt{1 - \rho^2}]^2 = \sigma^2(1 - \rho^2) \end{aligned} \tag{6.7}$$

Thus this model produces flows with the correct conditional mean and variance.

The unconditional mean of Q_{y+1} equals

$$E[Q_{y+1}] = \mu + \rho(E[Q_y] - \mu) + E[V_y]\sigma \sqrt{1 - \rho^2} \tag{6.8}$$

Noting that $E[V_y] = 0$ and that the distribution of streamflows is independent of time so that for all y, $E[Q_{y+1}] = E[Q_y] = E[Q]$, it is clear that

$$(1 - \rho)E[Q] = (1 - \rho)\mu \tag{6.9}$$

or

$$E[Q] = \mu \tag{6.10}$$

The unconditional variance of the annual flows is

$$\begin{aligned} E[(Q_{y+1} - \mu)^2] &= E[\{\rho(Q_y - \mu) + V_y \sigma \sqrt{1 - \rho^2}\}^2] \\ &= \rho^2 E[(Q_y - \mu)^2] + 2\rho\sigma \sqrt{1 - \rho^2}\, E[(Q_y - \mu)V_y] \\ &\quad + \sigma^2(1 - \rho^2)E[V_y^2] \end{aligned} \tag{6.11}$$

Because V_y is independent of Q_y (equation 6.5), the second term on the right-hand side of equation 6.11 is zero. Hence the unconditional variance of

Q satisfies

$$(1 - \rho^2)E[(Q - \mu)^2] = (1 - \rho^2)\sigma^2 \tag{6.12}$$

so that the unconditional variance is σ^2, as required. The covariance of consecutive flows is

$$\begin{aligned} E[(Q_{y+1} - \mu)(Q_y - \mu)] &= E\{[\rho(Q_y - \mu) + V_y\sigma\sqrt{1 - \rho^2}](Q_y - \mu)\} \\ &= \rho E[(Q_y - \mu)^2] = \rho\sigma^2 \end{aligned} \tag{6.13}$$

where $E[(Q_y - \mu)V_y] = 0$ because V_y and Q_y are independent (equation 6.5).

Another property of this model is that the covariance of flows in year y and $y + k$ is

$$E[(Q_{y+k} - \mu)(Q_y - \mu)] = \rho^k\sigma^2 \tag{6.14}$$

This equality can be proven by induction. It has already been shown for $k = 0$ and 1. If it is true for $k = l - 1$, then

$$\begin{aligned} E[(Q_{y+l} - \mu)(Q_y - \mu)] &= E\{[\rho(Q_{y+l-1} - \mu) \\ &\quad + V_{y+l-1}\sigma\sqrt{1 - \rho^2}](Q_y - \mu)\} \\ &= \rho[\rho^{l-1}\sigma^2] = \rho^l\sigma^2 \end{aligned} \tag{6.15}$$

where $E[(Q_y - \mu)V_{y+l-1}] = 0$. Hence equation 6.14 is true for any value of k.

It is important to note that the results in equations 6.6 to 6.15 do not depend on the assumption that the random variables Q_y and V_y are normally distributed. These relationships apply to all autoregressive Markov processes of the form in equation 6.3 regardless of the distributions of Q_y and V_y.

However, if the flow Q_y in year 1 is normally distributed with mean μ and variance σ^2 and if the V_y's are independent normally distributed random variables with mean zero and unit variance, then the generated Q_y's for $y \geq 1$ will also be normally distributed with mean μ and variance σ^2. The next section considers how this and other models can be used to generate streamflows which have other than a normal distribution.

6.4 REPRODUCING THE MARGINAL DISTRIBUTION

Most models for generating stochastic processes deal directly with normally distributed random variables. Unfortunately, flows are not always adequately described by the normal distribution. In fact, streamflows cannot really be normally distributed because of the impossibility of negative flows. In general, streamflow distributions are positively skewed having a lower bound of zero and, for practical purposes, an unbounded right-hand tail as shown in Figure 6.3.

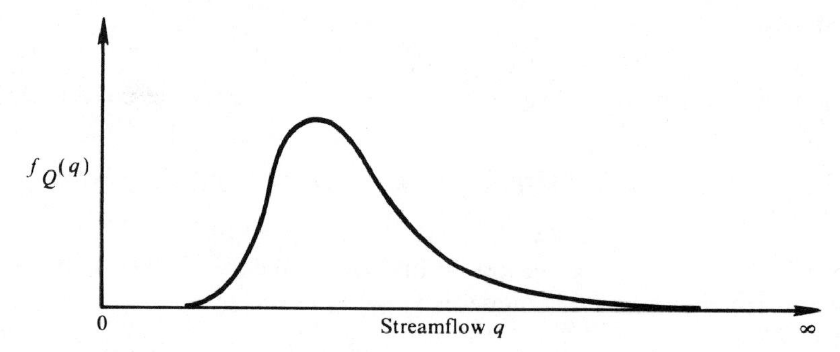

FIGURE 6.3. Typical positively skewed streamflow distribution.

The asymmetry of a distribution is often measured by its coefficient of skewness (introduced in Chapter 3). The skew coefficient is the value of the distribution's third moment scaled by the second moment so as to be dimensionless:

$$\gamma_Q = \frac{E[(Q - \mu)^3]}{\sigma^3} \tag{6.16}$$

In some streamflow models, the skew of the random elements V_y is adjusted so that the models produce flows with the desired mean, variance, and skew coefficient. For the autoregressive Markov model for annual flows

$$\begin{aligned} E[(Q_{y+1} - \mu)^3] &= E[\rho(Q_y - \mu) + V_y\sigma\sqrt{1 - \rho^2}]^3 \\ &= \rho^3 E[(Q_y - \mu)^3] + \sigma^3(1 - \rho^2)^{3/2}E[V_y^3] \end{aligned} \tag{6.17}$$

so that

$$\gamma_Q = \frac{E[(Q - \mu)^3]}{\sigma^3} = \frac{(1 - \rho^2)^{3/2}}{1 - \rho^3}E[V_y^3] \tag{6.18}$$

By appropriate choice of the skew of V_y, the desired skew coefficient of the annual flows is produced. This method has often been used to generate flows that have approximately a gamma distribution by using V_y's with a gamma distribution and the required skew. The resulting approximation is not always adequate [30].

The alternative and preferred method is to generate normal random variables and then transform these variates to streamflows with the desired marginal distribution. Common choices for the distribution of streamflows are the two-parameter and three-parameter lognormal or gamma distributions discussed in Section 3.3. If Q_y is a lognormally distributed random variable, then

$$Q_y = \tau + \exp(X_y) \tag{6.19}$$

where X_y is a normal random variable; the lower bound τ is zero if Q_y has a two-parameter lognormal distribution. Equation 6.19 transforms the normal variates X_y into lognormally distributed streamflows. The transformation is easily inverted to obtain

$$X_y = \ell n\,(Q_y - \tau) \tag{6.20}$$

where Q_y must be greater than its lower bound τ.

The mean, variance, and skewness of X_y and Q_y are related by the formulas [36]

$$\begin{aligned} \mu_Q &= \tau + \exp\,(\mu_X + \tfrac{1}{2}\sigma_X^2) \\ \sigma_Q^2 &= \exp\,(2\mu_X + \sigma_X^2)[\exp\,(\sigma_X^2) - 1] \\ \gamma_Q &= \frac{\exp\,(3\sigma_X^2) - 3\exp\,(\sigma_X^2) + 2}{[\exp\,(\sigma_X^2) - 1]^{3/2}} \end{aligned} \tag{6.21}$$

If normal variates X_y^s and X_y^u are used to generate lognormally distributed streamflows Q_y^s and Q_y^u at sites s and u, then the lag-k correlation of the Q_y's, $\rho_Q(k; s, u)$, is determined by the lag-k correlation of the X's, $\rho_X(k; s, u)$, and their variances $\sigma_X^2(s)$ and $\sigma_X^2(u)$, where

$$\rho_Q(k; s, u) = \frac{\exp\,[\rho_X(k; s, u)\sigma_X(s)\sigma_X(u)] - 1}{\{\exp\,[\sigma_X^2(s)] - 1\}^{1/2}\{\exp\,[\sigma_X^2(u)] - 1\}^{1/2}} \tag{6.22}$$

The correlations of the X_y^s's can be adjusted, at least in theory, to produce the observed correlations among the Q_y^s's. However, more efficient estimates of the true correlation of the Q_y^s's are obtained by transforming the historic flows q_y^s into their normal equivalent $x_y^s = \ell n\,(q_y^s - \tau)$ and using the historic correlations of these x_y^s's as estimates of $\rho_X(k; s, u)$ [50].

Some insight into the effect of this logarithmic transformation can be gained by considering the resulting model for annual flows at a single site. If the normal variates follow the simple autoregressive Markov model

$$X_{y+1} - \mu = \rho_X(X_y - \mu) + V_y\sigma_X\sqrt{1 - \rho_X^2} \tag{6.23}$$

then the corresponding Q_y's follow the model [36]

$$Q_{y+1} = \tau + D_y\{\exp\,[\mu_X(1 - \rho_X)]\}(Q_y - \tau)^{\rho_X} \tag{6.24}$$

where

$$D_y = \exp\,[(1 - \rho_X^2)^{1/2}\,\sigma_X V_y] \tag{6.25}$$

The conditional mean and standard deviation of Q_{y+1} given that $Q_y = q_y$ now depends on $(q_y - \tau)^{\rho_X}$. This is illustrated in Figure 6.4, which shows the conditional mean of Q_{y+1} and 90% confidence limits as a function of q_y. Because the conditional mean of Q_{y+1} is no longer a linear function of q_y, the streamflows are said to exhibit differential persistence. Low flows are now more likely to follow low flows than high flows are to follow high flows. This is a property often attributed to real streamflow distributions.

Another distribution often used to model streamflows is the gamma or Pearson Type III distribution. An easy exact method for transforming gamma

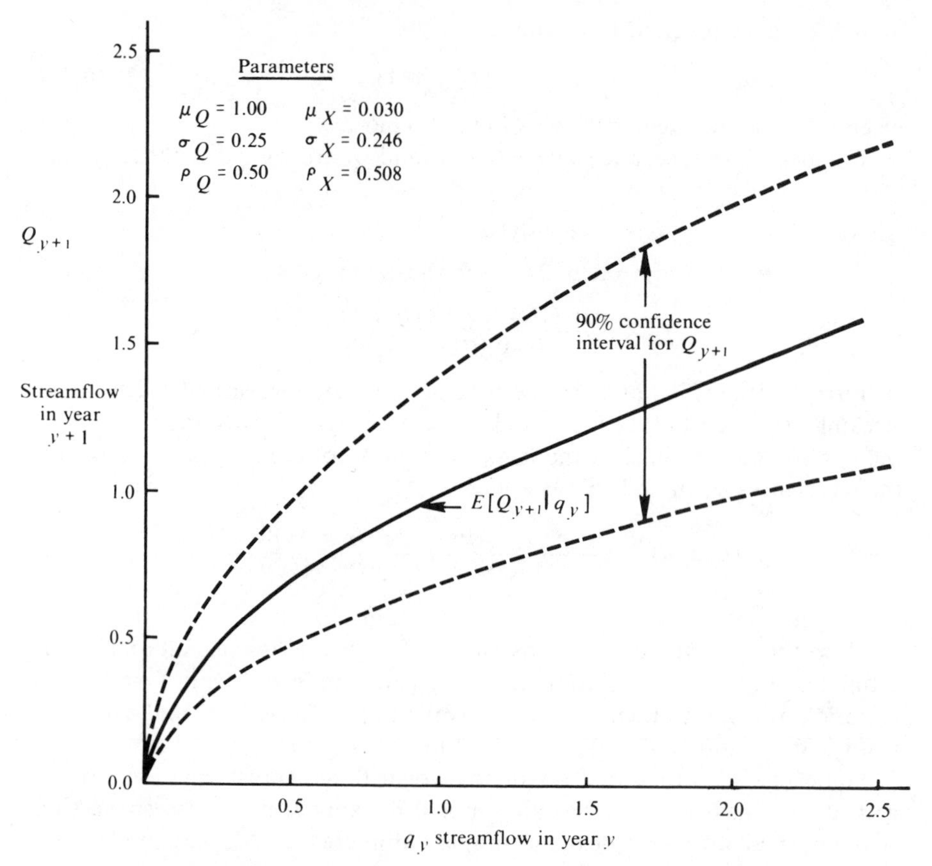

FIGURE 6.4. Conditional mean of Q_{y+1} given $Q_y = q_y$ and 90% confidence limits on value of Q_{y+1}.

variates to normal variates is not available. However, a good approximation for $|\gamma_Q| \leq 3$ is provided by the Wilson–Hilferty transformation,

$$Q_y = \mu_Q + \sigma_Q\left[\frac{2}{\gamma_Q}\left(1 + \frac{\gamma_Q X_y}{6} - \frac{\gamma_Q^2}{36}\right)^3 - \frac{2}{\gamma_Q}\right] \tag{6.26}$$

where X_y has a standard normal distribution and Q_y has a gamma distribution with mean μ_Q, variance σ_Q^2, and skew γ_Q. A modification of this transformation developed by Kirby [21] may be employed if $|\gamma_Q| > 3$.

As shown by Lettenmaier and Burges [30], for $1 \leq \gamma_Q \leq 3$ and $\sigma/\mu \leq 1$, the cumulative distribution function of the three-parameter lognormal and gamma distributions are quite close and the distributions will generally be indistinguishable in practice. However, for the two-parameter versions of the distributions or moments outside this range, the differences may be important.

6.5 AUTOREGRESSIVE-MOVING AVERAGE (ARMA) MODELS

The simple autoregressive Markov model presented in Section 6.3 is not always able to reproduce the observed persistance of streamflow sequences. It is, however, a simple example of the very flexible family of autoregressive-moving average (ARMA) time-series models [3] sometimes called Box–Jenkins models. These models have been widely used in the modeling and forecasting of time series in many fields besides water resources. This section discusses the available methods for selecting a member of this family for modeling an observed time series, estimating the chosen model's parameters, and checking to ensure that the model does provide an adequate fit to the observed sequence of values.

Let a sequence of stationary normally distributed zero-mean transformed streamflows be denoted $Z_1, \ldots, Z_n$. Then an autoregressive-moving average or ARMA model of this series might be

$$Z_{t+1} = \phi_1 Z_t + V_t - \theta_1 V_{t-1} \tag{6.27}$$

where the V_t's are independent identically distributed normal random variables. This particular ARMA model is denoted ARMA(1, 1); an ARMA(p, q) model has p autoregressive terms $\phi_1 Z_t, \ldots, \phi_p Z_{t-p+1}$ and q moving average terms $\theta_1 V_{t-1}, \ldots, \theta_q V_{t-q}$ in addition to the current V_t, so that

$$Z_{t+1} = \sum_{i=1}^{p} \phi_i Z_{t-i+1} + V_t - \sum_{j=1}^{q} \theta_j V_{t-j} \tag{6.28}$$

In many applications $p + q \leq 2$, although this is not necessarily the case.

In ARMA models restrictions must be placed on the allowable values of the parameters. To ensure that the Z_t's are stationary with finite variance, it is necessary that the autoregressive parameters are such that the roots r of the polynomial

$$1 - \sum_{i=1}^{p} \phi_i r^i = 0 \tag{6.29}$$

lie outside the unit circle. To ensure that the dependence of Z_t on Z_{t-k} for $k > 0$ decreases with increasing k, the moving-average parameters must also be such that the roots of the polynomial

$$1 - \sum_{j=1}^{q} \theta_j s^j = 0 \tag{6.30}$$

lie outside the unit circle.

Because transients in the generated time series will die out with time like r^{-t}, the values of r^{-1} describe the persistence of the generated series. If some r is near 1, then generated Z_t's may remain either above or below their mean value for considerable periods of time. The autoregressive model in Section

6.3 is an ARMA(1, 0) model for which $\phi_1 = \rho$. The one root of equation 6.29 satisfies $1/r = \rho$, showing the direct relationship between $1/r$ and ρ in this simple case.

The development of an ARMA model for a particular time series is often described as consisting of three steps: (1) *identification* of reasonable values of p and q; (2) *estimation* of the parameters $\phi_1, \ldots, \phi_p, \theta_1, \ldots, \theta_q$ and σ_V^2 for given values of p and q; and (3) *diagnostic checks* of the fitted model to ensure that it is an adequate model of the time series [3, 16]. However, in practice, these first two steps are often combined. To identify values of p and q that may produce a reasonable model of a time series, it is helpful to examine the autocorrelations r_k and the partial autocorrelations $\hat{\phi}_{kk}$ of a time series to see which values are significantly different from zero.

The theoretical autocorrelations of a stochastic process, introduced in Section 3.4, are given by

$$\rho_k = \frac{E[Z_{t+k} \cdot Z_k]}{E[Z_t^2]} \tag{6.31}$$

The autocorrelations of an observed time series $z_1, z_2, \ldots, z_n$ are generally estimated as

$$r_k = \frac{\sum_{t=1}^{n-k} z_{t+k} z_t}{\sum_{t=1}^{n} z_t^2} \tag{6.32}$$

where the mean or average of the z_t's is assumed to be zero. Under the assumption that $\rho_k = 0$ for $k > q$, r_k has approximately a normal distribution with mean zero and variance

$$\operatorname{Var}(r_k) = \frac{1}{n}[1 + 2(r_1^2 + r_2^2 + \ldots + r_q^2)] \qquad k > q \tag{6.33}$$

This allows specification of 90% or 95% confidence limits for r_k under the hypothesis that $\rho_l = 0$ for $l \geq k$ so that one can identify those r_k which are significantly different from zero.

The correlation function is particularly useful for identifying q in an ARMA(0, q) model, often denoted just MA(q). If

$$Z_{t+1} = V_t - \theta_1 V_{t-1} - \ldots - \theta_q V_{t-q} \tag{6.34}$$

then

$$\rho_k = \begin{cases} \dfrac{-\theta_k + \theta_1\theta_{k+1} + \ldots + \theta_{q-k}\theta_q}{1 + \theta_1^2 + \theta_2^2 + \ldots + \theta_q^2} & 1 \leq k \leq q \\ 0 & k > q \end{cases} \tag{6.35}$$

Hence if Z_t is an ARMA(0, q) process, its correlation function is nonzero for q lags, after which it is zero. If Z_t is an ARMA(p, 0) or ARMA(p, q) model, then its correlation function after initial transients decreases exponentially or as a dampened sine curve [3]. The relationship between the model's para-

TABLE 6.1. Some Relationships between Parameters and Correlations ρ_k of ARMA Processes

ARMA Model (p, q)	*Relationships*
(1, 0)	$\rho_k = (\phi_1)^k$
(2, 0)	$\rho_1 = \dfrac{\phi_1}{1 - \phi_2}$
	$\rho_2 = \phi_2 + \dfrac{\phi_1^2}{1 - \phi_2}$
	$\rho_k = \phi_1 \rho_{k-1} + \phi_2 \rho_{k-2}, \quad k \geq 2$
(1, 1)	$\rho_1 = \dfrac{(1 - \theta_1 \phi_1)(\phi_1 - \theta_1)}{1 + \theta_1^2 - 2\phi_1 \theta_1}$
	$\rho_k = \rho_1 \phi_1^{k-1}, \quad k \geq 2$
(0, 2)	$\rho_1 = \dfrac{-\theta_1(1 - \theta_2)}{1 + \theta_1^2 + \theta_2^2}$
	$\rho_2 = \dfrac{-\theta_2}{1 + \theta_1^2 + \theta_2^2}$
	$\rho_k = 0, \quad k \geq 3$
(0, 1)	$\rho_1 = \dfrac{-\theta_1}{1 + \theta_1^2}$
	$\rho_k = 0, \quad k \geq 2$

meters and the correlation function are given in Table 6.1 for five commonly used ARMA models.

While the correlation function is useful for identifying the value of q for an ARMA$(0, q)$ model, the partial autocorrelation function is particularly useful for identifying the value of p for an ARMA$(p, 0)$ model, often denoted just AR(p). If Z_t is an ARMA$(p, 0)$ process, so that

$$Z_{t+1} = \phi_1 Z_t + \dots + \phi_p Z_{t-p+1} + V_t \tag{6.36}$$

then by multiplying equation 6.36 by Z_{t-j+1} for $j = 1, 2, \dots, p$ and taking expectations, one obtains the relationships

$$\rho_j = \phi_1 \rho_{j-1} + \phi_2 \rho_{j-2} + \dots + \phi_p \rho_{j-p} \qquad \text{for } j = 1, \dots, p \tag{6.37}$$

Equation 6.37 can be used to estimate the ϕ_i's. Assuming that p equals some integer k, the resulting estimate of ϕ_i is the partial autocorrelation ϕ_{ik}. This value is defined by the equations

$$\begin{bmatrix} \rho_1 \\ \cdot \\ \cdot \\ \cdot \\ \rho_k \end{bmatrix} = \begin{bmatrix} 1 & \rho_1 & \cdots & \rho_{k-1} \\ \rho_1 & 1 & & \rho_{k-2} \\ \cdot & & \cdot & \\ \cdot & & & \cdot \quad \rho_1 \\ \rho_{k-1} & \cdot & \cdots \rho_1 & 1 \end{bmatrix} \begin{bmatrix} \phi_{1k} \\ \cdot \\ \cdot \\ \cdot \\ \phi_{kk} \end{bmatrix} \tag{6.38}$$

When fitting an ARMA model to an observed series, the values of ρ_i are not known so that sample estimates r_i of ρ_i are substituted into equation 6.38 to produce estimates $\hat{\phi}_{ik}$ of ϕ_{ik}.

For an ARMA(p, 0) process, the estimates $\hat{\phi}_{kk}$ for $k \geq p + 1$ are approximately independent normally distributed with mean zero and variance

$$\mathrm{Var}(\hat{\phi}_{kk}) = \frac{1}{n} \tag{6.39}$$

If Z_t is an ARMA(p, 0) process, then $\phi_{kk} = 0$ for $k \geq p + 1$ and in general $\phi_{kk} \neq 0$ for $k \leq p$. Hence the partial autocorrelation function can be used to identify p of an ARMA(p, 0) process just as the correlation function identifies q of an ARMA(0, q) process. If both p and q are nonzero, then both the correlation function ρ_k and partial correlation function ϕ_{kk} decay exponentially or as dampened sine curves. This behavior is summarized in Table 6.2.

TABLE 6.2. Behavior of Correlations ρ_k and Partial Autocorrelations ϕ_{kk} of ARMA Models [3]

ARMA Model	*Behavior of ρ_k*	*Behavior of ϕ_{kk}*	*Admissible Region*
(1, 0)	Decays exponentially	Only ϕ_{11} nonzero	$-1 < \phi_1 < 1$
(2, 0)	Mixture exponentials and dampened sine wave	Only ϕ_{11} and ϕ_{22} nonzero	$-1 < \phi_1 < 1$ $\phi_2 + \phi_1 < 1$ $\phi_2 - \phi_1 < 1$
(1, 1)	Decays exponentially from first lag	Dominated by exponential decay from first lag	$-1 < \phi_1 < 1$ $-1 < \theta_1 < 1$
(0, 2)	Only ρ_1 and ρ_2 nonzero	Dominated by exponentials or dampened sine wave	$-1 < \theta_2 < 1$ $\theta_2 + \theta_1 < 1$ $\theta_2 - \theta_1 < 1$
(0, 1)	Only ρ_1 nonzero	Exponential decay dominates	$-1 < \theta_1 < 1$

Examination of the autocorrelations and partial autocorrelations provides some insight into how many autoregressive and moving-average terms one might need to include in a good model of a time series. They do not indicate how many terms one should include, particularly when considering mixed models with $p, q \geq 1$. This information is generated by actually fitting models with different values of p and q and examining the significance of the estimated parameters and other statistics.

Maximum likelihood estimates of the parameters are generally employed when one has 50 or so observations. By assumption, the unobserved variables $V_1, V_2, \ldots, V_n$ are independent identically distributed zero-mean random variables whose joint density function equals

$$f_{V_1,\ldots,V_n}(v_1, \ldots, v_n) = (2\pi\sigma_V^2)^{-n/2} \exp\left(-\frac{1}{2\sigma_V^2}\sum_{t=1}^{n} v_t^2\right) \tag{6.40}$$

Given observed values z_t and estimates of the model's parameters, equation 6.28 is solved for values of v_t. Specific details of this procedure, including alternative ways of dealing with equation 6.28 when it involves either v_t's or z_t's for which $t < 0$, are given in Chapter 7 of Box and Jenkins [3].

Upon finding the maximum likelihood estimates of the ARMA parameters for given values of p and q, one also obtains the maximum likelihood estimate of σ_V^2,

$$\widehat{\sigma_V^2} = \frac{1}{n}\sum_{t=-\infty}^{n} v_t^2 \tag{6.41}$$

One can also obtain estimates of the standard errors of each of the parameters and, knowing that maximum likelihood parameter estimates are asymptotically normally distributed, can determine if the estimated parameters appear to be significantly different from zero at some reasonable confidence level. Those parameters that are not significantly different from zero are generally dropped from the model and p and q are adjusted accordingly. Still, it frequently happens that more than one ARMA model can be found all of whose parameters are significantly different from zero.

When one's objective is to identify a model that provides the best available description of an observed time series, the significance of the parameters of alternative models is not of fundamental importance. The real question is not whether an estimated parameter is significantly different from zero but whether including that term in the ARMA model results in a better description of the time series in question. By considering the model's total forecasting error due both to the residual variance of the V_t's and due to errors in the estimated parameters, one is led to selecting the time-series model that minimizes Akaike's information criterion

$$\text{AIC} = -2\,\ell\text{n}\,(\text{ML}_k^*) + 2k \tag{6.42}$$

where ML_k^* is the maximum value of the likelihood function (equation 6.40) for a model with k free parameters. In ARMA models, minimizing AIC is equivalent to minimizing

$$n\,\ell\text{n}\,(\widehat{\sigma_V^2}) + 2(p + q) \tag{6.43}$$

where $\widehat{\sigma_V^2}$ is given by equation 6.41. Use of the AIC criterion or equation 6.43 is to be recommended; still the user should be warned that local minima are common and that the modeler must still identify the sets of values of p and q to be investigated.

Finally, the adequacy of any time-series model should be checked by examination of the residuals $v_1, \ldots, v_n$ to ensure that they are indeed independent and normally distributed as assumed. Box and Jenkins [3], Hipel et al. [16], and McLeod et al. [40] discuss these model-fitting procedures in greater depth and include numerous examples.

As a simple example of this modeling procedure, consider the fitting of a model to the annual flows from 1926 to 1976 into New York City's reservoirs in the upper Delaware River basin in the northeastern United States. Figures 6.5 and 6.6 present the autocorrelations and partial autocorrelations of the annual flows along with 90% and 95% confidence intervals assuming that $\rho_k = 0$ for $k \neq 0$. The annual flows are well fitted by a normal distribution so that a transformation of the observed values was not required.

Examination of Figures 6.5 and 6.6 reveals that only r_1 and $\hat{\phi}_{11}$ are significantly different from zero while the general pattern of the functions suggest an ARMA(1, 0) model. Upon fitting ARMA(p, q) models for $p + q \leq 2$, the parameters of the ARMA(1, 0) and the ARMA(0, 1) models were found to be significantly greater than zero at the 95% level, while at least one of the parameters of each of the other models was not significant at this level.

Table 6.3 reports the values of the information criterion given in

TABLE 6.3. Value of Information Criterion Given by Equation 6.43 for Alternative ARMA (p, q) Models

q \ p	*0*	*1*	*2*
0	—	592.3	594.3
1	592.8	594.3	600.4
2	594.4	600.2	597.9

equation 6.42 for all the models tested. The values in the table indicate that the best model is ARMA(1, 0). Hence this is an example where the autoregressive Markov model presented in Section 6.3 is an appropriate model for an annual flow series. Note that the information criterion has a local minimum at ARMA(2, 2) in that dropping the last autoregressive or moving-average term increases the information criterion's value.

The example illustrates that significance tests and the information criterion are both blind to the design implications of the competing models. The parameters of both the ARMA(1, 0) and ARMA(0, 1) models are statistically significant and there is little difference in the corresponding values of the information criterion. Still, with the ARMA(0, 1) model

$$Z_{t+1} = V_t + 0.3375V_{t-1} \tag{6.44}$$

the autocorrelations of the modeled flows exhibit almost no persistence and $\rho_Z(k) = 0$ for $k \geq 2$. While with the ARMA(1, 0) model

$$Z_{t+1} = 0.356Z_t + V_t \tag{6.45}$$

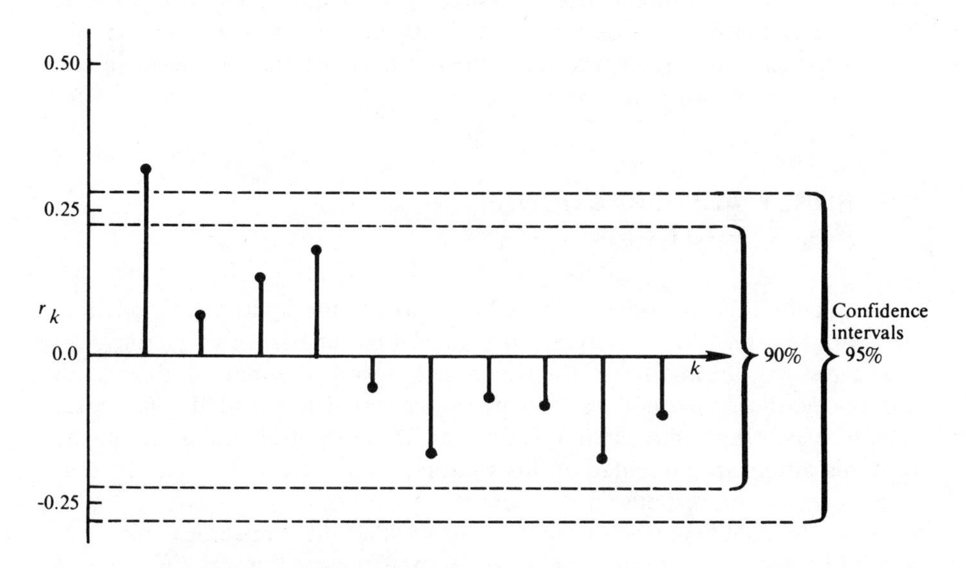

FIGURE 6.5. Autocorrelations of inflows to New York City reservoirs for various lags *k*.

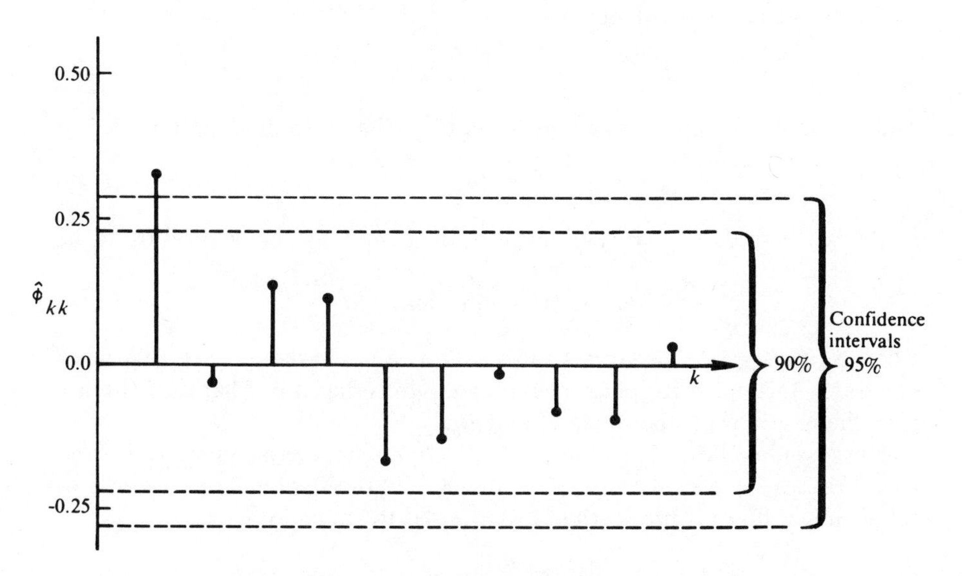

FIGURE 6.6. Partial autocorrelations of inflows to New York City reservoirs for various lags *k*.

the modeled flows will exhibit more persistence because $\rho_Z(k) = (0.356)^k$, thus perhaps decreasing the firm yield that a given size reservoir can provide. The systems analyst must be sensitive to these model differences and should use experience and judgment along with statistical criteria when selecting an appropriate model of hydrologic series.

6.6 HURST AND FRACTIONAL BROWNIAN NOISE

Long before the introduction of ARMA models into hydrologic modeling, the hydrologist H. E. Hurst [18, 19] studied the persistence of geophysical time-series data from all over the world. His work had a major influence on the models developed and used in hydrologic modeling and the statistical tests to which those models have been put. Because of his influence on the field, his results are presented in this section and the streamflow models that were developed in response are reviewed.

Hurst directed his attention to a statistic called the "range of cumulative departures from the mean," which equals the required storage volume of a reservoir which for a given inflow sequence can release in every year the mean inflow. Consider a sequence of inflows $\{q_1, \ldots, q_n\}$. Let the mean flow in the n-year period be denoted

$$\bar{q}_n = \frac{1}{n}\sum_{y=1}^{n} q_y \tag{6.46}$$

The accumulated departure of the flows from the mean flow after y years is

$$S_y^n = \sum_{y^*=1}^{y} (q_{y^*} - \bar{q}_n) \tag{6.47}$$

In the last period, $S_n^n = 0$. The range of the cumulative departures from the mean is

$$R_n = \max_y (S_y^n) - \min_y (S_y^n) = S_M^n - S_m^n \tag{6.48}$$

where S_M^n and S_m^n are the largest and smallest values in the set $\{S_y^n\}$. Figure 6.7 illustrates, through a Rippl or mass diagram introduced in Chapter 5 (Figure 5.4), the relationships between S_y^n and R_n.

Hurst studied how the average value of R_n changes as a function of n and found that the expected value of R_n divided by the standard deviation s_n of the n annual flows is proportional to n raised to some power H.

$$E\left[\frac{R_n}{s_n}\right] \sim n^H \tag{6.49}$$

The exponent H is called the Hurst coefficient, which he found to be between 0.69 and 0.80. Asymptotically, for independent normal random

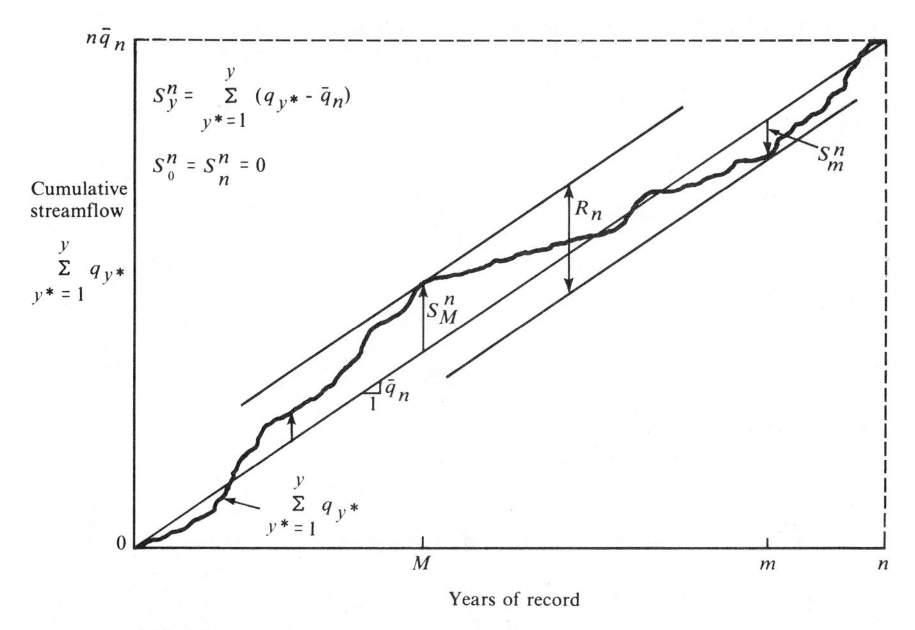

FIGURE 6.7. Mass diagram defines S_y^n and R_n for an n-year record of streamflows q_y^n having mean $\bar{q}_n$.

variables [11, 19],

$$E[R_n] = \left(\frac{\pi}{2}\right)^{0.5} \sigma n^{0.5} \tag{6.50}$$

The empirical value of H, equal to about 0.73, is larger than the value produced by simple autoregressive models, which exhibit values of H tending to 0.5 as n becomes large. Natural streams exhibit what is termed long-term persistence, long sequences of years in which streamflows remain higher or lower than the mean.

Modelers were at a loss for some time to develop a model that reproduced this phenomena until Mandelbrot and Van Ness [34] developed one that produces flows with a specified value of H. Their model is based on fractional Brownian motion which can be used to obtain fractional Gaussian noise.

A fractional Gaussian noise (FGN) process is a sequence of standard normal random variables $\{X_1, \ldots, X_y, \ldots\}$. The joint normal density of the random variables depends only on their means, variances, and covariances. Thus to finish the specification of the statistical properties of the sequence $\{X_1, X_2, \ldots\}$, it suffices to specify the correlation of any two which is given by their autocorrelations

$$\rho_X(k) = \tfrac{1}{2}[|k+1|^{2H} - 2|k|^{2H} + |k-1|^{2H}] \tag{6.51}$$

where H is the Hurst coefficient of the series. Table 6.4 contains values of

TABLE 6.4. Comparison of Correlations of Markov, Fractional Noise, and ARMA Processes (1, 1)

Lag[a]	*Markov Process* ($\rho_1 = 0.3755$)	*Fractional Noise* ($H = 0.73$)	*ARMA (1, 1) Process* ($\phi = 0.92, \theta = 0.706$)
1	0.376	0.376	0.376
2	0.141	0.235	0.346
4	0.020	0.160	0.293
8	4.0×10^{-4}	0.109	0.210
16	1.6×10^{-7}	0.075	0.108
32	2.4×10^{-14}	0.052	0.028
64	6.0×10^{-28}	0.036	0.002
125	6.8×10^{-54}	0.025	1.2×10^{-5}
250	5.0×10^{-107}	0.017	3.6×10^{-10}
500	2.0×10^{-213}	0.012	3.2×10^{-19}
1000	4.6×10^{-426}	0.008	2.5×10^{-32}

[a]A realistic planning period would not exceed 100 years, but very large lags are included to illustrate the differences between the models.

$\rho_X(k)$ for selected values of k for FGN with $H = 0.73$, a autoregressive Markov process with the same lag 1 autocorrelation and an ARMA(1, 1) process which also has the same lag 1 autocorrelation and whose generated flows over 25-year periods exhibit a Hurst coefficient on the order of 0.73 [47].

The definition of fractional Brownian motion involves an integral over an infinite time interval and hence is not convenient for the generation of FGN sequences. Thus Mandelbrot and Wallis [35] proposed two approximations, Type I and Type II, for the computer generation of flows. Type II was used in a number of studies by Mandelbrot and Wallis [35] and Wallis and Matalas [37, 56–58].

Subsequently, Mandelbrot [32] found that the popular and economical Type II approximation was inadequate and proposed a new approximation, fast fractional Gaussian noise (FFGN), which is more efficient and flexible. Fast fractional Gaussian noise allows the analyst to produce normally distributed random variables with a specified Hurst coefficient and a desired lag 1 correlation, within bounds determined by H [29, 32]. An exact method for producing short FGN noise sequences is given by McLeod and Hipel who also describe how to obtain maximum likelihood estimates of H [41].

Another model that has been proposed as an approximation to FGN is the broken-line model [43]. It is a continuous time process which can produce the Hurst effect, the lag 1 correlation $\rho(1)$ of flows and the second derivative of the correlation function at the origin $\rho''(0)$. The second derivative of the correlation function at the origin is related to crossing properties of the streamflow function, including the number of times the flow rate crosses a given level and the expected time between such crossings [26]. The model has several flaws, including being slightly non-Gaussian, that raise concerns about its ability to actually reproduce crossing properties [33] whose importance has not been demonstrated.

A good discussion of the geophysical implications of FGN models and possible physical explanations are given by Klemeš [22]. While the Hurst coefficient exhibited by ARMA models tends to 0.50 in very long sequences, their average value in annual flow sequences of 25, 50, or 100 years can be adjusted to exceed 0.80 [29, 47]. In addition, comparisons of the distributions of reservoir storage required to meet realistic targets over a 25- to 100-year period show that ARMA and FFGN models produce similar results [5, 29]. Careful comparisons of the fits of ARMA and fractional Gaussian noise models to long streamflow sequences using the Akaike information criterion (equation 6.42) indicate that the ARMA models are to be preferred [41].

6.7 MULTISITE MODELS

The previous sections developed models for generating annual flows at a single site. Simulations of complex multireservoir systems require the generation of concurrent flows at many sites. The appropriate complexity of multisite models depends on the availability of data and on the relationship between flows at the different sites. This section considers first the generation of flows in the tributaries of a major stream where simple multisite models may be sufficient. More complex multisite extensions of the models in Sections 6.5 and 6.6 are then considered.

6.7.1 A Simple Model

Figure 6.8 illustrates a simple river basin. Assume that a long historic record exists at gage site 1 on the river's main stem. Typically, projects may be proposed at sites 2 and 3, where only limited records, if any, may be available. Such short records, if they exist, do not provide reliable estimates of the mean and variance of flows at those sites. In these situations, it is sometimes appropriate for the purpose of generating concurrent streamflows at all three sites to assume that the flows Q_y^s at sites $s = 2$ or 3 are some fixed fraction α_s of the flow Q_y^1 at site 1.

$$Q_y^s = \alpha_s Q_y^1 \qquad \text{for } s = 2 \text{ or } 3 \tag{6.52}$$

A single-site model can generate flows at site 1 based on the record at gage 1, and these flows can then be used to generate synthetic flows at sites 2 and 3

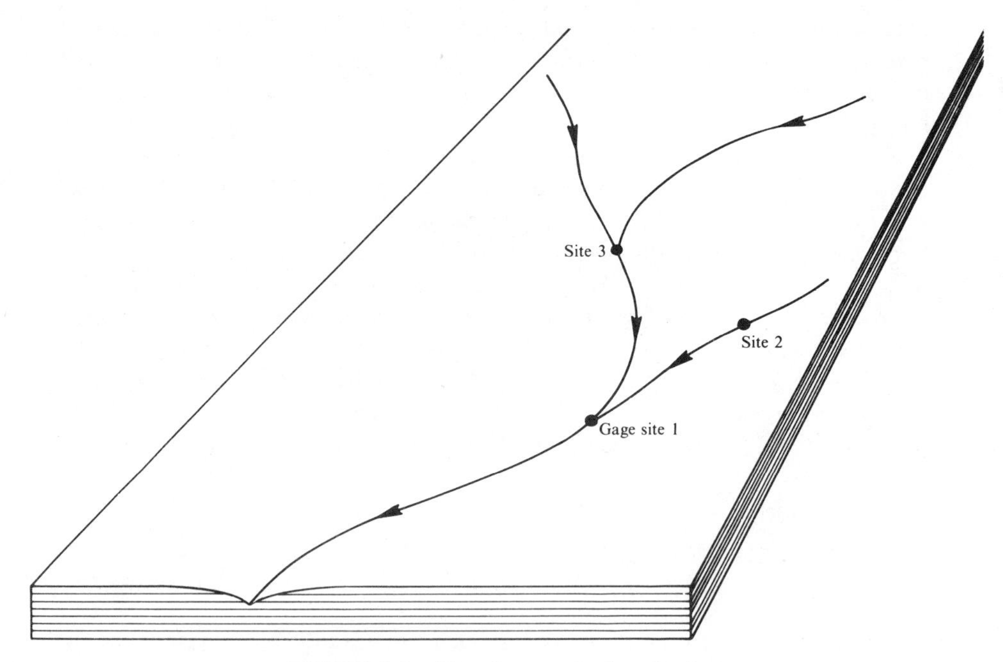

FIGURE 6.8. Sites in sample river basin.

using equation 6.52. The use of drainage areas to estimate α_s is discussed in Chapter 5.

6.7.2 General Multisite Models

If long concurrent streamflow records can be constructed at the several sites at which synthetic streamflows are desired, then ideally, a general multisite streamflow model could be employed. O'Connell [47], Ledolter [27], and Wilson [59] discuss multivariate ARMA models and parameter estimation. Unfortunately, parameter estimation and model identification is very difficult for the general multivariate ARMA model except for the simple AR(p) case or if special restrictions are placed on the models.

To illustrate the procedure by which these models can be generalized to the multivariate case, the multivariate generalization of the AR(1) or autoregressive Markov model is presented. This is followed by a discussion of how restricted versions of the ARMA or FGN models can be generalized to the multivariate case. Another solution to the multisite modeling problem is provided by the disaggregation model discussed in the next section.

For simplicity of presentation, vector notation is used. Let $\mathbf{Z}_y = (Z_y^1, \ldots, Z_y^n)^T$ be the transformed zero-mean annual flows at sites $s = 1, 2, \ldots, n$, so that

$$E[Z_y^s] = 0 \tag{6.53}$$

In addition, let $\mathbf{V}_y = (V_y^1, \ldots, V_y^n)^T$ be a column vector of standard normal random variables, where V_y^s is independent of V_w^r for $(r, w) \neq (s, y)$ and independent of past flows $\mathbf{Z}_w^r$ where $y \geq w$. A sequence of synthetic flows can be generated by the model

$$\mathbf{Z}_{y+1} = \mathbf{A}\mathbf{Z}_y + \mathbf{B}\mathbf{V}_y \tag{6.54}$$

where $\mathbf{A}$ and $\mathbf{B}$ are $(n \times n)$ matrices whose elements are chosen to reproduce the lag 0 and lag 1 cross covariances of the flows at each site. The lag 0 and lag 1 covariances and cross covariances can most economically be manipulated by use of the two matrices $\mathbf{S}_0$ and $\mathbf{S}_1$; $\mathbf{S}_0$ is defined as

$$\mathbf{S}_0 = E[\mathbf{Z}_y\mathbf{Z}_y^T] \tag{6.55a}$$

and has elements

$$S_0(i, j) = E[Z_y^i Z_y^j] \tag{6.55b}$$

Matrix $\mathbf{S}_1$ is defined as

$$\mathbf{S}_1 = E[\mathbf{Z}_{y+1}\mathbf{Z}_y^T] \tag{6.56a}$$

and has elements

$$S_1(i, j) = E[Z_{y+1}^i Z_y^j] \tag{6.56b}$$

The covariances do not depend on y because the streamflows are assumed to be stationary.

Matrix $\mathbf{S}_0$, often called the covariance matrix, contains the variances and lag 0 cross covariances. Matrix $\mathbf{S}_1$ contains the lag 1 covariances and lag 1 cross covariances. $\mathbf{S}_0$ is symmetric because the cross covariance $S_0(i, j)$ equals $S_0(j, i)$. In general, $\mathbf{S}_1$ is not symmetric. The equations that specify the values of $\mathbf{A}$ and $\mathbf{B}$ in terms of $\mathbf{S}_0$ and $\mathbf{S}_1$ are obtained by manipulations of equation 6.54. Multiplying both sides of equation 6.54 by $\mathbf{Z}_y^T$ and taking expectations yields

$$E[\mathbf{Z}_{y+1}\mathbf{Z}_y^T] = E[\mathbf{A}\mathbf{Z}_y\mathbf{Z}_y^T] + E[\mathbf{B}\mathbf{V}_y\mathbf{Z}_y^T] \tag{6.57}$$

The second term on the right-hand side vanishes because the components of $\mathbf{Z}_y$ and $\mathbf{V}_y$ are independent.

Now the first term

$$E[\mathbf{A}\mathbf{Z}_y\mathbf{Z}_y^T] \tag{6.58}$$

is a matrix whose (i, j)th element equals

$$E[\sum_{k=1}^{n} a_{ik} Z_y^k Z_y^j] \tag{6.59a}$$

which equals

$$\sum_{k=1}^{n} a_{ik} E[Z_y^k Z_y^j] \tag{6.59b}$$

The matrix with these elements is equivalent to the matrix

$$\mathbf{A}E[\mathbf{Z}_y\mathbf{Z}_y^T] \tag{6.59c}$$

Hence, $\mathbf{A}$—the matrix of constants—can be pulled through the expectation operator just as is done in the scalar case where $E[a\mathbf{Z}_y + b] = aE[\mathbf{Z}_y] + b$ for fixed constants a and b.

Substituting $\mathbf{S}_0$ and $\mathbf{S}_1$ for the appropriate expectations in equation 6.57 yields

$$\mathbf{S}_1 = \mathbf{A}\mathbf{S}_0 \quad \text{or} \quad \mathbf{A} = \mathbf{S}_1\mathbf{S}_0^{-1} \tag{6.60}$$

A relationship to determine the matrix $\mathbf{B}$ is obtained by multiplying both sides of equation 6.54 by its own transpose (this is equivalent to squaring both sides of the scalar equation $a = b$) and taking expectations to obtain

$$\begin{aligned} E[\mathbf{Z}_{y+1}\mathbf{Z}_{y+1}^T] = E[\mathbf{A}\mathbf{Z}_y\mathbf{Z}_y^T\mathbf{A}^T] + E[\mathbf{A}\mathbf{Z}_y\mathbf{V}_y^T\mathbf{B}^T] \\ + E[\mathbf{B}\mathbf{V}_y\mathbf{Z}_y\mathbf{A}^T] + E[\mathbf{B}\mathbf{V}_y\mathbf{V}_y^T\mathbf{B}^T] \end{aligned} \tag{6.61}$$

The second and third terms on the right-hand side vanish because $\mathbf{Z}_y$'s and $\mathbf{V}_y$'s components are independent. $E[\mathbf{V}_y\mathbf{V}_y^T]$ equals the identity matrix because the components of $\mathbf{V}_y$ are independently distributed standard normal random variables. Thus

$$\mathbf{S}_0 = \mathbf{A}\mathbf{S}_0\mathbf{A}^T + \mathbf{B}\mathbf{B}^T \quad \text{or} \quad \mathbf{B}\mathbf{B}^T = \mathbf{S}_0 - \mathbf{A}\mathbf{S}_0\mathbf{A}^T = \mathbf{S}_0 - \mathbf{S}_1\mathbf{S}_0^{-1}\mathbf{S}_1^T \tag{6.62}$$

The last equation results from substitution of the relationship for $\mathbf{A}$ given in equation 6.60 and the fact that $\mathbf{S}_0$ is symmetric; hence $\mathbf{S}_0^{-1}$ is symmetric. It should not be too surprising that the elements of $\mathbf{B}$ are not defined uniquely. The components of the random vector $\mathbf{V}_t$ may be combined in many ways to produce the desired covariances as long as $\mathbf{B}$ satisfies equation 6.62. A lower triangular matrix that satisfies equation 6.62 can be calculated easily by Cholesky decomposition [61] [exercise 6.2(a)]. The extension of this multivariate model to the AR(2) case is left to excercise 6.6.

Matalas and Wallis [38] call equation 6.54 the *lag-1 model.* They did not call it the Markov model because the streamflows at individual sites do not have the covariances of an autoregressive Markov process given in equation 6.14. They suggest an alternative model which they call the *Markov model.* It has the same structure as the lag-1 model except it does not preserve the lag-1 cross covariances. By relaxing this requirement, they obtain a simpler model that generates flows which have the covariances of an autoregressive Markov process at each site. In their Markov model, the new $\mathbf{A}$ matrix is simply a diagonal matrix whose diagonal elements are the lag-1 correlations of flows at each site:

$$\mathbf{A} = \begin{bmatrix} \rho(1;1,1) & & & 0 \\ & \rho(1;2,2) & & \\ & & \ddots & \\ 0 & & & \rho(1;n,n) \end{bmatrix} \tag{6.63}$$

The corresponding $\mathbf{B}$ matrix depends on the new $\mathbf{A}$ matrix and $\mathbf{S}_0$, where as before

$$\mathbf{B}\mathbf{B}^T = \mathbf{S}_0 - \mathbf{A}\mathbf{S}_0\mathbf{A}^T \tag{6.64}$$

This idea can be used to generalize univariate ARMA models to the multivariate case as it was by Matalas and Wallis [37] to generalize FGN to the multivariate case. Suppose that individual ARMA(p_s, q_s) models are fit to the transformed flows z_y^s at site s. This produces estimates of the residuals $\{v_1^s, \ldots, v_Y^s$ or estimates $\{\boldsymbol{v}_1, \ldots, \boldsymbol{v}_Y\}$ of the residual vectors. A reasonable way to model the statistical dependence of concurrent flows at the various sites is to reproduce the observed lag-0 covariances and cross covariances of these residuals [50]

$$\frac{1}{Y} \sum_{y=1}^{Y} v_y^r v_y^s \tag{6.65}$$

The resulting ARMA model would generate values of the $\mathbf{V}_y$ vectors with a model

$$\mathbf{V}_y = \mathbf{B}\mathbf{E}_y \tag{6.66}$$

where the vector $\mathbf{E}_y$ contains independent standard normal random variables and the $\mathbf{B}$ matrix satisfies

$$\mathbf{B}\mathbf{B}^T = \frac{1}{Y} \sum_{y=1}^{Y} \boldsymbol{v}_y \boldsymbol{v}_y^T \tag{6.67}$$

These V_y^s's would then be used in the ARMA models for each site. While this model provides a good representation of the flows at each site over time and a good model of the joint distribution of flows at all sites at the same time, no explicit attempt has been made to reproduce the joint distribution of flows at different sites at different times. However, it may well happen that the joint distribution of flows at different sites at different times is well described by the resulting model.

6.8 MULTISEASON MULTISITE MODELS

In most studies of surface water systems it is necessary to consider the variations of flows within each year. Streamflows in most areas have within-year variations, exhibiting wet and dry periods. Similarly, water demands for irrigation, municipal, and industrial uses also vary, and the variations in demand are generally out of phase with the variation in within-year flows; more water is usually desired when streamflows are low and less is desired when flows are high. This increases the stress on water delivery systems.

This section discusses two approaches to generating within-year flows. The first approach is based on the disaggregation of annual flows produced by an annual flow generator to seasonal flows. Thus the method allows for reproduction of both the annual and seasonal characteristics of streamflows. The second approach generates seasonal flows in a sequential manner, as was done for the generation of annual flows. Thus the models are a generalization of the annual flow models already discussed.

6.8.1 Disaggregation Model

The disaggregation model proposed by Valencia and Schaake [55, 52] and extended by Mejia and Rousselle [44] allows for the generation of synthetic flows which reproduce statistics both at the annual level and at the seasonal level. It can be used for either multiseason single-site or multisite streamflow generation. Annual flows for the several sites in question or the aggregate total annual flow at several sites are the input to the model. These must be generated by another model, such as those discussed in the previous sections. These annual flows or aggregated annual flows are then disaggregated to seasonal values.

Let $\mathbf{Z}_y = (Z_y^1, \ldots, Z_y^N)^T$ be the vector of N transformed normally distributed annual or aggregate annual flows. Let $\mathbf{X}_y = (X_{1,y}^1, \ldots, X_{T,y}^1, X_{1,y}^2, \ldots, X_{Ty}^2, \ldots, X_{1y}^n, \ldots, X_{Ty}^n)^T$ be the vector of nT transformed normally distributed seasonal flows X_{ty}^s for season t, year y, and site s. Assuming that the annual and seasonal series, Z_y^s and X_{ty}^s, have zero mean, the basic disaggregation model is

$$\mathbf{X}_y = \mathbf{A}\mathbf{Z}_y + \mathbf{B}\mathbf{V}_y \tag{6.68}$$

where $\mathbf{V}_y$ is a vector of nT independent standard normal random variables, and $\mathbf{A}$ and $\mathbf{B}$ are, respectively, $nT \times N$ and $nT \times nT$ matrices. The values of the elements of $\mathbf{A}$ and $\mathbf{B}$ can be selected so as to reproduce the observed correlations among the elements of $\mathbf{X}_y$ and between the elements of $\mathbf{X}_y$ and $\mathbf{Z}_y$. Alternatively, one could attempt to reproduce the observed correlations of the untransformed flows as opposed to the transformed flows, although this is not always possible [17] and often produces poorer estimates of the actual correlations of the flows [50]. The values of $\mathbf{A}$ and $\mathbf{B}$ are determined using the matrices

$$\begin{aligned}
\mathbf{S}_{ZZ} &= E[\mathbf{Z}_y\mathbf{Z}_y^T] \\
\mathbf{S}_{XX} &= E[\mathbf{X}_y\mathbf{X}_y^T] \\
\mathbf{S}_{XZ} &= E[\mathbf{X}_y\mathbf{Z}_y^T] \\
\mathbf{S}_{ZX} &= E[\mathbf{Z}_y\mathbf{X}_y^T]
\end{aligned} \tag{6.69}$$

where $\mathbf{S}_{ZZ}$ was called $\mathbf{S}_0$ in Section 6.7. Clearly, $\mathbf{S}_{XZ}^T = \mathbf{S}_{ZX}$. If $\mathbf{S}_{XZ}$ is to be reproduced, then by multiplying equation 6.68 on the right by $\mathbf{Z}_y^T$ and taking expectations, one sees that $\mathbf{A}$ must satisfy

$$E[\mathbf{X}_y\mathbf{Z}_y^T] = E[\mathbf{A}\mathbf{Z}_y\mathbf{Z}_y^T]$$

or

$$\mathbf{S}_{XZ} = \mathbf{A}\mathbf{S}_{ZZ} \tag{6.70}$$

so that upon solving for $\mathbf{A}$ one obtains

$$\mathbf{A} = \mathbf{S}_{XZ}\mathbf{S}_{ZZ}^{-1} \tag{6.71}$$

Likewise, multiplying both sides of equation 6.68 by their transpose and taking expectations yields

$$\mathbf{S}_{XX} = \mathbf{A}\mathbf{S}_{ZZ}\mathbf{A}^T + \mathbf{B}\mathbf{B}^T \tag{6.72}$$

or

$$\mathbf{B}\mathbf{B}^T = \mathbf{S}_{XX} - \mathbf{A}\mathbf{S}_{ZZ}\mathbf{A}^T \tag{6.73}$$

Equations 6.71 and 6.73 for determining **A** and **B** are completely analogous to equations 6.60 and 6.62, and equations 6.63 and 6.64 for the **A** and **B** matrices of the lag 1 and Markov models presented in Section 6.7.2. However, for the disaggregation model as formulated, $\mathbf{B}\mathbf{B}^T$ and hence the matrix **B** can actually be singular or nearly so. This occurs because the real seasonal flows sum to the observed annual flows, so that given the annual flow at a site and $(T - 1)$ of the seasonal flows, the value of the unspecified seasonal flow can be determined by subtraction.

If the seasonal variables X_{ty}^s correspond to nonlinear transformations of the actual flows Q_{ty}^s, then $\mathbf{B}\mathbf{B}^T$ is generally sufficiently non-singular that a **B** matrix can be obtained by Cholesky decomposition. On the other hand, when the model is used to generate values of X_{ty}^s to be transformed into synthetic flows Q_{ty}^s, the constraint that these seasonal flows should sum to the given value of the annual flow is lost. Thus the generated annual flows (equal to the sums of the seasonal flows) will deviate from the values which were to have been the annual flows. Some distortion of the specified distribution of the annual flows results. This small distortion can be ignored, or each year's seasonal flows can be scaled so that their sum equals the specified value of the annual flow. The latter approach eliminates the distortion in the distribution of the generated annual flows by distorting the distribution of the generated seasonal flows.

The disaggregation model has substantial data requirements. When the dimension of $\mathbf{Z}_y$ is N and the dimension of the generated vector $\mathbf{X}_y$ is M, the **A** matrix has MN elements. The lower diagonal **B** matrix and the symmetric $\mathbf{S}_{XX}$ matrix, upon which it depends, each have $0.5M(M + 1)$ nonzero or nonredundant elemets. For example, when disaggregating two aggregate annual flow series to monthly flows at five sites, $N = 2$ and $M = 12 \times 5 = 60$ so that **A** has 120 elements while **B** and $\mathbf{S}_{XX}$ each have 1830 nonzero or nonredundant parameters. As the number of sites included in the disaggregation increases the size of $\mathbf{S}_{XX}$ and **B** increases rapidly.

When flows at many sites or in many seasons are required, the size of the disaggregation model can be reduced by disaggregation of the flows in stages and not attempting to explicitly reproduce every season-to-season correlation [25]. Figure 6.9 illustrates how aggregate annual flows at sites 1 and 2 and sites 3, 4, and 5 could be disaggregated in stages so that the total size of the disaggregation model is kept within reasonable limits. If the two aggregate annual flows are disaggregated in one step to monthly flows at the five sites, there would be 1830 nonzero elements in the lower diagonal **B** matrix. A reasonable alternative is to jointly disaggregate, using equation 6.68, the aggregate annual flows to aggregate monthly flows at the two sets of sites.

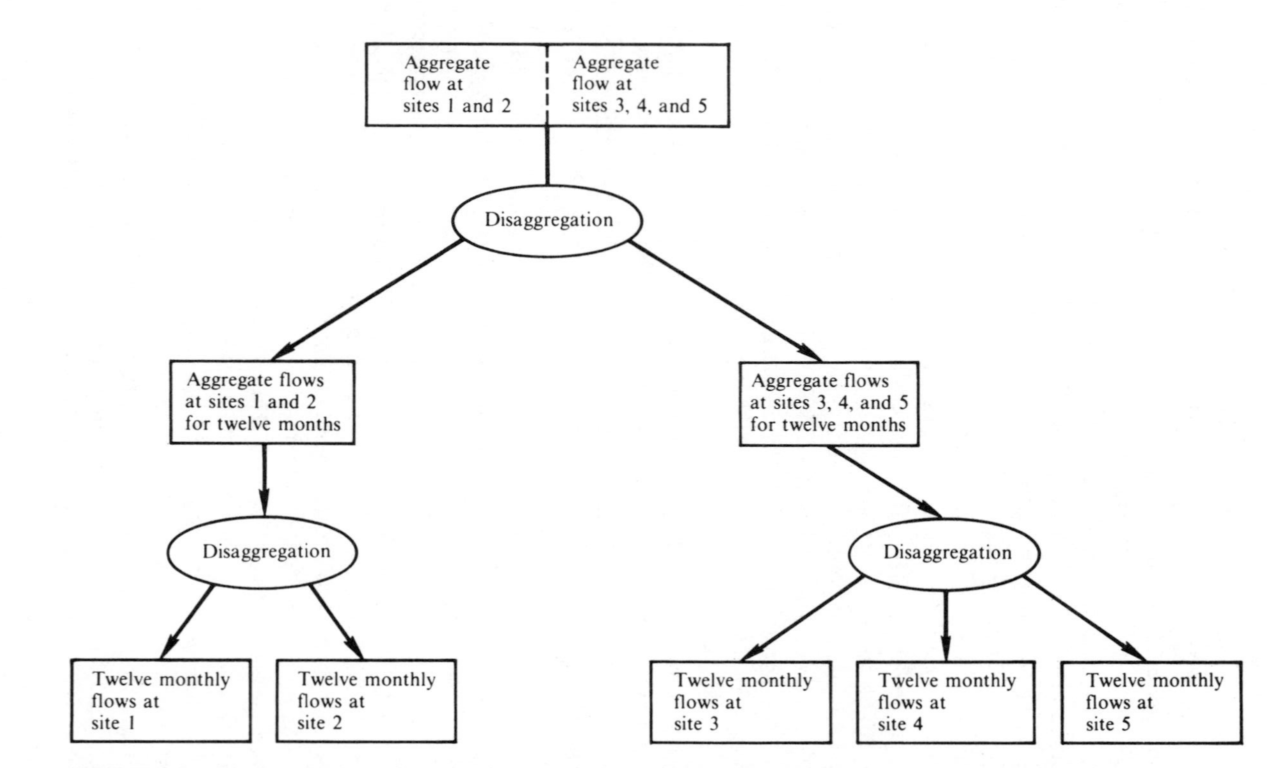

FIGURE 6.9. Disaggregation of aggregate annual flows in stages reduces the size of the disaggregation models.

These aggregate monthly flows can then be disaggregated individually to flows at each site while still preserving the month-to-month correlations of the flows at both sets of sites. The modified disaggregation model used in the second step at sites $s = 1$ and 2 is [44]

$$\begin{pmatrix} X_{ty}^{1} \\ X_{ty}^{2} \end{pmatrix} = \mathbf{A}_1 X_{ty}^{1,2} + \mathbf{A}_2 \begin{pmatrix} X_{t-1,y}^{1} \\ X_{t-1,y}^{2} \end{pmatrix} + \mathbf{BV}_{ty} \tag{6.74}$$

where $X_{ty}^{1,2}$ is the generated aggregate monthly flow at sites 1 and 2; $\mathbf{A}_1$, $\mathbf{A}_2$ and $\mathbf{B}$ are (2×1), (2×2) and (2×2) matrices; and $\mathbf{V}_{ty}$ is a vector of two standard normal random variables. Using the vectors $\mathbf{Y}_{ty} = (X_{ty}^{1,2}, X_{t-1,y}^{1}, X_{t-1,y}^{2})^T$ and $\mathbf{X}_{ty} = (X_{ty}^{1}, X_{ty}^{2})^T$ the models can be written

$$\mathbf{X}_{ty} = \mathbf{AY}_{ty} + \mathbf{BV}_{ty} \tag{6.75}$$

where $\mathbf{A} = (\mathbf{A}_1, \mathbf{A}_2)$. Using the same manipulations as before, one finds that

$$\begin{aligned} \mathbf{A} &= \mathbf{S}_{XY}\mathbf{S}_{YY}^{-1} \\ \mathbf{B} &= \mathbf{S}_{XX} - \mathbf{AS}_{YY}\mathbf{A}^T = \mathbf{S}_{XX} - \mathbf{S}_{XY}\mathbf{A}^T \end{aligned} \tag{6.76}$$

where all of these matrices are functions of t. This model preserves the variances, lag-zero covariances and lag-1 covariances of the monthly flows at sites 1 and 2 while also preserving their relationship with the aggregate annual and monthly flows at both sets of sites.

One should not attempt to reproduce the correlation of individual flows in the first month of each year X_{1y}^{s} ($s = 1$ or 2) with those in the last month of the previous year $X_{12,y}^{u}$ ($u = 1$ or 2) unless the correlation between the corresponding aggregate flow $X_{1y}^{1,2}$, and $X_{12,y-1}^{1,2}$ is also reproduced. The correlation of the last month and first month aggregate flows can be reproduced by the disaggregation model in equation 6.68 if the variables $X_{12,y-1}^{1,2}$ and $X_{12,y-1}^{3,4,5}$ are added as additional components of the $\mathbf{Z}_y$ vector. This is what was done in equation 6.75 to ensure that the disaggregation model reproduced the correlations of the transformed flows in consecutive months.

Table 6.5 summarizes the number of non-zero and non-redundant elements in the **A** and **B** matrices of the straightforward and the staged disaggregation models. Staged disaggregation allows over a 60% reduction in the number of model parameters and greatly reduces the size of the **B** matrices with which one must work. As this example shows, the disaggregation approach is very flexible. It can be adapted to meet the needs of many situations encountered in water resource simulation studies.

6.8.2 Extensions of Annual Flow Models

One approach that has been used to generate multiseason flows is to convert the time series of seasonal flows Q_{ty} into a homogeneous sequence of normally distributed zero-mean unit-variance random variables Z_{ty}.

TABLE 6.5. Comparison of Size of Straightforward and Staged Disaggregation Models

Problem

Disaggregate two aggregate annual flows to monthly flows at 5 sites

Solution 1: Straightforward Disaggregation

Size of **A** matrix = (5 × 12) × 2 = 120

Nonzero elements in **B** = (1/2)(5 × 12)[(5 × 12) + 1] = 1830

Total number of parameters = 120 + 1830 = 1950

Solution 2: Disaggregation in Stages

Stage 1: Disaggregate aggregate annual flows to aggregate monthly flows retaining correlation with aggregate flows in last month of previous year.

Size of **A** matrix = (2 × 12)(2 + 2) = 96

Nonzero elements in **B** = (1/2)(2 × 12)[(2 × 12) + 1] = 300

Stage 2: Disaggregate all aggregate monthly flows to individual flows at sites 1 and 2, and also 3, 4, and 5.

Elements in **A** matrices = 12[(2 × 3) + (3 × 4)] = 216

Nonzero elements in **B** matrices = $\left(\frac{12}{2}\right)[2(2 + 1) + 3(3 + 1)] = 108$

Total number of parameters = 96 + 300 + 216 + 108 = 720

These can then be modeled by an extension of the annual flow generators which have already been discussed. This transformation can be accomplished by fitting a reasonable marginal distribution to the flows in each season so as to be able to convert the observed flows q^s_{ty} into their transformed counterparts z^s_{ty}, and vice versa. The Young and Pisano algorithm [14, 46, 62] takes this approach and uses the lag 1 model of Section 6.7.2 to generate multisite multiseason streamflows. Others have suggested use of FGN[63] and ARMA models as models of the deseasonalized transformed streamflow series.

Particularly when shorter streamflow records are available, these simple approaches may yield a reasonable model of some streams for some studies. However, it implicitly assumes that the standardized series is stationary in the sense that the season-to-season correlations of the flows do not depend on the seasons in question. This assumption seems highly questionable. The work by Moss and Bryson [45] and the correlations of actual streamflow series suggest that the month-to-month correlation of monthly streamflows do, in fact, depend on the months in question.

This theoretical difficulty with the standardized series can be overcome by introducing a separate streamflow model for each month. For example, the classic Thomas–Fiering model [54] of monthly flows may be written

$$Z_{t+1,y} = \beta_t Z_{ty} + \sqrt{1 - \beta_t^2}\, V_{ty} \tag{6.77}$$

where the Z_{ty}'s are standard normal random variables corresponding to the streamflow in season t of year y, β_t is the season-to-season correlation of the standardized flows, and V_{ty} are independent standard normal random variables. Hoshi et al. [17] provides a comparison of the distribution of the reservoir storage required to regulate monthly flows so as to meet given target releases over a 40-year period when flows are generated by a Thomas–Fiering model and by AR(1) and ARMA(1, 1) annual flow models coupled with disaggregation. For an annual target release of 50% of the mean annual flow, the distribution of the required reservoir storage is model independent. However, for a target yield of 70% of the mean annual flow, the distribution of the required storage is model dependent, as illustrated by the results reproduced in Table 6.6.

TABLE 6.6. Storage Required[a] to Meet Monthly Target Releases Where Annual Target is 70% of Mean Annual Flow[b]

40-year Reliability (%)	*Thomas–Fiering Monthly Flow Model*	*AR*(1) *Annual Flow Generator*	*ARMA*(1, 1) *Annual Flow Generator*
80	0.31	0.37	0.38
95	0.34	0.50	0.61
98	0.36	0.62	0.83

[a]Storage reported as a fraction of mean annual flow.
[b]Values for Ishikari River, Japan, based on study reported in Hoshi et al. [17].

The divergence occurs, in part, because of the Thomas–Fiering model's inability to reproduce the persistence of observed flows. This difficulty might be overcome by introducing an ARMA model of the transformed flows Z_{ty} in each season t as a function of flows in previous seasons and perhaps even in the same season the previous year. Modeling the standardized monthly flows Z_{ty}^s by sets of simple models is an attractive alternative because of its simplicity. Still, such sets of models may not reproduce important statistics such as the variance and correlation of annual flow series.

6.9 MODEL SELECTION AND PARAMETER ESTIMATION

This chapter has only touched upon the problems of model selection and parameter estimation. As the reader can see, a large number of alternative models can be constructed, depending on which characteristics of the streamflows one wants to reproduce and how complex a model one is willing to construct. Still model selection is constrained by the availability of long high-quality streamflow records. Along this line, an interesting result reported by Slack [49] is that the sample error in observed correlations may make it impossible to fit a multisite annual flow model to a time series generated by that model. For example, FGN sequences must have a positive correlation between consecutive values, but the sample lag 1 correlation coefficient in short samples might be negative. A discussion of the difficulties that arise because of incomplete streamflow records or because the records at different sites are of different length is contained in Matalas and Wallis [38]. The importance and need for long high-quality streamflow records for the identification of an appropriate streamflow model and estimation of its parameters cannot be overemphasized.

Another important consideration is whether generated flows are consistent with the individual characteristics of the system or situation at hand. For example, unregulated flows at downstreams sties generally exceed the sum of the flows in upstream tributaries. Care must be exercised to include such constraints in statistical streamflow models. Also, streamflows at most locations must be nonnegative, a constraint that must often be enforced by setting generated negative flows to zero. This in itself is a questionable practice that illustrates the inexactness of the streamflow models and their potential physical unreasonableness. Every effort should be made to ensure that generated streamflow satisfy the physical constraints and characteristics of the modeled river basin.

Many decisions and assumptions are made in the process of selecting a streamflow model and estimating its parameters. To determine if the decisions and assumptions are statistically reasonable, that estimated model parameters are adequate, and that no gross errors have occurred, it is important to test the resultant model. At a minimum, a check should be made to ensure that generated streamflow sequences have the desired means, variances, covariances, and other characteristics that were to be reproduced.[1]

Reproduction of means, variances, and covariances alone does not ensure that a model is adequate for use in the design of water resource systems containing large storage volumes. The behavior of large reservoirs and groundwater aquifers will depend on the persistence of high and low flow sequences. While the historic record is only a single realization of a flow sequence, tests should be performed to ensure that modeled reservoir or aquifer behavior with synthetic flow sequences is consistent with the behavior that would occur if the historic flows were repeated. The adequacy of a particular streamflow model depends on the size of the projects under consid-

[1]This check can be complicated by the bias in small-sample estimates of the variances, skew coefficients, and covariances of streamflows. This problem can be overcome by use of the known mean of the annual or seasonal flows in the sample estimates of the indicated quantities; see Exercise 6.7.

eration and the length of the project planning period. This subject has not received enough attention. A reasonable test of model adequacy, along the lines suggested by Askew et al. [1], is discussed in Appendix 3B.

Omitted from this chapter is a discussion of methods for incorporating the imprecision of sample estimates of streamflow model parameters into the streamflow generation process. Sample estimates of means, variances, and other model parameters are only estimates. One should expect them to differ from the true means, variances, and parameter values which characterize the distribution from which future streamflow sequences will be drawn. The impact of this uncertainty on reservoir system simulation results and estimates of system reliability, given a 25 or 50 year historic record, is just now being realized and documented [24, 51, 42]. Given these preliminary results, the reader is advised to be aware of this omission and to be alert for future developments.

6.10 STREAMFLOW GENERATION FROM PRECIPITATION DATA

The preceding statistical methods for streamflow generation require historic streamflow records of sufficient length to provide reliable estimates of model parameters. In addition, they assume that past and future streamflows can be characterized as stationary stochastic processes. These conditions are not always met in practice. Flow records are sometimes very short or nonexistent. In addition, past and/or future modifications of a basin's drainage characteristics or climatic shifts may invalidate the assumption of stationarity. If streamflows are stationary, but the available record is short, it can sometimes be extended by use of correlation with flows at other sites at which longer records are available [12, 23, 39]. Alternatively, regional estimates of streamflow statistics can often be developed making use of watershed characteristics [2, 53]. Still another approach which is applicable if land-use changes have modified streamflow characteristics is to use rainfall–runoff models.

Rainfall–runoff models use precipitation and evapotranspiration data, either explicitly or implicitly, to calculate water balances and route water movements through a watershed. Runoff models are thus an example of the simulation process shown in Figure 6.1. In this case the simulation model of a watershed converts input sequences of precipitation and evapotranspiration into an output sequence of streamflows. The watershed's "management policy" and the drainage characteristics of the watershed are generally considered fixed. Dooge [10] and Chow [8] describe several of these models.

Generation of streamflow sequences using runoff models requires (1) selection of a runoff model that produces flows of a desired duration (hourly, daily, monthly, or annually), (2) estimation of the parameters of the runoff model, and (3) assembling a historic or synthetic precipitation sequence for input to the model. Rainfall–runoff models can be used in two different ways to obtain synthetic streamflows. The historic precipitation record may be used to produce a streamflow sequence of the same length as the historic precipitation record; this streamflow sequence can then be treated as a historic record and used directly in the development of streamflow generating models as was done in the preceding section. Thus the parameters (means, standard deviations, correlations, etc.) of streamflow models can be estimated using the "pseudo-historic" flows produced by a runoff model.

A second approach is to bypass the earlier streamflow generating models entirely and obtain synthetic flow sequences for river basin simulation directly from runoff models. In this case, synthetic precipitation sequences are generated with the characteristics of the historic precipitation record. Synthetic precipitation models (not discussed here) have different structures than streamflow generating models because of the large number of periods in which no precipitation occurs. For example, virtually all runoff models require at least daily precipitation data, and some require hourly or even every-5-minute precipitation depths. A generated sequence of daily precipitation must contain a large number of zeros, and special sampling schemes have been devised to produce this result [6, 7, 9]. Some models also account for the spatial distributed nature of rainfall, which must be considered if one has a detailed watershed model [28] or other special needs [4].

Simple rainfall–runoff models include the SCS equation [48], which gives the direct runoff from a storm as a function of precipitation and a soil moisture retention factor. The retention parameter is tabulated as a function of soil type, soil cover, season, and antecedent precipitation. With this equation, daily precipitation data are typically used to generate daily streamflows. More sophisticated rainfall–runoff models are often based on either a coarse or a fine description of the details of the hydrology of the watershed. Perhaps the most famous of these and also one of the more complex is the Stanford Watershed Model [31]. This model can consider time increments from 5 minutes to 6 hours and hence is often used to describe the continuous flood hydrograph of storm events. Haan has developed much simpler conceptual models to be used to estimate monthly or annual streamflows in small watersheds [15].

6.11 CONCLUSION

This chapter has introduced commonly used streamflow generating models. Since Thomas and Fiering first developed their autoregressive Markov model, interest in this subject has grown rapidly. Still, questions remain, including how best to estimate the parameters of alternative models and how to account for the uncertainty in model parameters. However, from a management or planning point of view, the models described in this chapter should be adequate in most situations, especially in view of the availability of streamflow data and the impact of other sources of uncertainty in the planning process [20]. Whether a particular generating model is adequate in a specific situation must be determined by the planning team. Their decision should be based on judgment, their understanding of the problem, and how sensitive system behavior is to the use of alternative models of the temporal and spatial distribution of streamflows.

EXERCISES

6-1. In Section 3.6.1 generated synthetic streamflow sequences were used to simulate a reservoir's operation. In the example, a Thomas–Fiering model was used to generate $\ell n\ Q_{1y}$ and $\ell n\ Q_{2y}$, the logarithms of the flows in year y and seasons 1 and 2, so as to preserve the season-to-season correlation of the untransformed flows. Noting that the annual flow is the sum of the untransformed seasonal flows Q_{1y} and Q_{2y}, calculate the correlation of annual flows produced by this model. The required data are given in Table 3.14. (*Hint:* You need to first calculate the covariance of $\ell n\ Q_{1y}$ and $\ell n\ Q_{1,y+1}$ and then of Q_{1y} and $Q_{2,y+1}$).

6-2. Part of New York City's municipal water supply is drawn from three parallel reservoirs in the upper Delaware River basin. The covariance matrix and lag-1 covariance matrix, as defined in equation 6.55, were estimated based on the 50-year flow record to be (in m³/s):

$$\mathbf{S}_0 = \begin{bmatrix} 20.002 & 21.436 & 6.618 \\ 21.436 & 25.141 & 6.978 \\ 6.618 & 6.978 & 2.505 \end{bmatrix} = [\text{Cov}(Q_y^i, Q_y^j)]$$

$$\mathbf{S}_1 = \begin{bmatrix} 6.487 & 6.818 & 1.638 \\ 7.500 & 7.625 & 1.815 \\ 2.593 & 2.804 & 0.6753 \end{bmatrix} = [\text{Cov}(Q_{y+1}^i, Q_y^j)]$$

Other statistics of the annual flows are:

Site	*Reservoir*	*Mean Flow*	*Standard Deviation*	r_1
1	Pepacton	20.05	4.472	0.3243
2	Cannonsville	23.19	5.014	0.3033
3	Neversink	7.12	1.583	0.2696

(a) Using these data, determine the values of the **A** and **B** matrices of the lag 1 model defined by equation 6.54. Assume that the flows are adequately modeled by a normal distribution. A lower triangular **B** matrix that satisfies $\mathbf{M} = \mathbf{BB}^T$ may be found by equating the elements of $\mathbf{BB}^T$ to those of M as follows:

$$M_{11} = b_{11}^2 \longrightarrow b_{11} = \sqrt{M_{11}}$$

$$M_{21} = b_{11}b_{21} \longrightarrow b_{21} = \frac{M_{21}}{b_{11}} = \frac{M_{21}}{\sqrt{M_{11}}}$$

$$M_{31} = b_{11}b_{31} \longrightarrow b_{31} = \frac{M_{31}}{b_{11}} = \frac{M_{31}}{\sqrt{M_{11}}}$$

$$M_{22} = b_{21}^2 + b_{22}^2 \longrightarrow b_{22}^2 = \sqrt{M_{22} - b_{21}^2} = \sqrt{M_{22} - M_{21}^2/M_{11}}$$

and so forth for M_{23} and M_{33}. Note that $b_{ij} = 0$ for $i < j$ and M must be symmetric because $\mathbf{BB}^T$ is necessarily symmetric.

(b) Determine **A** and $\mathbf{BB}^T$ for the Markov model which would preserve the variances and cross covariances of the flows at all sites at the same time and the lag 1 autocovariance of flows at each site, but not necessarily the lag 1 cross covariance of flows. Calculate the lag-1 cross covariances of flows generated with your calculated **A** matrix.

(c) Assume that some model has been built to generate the total annual flow into the three reservoirs. Construct and calculate the parameters of a disaggregation model that, given the total annual inflow to all three reservoirs, will generate annual inflows into each of the reservoirs preserving the variance and cross covariances of the flows. [*Hint:* The necessary statistics of the total flows can be calculated from those of the individual flows.]

6-3. Derive the variance of an ARMA(1, 1) process in terms of ϕ_1, θ_1, and σ_V^2. [*Hint:* Multiply both sides of equation 6.27 by V_t to obtain one equation and square both sides of the equation to obtain a second. Be careful to remember which V_t's are independent of which Z_t's.]

6-4. Show why minimizing the information criterion in equation 6.43 is equivalent to minimizing Akaike's information criterion, equation 6.42. Calculate the difference between the two criteria.

6-5. The accompanying table presents a 60-year flow record for the normalized flows of the Göta river near Sjötop-Vännersburg.

(a) Fit an autoregressive Markov model to the annual flow record.

(b) Using your model, generate a 50-year synthetic flow record. Demonstrate that the mean, variance, and correlation of your generated flows deviate from the specified values no more than would be expected as a result of sampling error.

(c) Calculate the autocorrelations and partial autocovariances of the annual flows for a reasonable number of lags. Calculate the standard errors of the calculated values. Determine reasonable values of p and q for an ARMA(p, q) model of the flows, perhaps by minimizing

Annual Flows, Göta River
Near Sjötop-Vännersburg, Sweden

1898	1.158	1918	0.948	1938	0.892
1899	1.267	1919	0.907	1939	1.020
1900	1.013	1920	0.991	1940	0.869
1901	0.935	1921	0.994	1941	0.772
1902	0.662	1922	0.701	1942	0.606
1903	0.950	1923	0.692	1943	0.739
1904	1.121	1924	1.086	1944	0.813
1905	0.880	1925	1.306	1945	1.173
1906	0.802	1926	0.895	1946	0.916
1907	0.856	1927	1.149	1947	0.880
1908	1.080	1928	1.297	1948	0.601
1909	0.959	1929	1.168	1949	0.720
1910	1.345	1930	1.218	1950	0.955
1911	1.153	1931	1.209	1951	1.186
1912	0.929	1932	0.974	1952	1.140
1913	1.158	1933	0.834	1953	0.992
1914	0.957	1934	0.638	1954	1.048
1915	0.705	1935	0.991	1955	1.123
1916	0.905	1936	1.198	1956	0.774
1917	1.000	1937	1.091	1957	0.769

Source: V. M. Yevdjevich, Fluctuations of Wet and Dry Years, Part I, Hydrology Paper No. 1, Colorado State University, Fort Collins, Colo., 1963.

the Akaike information criterion. Determine the parameter values for the selected model.

(d) Using the estimated model in (c), generate a 50-year synthetic streamflow flow record and demonstrate that the mean, variance, and first five autocorrelations of the synthetic flows deviate from the modeled values by no more than would be expected as a result of sampling error.

6-6. (a) Assume that one wanted to preserve the covariance matrices $\mathbf{S}_0$ and $\mathbf{S}_1$ of the flows at several sites $\mathbf{Z}_y$ by using the multivariate or vector ARMA(0, 1) model

$$\mathbf{Z}_{y+1} = \mathbf{A}\mathbf{V}_y - \mathbf{B}\mathbf{V}_{y-1}$$

where $\mathbf{V}_y$ contains n independent standard normal random variables. What is the relationship between the values of $\mathbf{S}_0$ and $\mathbf{S}_1$ and the matrices $\mathbf{A}$ and $\mathbf{B}$?

(b) Derive estimates of the matrices $\mathbf{A}$, $\mathbf{B}$, and $\mathbf{C}$ of the multivariate AR(2) model

$$\mathbf{Z}_{y+1} = \mathbf{A}\mathbf{Z}_y + \mathbf{B}\mathbf{Z}_{y-1} + \mathbf{C}\mathbf{V}_y$$

using the covariance matrices $\mathbf{S}_0$, $\mathbf{S}_1$ and $\mathbf{S}_2$.

6-7. When checking to ensure that generated streamflow sequences have the prescribed variance, skew coefficient, and covariances, the bias in small-sample estimates of these quantities can cause problems. For example, a biased estimator $\hat{\theta}$ of a quantity θ may be the best estimator of θ if one has only one sequence of observations with which to work. However, if one generates m independent sequences and obtains m estimates $\hat{\theta}_i$ of θ so as to estimate θ by

$$\hat{\theta}^* = \frac{1}{m}\sum_{i=1}^{m}\hat{\theta}_i$$

then

$$E[\hat{\theta}^*] = E[\hat{\theta}_i]$$

while

$$\text{Var}(\hat{\theta}^*) = \frac{1}{m}\text{Var}(\hat{\theta}_i)$$

Note that the variance of $\hat{\theta}^*$ decreases with m and thus can be made as small as required while its bias always equals the bias in $\hat{\theta}$. Thus when averaging many independent replicates of an estimator to obtain a better estimate of the quantity in question, it is often advantageous to start with an unbiased estimator.

(a) Generate 1000 20-year sequences of lognormally distributed synthetic annual flows with mean 1, variance $(0.25)^2$, skew 0.7656, and lag 1 covariance 0.03125. Calculate for each sequence i, each having n flows x_t^i, the sample estimate of the variance, skew, and lag 1 correlation using the standard estimates

$$\hat{\sigma}_i^2 = \frac{1}{n}\sum_{t=1}^{n}(x_t^i - \bar{x}_i)^2$$

$$\hat{\lambda}_i = \frac{1}{n}\sum_{t=1}^{n}(x_t^i - \bar{x}_i)^3$$

$$c_i = \frac{1}{n}\sum_{t=1}^{n-1}(x_{t+1}^i - \bar{x}_i)(x_t^i - \bar{x}_i)$$

where $n = 50$ and

$$\bar{x}_i = \frac{1}{n}\sum_{t=1}^{n}x_t^i \quad \text{and} \quad \hat{\gamma}_i = \frac{\hat{\lambda}_i}{\hat{\sigma}_i^3}$$

Show that the average of the 1000 averages $\bar{x}_i$'s indeed nearly equals μ_x. Show that the averages of $\hat{\sigma}_i^2$, $\hat{\gamma}_i$, and c_i deviate appreciably from the true values (due to the bias in the estimators). Show that this does not occur if one uses the estimators

$$\hat{\sigma}_i^2 = \frac{1}{n} \sum_{t=1}^{n} (x_t^i - \mu_x)^2$$

$$\hat{\gamma}_i = \frac{1}{n} \sum_{t=1}^{n} (x_t^i - \mu_x)^3 / \sigma_x^3$$

$$c_i = \frac{1}{n-1} \sum_{t=1}^{n-1} (x_{t+1}^i - \mu_x)(x_t^i - \mu_x)$$

with $n = 50$, $\mu_x = 1$, and $\sigma_x = 0.25$ in our case.

(b) Explain why these sample estimates of the variance, skew, and covariance yield unbiased estimates of those quantities. Is it acceptable in a check of a streamflow model's variance and covariances to use the known value of the mean flow rather than sample averages?

(c) How would you show that the differences in the first part of part (a) are statistically significant?

6-8. Formulate a model for the generation of monthly flows. The generated monthly flows should have the same marginal distributions as were fitted to the observed flows of record and should reproduce (i) the month-to-month correlation of the flows, (ii) the month-to-season correlation between each monthly flow and the total flow the previous season, and (iii) the month-to-year correlation between each monthly flow and the total 12-month flow in the previous year. Show how to estimate the model's parameters. How many parameters does your model have? How are values of the seasonal and annual flows obtained? Is this a reasonable model? How do you think that this model could be improved?

REFERENCES

1. ASKEW, A. J., W. W-G. YEH, and W. A. HALL,A Comparative Study of Critical Drought Simulation, *Water Resources Research*, Vol. 7, No. 1, 1971, pp. 52–62.
2. BENSON, M. A., and N. C. MATALAS, Synthetic Hydrology Based on Regional Statistical Parameters, *Water Resources Research*, Vol. 3, No. 4, 1967, pp. 931–945.
3. BOX, G. E. P., and G. M. JENKINS, *Times Series Analysis: Forecasting and Control*, Holden-Day, Inc., San Francisco, 1976.
4. BRAS, R. L., and R. COLÓN, Time-Average Areal Mean of Precipitation: Estimation and Network Design, *Water Resources Research*, Vol. 14, No. 5, 1978, pp. 878–888.

5. BURGES, S. J., and D. P. LETTENMAIER, Comparison of Annual Streamflow Models, *Journal of the Hydraulics Division, ASCE*, Vol. 99, No. HY9, 1973, pp. 1605–1622.

6. CAREY, D. I., and C. T. HAAN, Markov Processes for Simulating Daily Point Rainfall, *Journal of the Irrigation and Drainage Division, ASCE*, Vol. 103, No. IR1, 1978, pp. 111–125.

7. CHIN, E. H., Modeling Daily Precipitation Occurrence Process with Markov Chain, *Water Resources Research*, Vol. 13, No. 6, 1977, pp. 949–956.

8. CHOW, VEN TE (ed.), *Handbook of Applied Hydrology*, McGraw-Hill Book Company, New York, 1964, Secs. 14, 20, 21.

9. CLARK, R. T., Problems and Methods of Univariate Synthetic Hydrology, in *Mathematical Models for Surface Water Hydrology*, T. A. Ciriani, V. Maione, and J. R. Wallis (eds.), John Wiley & Sons, Inc., New York, 1977.

10. DOOGE, J. C. I., Problems and Methods of Rainfall-Runoff Modeling, in *Mathemactical Models for Surface Water Hydrology*, T. A. Ciriani, V. Maione, and J. R. Wallis (eds.), John Wiley & Sons, Inc., New York, 1977.

11. FELLER, W., The Asymptotic Distribution of the Range of Sums of Independent Random Variables, *Annals of Mathematical Statistics*, Vol. 22, 1951, pp. 427–432.

12. FIERING, M. B, On the Use of Correlation to Augment Data, *Journal of the American Statistical Association*, Vol. 57, March 1962, pp. 20–32.

13. FIERING, M. B, *Streamflow Synthesis*, Harvard University Press, Cambridge, Mass., 1967.

14. FIZZI, G., E. TODINI, and J. R. WALLIS, Comment upon Multivariate Synthetic Hydrology, *Water Resources Research*, Vol. 11, No. 6, 1975, pp. 844–850.

15. HAAN, C. T., Evaluation of a Monthly Yield Model, *Transactions of the American Society of Agricultural Engineers*, Vol. 19, No. 1, 1976, pp. 55–60.

16. HIPEL, K. W., A. I. MCLEOD, and W. C. LENNOX, Advances in Box–Jenkins Modeling. 1. Model Construction, *Water Resources Research*, Vol. 13, No. 3, 1977, pp. 567–576.

17. HOSHI, K., S. J. BURGES, and I. YAMAOKA, Reservoir Design Capacities for Various Seasonal Operational Hydrology Models, *Proceedings of the Japanese Society of Civil Engineers*, No. 273, 1978, pp. 121–134.

18. HURST, H. E., Long-Term Storage Capacity of Reservoirs, *Transactions of the ASCE*, Vol. 116, No. 776, 1951, pp. 770–799.

19. HURST, H. E., Methods of Using Long-Term Storage in Reservoirs, *Proceedings of the Institute of Civil Engineers*, Reprint 6059, 1956.

20. JAMES, I. C., B. T. BOWER, and N. C. MATALAS, Relative Importance of Variables in Water Resources Planning, *Water Resources Research*, Vol. 5, No. 6, 1969, pp. 1165–1173.

21. KIRBY, W., Computer Oriented Wilson–Hilferty Transformation That Preserves the First 3 Moments and Lower Bound of the Pearson Type 3 Distribution, *Water Resources Research*, Vol. 8, No. 5, 1972, pp. 1251–1254.

22. KLEMEŠ, V., The Hurst Phenomenon: A Puzzle? *Water Resources Research*, Vol. 10, No. 4, 1974, pp. 675–688.

23. KLEMEŠ, V., Applications of Hydrology to Water Resources Management, *Operational Hydrology Report*, No. 4, WMO-No. 356, 1973.

24. KLEMEŠ, V., and A. BULU, Limited Confidence in Confidence Limits Derived by Operational Stochastic Hydrologic Models, *Journal of Hydrology* Vol. 42, 1979, pp. 9–22 and comment by J. Stedinger, *Journal of Hydrology*, Vol. 43, 1980.

25. LANE, W., Applied Stochastic Techniques (Users Manual), Bureau of Reclamation, Engineering and Research Center, Denver, Colorado, December 1979.

26. LAWRENCE, A. J., and N. T. KOTTEGODA, Stochastic Modeling of Riverflow Time Series, *Journal of the Royal Statistical Society A*, Vol. 140, No. 1, 1977, pp. 1–47.

27. LEDOLTER, J., The Analysis of Multivariate Time Series Applied to Problems in Hydrology, *Journal of Hydrology*, Vol. 36, No. 3/4, 1978, pp. 327–352.

28. LENTON R. L., and I. RODRIGUEZ-ITURBE, A Multidimensional Model for the Synthesis of Processes of Areal Rainfall Averages, *Water Resources Research*, Vol. 13, No. 3, pp. 605–612, 1977.

29. LETTENMAIER, D. P., and S. J. BURGES, Operational Assessment of Hydrologic Models of Long-Term Persistence, *Water Resources Research*, Vol. 13, No. 1, 1977 pp. 113–124.

30. LETTENMAIER, D. P., and S. J. BURGES, An Operational Approach to Preserving Skew in Hydrologic Models of Long-Term Persistance *Water Resources Research*, Vol. 13, No. 2, 1977, pp. 281–290.

31. LINSLEY, R. K., Rainfall-Runoff Models, in *Systems Approach to Water Management*, A. K. Biswas (ed.), McGraw-Hill Book Company, New York, 1976, pp. 16–53.

32. MANDELBROT, B. B., A Fast Fractional Gaussian Noise Generator, *Water Resources Research*, Vol. 7, No. 3, 1971, pp. 543–553.

33. MANDELBROT, B. B., Broken Line Process Derived as an Approximation to Fractional Noise, *Water Resources Research*, Vol. 8, No. 5, 1972, pp. 1354–1356.

34. MANDELBROT, B. B., and J. W. VAN NESS, Fractional Brownian Motions, Fractional Brownian Noise and Applications, *SIAM Review*, Vol. 10, No. 4, 1968, pp. 422–437.

35. MANDELBROT, B. B., and J. R. WALLIS, Computer Experiments with Fractional Gaussian Noises, *Water Resources Research*, Vol. 5, No. 1, 1969, pp. 228–267.

36. MATALAS, N. C., Mathematical Assessment of Synthetic Hydrology, *Water Resources Research*, Vol. 3, No. 4, 1967, pp. 937–945.

37. MATALAS, N. C., and J. R. WALLIS, Statistical Properties of Multivariate Fractional Noise Processes, *Water Resources Research*, Vol. 7, No. 6, 1971, pp. 1460–1468.

38. MATALAS, N. C., and J. R. WALLIS, Generation of Synthetic Flow Sequences, in *Systems Approach to Water Management*, A. K. Biswas (ed.), McGraw-Hill Book Company, New York, 1976.

39. MATALAS, N. C., and B. JACOBS, A Correlation Procedure for Augmenting Hydrologic Data, U.S. Geological Survey, Professional Paper 434-E, 1964.

40. McLeod, A. I., K. W. Hipel, and W. C. Lennox, Advances in Box–Jenkins Modeling. 2. Applications, *Water Resources Research*, Vol. 13, No. 3, 1977, pp. 577–586.

41. McLeod, A. I., and K. W. Hipel, Preservation of the Rescaled Adjusted Range. 1. A Reassessment of the Hurst Phenomena, *Water Resources Research*, Vol. 14, No. 3, 1978, pp. 491–508.

42. McLeod, A. I., and K. W. Hipel, Simulation Procedures for Box–Jenkins Models, *Water Resources Research*, Vol. 14, No. 5, 1978, pp. 969–975.

43. Mejia, J. M., I. Rodriguez-Iturbe, and D. R. Dawdy, Streamflow Simulation. 2. The Broken Line Process as a Potential Model for Hydrologic Simulation, *Water Resources Research*, Vol. 8, No. 4, 1972, pp. 931–941.

44. Mejia, J. M., and J. Rousselle, Disaggregation Models in Hydrology Revisited, *Water Resources Research*, Vol. 12, No. 2, 1976, pp. 185–186.

45. Moss, M. E., and M. C. Bryson, Autocorrelation Structure of Monthly Streamflows, *Water Resources Research*, Vol. 10, No. 4, 1974, pp. 737–747.

46. O'Connell, P. E., Multivariate Synthetic Hydrology: A Correction, *Journal of the Hydraulics Division, ASCE*, Vol. 99, No. HY12, 1973, pp. 2393–2395.

47. O'Connell, P. E., ARMA Models in Synthetic Hydrology, in *Mathematical Models for Surface Water Hydrology*, T. A. Ciriani, V. Maione, and J. R. Wallis (eds.), John Wiley & Sons, Inc., New York, 1977.

48. Ogrosky, H. O., and V. Mackus, Hydrology of Agricultural Lands, in *Handbook of Applied Hydrology*, V. T. Chow (ed.), McGraw-Hill Book Company, New York, 1964.

49. Slack, J. R., I Would If I Could (Self-Denial of Conditional Models), *Water Resources Research*, Vol. 9, No. 1, 1973, pp. 247–249.

50. Stedinger, J. R., Estimating Correlations in Multivariate Streamflow Models, *Water Resources Research*, Vol. 17, 1981, to appear.

51. Stedinger, J. R., Parameter Estimation, Streamflow Model Validation and the Effects of Parameter Error and Model Choice on Derived Distributions, paper presented at Fall Meeting of the American Geophysical Union, San Francisco, 1979.

52. Tao, P. C., and J. W. Delleur, Multistation, Multiyear Synthesis of Hydrologic Time Series by Disaggregation, *Water Resources Research*, Vol. 12, No. 6, 1976, pp. 1303–1311.

53. Thomas, D. M., and M. A. Benson, Generalization of Streamflow Characteristics from Drainage-Basin Characteristics, U.S. Geological Survey Water-Supply Paper 1975, 1970.

54. Thomas, H. A., Jr., and M. B Fiering, Mathematical Synthesis of Streamflow Sequences for the Analysis of River Basins by Simulation, in *Design of Water Resources Systems*, A. Maass, M. M. Hufschmidt, R. Dorfman, H. A. Thomas, Jr., S. A. Marglin, and G. M. Fair (eds.), Harvard University Press, Cambridge, Mass., 1962.

55. Valencia, R., and J. C. Schaake, Jr., Disaggregation Processes in Stochastic Hydrology, *Water Resources Research*, Vol. 9, No. 3, 1973, pp. 580–585.

56. Wallis, J. R., and N. C. Matalas, Small Sample Properties of *H*- and *K*-Estimators of Hurst Coefficient, *h*, *Water Resources Research*, Vol. 6, No. 6, 1970, pp. 1583–1594.

57. WALLIS, J. R., and N. C. MATALAS, Correlogram Analysis Revisited, *Water Resources Research*, Vol. 7, No. 6, 1971, pp. 1448–1459.

58. WALLIS, J. R., and N. C. MATALAS, Sensitivity of Reservoir Design to the Generating Mechanism of Inflows, *Water Resources Research*, Vol. 8, No. 3, 1972, pp. 634–641.

59. WILSON, G. T., The Estimation of Parameters in Multivariate Time Series Models, *Journal of the Royal Statistical Society B*, Vol. 35, No. 1, 1973, pp. 76–85.

60. YEVJEVICH, V. M., *Stochastic Processes in Hydrology*, Water Resources Publications, Fort Collins, Colo., 1972.

61. YOUNG, G. K., Discussion of "Mathematical Assessment of Synthetic Hydrology" by N. C. Matalas, and Reply, *Water Resources Research*, Vol. 4, No. 3, 1968, pp. 681–683.

62. YOUNG, G. K., and W. C. PISANO, Operational Hydrology Using Residuals, *Journal of the Hydraulics Division, ASCE*, Vol. 94, No. HY4, 1968, pp. 909–923.

63. YOUNG, G. K., and R. V. JETTMAR, Modeling Monthly Hydrologic Persistence, *Water Resources Research*, Vol. 12, No. 5, 1976, pp. 829–835.

CHAPTER 7

Stochastic River Basin Planning Models

7.1 INTRODUCTION

The streamflow generation methods presented in Chapter 6 provide a reasonable input to simulation models for estimating the likely future performance of any particular water resource system plan. But such simulations do not provide a very effective means of choosing among all plans, those designs and operating policies that will maximize system performance indices. For this task it is once again appropriate to examine the use of optimization models, such as those developed for multipurpose river basin planning in Chapter 5.

The river basin planning model developed in Chapter 5 is deterministic. It assumes that the unregulated streamflow at any site in the basin in any time period equals the historical average or some critical value for that site and period. This assumption ignores the natural variability of such flows and hence the need to consider over-year as well as within-year active reservoir storage capacity requirements. This streamflow variability often justifies active reservoir storage capacity even though it is not required to regulate average within-year flows.

Deterministic models based on average or mean values of inputs, such as streamflows, are usually optimistic. System benefits are overestimated, and costs and losses are underestimated, if they are based only on the expected values of each input variable instead of the probability distributions describing those variables. Hence, even for the preliminary identification of efficient project designs and operating policies prior to a detailed simulation study, deterministic models are of limited value. It is because of these limitations that many stochastic planning models have been developed to account for hydrologic uncertainty. In spite of this added complexity, these models still serve only as a means of screening a wide range of design and operating policy alternatives prior to a more detailed simulation study.

The purpose of these stochastic design and operating models is not to identify the single best solution, even if a single well defined planning objective could be agreed upon. Rather they are intended to eliminate those alternatives that are clearly inferior. This is the criterion that should be used when judging the relative advantages and limitations of various types of deterministic and stochastic water resource optimization models.

In this chapter three types of stochastic river basin planning models are discussed. Each incorporates hydrologic variability and uncertainty and each is structured for solution by either linear or dynamic programming techniques. These model types include (1) models that define a number of possible discrete streamflows and storage volumes, and their probabilities, in each time interval and at each site; (2) models that identify an annual "firm" water yield, its within-year distribution, and its reliability; and (3) chance-constrained models which have rules that express the unknown reservoir storage volume and release probability distributions as linear functions of the unknown unregulated streamflows.

This chapter begins with a discussion of stochastic models for the operation of a single reservoir whose capacity, release targets, and storage volume targets are fixed. Following this, models for single- and multiple-reservoir design and operation are examined. Of primary interest will be the estimation of required active reservoir storage capacities, release and storage targets, and operating policies that maximize system performance given the distribution of inflows and the performance criteria. Those components of river basin models, discussed in Chapter 5, that identify dead storage and flood storage capacities, and downstream channel capacity improvements, are similar for all these multipurpose river basin planning models and therefore are not discussed in this chapter. However, they could be included in any of the multipurpose river basin planning models in this chapter.

7.2 RESERVOIR OPERATION

In Chapter 2 a number of models are presented for determining the optimal sequence of releases from a single reservoir given a series of known inflows. Let Q_t represent the *random* unregulated inflow into a reservoir of active capacity K_a in within-year period t. Assume that the probability distribution of the random inflow is known. Let S_t and R_t be the corresponding random initial reservoir storage volumes and releases in each period t. Clearly, the probability distributions of these random variables are not yet known, since they depend in part on the unknown reservoir operating policy. The deterministic models of Chapter 2 are often used with the expected values of each of these inflows. They can then identify the optimal reservoir releases given the expected inflows in each period. Such a deterministic operating policy is not very helpful in deciding what to release, given other than the expected or average inflows and the storage volumes that would have resulted.

To define reservoir operating policies that specify the desired reservoir release as a function of the initial storage volume and inflow in each period, several discrete optimization models may be used. The development of such models requires an analysis of streamflow data and some new notation.

First consider the random inflows Q_t. Restrict the possible inflows in each period to a number of discrete values, ordered by the index i. Let PQ_{it} be the probability that the inflow in period t will be within an interval represented by the discrete value Q_{it}. Note that these discrete inflows Q_{it} should have the same first several moments as the historical inflow probability distribution. Hence, instead of representing the entire range of possible inflows in each period t by its expected value, the range will be represented by a number of discrete inflows Q_{it}, each having a probability PQ_{it} of occurring. Similarly, restrict the range of possible initial reservoir storage volumes S_t to various discrete values, each denoted by the index k. In this case each S_{kt} is known, but each PS_{kt}, the probability that the storage volume will be within an interval represented by S_{kt} in period t, is unknown, since the probability of any particular S_{kt} is a function of the unknown operating policy.

Having defined indices i and k associated with inflow and storage volumes in period t, let indices j and l denote the corresponding inflow and storage volumes in period $t + 1$. This notation is illustrated in Figure 7.1.

Given the initial storage volumes S_{kt}, the inflows Q_{it}, and final storage volumes $S_{l,t+1}$ in period t, the releases R_{kilt} are determined by the continuity requirements:

$$R_{kilt} = S_{kt} + Q_{it} - E_{klt} - S_{l,t+1} \tag{7.1}$$

where E_{klt} is the possible evaporation and seepage losses based on the initial and final storage volumes in period t.

Let PR_{kilt} be the unknown probability of a release R_{kilt}. This equals the probability that the initial storage volume will be S_{kt}, the inflow will be Q_{it}, and the final storage volume will be $S_{l,t+1}$ in period t (which of course equals the initial storage volume in period $t + 1$). One method of determining the optimal operating policy is to develop a model for finding these joint probabilities. For each S_{kt} and Q_{it} in each period t there will be a unique $S_{l,t+1}$

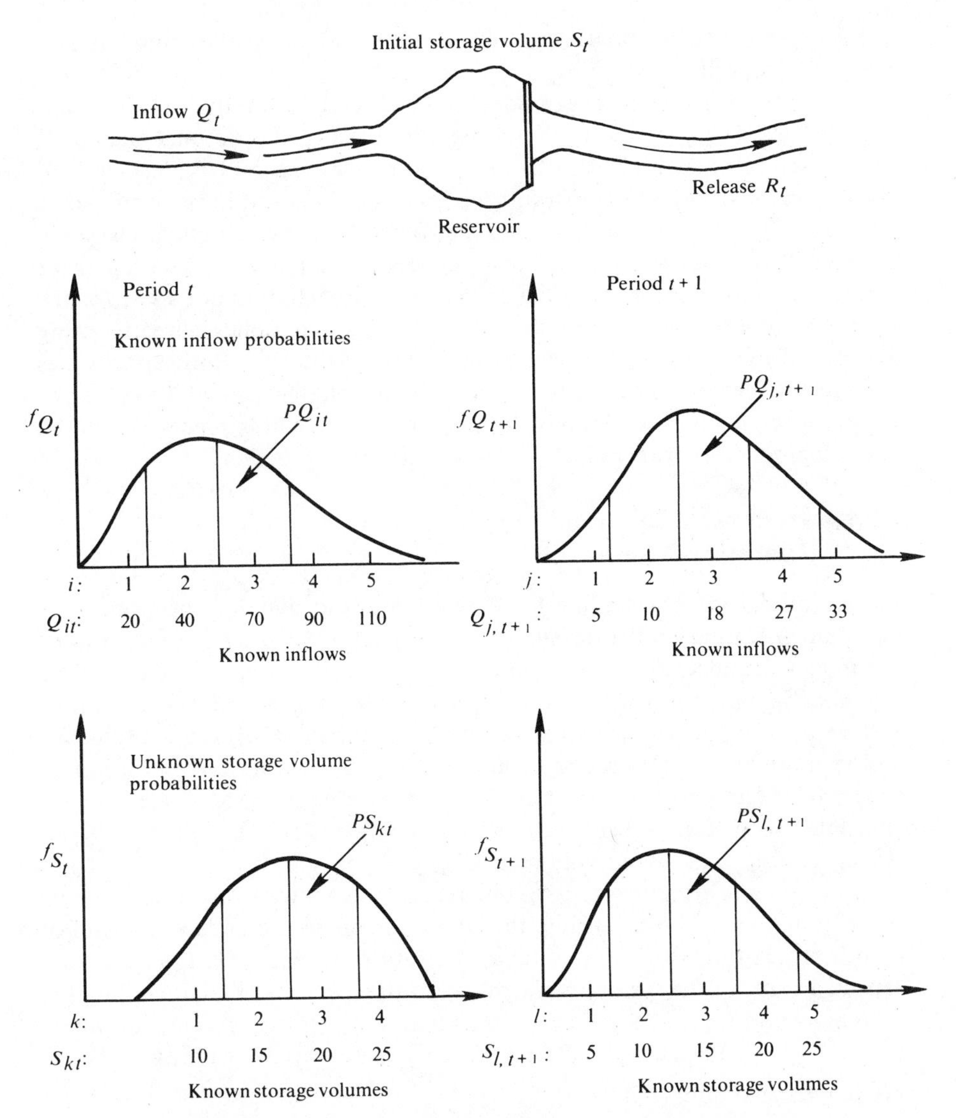

FIGURE 7.1. Discretization of random inflows Q_t and initial storage volumes S_t in each period t.

that is optimal. As a result the joint probabilities PR_{kilt} will all equal zero, except for the one corresponding to the unique optimal final volume $S_{l,t+1}$ for each combination of S_{kt}, Q_{it}, and t. Stated another way, the conditional probability of $S_{l,t+1}$ given S_{kt} and Q_{it}, which equals

$$\Pr[l \mid k, i, t] = \frac{\mathrm{PR}_{kilt}}{\sum_l \mathrm{PR}_{kilt}} \tag{7.2}$$

will, if it exists, be zero for all nonoptimal final storage volumes $S_{l,t+1}$ and will equal 1 for the optimal $S_{l,t+1}$ given the initial storage volume S_{kt} and inflow Q_{it} in period t.

It is possible to develop a model to solve directly for the optimal joint probabilities PR_{kilt} of reservoir releases, and hence the optimal operating policy itself. One such model was proposed by Thomas and Watermeyer [9] and others [1, 4]. It has a constraint set structure similar to that used to solve for the steady-state probabilities of a first-order Markov process discussed in Chapter 3. This model, presented in Appendix 7A, can be solved with a linear programming procedure. Another modeling approach does not solve for the joint probabilities PR_{kilt} but identifies the optimal policy directly using dynamic programming as a solution procedure [1, 5, 30]. Both approaches maximize expected system performance based on the probabilities of the various discrete inflows. Happily, both approaches result in identical optimal operating policies for any given reservoir operation problem.

7.2.1 Stochastic Dynamic Programming Operating Model

This model is very similar to the deterministic dynamic programming reservoir operating model developed and solved in Chapter 2. One form of the model has already been outlined in Chapter 3. At each stage or time period t, the optimal reservoir release R_{kilt}, or equivalently the final storage volume $S_{l,t+1}$, depends on two state variables: the initial storage volume S_{kt} and current inflow Q_{it}. Assume that the objective is the maximization of expected system performance. Let B_{kilt} be the value of system performance associated with an initial reservoir storage volume of S_{kt}, an inflow of Q_{it}, a release of R_{kilt}, and a final volume of $S_{l,t+1}$.

To begin the development of the recursive backward-moving dynamic programming algorithm, assume that reservoir operation ends at the end of month or season $t = T$ some year in the future. Define $f_t^n(k, i)$ as the total expected value of the system performance with n periods to go, including the current period t, given that in period t the initial storage volume and inflow are S_{kt} and Q_{it}, respectively. Then with only one period remaining,

$$f_T^1(k, i) = \underset{l}{\text{maximum}}\,(B_{kilt}) \qquad \forall k, i; \quad l \text{ must be feasible given } k \text{ and } i \text{ and } t \tag{7.3}$$

With two periods remaining before the end of reservoir operation, the maximum expected value of system performance can be calculated as

$$f_{T-1}^2(k, i) = \underset{l}{\text{maximum}}\,\Big(B_{k,i,l,T-1} + \sum_j P_{ij}^{T-1} f_T^1(l, j)\Big) \qquad \forall k, i; \quad l \text{ feasible} \tag{7.4}$$

where P_{ij}^{T-1} is the probability of inflow $Q_{j,T}$ in period T, when the inflow in period $T - 1$ equals $Q_{i,T-1}$.

The function $f_{T-1}^{2}(k, i)$ is the expected performance in the final two periods since the inflow in period T is not known with certainty in period $T - 1$. The probability of any flow Q_{jT} in period T depends on the flow $Q_{i,T-1}$ in period $T - 1$, and is given by the transition probability P_{ij}^{T-1}. Note also that $f_T^1(k, i)$ in equation 7.3 is a function of the initial volume S_{kT} and inflow Q_{iT} in that period T. When in period $T - 1$, these same volumes for the following period T are denoted by indices l and j, respectively.

The recursive relationships defined by equation 7.4 can be generalized:

$$f_t^n(k, i) = \underset{l}{\text{maximum}}\,[B_{kilt} + \sum_j P_{ij}^t f_{t+1}^{n-1}(l, j)] \qquad \forall k, i; \quad l \text{ feasible} \qquad (7.5a)$$

where t refers to the within-year period and n to the total number of periods remaining before reservoir operation terminates. As these recursive equations are solved for each period in successive years, the policy $l(k, i, t)$ defined in each particular period t will, relatively quickly, repeat in each successive year. When this condition is satisfied, and when the expected annual performance, $f_t^{n+T}(k, i) - f_t^n(k, i)$, is constant for all states k, i and all periods t within a year, the policy has reached a steady state.

This steady-state condition is attained only if the constants B_{kilt} and the transition probabilities P_{ij}^t do not change from one year to the next. Because it is this steady-state operating policy that is desired, the selection of some period and year in the future to begin writing and solving this recursive system of equations is arbitrary. This future date does not affect the steady-state reservoir operating policy.

Another common situation is when the present value of the discounted net benefits is to be maximized. In this case, the undiscounted benefit functions are assumed to be periodic so that $B_{kilt} = B_{kilt+T}$ where there are T periods per year. Then the maximum discounted net benefits can be found by solution of

$$f_t^n(k, i) = \underset{l}{\text{maximum}}\,[B_{kilt} + (1 + r)^{-1} \sum_j P_{ij}^t f_{t+1}^{n-1}(l, j)] \qquad (7.5b)$$
$$\forall k, i; \quad l \text{ feasible}$$

where now $f_t^n(k, i)$ is the present value in period t of the maximum expected discounted net benefits obtainable in the next n periods when one starts in state (k, i). If equations 7.5b are solved recursively working backward in time, a situation is eventually reached where $f_{t-T}^{n+T}(k, i)$ equals $f_t^n(k, i)$ for all k, i, and t and the steady-state solution to the problem has been found. At this point the assumed termination of the reservoir's operation after n or $n + T$ periods is so far in the future that it has no affect on either the current discounted net benefits or the reservoir's optimal operating policy. When discounting benefits, the optimal steady-state policy and the infinite-horizon discounted net benefits can also be found with another technique, called policy iteration, discussed in texts on operations research and in the water resources literature [5].

7.2.2 Probability Distributions of Storage Volumes and Releases

The steady-state operating policy derived from the solution of the dynamic programming model defines the optimal final storage volume $S_{l,t+1}$ for each initial storage volume S_{kt} and inflow Q_{it} in each period t. Let this policy be denoted by $l = l(k, i, t)$. Assuming a *pure operating strategy* (i.e., a unique and now known $S_{l,t+1}$ for each possible combination of initial storage volume S_{kt} and inflow Q_{it} in each period t), the l index is not needed for the definition of the reservoir releases and their joint probabilities. It is specified by the policy $l(k, i, t)$ for any value of k, i, and t.

Thus, the continuity equation 7.1 can be written

$$R_{kit} = S_{kt} + Q_{it} - E_{klt} - S_{l,t+1} \qquad \forall k, i, t; \quad l = l(k, i, t) \tag{7.6}$$

and PR_{kit} can be defined as the joint probability of the initial storage volume S_{kt}, the inflow Q_{it}, and also the final storage volume $S_{l,t+1}$ which is specified for each k, i, and t by the function $l(k, i, t)$. This probability PR_{kit} times the streamflow transition probability P_{ij}^t is the joint probability of S_{kt}, Q_{it}, and $Q_{j,t+1}$. Summing all such joint probabilities $\text{PR}_{kit}P_{ij}^t$ over all k and i that result in the same $S_{l,t+1}$, as defined by the function $l(k, i, t)$, results in the joint probability of $S_{l,t+1}$ and $Q_{j,t+1}$. This joint probability is also denoted as $\text{PR}_{l,j,t+1}$. In addition, the joint probabilities PR_{kit} must sum to 1 in each period t. Hence knowing the values of P_{ij}^t and the functions $l(k, i, t)$, the simultaneous solution of the following sets of equations:

$$\text{PR}_{l,j,t+1} = \sum_{\substack{k \\ l=l(k,i,t)}} \sum_{i} \text{PR}_{kit}P_{ij}^t \qquad \forall l, j, t \tag{7.7}$$

$$\sum_k \sum_i \text{PR}_{kit} = 1 \qquad \forall t \tag{7.8}$$

will yield the values of all PR_{kit}.

These equations are the familiar equations for finding the steady-state probabilities of a two-state first-order Markov chain. They are similar in form to the steady-state equations 3.74 and 3.81a. The right-hand side of equation 7.7 is a selective summation over only those initial storage and inflow indices k and i that result in a final volume having the index l. The P_{ij}^t's are the known streamflow transition probabilities.

One equation in 7.7 is redundant in each period t; hence the number of independent equations 7.7 and 7.8 equals the number of unknowns. A unique solution can be obtained by solving the simultaneous set of those equations. The values of the unknown variables PR_{kit} are the steady-state joint probabilities of initial storage volumes, inflows, and final storage volumes corresponding to indices k, i, and also their l. The corresponding marginal probability distributions of initial storage volumes, inflows, and final storage volumes associated with any operating policy $l = l(k, i, t)$ are

$$\mathrm{PS}_{kt} = \sum_{i} \mathrm{PR}_{kit} \qquad \forall k, t \tag{7.9}$$

$$\mathrm{PQ}_{it} = \sum_{k} \mathrm{PR}_{kit} \qquad \forall i, t \tag{7.10}$$

$$\mathrm{PS}_{l,t+1} = \sum_{\substack{k \\ l=l(k,i,t)}} \sum_{i} \mathrm{PR}_{kit} \qquad \forall l, t \tag{7.11}$$

7.2.3 Policy Implementation

Throughout this discussion of operating policy models, there has been the assumption that the current period's inflow is known. The generated sequential operating policies define the release or final storage volume as a function of the initial storage volume, which can be observed, and the current period's inflows, which cannot be observed until the end of the period. Hence the final task in defining a useful sequential operating policy is to express it in a manner that does not depend on unknown future inflows. This is necessary if the policy is to be implemented starting at the beginning of each period prior to a knowledge of the period's streamflow. It can be done by identifying either a final storage volume target, subject to limitations on the releases, or reservoir release targets subject to limitations on the final storage volumes, respectively, in each period t. The implementation of such an operating policy cannot guarantee constant reservoir releases throughout a period but can be used by reservoir operators as a guide at the beginning of a period prior to their knowledge of the period's inflow. How this may be done can be illustrated with a simplified example reservoir operating problem. This example problem also serves to illustrate the development and solution of the stochastic reservoir operating policy model just discussed.

7.2.4 Example Reservoir Operating Problem

Consider a two-period problem, each period having only two possible discrete inflows Q_{it} and two possible initial storage volumes for S_{kt} and $S_{l,t+1}$. The objective is to minimize the total expected sum of squared deficit and excess deviations from a constant storage target T^S, and the squared deficit deviations from a constant release target T^R, in each period t. Table 7.1 lists the discrete storage volumes, inflows, evaporation losses, releases, targets, and the resulting squared deviations B_{kilt} from the desired storage volume and release targets. This is the measure of system performance in this example problem.

The first six columns of Table 7.1 describe the states and transitions allowed in the model. Next, evaporation and seepage losses must be com-

TABLE 7.1. Calculation of Discrete System Performance Values

INITIAL STORAGE		INFLOW		FINAL STORAGE		*Evaporation Volume*	*Release*	*Squared Deviations from Storage*	*Squared Deficit Deviations from Release*	*Sum of Squared*
Index, k	*Volume,* S_{kt}	*Index,* i	*Volume,* Q_{it}	*Index,* l	*Volume,* $S_{l,t+1}$	*Loss,* E_{klt}	*Volume,* R_{kilt}	*Target,* $T^S = 20$	*Target,* $T^R = 30$	*Deviations,* B_{kilt}
					Period $t = 1$					
1	20	1	10	1	10	0	20	0	100	100
1	20	1	10	2	20	1	9	0	441	441
1	20	2	20	1	10	0	30	0	0	0
1	20	2	20	2	20	1	19	0	121	121
2	30	1	10	1	10	1	29	100	1	101
2	30	1	10	2	20	2	18	100	144	244
2	30	2	20	1	10	1	39	100	0	100
2	30	2	20	2	20	2	28	100	4	104
					Period $t = 2$					
1	10	1	30	1	20	0	20	100	100	200
1	10	1	30	2	30	0	10	100	400	500
1	10	2	40	1	20	0	30	100	0	100
1	10	2	40	2	30	0	20	100	100	200
2	20	1	30	1	20	0	30	0	0	0
2	20	1	30	2	30	1	19	0	121	121
2	20	2	40	1	20	0	40	0	0	0
2	20	2	40	2	30	1	29	0	1	1

puted. The remaining columns in Table 7.1 are computed using the information in these first eight columns. In this example, squared deviations from the storage target and squared deficits from the release targets are considered equally important in each time period. If they are not, they may be weighted, as is appropriate.

Note that for this example problem, all discrete combinations of reservoir storage volumes and inflows result in nonnegative releases. In some situations this may not occur, and those particular combinations would be infeasible.

Table 7.2 defines the streamflow transition probabilities P^t_{ij}, the probability of a streamflow $Q_{j,t+1}$ in period $t+1$ given an inflow Q_{it} in period t. Using

TABLE 7.2. Streamflow Transition Probabilities

			PERIOD $t = 2$	
		j:	1	2
	i	Q_{i1} \ Q_{j2}:	30	40
PERIOD $t = 1$	1	10	0.7	0.3
	2	20	0.2	0.8

			PERIOD $t = 1$	
		j:	1	2
	i	Q_{i2} \ Q_{j1}:	10	20
PERIOD $t = 2$	1	30	0.6	0.4
	2	40	0	1.0

the dynamic programming model to solve this example problem, let the terminal period t be 2. In this case, the minimum squared deviations $f^1_2(k, i)$ associated with each possible initial state vector k and i, with only one period remaining, are equal to

$$f^1_2(k, i) = \underset{l}{\text{minimum}}\,(B_{kil2}) \tag{7.12}$$

Once this is solved for each k and i in period $t = 2$, the next recursive equation for the expected sum of squared deviations with two remaining periods can be solved for each k and i in period $t = 1$:

$$f^2_1(k, i) = \underset{l}{\text{minimum}}\left[B_{kil1} + \sum_{j=1}^{2} P^1_{ij} f^1_2(l, j)\right] \tag{7.13}$$

Table 7.3 shows the computations associated with recursive equations 7.12 and 7.13, and the computations for the following eight stages using recursive equations of the general form:

$$f^n_t(k, i) = \underset{l}{\text{minimum}}\left[B_{kilt} + \sum_{j=1}^{2} P^t_{ij} f^{n-1}_{t+1}(l, j)\right] \tag{7.14}$$

After 10 stages a steady-state solution is obtained. A steady-state policy $l = l(k, i, t)$ is reached when the difference between $f^{n+2}_t(k, i)$ and $f^n_t(k, i)$ is essentially constant for all k, i, and t. This constant is the expected annual sum of squared deviations. For this example problem, the expected annual sum of squared deviations equals 135.9.

TABLE 7.3. Dynamic Programming Calculations for Example Problem $[f_t^n(k, i, l) = B_{kilt} + \sum_j P_{ij}^t f_{t+1}^{n-1}(l, j)]$

Stage 1: $t = 2, n = 1$

STATE		B_{kil}			
k	i	l: 1	2	$f_2^1(k, i)$	l^*
1	1	200	500	200	1
1	2	100	200	100	1
2	1	0	121	0	1
2	2	0	1	0	1

Stage 2: $t = 1, n = 2$

STATE		$f_1^2(k, i, l)$			
k	i	l: 1	2	$f_1^2(k, i)$	l^*
1	1	270	441	270	1
1	2	120	121	120	1
2	1	271	244	244	2
2	2	220	104	104	2

Stage 3: $t = 2, n = 3$

STATE		$f_2^3(k, i, l)$			
k	i	l: 1	2	$f_2^3(k, i)$	l^*
1	1	410	688	410	1
1	2	220	304	220	1
2	1	210	309	210	1
2	2	120	105	105	2

Stage 4: $t = 1, n = 4$

STATE		$f_1^4(k, i, l)$			
k	i	l: 1	2	$f_1^4(k, i)$	l^*
1	1	453	619.5	453	1
1	2	258	247	247	2
2	1	454	422.5	422.5	2
2	2	358	230	230	2

Stage 5: $t = 2, n = 5$

STATE		$f_2^5(k, i, l)$			
k	i	l: 1	2	$f_2^5(k, i)$	l^*
1	1	570.6	845.5	570.6	1
1	2	347.0	430.0	347.0	1
2	1	370.6	466.5	370.6	1
2	2	247.0	231.0	231.0	2

Stage 6: $t = 1, n = 6$

STATE		$f_1^6(k, i, l)$			
k	i	l: 1	2	$f_1^6(k, i)$	l^*
1	1	603.5	769.7	603.5	1
1	2	391.7	379.9	379.9	2
2	1	604.5	572.7	572.7	2
2	2	491.7	362.9	362.9	2

Stage 7: $t = 2, n = 7$

STATE		$f_2^7(k, i, l)$			
k	i	l: 1	2	$f_2^7(k, i)$	l^*
1	1	714.1	988.8	714.1	1
1	2	479.9	562.9	479.9	1
2	1	514.1	609.8	514.1	1
2	2	379.9	363.9	363.9	2

Stage 8: $t = 1, n = 8$

STATE		$f_1^8(k, i, l)$			
k	i	l: 1	2	$f_1^8(k, i)$	l^*
1	1	743.8	910.0	743.8	1
1	2	526.8	515.0	515.0	2
2	1	744.8	713.0	713.0	2
2	2	626.7	498.0	498.0	2

Stage 9: $t = 2, n = 9$

STATE		$f_2^9(k, i, l)$			
k	i	l: 1	2	$f_2^9(k, i)$	l^*
1	1	852.3	1127.0	852.3	1
1	2	616.0	698.0	616.0	1
2	1	652.3	748.0	652.3	1
2	2	515.0	499.0	499.0	2

Stage 10: $t = 1, n = 10$

STATE		$f_1^{10}(k, i, l)$			
k	i	l: 1	2	$f_1^{10}(k, i)$	l^*
1	1	881.4	1047.3	881.4	1
1	2	663.3	650.7	650.7	2
2	1	882.4	850.3	850.3	2
2	2	763.3	633.7	633.7	2

Examining the last few stages in Table 7.3, it is apparent that the optimal steady-state operating policy is as defined in Table 7.4. The policy is a sequential one, dependent on each initial storage volume and the current inflow. Since the current inflow is not known before the end of the period, it is desirable to state the policy in terms of a final storage volume target, or a release target, possibly subject to lower and upper bounds on the releases or final storage volumes, respectively. (In this example these bounds are not necessary, but for other problems they may be [4].) This is done in Table 7.5. Note that the policy specified in Table 7.5 is not dependent on a knowledge of the current inflow. If the policy is followed, the policy specified in Table 7.4 (which is dependent on a knowledge of the current inflow) will also be implemented.

TABLE 7.4. Optimal Operating Policy Identifying Final Storage Volumes, $S_{l,t+1}$, and Index l for Each Period t

PERIOD $t = 1$				PERIOD $t = 2$			
Q_{i1}:		10	20	Q_{i2}:		30	40
i:		1	2	i:		1	2
S_{k1}	k	S_{l2} (l)		S_{k2}	k	S_{l1} (l)	
20	1	10(1)	20(2)	10	1	20(1)	20(1)
30	2	20(2)	20(2)	20	2	20(1)	30(2)

TABLE 7.5. Reservoir Operating Policy Stated Independently of Current Inflow

Period, t	Initial Volume, S_{kt}	Final Volume Target	RELEASE LIMITS Lower	RELEASE LIMITS Upper	Release Target	FINAL VOLUME LIMITS Lower	FINAL VOLUME LIMITS Upper
1	20	—	—	—	20	—	—
1	30	20	—	—	—	—	—
2	10	20	—	—	—	—	—
2	20	—	—	—	30	—	—

This is admittedly a very simple example problem, but even for reservoir operating problems having many discrete inflow and storage volume states, the resulting discrete policies defined in tables similar to Tables 7.4 and 7.5 serve only as guides for reservoir operation. Occasionally, the uneven distribution of inflow in a period or temporary changes in the operating policy objectives may justify deviations from such policies.

Having determined the optimal sequential steady-state operating policy

$l^* = l^*(k, i, t)$, it is possible to solve equations 7.7 and 7.8 for the resulting optimal steady-state probabilities of reservoir releases PR_{kit}, and subsequently the steady-state probabilities of storage volumes and inflows using equations 7.9 and 7.10. Table 7.6 lists these probabilities for the example problem. These probabilities are also the solution to the stochastic linear programming model, equations 7A.1, 7A.3, and 7A.4, as defined for the example problem in Figure 7A.1 in this chapter's Appendix.

TABLE 7.6. Steady-State Probabilities for Example Problem

Period, t	*Initial Volume Index,* k	*Inflow Index,* i	*Optimal Final Value Index,* l	STEADY-STATE PROBABILITIES PS_{kt}	PQ_{it}	$PR_{kit} = PR_{kilt}$ (*l optimal*)
1	1			0.337		
1	2			0.663		
1		1			0.171	
1		2			0.829	
1	1	1	1			0.171
1	1	2	2			0.166
1	2	1	2			0.0
1	2	2	2			0.663
2	1			0.171		
2	2			0.829		
2		1			0.286	
2		2			0.714	
2	1	1	1			0.120
2	1	2	1			0.051
2	2	1	1			0.166
2	2	2	2			0.663

Note that the steady-state joint probability of an initial storage volume corresponding to $k = 2$, and an inflow corresponding to $i = 1$ in period $t = 1$, will be zero if the optimal operating policy, defined in Table 7.4, is followed. Such transient states are certainly possible, and if they occur, the conditional probabilities (equation 7.2) for these transient states are undefined. Nevertheless, the dynamic programming solution procedure identifies an optimal final volume for each initial volume and inflow state.

In this example problem it is easy to see why state $k = 2$, $i = 1$, will not occur in period $t = 1$ once the system has occupied a nontransient state. A final storage volume corresponding to $l = 2$ in period $t = 2$ can result only if the inflow in that second period corresponds to $j = 2$. If the inflow in period 2 is Q_{22}, then from Table 7.2, the inflow in the next period $t = 1$ is certain to be Q_{21}. Hence Q_{11} will never be observed together with S_{21} in period $t = 1$ once the reservoir has reached one of the optimal stationary states and is operated using the policy defined in Table 7.4.

7.3 SINGLE RESERVOIR DESIGN AND OPERATION

So far this chapter has focused on reservoir operation models. The remaining discussion will outline three alternative approaches for including hydrologic uncertainty in models that consider both design and operating policy variables. These design variables will include reservoir capacities, reservoir storage and/or release or diversion targets, and water allocation targets. The methods discussed in Chapter 5 for including flood storage capacities and dead storage within any model still apply and can be used with each of these modeling approaches.

7.3.1 Stochastic Linear Programming Design Model

In the previous section on operating models, equations 7.7 through 7.9 were used to define the steady-state probabilities of reservoir inflows, storage volumes, and releases. These probabilities were dependent on an operating policy $l = l(k, i, t)$ that specified the final volume $S_{l,t+1}$ given an initial volume S_{kt} and an inflow Q_{it} in each period t. For these reservoir operation models, the storage volumes S_{kt} and $S_{l,t+1}$ were given and were assumed to represent the range of all possible volumes and inflows. In this reservoir design model, the discrete volumes S_{kt} and $S_{l,t+1}$ will be considered as unknowns. Not knowing the actual storage volumes represented by each value of the index k or l does not affect the solution of equations 7.7 through 7.10. However, the functions $l(k, i, t)$ by themselves no longer define an operating policy. Without a knowledge of the actual values represented by the indices k or l in each period t, any functional relationship among the indices k, i, and l in each period t [i.e., $l(k, i, t)$] does not define the actual storage volumes or releases that should be achieved. Yet, using equations 7.7 and 7.8 together with some functions $l(k, i, t)$ does permit an estimation of the steady-state probabilities of various undetermined storage volumes and releases.

One approach for estimating the values of storage volumes, storage capacity, and releases that maximize some measure of expected system performance is to (1) select a reasonable set of functions $l(k, i, t)$, (2) solve for the steady-state probabilities PS_{kt} and PR_{kit} of all storage volumes S_{kt} and releases R_{kit}, and then (3) solve for the actual values of the volumes S_{kt} and releases R_{kit} and associated targets and capacities. Given these design and operating policy variable values, one can compute the system performance B_{kilt} associated with all feasible combinations of S_{kt} and R_{kit}. Then using the operating policy model just described, one can compute the optimal operating policy functions $l(k, i, t)$ for those design and operating policy variable values. One can then return to step (3) and repeat the process until no further improvement in expected system performance is obtained.

Experience in the actual application of these models indicates that good solutions, appropriate for further study using simulation rather than optimization methods, can be obtained if only steps 1 through 3 are completed [3]. An example problem serves to illustrate this modeling approach.

Example Problem. Consider once again the single-reservoir problem involving only two within-year periods, each having only two possible inflows. These inflows Q_{it} and their transition probabilities P_{ij}^t are defined in Table 7.2. Allow for two discrete unknown initial storage volumes S_{kt} $(k = 1, 2)$ in each period t. Of interest will be the estimation of the maximum expected annual net benefits from reservoir recreation and downstream releases or diversions less the annual cost $C(K)$ of total reservoir capacity K.

The expected net benefits of reservoir releases will depend on the long-run annual benefits $B^R(T^R)$, derived from an assumed constant but unknown release target T^R, less the expected losses, $E[L_{ikt}^R(D_{kit})]$, associated with any deficit D_{kit}. This deficit is simply the positive difference between the actual release R_{kit} and the target release T^R. Assume that excesses E_{kit} result in no loss but should not exceed a fixed amount. The deficit and excesses can be defined by:

$$R_{kit} = T^R - D_{kit} + E_{kit} \qquad \forall k, i, t \tag{7.15}$$

where D_{kit} and E_{kit} will not both be nonzero. The target, although unknown, is not a random variable; thus the probability of each deficit D_{kit} is equal to the probability of each release R_{kit}, namely PR_{kit}.

The expected annual recreation benefits can be expressed as a function $B^S(T^S)$ of the storage volume target T^S during the recreation period, less the expected losses $E[L_{kt}^S(D_{kt}, E_{kt})]$ resulting from storage volume deficits D_{kt} or surpluses E_{kt} in the recreation period. These deficits and excesses can be defined by the equation

$$S_{kt} = T^S - D_{kt} + E_{kt} \qquad \forall k, t \tag{7.16}$$

Since the target storage volume T^S is not a random variable, the probabilities of D_{kt} and E_{kt} are the same as the probabilities of S_{kt}, namely PS_{kt}.

Figure 7.2 defines the piecewise-linear approximations of the reservoir capacity cost function and the benefit and loss functions for reservoir storage and release volumes used in this example problem. Also defined are the area storage volume function and the evaporation rates $E_t\,(L)$, both of which are required for estimating evaporation losses $E_{klt}\,(L^3)$. The selected functions $l(k, i, t)$ indicating the final storage volume $S_{l,t+1}$, given an initial storage volume S_{kt} and inflow Q_{it}, are also defined in Figure 7.2.

The storage volumes represented by all k and l have yet to be determined. Since the streamflow transition probabilities and the functions $l(k, i, t)$ are the same as those associated with the steady-state probabilities listed in

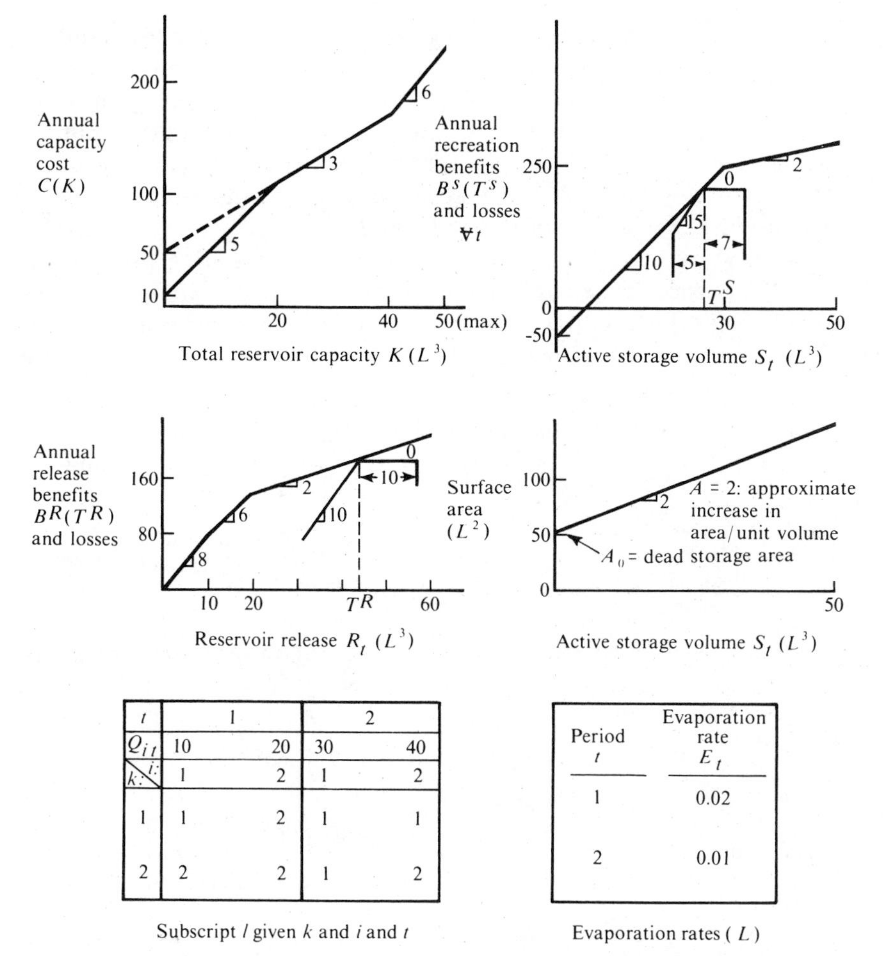

t	1		2	
Q_{it}	10	20	30	40
i: / k:	1	2	1	2
1	1	2	1	1
2	2	2	1	2

Subscript l given k and i and t

Period t	Evaporation rate E_t
1	0.02
2	0.01

Evaporation rates (L)

FIGURE 7.2. Data for example reservoir design problem.

Table 7.6, these steady-state joint probabilities will apply to this example problem. [If other functions $l(k, i, t)$ were selected, equations 7.7, 7.8, and 7.9 would have to be solved to obtain the probabilities PS_{kt} and PR_{kit}.] Given any index policy $l(k, i, t)$, the probabilities of various storage volumes and releases can be determined prior to a knowledge of the storage volumes and releases themselves.

Some care must be used in selecting the initial subscript relationships $l(k, i, t)$ or the stochastic problem may be forced into a deterministic one. Assuming, without loss of generality, that each

$$Q_{it} < Q_{i+1,t} \qquad \text{for all } i \text{ and } t \tag{7.17}$$

and

$$S_{kt} \leq S_{k+1,t} \qquad \text{for all } k \text{ and } t \tag{7.18}$$

then it also seems reasonable to assume that associated with a low-valued k and i would be the low-valued l, and vice versa. This assumption does not restrict S_{kt} to be more or less than $S_{k,t+1}$ for any k and t. Defining one more discrete storage volume than discrete inflow, or vice versa, would permit l to equal the largest integer in $(k + i)/2$. Such policy functions have been used with some success for problems that are larger than this example, but of course this or any other reasonable function $l(k, i, t)$ may not be optimal [3]. Any assumed or derived function $l(k, i, t)$ and the resulting probabilities PR_{kit} will influence the values of the reservoir releases and storage volumes.

For this model, the objective function and the constraints can be written in the following general form. The objective selected for this problem is the maximization of expected net benefits.

$$\text{maximize } B^S(T^S) + B^R(T^R) - \sum_k \sum_t \{(PS_{kt} L^S_{kt}(D_{kt}, E_{kt}) + \sum_i PR_{kit} L^R_{kit}(D_{kit})\} - C(K) \tag{7.19}$$

$$\begin{bmatrix}\text{annual benefits}\\ \text{from target}\\ \text{storage and}\\ \text{releases}\end{bmatrix} - \begin{bmatrix}\text{expected annual losses}\\ \text{from deficit or excess}\\ \text{storage volumes and}\\ \text{deficit releases}\end{bmatrix} - \begin{bmatrix}\text{annual}\\ \text{capacity}\\ \text{cost}\end{bmatrix}$$

This objective is to be maximized subject to the following set of constraints.

1. Continuity of storage:

$$S_{kt} + Q_{it} - R_{kit} - E_{klt} = S_{l,t+1} \qquad \forall k, i, t; \quad l = l(k, i, t) \tag{7.20}$$

$$\begin{bmatrix}\text{initial}\\ \text{storage}\\ \text{volume}\end{bmatrix} + \begin{bmatrix}\text{inflow}\end{bmatrix} - \begin{bmatrix}\text{release}\end{bmatrix} - \begin{bmatrix}\text{evaporation}\\ \text{losses}\end{bmatrix} = \begin{bmatrix}\text{final}\\ \text{storage}\\ \text{volume}\end{bmatrix}$$

$$(A_0 E_t)z + \frac{(A)(E_t)(S_{kt} + S_{l,t+1})}{2} = E_{klt} \qquad \forall k, l, t \tag{7.21}$$

$$\begin{bmatrix}\text{fixed}\\ \text{losses}\\ \text{if}\\ \text{capacity}\\ K > 0\end{bmatrix} + \begin{bmatrix}\text{evaporation}\\ \text{loss per}\\ \text{unit volume}\end{bmatrix} \cdot \begin{bmatrix}\text{average}\\ \text{storage}\\ \text{volume}\end{bmatrix} = \begin{bmatrix}\text{total}\\ \text{loss}\end{bmatrix}$$

$$z = \begin{cases} 1 & \text{if } K > 0 \\ 0 & \text{otherwise} \end{cases} \tag{7.22}$$

2. Reservoir capacity:

$$S_{kt} + (K_d)z + K_{ft} \leq K \qquad \forall k, t \tag{7.23}$$

$$\begin{bmatrix}\text{initial}\\ \text{storage}\\ \text{volume}\end{bmatrix} + \begin{bmatrix}\text{dead storage}\\ \text{volume}\\ \text{if } K > 0\end{bmatrix} + \begin{bmatrix}\text{flood storage}\\ \text{volume required}\\ \text{in period } t\end{bmatrix} \leq \begin{bmatrix}\text{total}\\ \text{capacity}\end{bmatrix}$$

3. Definition of deficits and excesses:

$$S_{kt} = T^S - D_{kt} + E_{kt} \qquad \forall k, t \tag{7.24}$$

$$R_{kit} = T^R - D_{kit} + E_{kit} \qquad \forall k, i, t \tag{7.25}$$

Using the data defined in Table 7.6 and Figure 7.2, this model can be written as shown in Figure 7.3:

maximize

$$\begin{aligned}
\text{ROBJ: } & (-50T_0^S + 250T_1^S + 290T_2^S) + (80T_1^R + 140T_2^R + 220T_3^R) \\
& - 5.055D_{11} - 9.945D_{21} - 2.565D_{12} - 12.435D_{22} \\
& - 1.71D_{111} - 1.66D_{121} - 0.0D_{211} - 6.33D_{221} \\
& - 1.20D_{112} - 0.51D_{122} - 1.66D_{212} - 6.63D_{222} \\
& - (10z_1 + 5K_1 + 50z_2 + 3K_2 + 6K_3)
\end{aligned}$$

subject to

1. Definition of piecewise-linear variables:
   (a) For storage target T^S

$$\text{ROW1: } T_0^S + T_1^S + T_2^S = 1$$

$$\text{ROW2: } 30T_1^S + 50T_2^S - T^S = 0$$

   (b) For release target T^R

$$\text{ROW3: } T_1^R + T_2^R + T_3^R \leq 1$$

$$\text{ROW4: } 10T_1^R + 20T_2^R + 60T_3^R - T^R = 0$$

   (c) For reservoir capacity K

$$\text{ROW5: } K_1 - 20z_1 \leq 0$$

$$\text{ROW6: } K_2 - 40z_2 \leq 0$$

$$\text{ROW7: } K_3 - (50 - 40)z_2 \leq 0$$

$$\text{ROW8: } K_1 + K_2 + K_3 - K = 0$$

$$\text{ROW9: } z_1 + z_2 - z = 0 \qquad z_1, z_2 \text{ integers}$$

2. Definition of deficit and excess allocations and bounds:

$$\text{ROW10–ROW17: } R_{kit} - T^R + D_{kit} - E_{kit} = 0 \qquad \forall k, i, t$$

$$\text{ROW18–ROW21: } S_{kt} - T^S + D_{kt} - E_{kt} = 0 \qquad \forall k, t$$

$$\text{BOUNDS: } D_{kt} \leq 5 \qquad \forall k, t$$

$$\text{BOUNDS: } E_{kt} \leq 7 \qquad \forall k, t$$

$$\text{BOUNDS: } E_{kit} \leq 10 \qquad \forall k, i, t$$

FIGURE 7.3. (a) Stochastic mixed-integer linear programming model for reservoir planning.

3. Continuity of storage:

$$\text{ROW22:} \quad -(1-0.02)S_{11} + R_{111} + 1.0z + (1+0.02)S_{12} = 10$$
$$\text{ROW23:} \quad -(1-0.02)S_{21} + R_{211} + 1.0z + (1+0.02)S_{22} = 10$$
$$\text{ROW24:} \quad -(1-0.02)S_{11} + R_{121} + 1.0z + (1+0.02)S_{22} = 20$$
$$\text{ROW25:} \quad -(1-0.02)S_{21} + R_{221} + 1.0z + (1+0.02)S_{22} = 20$$
$$\text{ROW26:} \quad -(1-0.01)S_{12} + R_{112} + 0.5z + (1+0.01)S_{11} = 30$$
$$\text{ROW27:} \quad -(1-0.01)S_{22} + R_{212} + 0.5z + (1+0.01)S_{11} = 30$$
$$\text{ROW28:} \quad -(1-0.01)S_{12} + R_{122} + 0.5z + (1+0.01)S_{11} = 40$$
$$\text{ROW29:} \quad -(1-0.01)S_{22} + R_{222} + 0.5z + (1+0.01)S_{21} = 40$$

BOUNDS: $z \leq 1$

4. Reservoir capacity:
(Since period 2 has the higher inflows, only the initial storage in period 1 need be constrained not to exceed the maximum storage capacity. Assume that a dead storage K_d of 10 units is required if the reservoir is built and no flood storage capacity is required.)

$$\text{ROW30:} \quad S_{11} + 10z - K \leq 0$$
$$\text{ROW31:} \quad S_{21} + 10z - K \leq 0$$
$$\begin{bmatrix}\text{storage}\\\text{volume}\end{bmatrix} + \begin{bmatrix}\text{dead}\\\text{storage}\\\text{if } K > 0\end{bmatrix} - \begin{bmatrix}\text{total}\\\text{reservoir}\\\text{capacity}\end{bmatrix} \leq 0$$

FIGURE 7.3. (a) (continued)

The model illustrated in Figure 7.3a can be solved by any of a variety of mixed-integer linear programming procedures. Using IBM's version, MPSX-MIP, Table 7B.1 illustrates the input data and Figure 7.3b and Table 7B.2 contain the resulting solution. In this simple example, the two possible storage volumes in each period take the same value. In models that include more discrete reservoir and inflow volumes, the discrete storage volume variables in each period would usually have different values.

In some planning situations, economic benefit and loss functions are not readily available. In these cases the objective function might be one of cost minimization subject to constraints on acceptable storage and release targets and the expected deviations from them. These and other possible modifications of the foregoing example problem will be explored in the exercises at the end of the chapter. Also discussed in a later section of the chapter will be some procedures that extend the model to multiple reservoir and water allocation planning problems. At this point, however, it is appropriate to examine some other types of stochastic design and operating models for single reservoirs.

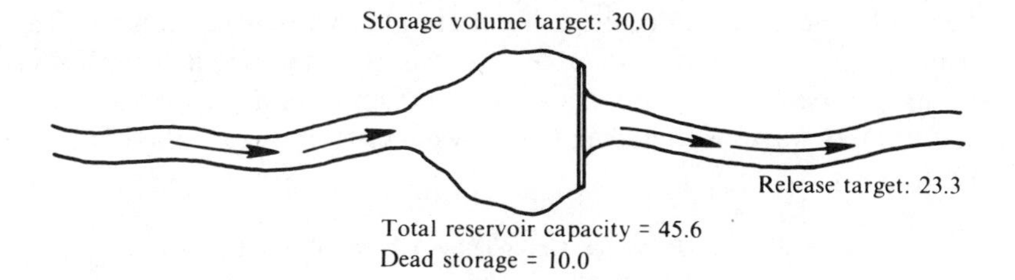

Maximum annual expected net benefits: 176

Reservoir storage	Volumes	Probabilities
Period t = 1	35.6	1.0
Period t = 2	30.0	1.0

Sequential reservoir operating policy

Period 1

Storage volume	Inflows 10.0	Inflows 20.0
35.6	13.3	23.3
	Releases	

Period 2

Storage volume	Inflows 30.0	Inflows 40.0
30.0	23.3	33.3
	Releases	

(b)

FIGURE 7.3. (b) Solution to example stochastic reservoir planning model.

7.3.2 Yield Models for Design and Operation

An alternative modeling approach for reservoir planning and operation is one that emphasizes the yields that can be achieved, and their reliabilities, with a given streamflow sequence. Yields refer to flows having a relatively high reliability or probability of being equaled or exceeded in future periods. The models to be introduced in this section can be used to estimate the storage capacity required to deliver various yields with given probabilities.

This section is divided into four subsections. The first outlines a method for estimating the reliabilities of various yields. The second discusses a modeling approach for estimating over-year and within-year active storage requirements to deliver a given yield with a specified reliability. The third and fourth subsections expand this modeling approach to include multiple yields, evaporation losses, and the construction of reservoir operation rule curves.

Throughout this discussion it will be convenient to illustrate the various yield models and their solutions using a simple two-season example. Table 7.7 lists the nine years of available streamflow data for each within-year season at a potential reservoir site. These streamflows are used to compare the solutions of various yield models as well as to illustrate the concept of yield reliability.

TABLE 7.7. Recorded Unregulated Historical Streamflows at a Reservoir Site

year,	WITHIN-YEAR PERIOD FLOW, Q_{ty}		*Annual Flow,*
y	Q_{1y}	Q_{2y}	Q_y
1	1.0	3.0	4.0
2	0.5	2.5	3.0
3	1.0	2.0	3.0
4	0.5	1.5	2.0
5	0.5	0.5	1.0
6	0.5	2.5	3.0
7	1.0	5.0	6.0
8	2.5	5.5	8.0
9	1.5	4.5	6.0
Totals	9.0	27.0	36.0
Average flow	1.0	3.0	4.0

(1) **Reliability of Annual Yields.** The maximum flow that can be made available at a specific site by the regulation of the historic streamflows from a reservoir of a given size is often referred to as the "firm yield" or "safe yield." These terms imply that the firm or safe yield is that yield which the reservoir will always be able to provide and that larger yields are unsafe in the sense that they cannot always be met. Of course, this is not true.

The firm or safe yield is 100% reliable only if in future years of reservoir operation no low flow periods will occur which are more extreme than those which occurred in the historic record. Clearly, this is not likely to be the case. Hence associated with any historic yield is a probability that that yield can be provided in any future year by a given-size reservoir with a particular operating policy.

Referring to the nine-year streamflow record listed in Table 7.7, if no reservoir is built to increase the yields downstream of the reservoir site, the historic firm yield is the lowest flow on record, namely 1.0 in year 5. The reliability of this annual yield is the probability that the streamflow in any year is greater than or equal to this value. In other words, it is the probability that this flow is exceeded. The expected value of the exceedence probability of the lowest flow in an n-year record is approximately $n/(n + 1)$, which for the 9-year record is $9/(9 + 1)$, or 0.90.

In general, the expected probability p that any flow of rank m (the largest flow having a rank of $m = 1$ and the lowest flow having a rank of $m = n$) will be equaled or exceeded in any year is approximately $m/(n + 1)$. An annual yield having a probability p of exceedence will be denoted as Y_p.

For independent events, the expected number of years until a flow of rank m is equaled or exceeded is the reciprocal of its probability of exceedence p, namely $1/p = (n + 1)/m$. The recurrence time or expected time until a failure (a flow less than that of rank m) is the reciprocal of the probability of failure in any year. Thus the expected recurrence time T_p associated with a flow having an expected probability p of exceedence is $1/(1 - p)$.

Estimation of Active Reservoir Storage Capacities Required for Increased Yields. A reservoir with active overyear storage capacity provides a means of increasing the magnitude and/or the reliabilities of various annual yields. For example, the sequent peak algorithm presented in Chapter 5 (equation 5.8) provides a means of estimating the reservoir storage volume capacity required to meet various "firm" yields $Y_{0.9}$, associated with the nine annual flows presented in Table 7.7. Ignoring the within-year distribution of the yields for the moment and assuming no evaporation or seepage losses, Figure 7.4 illustrates the annual yields that result from various active over-year reservoir storage capacities K_a^o. This is called a storage-yield function.

The same storage yield function could be obtained by specifying various yields $Y_{0.9}$, and for each yield finding the minimum active over-year reservoir

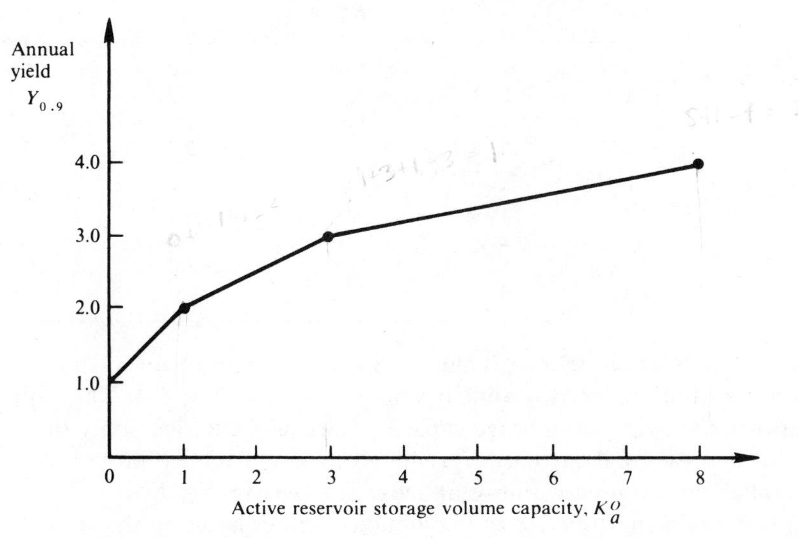

FIGURE 7.4. Active reservoir storage capacity required to provide various yields, $Y_{0.9}$, based on annual flows in Table 7.7.

capacity K_a^o required to satisfy continuity and reservoir capacity constraint equations.

$$S_y \quad + \quad Q_y \quad - \quad Y_{0.9} \quad - \quad R_y \quad = \quad S_{y+1} \tag{7.26}$$

$$\begin{bmatrix}\text{initial}\\ \text{storage}\\ \text{volume}\\ \text{year } y\end{bmatrix} + \begin{bmatrix}\text{inflow}\\ \text{year } y\end{bmatrix} - \begin{bmatrix}\text{annual}\\ \text{yield}\\ (p = 0.9)\end{bmatrix} - \begin{bmatrix}\text{excess}\\ \text{release}\\ \text{if any}\end{bmatrix} = \begin{bmatrix}\text{final}\\ \text{storage}\\ \text{volume}\\ \text{year } y\end{bmatrix}$$

$$S_y \leq K_a^o \tag{7.27}$$

$$\begin{bmatrix}\text{storage}\\ \text{volume}\end{bmatrix} \leq \begin{bmatrix}\text{overyear}\\ \text{capacity}\end{bmatrix}$$

Once again, if $y = 9$, the last year of record, then $y + 1$ is assumed to equal 1.

So far, only annual flows, annual yields, and over-year active storage volumes have been considered. Within-year periods requiring specific yields y_{pt} that sum to the annual yield Y_p may also be considered in the estimation of the required active storage capacity. Any distribution of within-year yields that differs from the distribution of the inflows may require additional active reservoir storage capacity. Given the flows in Table 7.7, Table 7.8 illustrates

TABLE 7.8. Active Storage Requirements for Various Within-Year Yields

Annual Yield, $Y_{0.9}$	WITHIN-YEAR YIELDS		REQUIRED ACTIVE STORAGE VOLUME CAPACITY		
	$t = 1$	$t = 2$	*Within-Year Capacity,* K_a^w	*Overyear Capacity,* K_a^o	*Total Capacity,* K_a
3	0	3	1.0	3.0	4.0
	1	2	0.5	3.0	3.5
	2	1	1.5	3.0	4.5
	3	0	2.5	3.0	5.5
4	0	4	1.0	8.0	9.0
	1	3	0.0	8.0	8.0
	2	2	1.0	8.0	9.0
	3	1	2.0	8.0	10.0
	4	0	3.0	8.0	11.0

the different active storage volume capacities required for various within-year distributions of two annual firm yields $Y_p = 3$ and 4. The difference between the over-year storage capacity K_a^o required to meet only the annual yields and the total capacity K_a required to meet each within-year yield can be called the required within-year active storage capacity K_a^w.

The results in Table 7.8 can be obtained either by using the sequent peak algorithm as discussed in Chapter 5 (equation 5.8), or by minimizing the total capacity K_a subject to continuity and capacity constraints for every within-year period in every year. This model is defined by equations 7.28 through 7.30 for each period t in each year y:

$$\text{minimize } K_a \tag{7.28}$$

subject to

$$S_{ty} + Q_{ty} - y_{0.9,t} - R_{ty} = S_{t+1,y} \tag{7.29}$$

$$\begin{bmatrix}\text{initial}\\ \text{storage}\\ \text{volume}\\ \text{in period}\\ t\text{, year } y\end{bmatrix} + \begin{bmatrix}\text{inflow}\\ \text{in}\\ \text{period}\\ t\text{, year } y\end{bmatrix} - \begin{bmatrix}\text{period}\\ \text{yield}\end{bmatrix} - \begin{bmatrix}\text{excess}\\ \text{release}\\ \text{if any}\end{bmatrix} = \begin{bmatrix}\text{final}\\ \text{storage}\\ \text{volume in}\\ \text{period } t\text{,}\\ \text{year } y\end{bmatrix}$$

$$S_{ty} \leq K_a \tag{7.30}$$

$$\begin{bmatrix}\text{initial}\\ \text{storage}\\ \text{volume}\end{bmatrix} \leq \begin{bmatrix}\text{total}\\ \text{capacity}\\ \text{active}\end{bmatrix}$$

In equation 7.29, if t is the final period in year y, the next period is $t = 1$ in year $y + 1$, or 1 in year 1 if y is the last year of record.

Clearly, the number of continuity and reservoir capacity constraints in the model just developed can become very large when a large number of years and within-year periods are considered. This is especially true if a number of reservoir sites are being considered, since each reservoir site requires an additional set of constraints. However, examination of solutions from reservoir storage models shows that it is only a relatively short sequence of flows within the total record of flows that generally determines the required active storage capacity K_a in a reservoir. This *critical drought period* is often used in engineering studies to estimate the "firm" or "safe" yield of any particular reservoir or system of reservoirs and other installations. Even though the severity of future droughts is unknown, many people accept the traditional practice of using the critical drought period for reservoir design and operation studies on the assumption that having observed such an event in the past, it is certainly possible to experience similar conditions in the future. (For a more complete discussion of critical period analyses, see Hall and Dracup [31].)

Since reservoir storage requirements are determined from critical periods of record, this suggests that it may not be necessary to include every period of every year in a reservoir storage yield model such as that defined by equations 7.28 through 7.30. Indeed, a simpler model can be constructed.

To begin the development of a simpler model, consider a simulation of a reservoir over the 9 years of inflows listed in Table 7.7. Assume that the desired firm yield $y_{0.9,1}$ in the dry period ($t = 1$) of each year is 3 and that no yield is desired in the wet period ($t = 2$); hence $y_{0.9,2} = 0$. Figure 7.5 is a plot of the simulated storage volumes at the beginning and end of each of the two periods, for the 9 years of record. Note that the total active storage

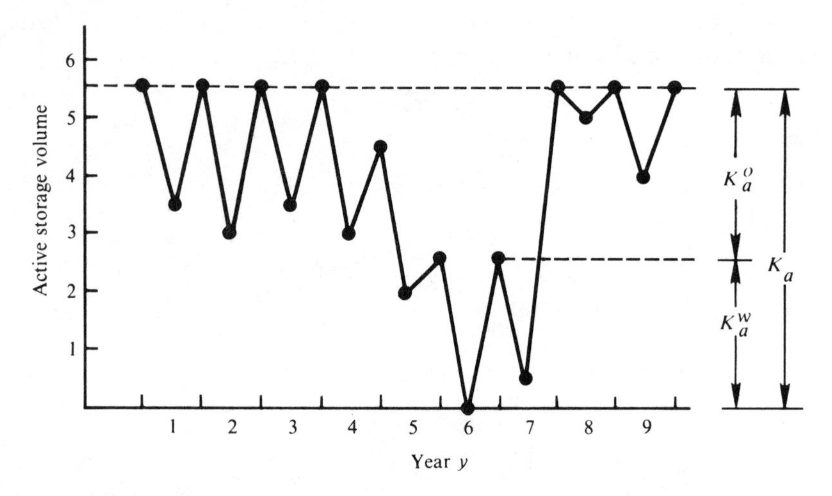

FIGURE 7.5. Sequence of storage volumes for a reservoir having sufficient capacity to supply a "firm" yield of 3 in period 1 given the inflows listed in Table 7.7.

requirement K_a is 5.5. As expected, the storage volumes S_{1y} at the end of the wet period (beginning of dry period) each year equal the active storage capacity K_a except during the years of the critical drought period. These years of storage drawdown are those years in which the total annual flow is less than the annual yield.

From Figure 7.5 note that the range of volumes S_{1y} at the beginning of each year y defines the over-year storage requirement K_a^o. The remaining storage requirement is the within-year storage capacity K_a^w needed to get through the critical year. This critical year is generally at the end of a sequence of years having annual flows less than the annual yields. Thus the reservoir storage capacity derived from solving equations 7.28 to 7.30 can also be obtained from a model having year-to-year or over-year continuity constraints to define each year's initial storage volume S_{1y}, or simply S_y, plus a set of within-year continuity constraints for the critical year.

This reduces the number of storage continuity constraints and storage capacity constraints from twice the product of the number of years times the number of within-year periods (equations 7.29 and 7.30) to twice the sum of the number of years plus the number of within-year periods (equations 7.26 and 7.27 for over-year constraints plus another set of within-year continuity and capacity constraints).

Writing the within-year continuity constraints for the critical year requires the identification of the critical year and its inflows. In studies where the yields y_{pt} in each period of each year are to be determined, as they are in design models being developed here, it is not possible to identify the critical year at the time of model development. This is because the critical year depends in part on the values of the annual and within-year yields. However, good results are generally obtained by letting some appropriate fraction β_t of the total annual yield Y_p be the inflow in each period t within the critical year. Hence $\sum_t \beta_t = 1$. A good choice for β_t is the ratio of the inflow in period t of the driest year of record to the total inflow that year. Each β_t thus reflects the relative proportion of the critical year's inflow that is likely to occur in period t.

The within-year continuity constraints for a single yield can be written

$$s_t + \beta_t Y_p - y_{pt} = s_{t+1} \qquad \forall t \tag{7.31a}$$

Because the β_t's sum to one these constraints ensure that $\sum_t y_{pt}$ equals the annual yield Y_p.

If the within-year yields y_{pt} are constrained to equal some predetermined fraction f_t of the annual yield Y_p ($\sum_t f_t = 1$), the within-year constraints can be written

$$s_t + \beta_t Y_p - f_t Y_p = s_{t+1} \qquad \forall t \tag{7.31b}$$

In both equations 7.31a and 7.31b, the inflows and required releases are just in balance, so that the reservoir neither fills nor empties during the modeled critical year. This is similar to what would be expected in the critical year that generally occurs at the end of a drawdown period.

The within-year capacity K_a^w is the maximum of all within-year storage volumes s_t:

$$K_a^w \geq s_t \qquad \forall t \tag{7.32}$$

The total active storage capacity is simply the sum of the over-year storage and within-year storage capacities (Figure 7.5):

$$K_a = K_a^o + K_a^w \tag{7.33}$$

Combining equations 7.32 and 7.33,

$$K_a \geq K_a^o + s_t \qquad \forall t \tag{7.34}$$

The complete yield model defined by equations 7.28 through 7.30 can be compared with the approximate yield model defined by equations 7.26, 7.27, 7.28, 7.31, and 7.34. For a hydrologic record of n years, each having T periods, the number of constraint equations is reduced from $2nT$ to $2(n + T)$. The number of variables is reduced from $2nT + T + 2$ to $2n + 2T + 3$.

Table 7.9 presents the solutions obtained from the approximate model for the example problem whose streamflows are defined in Table 7.7. In this example, $\beta_t = \frac{1}{2}$ for $t = 1$ and 2. The active storage volume capacity estimates shown in Table 7.9 can be compared to those derived from the complete model (equations 7.28 through 7.30, or the sequent peak algorithm) that are listed in Table 7.8.

Comparing Tables 7.8 and 7.9 shows that the approximate model may

TABLE 7.9. Active Storage Requirements for Various Within-Year Yields Based on Approximate Yield Models

	WITHIN-YEAR PERIOD YIELDS		ESTIMATED REQUIRED ACTIVE STORAGE VOLUME CAPACITIES		
Annaul Yield, $Y_{0.9}$	$y_{0.9,1}$	$y_{0.9,2}$	*Within-Year,* K_a^w	*Over-Year,* K_a^o	*Total Capacity,* K_a
3	0	3	1.5	3.0	4.5
	1	2	0.5	3.0	3.5
	2	1	0.5	3.0	3.5
	3	0	1.5	3.0	4.5
4	0	4	2.0	8.0	10.0
	1	3	1.0	8.0	9.0
	2	2	0.0	8.0	8.0
	3	1	1.0	8.0	9.0
	4	0	2.0	8.0	10.0

over- or underestimate the storage required to provide a specified yield with a particular hydrologic flow sequence. A more thorough comparison of actual and estimated storage requirements is presented in Figure 7.6.

Figure 7.6 shows the results of solving for the active storage required to provide a constant monthly yield equal to one-twelfth of the annual yield. Annual yields equal to 50, 75, and 90% of the mean annual flow μ are considered. A solution was obtained for each of 2000 50-year synthetic monthly streamflow sequences.

Three different models were used to determine or estimate the required active storage capacity. The first, denoted as SP on Figure 7.6, represents solutions to the complete model (equations 7.28 through 7.30) or, equivalently, the sequent peak procedure, equation 5.8. The approximate model was solved twice for each flow sequence, once using β_t's based on the inflow distribution of the driest flow year, and again using $\bar{\beta}_t$ based on the inflow distribution of the average monthly inflows. Note the wide range in the required active storage capacities K_a. Overall, the use of β_t based on the driest year of record provides as reasonable an estimate of the future storage requirements as does the more complete and larger model.

Simulation studies with other within-year yield distributions produce results similar to those in Figure 7.6, except when the β_t's representing the inflow distribution closely correspond to the within-year distribution of the yields y_{pt}/Y_p. Then the yield model tends to underestimate within-year storage requirements, especially if the level of development is low. Fortunately, this situation is not commonly encountered in practice, since demands for water generally increase during periods of low natural flows.

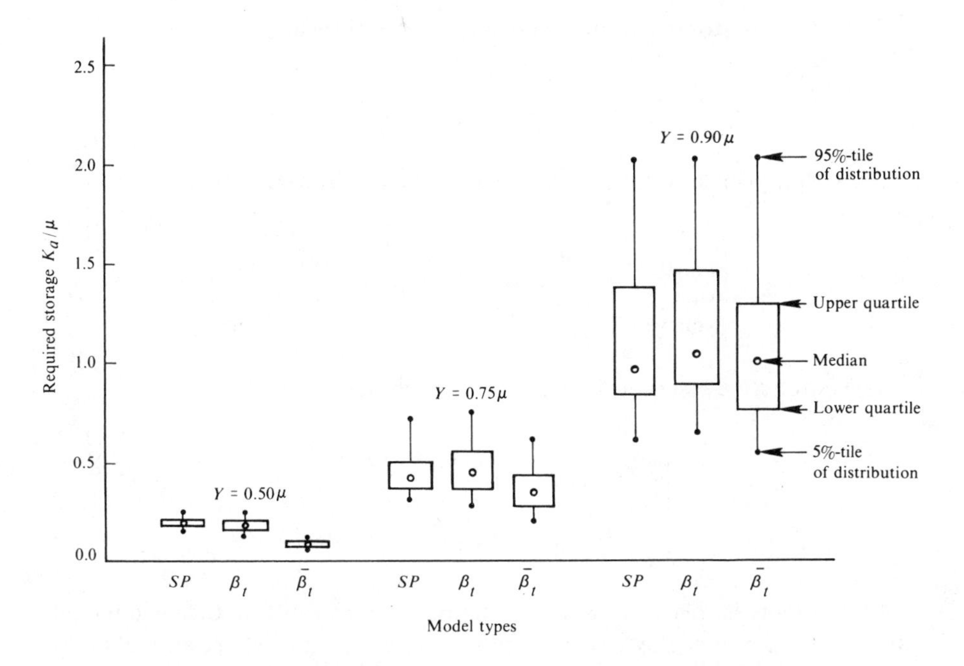

FIGURE 7.6. Comparison of required storage capacities K_a and estimates of required capacities obtained with yield models using driest-year inflow distribution (β_t) and average inflow distribution ($\bar{\beta}_t$). Results derived from 2000 fifty-year monthly streamflow sequences and expressed as fractions of true mean annual flow μ. Points (•) correspond to 95%-tiles and 5%-tiles; circles (∘) correspond to median, and box ranges from upper to lower quartile and hence contains 50% of the observations.

The simple yield model with an objective of minimizing the total active storage capacity K_a required to ensure known within-year yields y_{pt} can be written as:

$$\text{minimize} \quad \underset{\begin{bmatrix}\text{total active}\\ \text{capacity}\end{bmatrix}}{K_a} \tag{7.35a}$$

subject to

1. Over-year storage continuity. For each year y:

$$\underset{\begin{bmatrix}\text{initial}\\ \text{storage}\\ \text{volume}\\ \text{in year } y\end{bmatrix}}{S_y} + \underset{\begin{bmatrix}\text{inflow}\\ \text{in year } y\end{bmatrix}}{Q_y} - \underset{\begin{bmatrix}\text{yield}\\ p = 0.9\end{bmatrix}}{Y_p} - \underset{\begin{bmatrix}\text{excess}\\ \text{release,}\\ \text{if any}\end{bmatrix}}{R_y} = \underset{\begin{bmatrix}\text{final}\\ \text{storage}\\ \text{volume}\\ \text{in year } y\end{bmatrix}}{S_{y+1}} \tag{7.35b}$$

2. Over-year storage volume capacity. For each year y:

$$S_y \leq K_a^o \tag{7.35c}$$

$$\begin{bmatrix}\text{storage}\\ \text{volume}\end{bmatrix} \leq \begin{bmatrix}\text{over-year}\\ \text{capacity}\end{bmatrix}$$

3. Within-year storage continuity. For each within-year period t:

$$s_t + \beta_t Y_p - y_{pt} = s_{t+1} \tag{7.35d}$$

$$\begin{bmatrix}\text{initial}\\ \text{storage}\\ \text{volume}\end{bmatrix} + \begin{bmatrix}\text{inflow}\\ \text{proportion}\\ \text{of annual}\\ \text{yield}\end{bmatrix} - \begin{bmatrix}\text{yield}\end{bmatrix} = \begin{bmatrix}\text{final}\\ \text{storage}\\ \text{volume}\end{bmatrix}$$

4. Total active reservoir capacity. For each period t:

$$K_a^o + s_t \leq K_a \tag{7.35e}$$

$$\begin{bmatrix}\text{over-year}\\ \text{storage}\\ \text{capacity}\end{bmatrix} + \begin{bmatrix}\text{within-}\\ \text{year}\\ \text{storage}\\ \text{volume}\end{bmatrix} \leq \begin{bmatrix}\text{total}\\ \text{active}\\ \text{capacity}\end{bmatrix}$$

Without any increase in error, the number of constraints in the yield model can be further reduced by grouping all sequences of years whose annual flows Q_y exceed the mean annual flow or maximum yield, whichever is smaller. During these years the reservoir will fill so that no shortages need occur. Referring to the example 9-year flow sequence presented in Table 7.7, this would permit years 7, 8, 9, and 1 to be combined if the annual yield is 4. There would then be 12 over-year constraints rather than 18. The resulting over-year constraints are

$$S_y + Q_y - Y_p - R_y = S_{y+1} \qquad y = 2, 3, \ldots, 6 \tag{7.36a}$$

$$S_7 + Q_7 + Q_8 + Q_9 + Q_1 - 4Y_p - R_7 = S_2 \tag{7.36b}$$

$$S_y \leq K_a^o \qquad y = 2, 3, \ldots, 7 \tag{7.37}$$

If the annual yield were 3, then years 6 through 3 (assuming again that year 1 follows year 9) can be combined, representing an additional reduction of six constraints. Furthermore, not all reservoir capacity constraint equations 7.35c or 7.37 are needed, since the initial storage volumes in the years following low flows will probably be less than the over-year capacity K_a^o. Note, in equation 7.36b, R_7 corresponds to an aggregate release rather than the release in just the seventh year.

The approximate yield model can be used to estimate active storage volume capacity requirements associated with yields having less than the maximum estimated probability of exceedence p. In these cases a yield failure is permitted. The number of years of yield failure determines the estimated reliability of each yield. An annual yield that fails in f years has an estimated probability $(n - f)/(n + 1)$ of being equaled or exceeded in any future year. Once the desired reliability of a yield is known, the problem is to select the appropriate number f of failure years and the specific failure years themselves, and to specify the permissible extent of failure in those failure years.

Let α_{py} be defined as follows:

$$\alpha_{py} = \begin{cases} 1 & \text{if the yield } Y_p \text{ is to be provided in year } y \\ < 1 & \text{if the yield } Y_p \text{ need not necessarily be provided in year } y \end{cases} \tag{7.38}$$

The value of α_{py}, when less than 1, indicates the extent of permissible yield failure. Its value is in part dependent on the consequences of failure and on the ability to forecast when a failure may occur and to adjust the reservoir operating policy accordingly.

The over-year storage continuity constraints can now be written in a form appropriate for identifying any single annual yield Y_p having an exceedence probability p.

$$S_y + Q_y - \alpha_{py} Y_p - R_y = S_{y+1} \quad \forall y \tag{7.39}$$

$$\begin{bmatrix}\text{initial}\\ \text{storage}\\ \text{volume}\end{bmatrix} + \begin{bmatrix}\text{annual}\\ \text{inflow}\end{bmatrix} - \begin{bmatrix}\text{yield}\\ \text{release}\end{bmatrix} - \begin{bmatrix}\text{excess}\\ \text{release}\end{bmatrix} = \begin{bmatrix}\text{final}\\ \text{storage}\\ \text{volume}\end{bmatrix}$$

When writing equations 7.39, the failure year or years should be selected from among those in which permitting a failure decreases the required reservoir capacity K_a. If a failure year is selected in which an excess release would be made anyway, no reduction in the required active storage capacity will result, and the reliability of the yield may be higher than intended.

The critical year or years that determine the required active storage volume capacity may be dependent on the yield itself. Consider, for example, the 7-year sequence of annual flows (4, 3, 3, 2, 8, 1, 7) whose mean is 4. If a yield of 2 is desired in each of the 7 years, the critical year requiring reservoir capacity is year 6. If a yield of 4 is desired (again assuming no losses), the critical years are years 2 through 4. The stream-flows and yields in these *critical years* determine the required over-year storage capacity. The failure years, if any, must be selected from within the critical drought periods for the desired yield.

When the magnitudes of the yields are unknown, some trial-and-error procedures may be necessary to ensure that any failure years are within the critical period of years for the associated yields. To ensure a wider range of applicable yield magnitudes, the year having the lowest flow within the critical period should be selected as the failure year if only one failure year is selected. Even though the actual failure year may follow that year, the resulting required reservoir storage volume capacity will be the same.

(3) **Multiple Yields and Evaporation Losses.** The yield models developed so far define only single annual and within-year yields. Incremental secondary yields having lower reliabilities can also be included in the model. Referring to the 9-year streamflow record in Table 7.9, assume that two yields are desired, one 90% reliable and the other 70% reliable. Let $Y_{0.9}$ and $Y_{0.7}$ represent those annual yields having reliabilities of 0.9 and 0.7, respectively. The incremental secondary yield $Y_{0.7}$ represents the amount in addition to $Y_{0.9}$ that is only 70% reliable. Assume that the problem is one of estimating the appropriate values of $Y_{0.9}$ and $Y_{0.7}$, their respective within-year distributions y_{pt} and the total reservoir capacity K that maximizes some function of these yield and capacity variables.

In many regions of the world, it is necessary to include evaporation losses in these analyses to obtain accurate estimates of the required storage. Since the approximate yield model (e.g., equations 7.35a to 7.35e) does not identify the exact storage volumes at the beginning of each period in each year, evaporation losses must be based on an expected storage volume in each period and year. The approximate storage volume in any period t in year y can be defined as the initial overyear volume S_y, plus the estimated average within-year volume $(s_t + s_{t+1})/2$. The annual evaporation volume loss E_y in each year y can be based on these estimated average storage volumes. This loss will be approximately equal to the average annual fixed loss E_0 (see Figure 5.5), plus the sum of each period's volume loss per unit of active storage volume times the expected storage volume in the period. Letting γ_t be the fraction of the annual evaporation volume loss that occurs in period t and E be the average annual evaporation volume loss rate per unit of active storage volume, the annual evaporation loss in year y equals

$$E_y = \sum_t \left[\gamma_t E_0 + \left(S_y + \frac{s_t + s_{t+1}}{2} \right) \gamma_t E \right] \tag{7.40}$$

Since the sum of all fractions γ_t equals 1, equation 7.40 can be simplified to

$$E_y = E_0 + \left[S_y + \sum_t \left(\frac{s_t + s_{t+1}}{2} \right) \gamma_t \right] E \tag{7.41}$$

The within-year evaporation loss in each period t of the critical year is approximately

$$e_t = \gamma_t E_0 + \left(\frac{s_t + s_{t+1}}{2} \right) \gamma_t E \tag{7.42}$$

assuming that the initial over-year storage in the critical year is 0. (If upon the solution of the model, the minimum value $S_y^{\min}$ of over-year storage S_y is substantially greater than zero due to, say, the benefits derived from increased inactive storage for recreation or hydropower, the problem should be resolved with the added term $\gamma_t E S_y^{\min}$ in equation 7.42.)

The yield model can now be written to include the two desired yields and the evaporation losses for the example problem. Assume that $B_{pt}(y_{pt})$ are the benefits derived from a yield y_{pt}, their sum Y_p having a probability p of being equaled or exceeded in each year. Let $C(K)$ be the annual cost of the total reservoir capacity K. This capacity includes dead storage K_d and flood control storage capacity K_{ft} required at the beginning of period t. Let P denote the set of exceedence probabilities p of interest, namely 0.9 and 0.7. The objective will be to find the yields y_{pt} $(p \in P)$ that maximize net benefits. While active capacity is an unknown variable, it is assumed to be greater than zero and hence dead storage K_d will be required and fixed evaporation losses will occur.

The yield model for this problem is as follows:

$$\text{maximize} \quad \sum_t \sum_{p \in P} B_{pt}(y_{pt}) - C(K) \tag{7.43}$$

$$\begin{bmatrix}\text{total annual}\\ \text{benefits from}\\ \text{yields } y_{pt},\\ p = 0.9 \text{ and } 0.7\end{bmatrix} - \begin{bmatrix}\text{annual}\\ \text{cost of}\\ \text{capacity}\end{bmatrix}$$

subject to

1. Over-year storage continuity. For each year y:

$$S_y + Q_y - Y_{0.9} - \alpha_{0.7,y}Y_{0.7}$$

$$\begin{bmatrix}\text{initial}\\ \text{storage}\\ \text{volume}\end{bmatrix} + \begin{bmatrix}\text{annual}\\ \text{inflow}\\ \text{(Table 7.7)}\end{bmatrix} - \begin{bmatrix}\text{firm}\\ \text{yield}\end{bmatrix} - \begin{bmatrix}\text{secondary}\\ \text{incremental}\\ \text{yield}\end{bmatrix}$$

$$- E_y - R_y = S_{y+1} \tag{7.44}$$

$$- \begin{bmatrix}\text{evaporation}\\ \text{loss}\end{bmatrix} - \begin{bmatrix}\text{excess}\\ \text{release}\end{bmatrix} = \begin{bmatrix}\text{final}\\ \text{storage}\\ \text{volume}\end{bmatrix}$$

where

$$\alpha_{0.7,y} = \begin{cases} 0 & \text{for failure years} \\ 1 & \text{otherwise} \end{cases} \tag{7.45}$$

Note that if y is the last year of record, then $y + 1 = 1$. Also, for multiple yield problems, failure fractions $(\alpha_{0.7,y})$ for incremental secondary yields are zero. Otherwise, the firm yield (e.g., $Y_{0.9}$) is essentially increased by $\alpha_{0.7,y}Y_{0.7}$.

2. Over-year active storage volume capacity. For each year y:

$$S_y \leq K_a^o \tag{7.46}$$

$$\begin{bmatrix}\text{storage}\\ \text{volume}\end{bmatrix} \leq \begin{bmatrix}\text{active}\\ \text{over-year}\\ \text{capacity}\end{bmatrix}$$

3. Within-year storage continuity. For each period t:

$$s_t + \beta_t\left(\sum_{p \in P} Y_p + \sum_t e_t\right) - \sum_{p \in P} y_{pt} - e_t = s_{t+1} \tag{7.47}$$

$$\begin{bmatrix}\text{initial} \\ \text{within-year} \\ \text{storage} \\ \text{volume}\end{bmatrix} + \begin{bmatrix}\text{assumed} \\ \text{critical period} \\ \text{inflow}\end{bmatrix} - \begin{bmatrix}\text{firm and} \\ \text{incremental} \\ \text{secondary} \\ \text{yields}\end{bmatrix} - \begin{bmatrix}\text{estimated} \\ \text{evaporation} \\ \text{loss}\end{bmatrix} = \begin{bmatrix}\text{final} \\ \text{storage} \\ \text{volume}\end{bmatrix}$$

where $t + 1 = 1$ if t is last period in the year.

4. Within-year yields must sum to the respective over-year yields. If there is only one yield Y_p defined, this constraint will be redundant. However, for more than one yield, for all but one p,

$$\sum_t y_{pt} = Y_p \tag{7.48}$$

$$\begin{bmatrix}\text{sum of} \\ \text{within-year} \\ \text{yields}\end{bmatrix} = \begin{bmatrix}\text{annual} \\ \text{yield}\end{bmatrix}$$

5. Definition of estimated evaporation losses. For each year y:

$$E_y = E_0 + \left[S_y + \sum_t \left(\frac{s_t + s_{t+1}}{2}\right)\gamma_t\right]E \tag{7.49}$$

$$\begin{bmatrix}\text{annual} \\ \text{evaporation} \\ \text{loss}\end{bmatrix} = \begin{bmatrix}\text{annual fixed} \\ \text{evaporation} \\ \text{loss}\end{bmatrix} + \begin{bmatrix}\text{annual evaporation} \\ \text{loss from active} \\ \text{storage}\end{bmatrix}$$

and (again assuming the minimum over-year storage is zero) for each period t of the critical year:

$$e_t = \gamma_t E_0 + \tfrac{1}{2}(s_t + s_{t+1})\gamma_t E \tag{7.50}$$

$$\begin{bmatrix}\text{evaporation} \\ \text{loss in critical} \\ \text{period}\end{bmatrix} = \begin{bmatrix}\text{fixed} \\ \text{evaporation} \\ \text{loss}\end{bmatrix} + \begin{bmatrix}\text{evaporation} \\ \text{loss from active} \\ \text{storage}\end{bmatrix}$$

6. Total reservoir capacity. For each period t:

$$K_d + K_a^o + s_t + K_{ft} \le K \tag{7.51}$$

$$\begin{bmatrix}\text{dead} \\ \text{storage} \\ \text{capacity}\end{bmatrix} + \begin{bmatrix}\text{over-year} \\ \text{storage} \\ \text{capacity}\end{bmatrix} + \begin{bmatrix}\text{within-} \\ \text{year} \\ \text{storage} \\ \text{volume}\end{bmatrix} + \begin{bmatrix}\text{flood} \\ \text{storage} \\ \text{capacity}\end{bmatrix} \le \begin{bmatrix}\text{total} \\ \text{reservoir} \\ \text{capacity}\end{bmatrix}$$

where

$$K_{ft} = \begin{cases} K_f & \text{in flood season} \\ 0 & \text{otherwise} \end{cases}$$

This model can be solved with specific benefit and cost functions, values of evaporation loss rates, dead and flood storage capacities, and permissible yield failure fractions $\alpha_{0.7,y}$. The solution will provide estimates of the yields and the required over-year and within-year active storage capacities. The model can be expanded to include flood storage capacity as an unknown endogenous variable, using any of the methods discussed in Chapter 5.

Since actual reservoir storage volumes in each period t of each year y are not identified in this model, measures of system performance that are functions of those storage volumes, such as hydroelectric energy or reservoir recreation, are only approximate, especially when compared to the stochastic linear programming design models discussed earlier. Thus as with any of these stochastic screening models, any set of solutions should be evaluated and further improved using more precise simulation methods. The information provided by the solution of the yield model can aid in defining a reservoir operating policy for such simulation studies.

(4) **Reservoir Operation Rules.** Reservoir operation rules are guides for those responsible for reservoir operation. They apply to reservoirs being operated in a steady-state condition (i.e., not filling up immediately after construction or being operated to meet a set of new and temporary objectives). There are several types of rules but each indicates the desired or required reservoir release or storage volumes at any particular time of year. Some rules identify storage volume targets ("rule curves") that the operator is to maintain, if possible, and others identify storage zones, each associated with a particular release policy. This latter type of rule can be developed from the solution of the yield model.

To construct an operation rule that identifies storage zones, each having a specific release policy, the values of the dead and flood storage capacities K_d and K_{ft} are needed together with the over-year storage capacity K_a^o and within-year storage volumes s_t in each period t. Since both K_a^o and all s_t derived from the yield model are for all yields being considered (i.e., for all $p \in P$), it is necessary to determine the over-year capacities and within-year volumes required to provide each separate yield y_{pt}.

For example, assume that the total over-year capacity required to supply yields $Y_{0.9}$ and $Y_{0.7}$ is K_a^o. These yields and capacities are obtained from the solution of the yield model. Using any appropriate method, the over-year storage $\hat{K}_a^o$ required only for $Y_{0.9}$ can be computed. Similarly, the within-year storage $\hat{s}_t$ required at the beginning of each period t for yields $y_{0.9,t}$

can be computed. The sum of this over-year capacity and within-year volume $\hat{K}_a^o + \hat{s}_t$ in each period t defines a zone of active storage volumes for each period t required to supply the firm yields $y_{0.9,t}$. If at any time t the actual reservoir storage volume is within this zone, then reservoir releases should not exceed those required to meet the firm yield $y_{0.9,t}$ if the reliability of this yield is to be maintained.

The difference between the sum of total active over-year capacity K_a^o plus the total within-year storage volume s_t and the corresponding storage volume $\hat{K}_a^o + \hat{s}_t$ required for yields $y_{0.9,t}$ is the volume required to deliver the 70% reliable incremental secondary yields $y_{0.7,t}$. If at any time t the actual storage volume is greater than that needed for only the firm yields, namely $\hat{K}_a^o + \hat{s}_t$, the releases should meet both yields $y_{0.9,t}$ and $y_{0.7,t}$. If the actual storage volume is greater than the total required over-year storage capacity K_a^o plus the within-year volume s_t, then a release can be made to satisfy any downstream demand. However, if the actual storage volume is within the flood control zone in the flood season, releases should be made to reduce the actual storage to a volume no greater than the total capacity K less the flood storage capacity K_{ft}.

An operating rule containing these four storage zones for a 12 within-year period problem is illustrated in Figure 7.7. Each zone identifies a specific

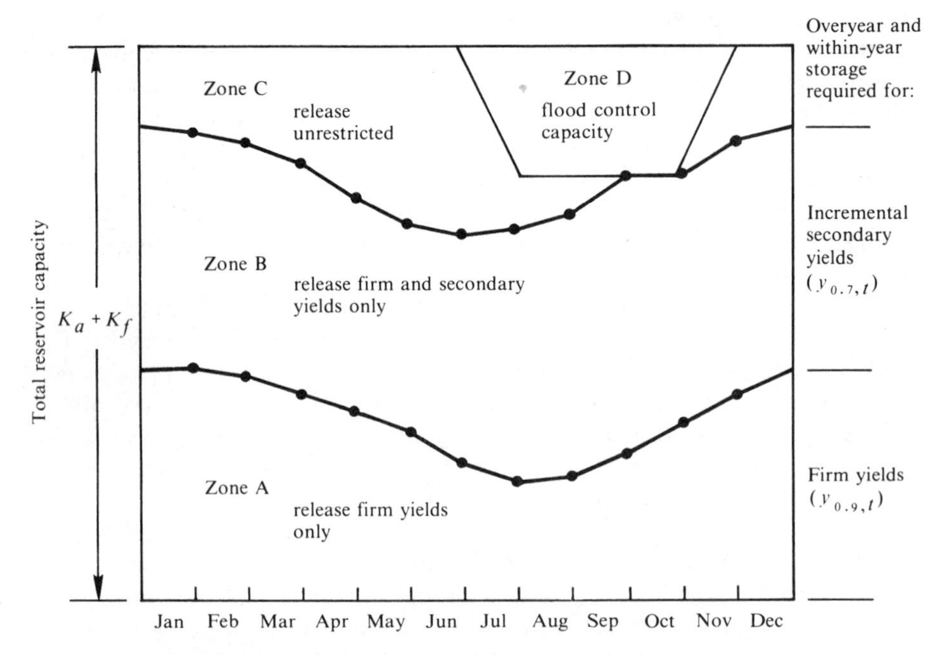

FIGURE 7.7. Reservoir operation rules for firm and secondary yields in each of 12 periods within the year.

release policy. If the actual storage volume is within zone C, any downstream release may be made. If the actual storage volume is within the flood control storage zone D, releases should be made to reduce the storage volume to a zone outside the flood control storage zone. If the actual storage volume is within zone B, releases should be restricted to only the firm and secondary yields (e.g., $y_{0.9,t}$ and $y_{0.7,t}$ in this example). If in a drought condition indicated by a storage volume within zone A, only firm yields $y_{0.9,t}$ should be released.

Reservoir rules developed from simplified models such as this yield model are only guides, and once developed they should be simulated, evaluated, and refined prior to their actual adoption.

7.3.3 Chance-Constrained Models

Chance-constrained models have been proposed for the preliminary estimation of cost-effective reservoir designs (capacities, targets) and operating policies. The solution of these models identifies the reservoir capacity and operating policy that should ensure a prespecified range of reservoir releases and reservoir storage volumes at given levels of reliability. The models can also be used to identify storage volume and release targets and operating policies which, at given levels of reliability, ensure that any deviations from these targets will be within prespecified ranges or limits.

Before developing these models, a brief review of chance constraints and their deterministic equivalents may be helpful. Recall from Chapter 3 that chance constraints are of the form

$$\Pr[g_i(\mathbf{X}) \leq B_i] \geq P_i \tag{7.52}$$

The function $g_i(\mathbf{X})$ contains the nonrandom decision vector $\mathbf{X}$, and B_i is a random variable whose distribution function $F_{B_i}(b_i)$ is illustrated in Figure 7.8.

In Figure 7.8 the values $b_i^{(1-P_i)}$ and $b_i^{(P_i)}$ are particular values of the random variable B_i. The superscripts $1 - P_i$ and P_i denote the probabilities that the random variable will be less than or equal to the values $b_i^{(1-P_i)}$ and $b_i^{(P_i)}$, respectively. Stated another way, the probability that the random variable B_i will exceed either $b_i^{(P_i)}$ or $b_i^{(1-P_i)}$ is $1 - P_i$ and P_i, respectively.

Chance constraints 7.52 require that the function $g_i(\mathbf{X})$ be no greater than the random variable B_i with at least probability P_i. Since the value $b_i^{(1-P_i)}$ will be no greater than the random variable B_i with probability P_i, it is sufficient to require that the function $g_i(\mathbf{X})$ be no greater than $b_i^{(1-P_i)}$.

Hence the chance constraints

$$\Pr[g_i(\mathbf{X}) \leq B_i] \geq P_i \tag{7.52}$$

or

$$\Pr[g_i(\mathbf{X}) \geq B_i] \leq 1 - P_i \tag{7.53}$$

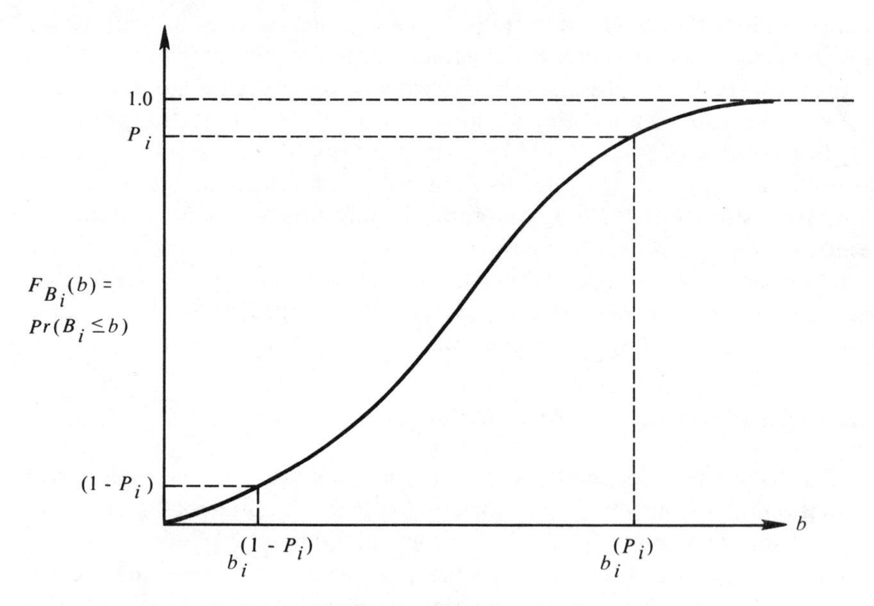

FIGURE 7.8. Distribution function of random variable B_i.

are equivalent to the deterministic constraints

$$g_i(\mathbf{X}) \leq b_i^{(1-P_i)} \tag{7.54}$$

These deterministic constraints will ensure that the functions $g_i(\mathbf{X})$ will be no greater than the random variable B_i, at least with probability P_i. Equation 7.54 is the deterministic equivalent of chance-constraint equation 7.52 or 7.53.

Similarly, the deterministic equivalents of chance constraints

$$\Pr[g_i(\mathbf{X}) \geq B_i] \geq P_i \quad \text{or} \quad \Pr[g_i(\mathbf{X}) \leq B_i] \leq 1 - P_i \tag{7.55}$$

are

$$g_i(\mathbf{X}) \geq b_i^{(P_i)} \tag{7.56}$$

where once again B_i has probability P_i of being less than or equal to $b_i^{(P_i)}$. In the models that follow, equations 7.52 to 7.56 may serve as a guide for the development of deterministic equivalents of chance constraints.

Consider now a single reservoir in which it is desired to know the total capacity K required to satisfy the following chance constraints limiting initial storage volumes S_t and releases R_t in each period t.

$$\Pr[S_t \leq S_t^{\max}] \geq P_{1t} \tag{7.57}$$

$$\Pr[R_t \leq R_t^{\max}] \geq P_{2t} \tag{7.58}$$

$$\Pr[R_t \geq R_t^{\min}] \geq P_{3t} \tag{7.59}$$

The reservoir release limits, $R_t^{\min}$ and $R_t^{\max}$, and the probabilities P_{it} ($i = 1$,

2, 3), must be specified and thus are known parameters. S_t^{max} will be an unknown but nonrandom variable that may equal the total active storage capacity K_a required in period t. Clearly, if no flood storage capacity is required in period t, S_t^{max} will equal the total active capacity K_a, and the constraint should always be met. If flood storage is constant and required in each period t, it can be ignored as can the dead storage capacity. Constant volume requirements in each period t have no affect on the active capacity K_a required to satisfy equations 7.57 to 7.59.

The initial storage volumes S_t and releases R_t are random variables whose distributions are unknown, since they are functions of the storage capacity limits S_t^{max} and the reservoir operating policy, which are to be determined. Therefore, it is necessary to express the random variables S_t and R_t as functions of nonrandom variables or random variables whose distributions are known. If these functions are linear, the resulting deterministic equivalents of equations 7.57 to 7.59 will be linear, and the model can be solved using linear optimization methods.

Consider the following linear release rule,

$$R_t = S_t + Q_t - E_t - b_t \tag{7.60}$$

in which each E_t is the evaporation loss and b_t is a nonrandom, nonnegative unknown operating policy parameter. This release rule indicates a release equal to the total available quantity $S_t + Q_t - E_t$, less some fixed amount b_t. Substituting the linear release rule into the continuity equation

$$S_t + Q_t - E_t - R_t = S_{t+1} \tag{7.61}$$

results in the linear storage rule

$$S_{t+1} = b_t \tag{7.62}$$

which equates the random end of period storage volume to a nonrandom parameter b_t. Using this rule, the variance in the inflows will be transferred directly to the releases.

Again assume that the evaporation loss E_t is a linear function, $e_{0t} + (e_t)(S_t + S_{t+1})/2$, of the average storage volume in each period (similar to equation 7.42). Substituting equation 7.62 into equation 7.60, the linear release rule in equation 7.61 becomes:

$$R_t = Q_t - e_{0t} + \left(1 - \frac{e_t}{2}\right)b_{t-1} - \left(1 + \frac{e_t}{2}\right)b_t \tag{7.63}$$

where e_{0t} is the fixed evaporation volume loss and e_t is the evaporation volume loss per unit average storage volume in period t. Equations 7.62 and 7.63 define linear rules for storage volumes and releases in a form that will permit a linear programming solution of a chance-constrained problem.

A linear model can now be developed for estimating total reservoir capacity K required for meeting various chance constraints on reservoir releases and storage volumes. The objective will be to

$$\text{minimize } K \tag{7.64}$$

subject to

$$\Pr[S_t \leq K] \geq P_{1t} \equiv 1 \qquad \forall t \text{ not in flood season} \tag{7.65}$$

$$\Pr[S_t \leq K - K_f] \geq P_{1t} \qquad \forall t \text{ in flood season} \tag{7.66}$$

$$\Pr[R_t \leq R_t^{\max}] \geq P_{2t} \qquad \forall t \tag{7.67}$$

$$\Pr[R_t \geq R_t^{\min}] \geq P_{3t} \qquad \forall t \tag{7.68}$$

The deterministic equivalents of chance constraints 7.65 to 7.68 can be derived by substituting the linear release and storage rules (equations 7.62 and 7.63) into chance constraints 7.65 to 7.68, as appropriate, rearranging the equations into a form similar to equation 7.52 or 7.55, and then writing the deterministic equivalents using as a guide equation 7.54 or 7.56, respectively. Using equations 7.62 and 7.63 to replace the variables S_t and R_t, the model becomes

$$\text{minimize } K \tag{7.64}$$

subject to

$$\Pr[b_{t-1} - K \leq 0] \geq P_{1t} \qquad \forall t \text{ not in flood season} \tag{7.65a}$$

$$\Pr[b_{t-1} - K + K_f \leq 0] \geq P_{1t} \qquad \forall t \text{ in flood season} \tag{7.66a}$$

$$\Pr\left[e_{0t} - \left(1 - \frac{e_t}{2}\right)b_{t-1} + \left(1 + \frac{e_t}{2}\right)b_t + R_t^{\max} \geq Q_t\right] \geq P_{2t} \qquad \forall t \tag{7.67a}$$

$$\Pr\left[e_{0t} - \left(1 - \frac{e_t}{2}\right)b_{t-1} + \left(1 - \frac{e_t}{2}\right)b_t + R_t^{\min} \leq Q_t\right] \geq P_{3t} \qquad \forall t \tag{7.68a}$$

Note that the first two chance constraints (7.65 and 7.66a) contain no random variables; hence if $P_{1t} > 0$, the deterministic equivalents are

$$b_{t-1} - K \leq 0 \qquad \forall t \text{ not in flood season} \tag{7.65b}$$

$$-b_{t-1} + K \geq K_f \qquad \forall t \text{ in flood season} \tag{7.66b}$$

If flood storage capacity is not required in any period, the foregoing two constraints can be written simply as $b_t - K \leq 0$. Since b_t represents the storage at the end of period t (equation 7.62), equations 7.66a and 7.66b must also be specified for the period just preceding the flood season if flood storage K_f is required throughout the flood season.

Using equation 7.56 as a guide, the deterministic equivalent of equation 7.67a is

$$\left(1 + \frac{e_t}{2}\right)b_t - \left(1 - \frac{e_t}{2}\right)b_{t-1} \geq q_t^{(P_{2t})} - R_t^{\max} - e_{0t} \qquad \forall t \tag{7.67b}$$

where $q_t^{(P_{2t})}$ has a probability P_{2t} of equaling or exceeding the random streamflow Q_t.

Using equation 7.54 as a guide, the deterministic equivalent of equation 7.68a is

$$\left(1 + \frac{e_t}{2}\right)b_t - \left(1 - \frac{e_t}{2}\right)b_{t-1} \leq q_t^{(1-P_{3t})} - R_t^{\min} - e_{0t} \qquad \forall t \tag{7.68b}$$

where $q_t^{(1-P_{3t})}$ has a probability $1 - P_{3t}$ of equaling or exceeding Q_t.

Examining equations 7.67b and 7.68b for each period t, it is clear that since the left-hand sides of each equation are identical, the right-hand side of equation 7.67b must be no greater than the right-hand side of equation 7.68b.

$$q_t^{(P_{2t})} - R_t^{\max} \leq q_t^{(1-P_{3t})} - R_t^{\min} \tag{7.69}$$

Furthermore, if all the evaporation rates e_t are zero, summing each set of equations 7.67b and 7.68b over all periods t results in two other feasibility conditions.

$$\sum_t q_t^{(P_{2t})} \leq \sum_t R_t^{\max} \tag{7.70}$$

$$\sum_t q_t^{(1-P_{3t})} \geq \sum_t R_t^{\min} \tag{7.71}$$

These mathematical restrictions on the values of the bounds $R_t^{\max}$ and $R_t^{\min}$ and on the probabilities P_{2t} and P_{3t} limit the range of solutions that may be feasible. Some mathematically infeasible solutions may prove to be physically feasible, and possibly preferable, in a detailed simulation analysis.

Example Problem. Table 7.10 presents the data for a particular single-reservoir design and operating problem to be solved with the model defined by equations 7.64 through 7.68. Substituting the parameters given in Table 7.10 into the deterministic equivalent of that model, equations 7.64 and 7.65b to 7.68b results in the linear programming problem shown in Figure 7.9. The solution of this problem is summarized in Table 7.11.

To implement or simulate the reservoir release policy defined in Table 7.11 requires a knowledge of the current inflow Q_t. This may be difficult to forecast. An estimate of the additional active reservoir capacity required for a release policy that is independent of the current inflow can be obtained by omitting the current inflow term Q_t from the reservoir release rule, equation 7.60. If this is done, the reservoir release rule becomes dependent on the previous period's inflow Q_{t-1}, and the storage volume at the end of the period becomes dependent on the current inflow. Now the inflow variance is transferred to both the storage volumes and the releases.

Ignoring evaporation losses, the revised linear reservoir release and storage volume rules are

$$R_t = S_t - b_t = Q_{t-1} + b_{t-1} - b_t \tag{7.72}$$

$$S_{t+1} = Q_t + b_t \tag{7.73}$$

TABLE 7.10. Data for Chance-Constrained Example Problem

Period, t	*Required Flood Capacity during Period,* t	$R_t^{\min}$	$R_t^{\max}$	CHANCE-CONSTRAINT PROBABILITIES, P_{it}			STREAMFLOW VALUES				*Fixed Losses,* e_{0t}	*Evap. Loss Rates,* e_t
				i: 1	*2*	*3*	$q_t^{(0.05)}$	$q_t^{(0.10)}$	$q_t^{(0.90)}$	$q_t^{(0.95)}$		
1	0	20	70	1.00	0.95	0.95	33	45	50	80	3	0.02
2	0	20	70	0.90	0.95	0.95	18	38	50	65	5	0.10
3	0	20	70	0.90	0.95	0.95	15	20	45	63	6	0.10
4	40	20	70	1.00	0.90	0.95	35	42	83	90	3	0.02

in which each b_t may be positive or negative. Using these decision rules in a chance-constrained reservoir model results in a larger required active storage capacity. When b_t are allowed to take negative values, constraints should be added to the model to insure that all S_t are nonnegative with a reasonable probability [23].

$$\text{minimize } K \text{ (storage capacity)}$$

subject to

(a) Storage capacity constraints for periods $t = 2, 3, 4, 1$

$$b_1 - K \leq 0$$
$$b_2 - K \leq 0$$
$$\left.\begin{array}{l} -b_3 + K \geq 40 \\ -b_4 + K \geq 40 \end{array}\right\} = \text{flood storage requirements for period 4}$$

(b) Upper limits on reservoir releases for periods $t = 1, 2, 3, 4$

$$1.01b_1 - 0.99b_4 \geq 80 - 70 - 3 = 7$$
$$1.05b_2 - 0.95b_1 \geq 65 - 70 - 5 = -10$$
$$1.05b_3 - 0.95b_2 \geq 63 - 70 - 6 = -13$$
$$1.01b_4 - 0.99b_3 \geq 83 - 70 - 3 = 10$$

(c) Lower limits on reservoir releases for periods $t = 1, 2, 3, 4$

$$1.01b_1 - 0.99b_4 \leq 33 - 20 - 3 = 10$$
$$1.05b_2 - 0.95b_1 \leq 18 - 20 - 5 = -7$$
$$1.05b_3 - 0.95b_2 \leq 15 - 20 - 6 = -11$$
$$1.01b_4 - 0.99b_3 \leq 35 - 20 - 3 = 12$$

FIGURE 7.9. Numerical example of chance-constrained linear optimization problem, Equations 7.64, 7.65(b) to 7.68(b).

TABLE 7.11. Solution of Chance-Constrained Example Problem

Required storage capacity, $K = 49.9$

Period, t	*Operating Policy Parameters,* b_t	=	*Final Storage Volumes,* S_{t+1}
1	19.6	=	S_2
2	8.2	=	S_3
3	0	=	S_4
4	9.9	=	S_1

A compromise solution could be obtained using some combination of rules 7.60 and 7.72. Assuming some fractions λ_t $(0 \leq \lambda_t \leq 1)$ for each period t and ignoring evaporation losses, the compromise release and storage rules are

$$R_t = (S_t + \lambda_t Q_t - b_t) = (1 - \lambda_{t-1})Q_{t-1} + \lambda_t Q_t + b_{t-1} - b_t \tag{7.74}$$

$$S_{t+1} = (1 - \lambda_t)Q_t + b_t \tag{7.75}$$

The reservoir capacities associated with various fractions λ_t will range from the least conservative estimate obtained when all $\lambda_t = 1$ to the most conservative estimate when all $\lambda_t = 0$.

Chance-Constrained Models with Targets. An alternative approach to restricting the range of possible releases and storage volumes, equations 7.57 to 7.59, is one that identifies storage and release targets and an operating policy that will ensure chance-constrained target deficits or excesses. Such an approach involves the development of what can be termed target models.

In situations where it is desired to examine trade-offs among constant storage targets T^S and release targets T^R as well as reservoir capacity K, these targets can be included as unknowns in a chance-constrained model. Any deficits D_t^S or D_t^R or excesses E_t^S or E_t^R will be random variables and hence may be bounded using chance constraints. The deficits and excesses are defined by the following equations:

$$R_t = T^R - D_t^R + E_t^R \qquad \forall t \tag{7.76}$$

$$S_t = T^S - D_t^S + E_t^S \qquad \forall t \tag{7.77}$$

If one wants to ensure that release deficits D_t^R, and storage deficits D_t^S and excesses E_t^S are within some fraction f_t of the respective target T with at least probability P_{it}, they can require that

$$\Pr[D_t^R \leq f_t T^R] \geq P_{1t} \qquad \forall t \tag{7.78}$$

$$\Pr[D_t^S \leq f_t T^S] \geq P_{2t} \qquad \forall t \tag{7.79}$$

$$\Pr[E_t^S \leq f_t T^S] \geq P_{3t} \qquad \forall t \tag{7.80}$$

These fractions f_t could differ for different probabilities P_{it}, the larger the permissible fraction the larger the probability P_{it}. Alternatively, the permissible deficits or excesses could be fixed and independent of the target.

For example, any or all of the following three sample chance constraints might simultaneously apply.

$$\Pr[D_t^R \leq 0.1T^R] \geq 0.95 \tag{7.81}$$

$$\Pr[D_t^R \leq 0.2T^R] \geq 0.98 \tag{7.81a}$$

$$\Pr[D_t^R \leq 20] \geq 0.99 \tag{7.81b}$$

Chance constraints containing random deficits and excesses can be converted to their deterministic equivalents by noting that if $D_t > 0$, then $E_t = 0$, and vice versa. Hence, if $D_t \leq f_t T$, then certainly it is also true that $-E_t + D_t \leq$

f_tT. Similarly, if $E_t \leq f_tT$, then certainly $-D_t + E_t \leq f_tT$. Thus the chance constraints 7.78 to 7.80 can be written

$$\Pr[D_t^R - E_t^R \leq f_tT^R] \geq P_{1t} \tag{7.78a}$$

$$\Pr[D_t^S - E_t^S \leq f_tT^S] \geq P_{2t} \tag{7.79a}$$

$$\Pr[-D_t^S + E_t^S \leq f_tT^S] \geq P_{3t} \tag{7.80a}$$

The task now is to express the random variables D_t and E_t, whose distributions are unknown, as functions of unknown parameters and random variables whose distributions are known. Again ignoring evaporation losses and incorporating the rules defined by equations 7.74 and 75 into equations 7.76 and 77, one obtains

$$D_t^R - E_t^R = T^R - R_t = T^R - (1 - \lambda_{t-1})Q_{t-1} - \lambda_t Q_t - b_{t-1} + b_t \tag{7.82}$$

$$D_t^S - E_t^S = T^S - S_t = T^S - (1 - \lambda_{t-1})Q_{t-1} - b_{t-1} \tag{7.83}$$

Substituting the right-hand sides of equations 7.82 and 7.83 in place of the respective left-hand terms found in equations 7.78a to 7.80a results in the following chance constraints:

$$\Pr[(1 - f_t)T^R - b_{t-1} + b_t \leq (1 - \lambda_{t-1})Q_{t-1} + \lambda_t Q_t] \geq P_{1t} \tag{7.78b}$$

$$\Pr[(1 - f_t)T^S - b_{t-1} \leq (1 - \lambda_{t-1})Q_{t-1}] \geq P_{2t} \tag{7.79b}$$

$$\Pr[(1 + f_t)T^S - b_{t-1} \geq (1 - \lambda_{t-1})Q_{t-1}] \geq P_{3t} \tag{7.80b}$$

Their deterministic equivalents for each period t are

$$(1 - f_t)T^R - b_{t-1} + b_t \leq [(1 - \lambda_{t-1})Q_{t-1} + \lambda_t Q_t]^{(1-P_{1t})} \tag{7.78c}$$

$$(1 - f_t)T^S - b_{t-1} \leq (1 - \lambda_{t-1})q_{t-1}^{(1-P_{2t})} \tag{7.79c}$$

$$(1 + f_t)T^S - b_{t-1} \geq (1 - \lambda_{t-1})q_{t-1}^{(P_{3t})} \tag{7.80c}$$

The right-hand term of equation 7.78c is the value of the joint distribution of $[(1 - \lambda_{t-1})Q_{t-1} + \lambda_t Q_t]$ that is exceeded with a probability of P_{1t}.

A fourth set of constraints ensures that the storage volume is no greater than the available active capacity with probability P_{4t}. It completes the constraint set of the model. Assuming a flood storage requirement of K_f that must be provided with probability P_{4t}, this constraint can be written

$$\Pr[S_t + K_f \leq K] \geq P_{4t} \tag{7.84}$$

Using equation 7.75 to replace the random storage term S_t in the equation above and using equation 7.56 as a guide, chance constraint 7.84 has as its deterministic equivalent

$$K - K_f - b_{t-1} \geq (1 - \lambda_{t-1})q_{t-1}^{(P_{4t})} \tag{7.84a}$$

An objective function for estimating efficient (Pareto-optimal) combinations of targets T^S and T^R and capacity K might include relative weights w^S and w^R that can be varied:

$$\text{maximize } w^ST^S + w^RT^R - K \tag{7.85}$$

Alternatively, the capacity K could be minimized subject to lower bounds on the storage and release targets T^S and T^R, respectively:

$$\text{minimize } K$$

subject to

$$T^S \geq \text{minimum storage target value} \tag{7.86}$$

$$T^R \geq \text{minimum release target value} \tag{7.87}$$

$$\text{Constraints on storage capacities, as required} \tag{7.88}$$

These two alternative multiobjective approaches are discussed in greater detail in Chapter 4.

Other modifications are possible [13, 16, 17, 19, 24]. No matter what the modification, even when using the least conservative of operating rules, the solutions will tend to be conservative. This stems, in part, from the fact that only the relatively high or low streamflows ($q^{(P_{it})}$ or $q_t^{(1-P_{it})}$) are contained within the model in each period t, and the joint probability of a succession of such critical flow conditions is very unlikely. Hence it is important that the solutions obtained from these models be simulated to obtain improved estimates of appropriate design parameters and operating policies.

7.4 MULTIPLE-SITE RIVER BASIN PLANNING MODELS

Each of the design models just discussed can be extended to analyze river basin problems involving multiple reservoir and water allocation or use sites. To illustrate how this can be done, consider the three-reservoir, two-use river basin planning problem illustrated in Figure 7.10. Sites 1, 2, and 3 are potential reservoir sites, and sites 4 and 5 are water-use sites. Outlined in this final section of the chapter are the constraints for each type of model needed to describe this example problem. To simplify this illustration, evaporation and dead and flood storage capacities will be ignored. These variables and parameters can be added when appropriate, using any of the approaches that have been discussed. Planning objectives will be functions of the variables defined in the constraints.

First the single-reservoir stochastic design model will be expanded for use in analyzing the multiple-reservoir planning problem. This discussion will be followed by a similar expansion of the chance-constrained models and yield models. Only the fundamental assumptions and techniques will be presented. Additional information on each of these techniques may be found in the literature listed under the appropriate reference section at the end of this chapter.

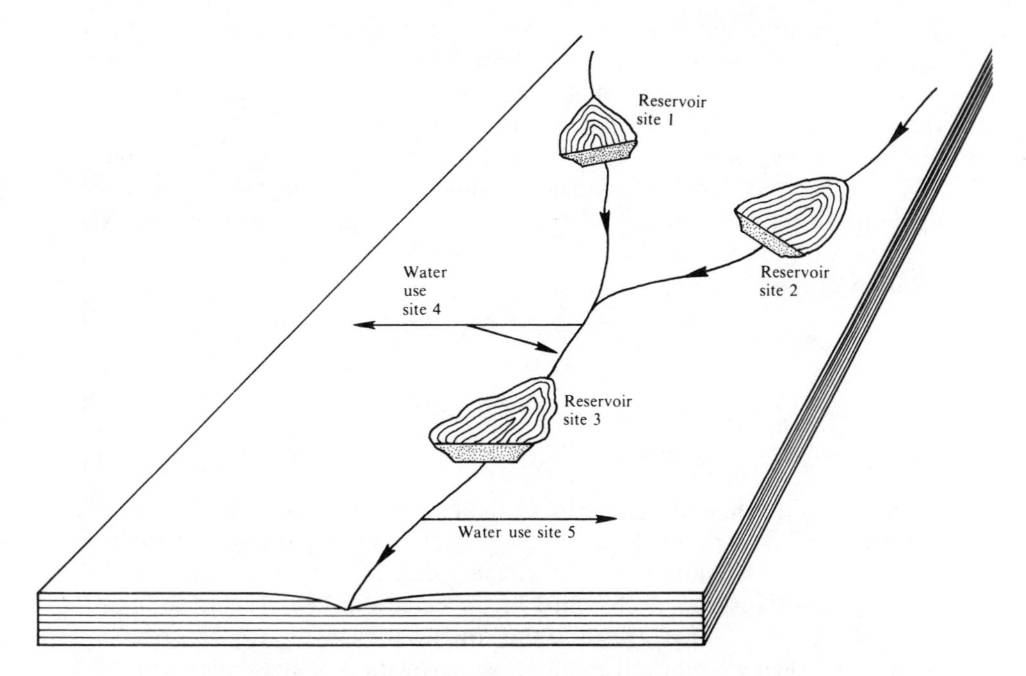

FIGURE 7.10. Multiple-site river basin planning problem.

7.4.1 Stochastic Linear Programming Design Models

Stochastic design models are appropriate for analyzing relatively small multireservoir river basin systems having streamflows that are highly cross-correlated. This assumption plus the imposition of the same final volume subscript l associated with particular values of the initial reservoir storage volume subscript k and inflow subscripts i at each reservoir site simplifies the problem considerably [2, 3].

For multisite problems, define Q_{it}^{s} as the discrete unregulated streamflow corresponding to flow index i at site s in period t. Define the ith flow at each site so that the unconditional probability PQ_{it} of a streamflow corresponding to index i in period t at any site s is the same for all sites s. Defining the discrete streamflows Q_{it}^{s} such that their probabilities PQ_{it} are the same at all sites s does not imply that the flows represented by any particular index i at each site will be the same at each site. In most cases they will differ.

Next, assume that the flows within the multisite basin are highly cross-correlated, so that the joint probabilities of observing flows corresponding to different i's at different sites in the same time period are zero. Finally, assume that the subscript policy $l = l(k, i, t)$ is the same at each reservoir site. Once again, this does not define the operating policy itself, only the relationships between the initial storage volume subscripts k, the inflow subscripts i, and the final storage volume subscripts l, in each period t.

These assumptions allow calculation of the steady-state joint probabilities PS_{kt} and PR_{kit} of storage volumes S_{kt} and reservoir releases R_{kit} that will apply to all reservoir sites. For transition probabilities P_{ij}^t at every site, the steady-state joint probabilities, PR_{kit} can be derived from the simultaneous solution of the following equations.

$$\mathrm{PR}_{l,j,t+1} = \sum_{k} \sum_{\substack{i \\ l=l(k,i,t)}} \mathrm{PR}_{kit} P_{ij}^t \qquad \forall l, j, t \tag{7.89}$$

$$\sum_{k} \sum_{i} \mathrm{PR}_{kit} = 1 \qquad \forall t \tag{7.90}$$

$$\sum_{i} \mathrm{PR}_{kit} = \mathrm{PS}_{kt} \qquad \forall k, t \tag{7.91}$$

If there is more than one gage site in the basin, it may be difficult to define the discrete flows Q_{it}^s and $Q_{j,t+1}^s$ so that the observed transition probabilities P_{ij}^t are the same at all sites s. In this case it may be simpliest to substitute in place of the transition probability P_{ij}^t in equation 7.89 the unconditional probabilities $P_{j,t+1}$ that are the same for each flow $Q_{j,t+1}^s$. The resulting screening model will not reflect the serial correlation of streamflows but will still be adequate for preliminary estimates of various design variables [3].

Use of the same subscript policy $l = l(k, i, t)$ for each reservoir in addition to the common streamflow probabilities PQ_{it} or P_{ij}^t eliminates the need to consider the possibility of having different storage volume indices and different flow indices at different sites at any particular time. The joint probabilities of such combinations of different subscripts k or i will be zero. Again, the values represented by any particular k or i may differ at each site s. This is not prevented by assuming that the reservoir storage subscripts k will be the same, or that the streamflow subscripts i will be the same, at all sites in any particular period.

Another possible simplification when applying these stochastic design models to multisite problems is the reduction of the number of reservoir releases R_{kit} to the number of streamflows Q_{it}. In order to add the resulting expected reservoir release R_{it} to the corresponding natural or unregulated streamflow Q_{it}, their probabilities must be the same. These conditions can be satisfied by the addition of the following constraints at each reservoir site s:

$$R_{it}^s = \frac{\sum_{k} \mathrm{PR}_{kit} R_{kit}^s}{\mathrm{PQ}_{it}} \tag{7.92}$$

The equation above defines an expected reservoir release whose probability is PQ_{it}. This can be added to the corresponding natural streamflow, or incremental flow Q_{it}^s, at any site s downstream from the reservoir to obtain the total flow at that site having probability PQ_{it}.

The assumptions just discussed are incorporated into the constraints below for the multisite problem defined in Figure 7.10.

1. Continuity of storage at each reservoir site ($s = 1, 2, 3$). For each storage volume index k and streamflow index i in each period t:

$$S_{kt}^s + Q_{it}^s - R_{kit}^s = S_{l,t+1}^s \qquad s = 1, 2 \tag{7.93}$$

$$\begin{bmatrix}\text{initial}\\ \text{storage}\\ \text{volume}\end{bmatrix} + \begin{bmatrix}\text{inflow}\end{bmatrix} - \begin{bmatrix}\text{release}\end{bmatrix} = \begin{bmatrix}\text{final}\\ \text{storage}\\ \text{volume}\end{bmatrix}$$

$$S_{kt}^3 + (R_{kit}^1 + R_{kit}^2) + (Q_{it}^3 - Q_{it}^1 - Q_{it}^2) - \delta_t^4 A_{it}^4$$

$$\begin{bmatrix}\text{initial}\\ \text{storage}\\ \text{volume}\end{bmatrix} + \begin{bmatrix}\text{upstream}\\ \text{reservoir}\\ \text{releases}\end{bmatrix} + \begin{bmatrix}\text{unregulated}\\ \text{incremental}\\ \text{flow}\end{bmatrix} - \begin{bmatrix}\text{consumption}\\ \text{at site 4}\end{bmatrix}$$

$$R_{kit}^3 = S_{l,t+1}^3 \tag{7.94}$$

$$- \begin{bmatrix}\text{release}\end{bmatrix} = \begin{bmatrix}\text{final}\\ \text{storage}\\ \text{volume}\end{bmatrix}$$

(δ_t^4 is the fraction of the allocation consumed.)

2. Reservoir capacity requirements. For each discrete storage volume index k in each period t at all reservoir sites $s = 1, 2, 3$:

$$S_{kt}^s \leq K_a^s \tag{7.95}$$

$$\begin{bmatrix}\text{initial}\\ \text{storage}\\ \text{volume}\end{bmatrix} \leq \begin{bmatrix}\text{active}\\ \text{storage}\\ \text{capacity}\end{bmatrix}$$

3. Expected reservoir releases. For each flow index i in each period t at each reservoir site s:

$$R_{it}^s = \frac{\sum_k PR_{kit} R_{kit}^s}{PQ_{it}} \tag{7.96}$$

4. Allocation at site 4. For each flow index i in each period t:

$$A_{it}^4 \leq R_{it}^1 + R_{it}^2 + Q_{it}^4 - Q_{it}^1 - Q_{it}^2 \tag{7.97}$$

$$\begin{bmatrix}\text{allocation}\\ \text{to site 4}\end{bmatrix} \leq \begin{bmatrix}\text{upstream}\\ \text{reservoir}\\ \text{releases}\end{bmatrix} + \begin{bmatrix}\text{unregulated}\\ \text{incremental}\\ \text{flow}\end{bmatrix}$$

5. Allocation at site 5: For each flow index i in each period t:

$$A_{it}^5 \leq R_{it}^3 + Q_{it}^5 - Q_{it}^3 \tag{7.98}$$

$$\begin{bmatrix}\text{allocation}\\ \text{at site 5}\end{bmatrix} \leq \begin{bmatrix}\text{reservoir}\\ \text{release}\end{bmatrix} + \begin{bmatrix}\text{unregulated}\\ \text{flow}\end{bmatrix}$$

The probabilities PR_{kit} would be determined from the solution of equations 7.89 and 7.90 prior to their incorporation into any objective function as well as equation 7.96.

7.4.2 Yield Models for Design and Operation

Compared to the other two types of stochastic models, yield models are perhaps the easiest to extend to multisite planning problems. A special requirement, however, is that the yield failure year or years must be the same at all allocation sites throughout the basin. For basins having multiple gage sites, the identification of the failure years may be difficult, especially if the annual flows at different sites are not highly, and positively, cross-correlated. Another requirement is that the incremental flow yields must be of the same reliability as the reservoir release yields if they are to be added to define the yield available at any point downstream from one or more reservoirs.

Once again referring to the three-reservoir, two-use basin planning problem in Figure 7.10, the basic constraints for defining active reservoir capacities, and yields having the same reliability throughout the basin, are as follows:

1. Continuity of storage at each reservoir site.
   a. Over-year storage continuity. For each year y:

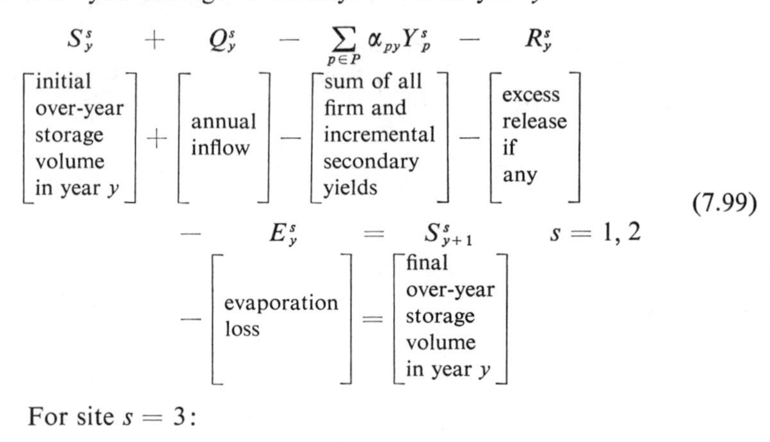

$$S_y^s + Q_y^s - \sum_{p \in P} \alpha_{py} Y_p^s - R_y^s - E_y^s = S_{y+1}^s \qquad s = 1, 2 \tag{7.99}$$

[initial over-year storage volume in year y] + [annual inflow] − [sum of all firm and incremental secondary yields] − [excess release if any] − [evaporation loss] = [final over-year storage volume in year y]

For site $s = 3$:

$$S_y^3 + (R_y^1 + R_y^2) + (Q_y^3 - Q_y^1 - Q_y^2) - E_y^3 - \sum_{p \in P} \alpha_{py} Y_p^3 - R_y^3 = S_{y+1}^3 \tag{7.100}$$

[initial over-year storage volume] + [excess releases from upstream reservoirs] + [incremental flow] − [evaporation loss] − [sum of all firm and incremental secondary yields] − [excess release if any] = [final over-year storage volume]

Note that the annual yields Y_p^3 defined by this equation do not include any portions of the upstream yields, Y_p^1 and Y_p^2.

b. Within-year storage continuity. For each period t within the year:

$$\underset{\begin{bmatrix}\text{initial}\\ \text{within-year}\\ \text{storage volume}\end{bmatrix}}{s_t^s} + \underset{\begin{bmatrix}\text{assumed critical}\\ \text{period inflow}\end{bmatrix}}{\beta_t^s\left(\sum_{p \in P} Y_p^s + \sum_t e_t^s\right)} - \underset{\begin{bmatrix}\text{yield releases}\end{bmatrix}}{\sum_{p \in P} y_{pt}^s} - \underset{\begin{bmatrix}\text{evaporation}\\ \text{loss}\end{bmatrix}}{e_t^s} = \underset{\begin{bmatrix}\text{final}\\ \text{storage}\\ \text{volume}\end{bmatrix}}{s_{t+1}^s} \qquad s = 1, 2 \tag{7.101}$$

For site $s = 3$:

$$\underset{\begin{bmatrix}\text{initial}\\ \text{storage}\\ \text{volume}\end{bmatrix}}{s_t^3} + \underset{\begin{bmatrix}\text{assumed critical}\\ \text{period inflow}\end{bmatrix}}{\beta_t\left(\sum_{p \in P} Y_p^3 + \sum_t e_t^3\right)} + \underset{\begin{bmatrix}\text{upstream yield}\\ \text{releases less}\\ \text{consumption}\\ \text{at site 4 and}\\ \text{yield releases}\\ \text{at site 3}\end{bmatrix}}{\sum_{p \in P}\left[\left(\sum_{s=1}^{2} y_{pt}^s\right) - \delta_t^4 A_{pt}^4 - y_{pt}^3\right]} - \underset{\begin{bmatrix}\text{evaporation}\\ \text{loss}\end{bmatrix}}{e_t^3} = \underset{\begin{bmatrix}\text{final}\\ \text{storage}\\ \text{volume}\end{bmatrix}}{s_{t+1}^3} \tag{7.102}$$

Note that unlike the annual incremental yields Y_p^3 at site 3, the within-year yields y_{pt}^3 are the total yields at that reservoir site in each period t. They include the upstream yields that flow into the reservoir at site 3. The reason yields Y_p^1 and Y_p^2 are not included in the yield Y_p^3, in the over-year continuity constraint at reservoir site 3, is to make it possible to define the within-year inflow distribution of the incremental yield Y_p^3. The within-year inflow distribution of the natural incremental yield Y_p^3 defined by β_t's is not likely to be the same as the controlled within-year outflow distributions of yields Y_p^1 and Y_p^2 from the upstream reservoirs at sites 1 and 2.

2. Continuity of yields at each reservoir site. If the model includes multiple yields, it will be necessary to equate the sum of within-year period yields to the appropriate annual yield. Otherwise, it is unnecessary, since such constraints will be redundant. Assuming multiple yields Y_p^s at each reservoir site s and for each probability

p of interest ($p \in P$):

$$\sum_t y^s_{pt} = Y^s_p \qquad s = 1, 2 \tag{7.103a}$$

$$\begin{bmatrix}\text{sum of}\\ \text{within-year}\\ \text{yields}\end{bmatrix} = \begin{bmatrix}\text{annual}\\ \text{yield}\end{bmatrix}$$

$$\sum_t y^3_{pt} = Y^3_p + \sum_{s=1}^{2} Y^s_p - \sum_t \delta^4_t A^4_{pt} \tag{7.103b}$$

$$\begin{bmatrix}\text{sum of}\\ \text{within-year}\\ \text{total yields}\end{bmatrix} = \begin{bmatrix}\text{annual}\\ \text{incremental}\\ \text{yield}\end{bmatrix} + \begin{bmatrix}\text{net annual}\\ \text{yields from}\\ \text{upstream reservoirs}\end{bmatrix}$$

3. Over-year storage capacity and total active capacity: For each year y, or each period t, at each reservoir site:

$$S^s_y \leq K^{os}_a \tag{7.104}$$

$$\begin{bmatrix}\text{over-year}\\ \text{storage}\\ \text{volume}\end{bmatrix} \leq \begin{bmatrix}\text{over-year}\\ \text{storage}\\ \text{capacity}\end{bmatrix}$$

$$K^{os}_a + s^s_t \leq K^s_a \tag{7.105}$$

$$\begin{bmatrix}\text{over-year}\\ \text{storage}\\ \text{capacity}\end{bmatrix} + \begin{bmatrix}\text{within-year}\\ \text{storage}\\ \text{volume}\end{bmatrix} \leq \begin{bmatrix}\text{total}\\ \text{active}\\ \text{capacity}\end{bmatrix}$$

4. Allocation of yields: For each period t and probability $p \in P$, at water use sites 4 and 5:

$$A^4_{pt} \leq y^1_{pt} + y^2_{pt} + I^4_{pt} \tag{7.106}$$

$$\begin{bmatrix}\text{allocation}\\ \text{to site 4}\end{bmatrix} \leq \begin{bmatrix}\text{upstream}\\ \text{reservoir}\\ \text{yields}\end{bmatrix} + \begin{bmatrix}\text{incremental}\\ \text{flow having}\\ \text{probability } p\end{bmatrix}$$

$$A^5_{pt} \leq y^3_{pt} + I^5_{pt} \tag{7.107}$$

$$\begin{bmatrix}\text{allocation}\\ \text{to site 5}\end{bmatrix} \leq \begin{bmatrix}\text{upstream}\\ \text{reservoir}\\ \text{yield}\end{bmatrix} + \begin{bmatrix}\text{incremental}\\ \text{flow having}\\ \text{probability } p\end{bmatrix}$$

5. Evaporation losses: The estimated annual evaporation loss in each year y at each reservoir site s is:

$$E^s_y = E^s_0 + \left[S^s_y + \sum_t \left(\frac{s^s_t + s^s_{t+1}}{2}\right)\gamma^s_t\right]E^s \tag{7.108}$$

$$\begin{bmatrix}\text{total}\\ \text{evaporation}\\ \text{loss}\end{bmatrix} = \begin{bmatrix}\text{fixed}\\ \text{evaporation}\\ \text{loss}\end{bmatrix} + \begin{bmatrix}\text{evaporation loss}\\ \text{from active}\\ \text{storage volumes}\end{bmatrix}$$

For each period t in the critical year, assuming an initial over-year storage volume of zero:

$$e^s_t = \gamma^s_t E^s_0 + \left(\frac{s^s_t + s^s_{t+1}}{2}\right)\gamma^s_t E^s \tag{7.109}$$

$$\begin{bmatrix}\text{total}\\ \text{evaporation}\\ \text{loss}\end{bmatrix} = \begin{bmatrix}\text{fixed}\\ \text{evaporation}\\ \text{loss}\end{bmatrix} + \begin{bmatrix}\text{evaporation loss}\\ \text{from storage}\\ \text{in period } t\end{bmatrix}$$

This completes the basic constraint set of the yield model for the multiple-site planning problem. Just as is the case with the previous model, there are many possible modifications of this modeling approach that might be considered in any particular situation. The purpose here has been merely to introduce and emphasize the modeling procedure and not to discuss the variety of modifications that might be desired for specific river basin studies.

7.4.3 Chance-Constrained Models

Required prior to developing a multisite, chance-constrained model is the definition of a linear allocation rule. This rule together with the linear release and storage rules are required in order to write linear deterministic equivalents of any chance constraints on the reservoir storage volumes, releases, and/or allocations.

In this example consider the linear operating rules defined by equations 7.74 and 7.75, in which all λ_t equal 1. These rules will apply for all reservoir releases R_t^s and storage volumes S_t^s at reservoir sites s.

$$R_t^s = \hat{Q}_t^s + b_{t-1}^s - b_t^s \tag{7.110}$$

$$\begin{bmatrix}\text{reservoir}\\ \text{release}\end{bmatrix} = \begin{bmatrix}\text{current}\\ \text{inflow}\end{bmatrix} + \begin{bmatrix}\text{nonrandom}\\ \text{release-rule}\\ \text{parameters}\end{bmatrix}$$

$$S_{t+1}^s = b_t^s \tag{7.111}$$

$$\begin{bmatrix}\text{final}\\ \text{storage}\\ \text{volume}\\ \text{in } t\end{bmatrix} = \begin{bmatrix}\text{storage}\\ \text{rule}\\ \text{parameter}\end{bmatrix}$$

Let the allocation rule for each period t at each allocation site s be

$$A_t^s = \beta_t \hat{Q}_t^s - a_t^s \tag{7.112}$$

$$\begin{bmatrix}\text{water}\\ \text{allocation}\\ \text{to use site}\end{bmatrix} = \begin{bmatrix}\text{fraction of}\\ \text{flow at}\\ \text{use site}\end{bmatrix} - \begin{bmatrix}\text{nonrandom}\\ \text{allocation}\\ \text{parameter}\end{bmatrix}$$

In the foregoing rules, the flows $\hat{Q}_t^s$ denote the total available (regulated plus unregulated) flow at site s in period t. Flows Q_t^s will denote the unregulated natural flows at site s in period t.

Given these linear operating rules, a chance-constrained model of any multiple-reservoir, multiple-use, river basin planning problem can be converted to an equivalent deterministic model. Required are the joint distributions of various combinations of fractions of streamflows at various sites.

To illustrate one such model, consider the following set of chance constraints on the releases and allocations in the basin shown in Figure 7.10, for each period t:

$$\Pr[R_t^s \geq R_t^{\min, s}] \geq P_{1t}^s \qquad s = 1, 2, 3 \tag{7.113}$$

$$\Pr[R_t^s \leq R_t^{\max, s}] \geq P_{2t}^s \qquad s = 1, 2, 3 \tag{7.114}$$

$$\Pr[A_t^s \geq A_t^{\min, s}] \geq P_{3t}^s \qquad s = 4, 5 \tag{7.115}$$

The lower and upper limits defined in each of the equations 7.113 to 7.115 are assumed known, as are the probabilities P^s_{it}.

Using the appropriate rules defined by equations 7.110 to 7.112 yields the following deterministic equivalents of equations 7.113 and 7.114 for the upstream reservoir sites $s = 1, 2$ in each period t:

$$R_t^{\min, s} - b^s_{t-1} + b^s_t \leq q_t^{s, (1-P^s_{1t})} \tag{7.116}$$

$$R_t^{\max, s} - b^s_{t-1} + b^s_t \geq q_t^{s, (P^s_{2t})} \tag{7.117}$$

The right-hand terms of equations 7.116 and 7.117 represent particular values of the unregulated random streamflows whose probabilities of being exceeded are P^s_{1t} and $1 - P^s_{2t}$, respectively.

Chance-constrained equation 7.115 applied to site 4 can be rewritten

$$\Pr[\beta^4_t \hat{Q}^4_t - a^4_t \geq A_t^{\min, 4}] \geq P^4_{3t} \tag{7.118}$$

The total streamflow $\hat{Q}^4_t$ at site 4 equals the releases $R^1_t + R^2_t$ from the upstream reservoirs at sites $s = 1$ and 2 plus the difference $Q^4_t - Q^1_t - Q^2_t$.

$$\hat{Q}^4_t = R^1_t + R^2_t + Q^4_t - Q^1_t - Q^2_t \tag{7.119}$$

Substituting the linear release rule, equation 7.110, results in an expression for the total flow at site 4 in terms of the unregulated flow Q^4_t at that site.

$$\hat{Q}^4_t = Q^4_t + \sum_{s=1}^{2} (b^s_{t-1} - b^s_t) \tag{7.120}$$

Thus the chance-constrained equation 7.118 becomes

$$\Pr\left[A_t^{\min, 4} + a^4_t - \beta^4_t \sum_{s=1}^{2} (b^s_{t-1} - b^s_t) \leq \beta^4_t Q^4_t\right] \geq P^4_{3t} \tag{7.121}$$

which is equivalent to

$$A_t^{\min, 4} + a^4_t - \beta^4_t \sum_{s=1}^{2} (b^s_{t-1} - b^s_t) \leq \beta^4_t q_t^{4, (1-P^4_{3t})} \tag{7.122}$$

The right-hand term $q_t^{4, (1-P^4_{3t})}$ is the value of the random streamflow Q^4_t exceeded with a probability of P^4_{3t}.

The total streamflow at reservoir site 3, $\hat{Q}^3_t$, equals the releases from reservoirs 1 and 2, plus the incremental flow, less the consumption $\delta^4_t A^4_t$ at site 4.

$$\hat{Q}^3_t = R^1_t + R^2_t + Q^3_t - Q^1_t - Q^2_t - \delta^4_t A^4_t \tag{7.123}$$

Substituting the appropriate release rules for the releases R^1_t, R^2_t and the allocation rule for A^4_t results in

$$\hat{Q}^3_t = (1 - \delta^4_t \beta^4_t) \sum_{s=1}^{2} (b^s_{t-1} - b^s_t) + Q^3_t - \delta^4_t(\beta^4_t Q^4_t - a^4_t) \tag{7.124}$$

The reservoir release R^3_t at site 3 is a function of this total streamflow

$$R^3_t = \hat{Q}^3_t + b^3_{t-1} - b^3_t \tag{7.125}$$

or

$$R^3_t = (1 - \delta^4_t \beta^4_t) \sum_{s=1}^{2} (b^s_{t-1} - b^s_t) - \delta^4_t(\beta^4_t Q^4_t - a^4_t) + Q^3_t + b^3_{t-1} - b^3_t \tag{7.126}$$

Substituting the right-hand side of this equation into chance-constrained equations 7.113 and 7.114 results in the following deterministic equivalents for site 3:

$$(1 - \delta_t^4\beta_t^4)\sum_{s=1}^{2}(b_{t-1}^s - b_t^s) + \delta_t^4 a_t^4 + b_{t-1}^3 - b_t^3 - R_t^{\min,3} \geq (\beta_t^4\delta_t^4 Q_t^4 - Q_t^3)^{(P_{1t}^3)} \tag{7.127}$$

$$(1 - \delta_t^4\beta_t^4)\sum_{s=1}^{2}(b_{t-1}^s - b_t^s) + \delta_t^4 a_4 + b_{t-1}^3 - b_t^3 - R_t^{\max,3} \leq (\beta_t^4\delta_t^4 Q_t^4 - Q_t^3)^{(1-P_{2t}^3)} \tag{7.128}$$

The right-hand sides of equations 7.127 and 7.128 represent particular values of the joint distributions of $\beta_t^4\delta_t^4 Q_t^4 - Q_t^3$ that are exceeded with probabilities $1 - P_{1t}^3$ and P_{2t}^3, respectively. These values can be estimated from records of the unregulated streamflows Q_t^3 and Q_t^4, at sites 3 and 4.

Finally, the allocation at site 5 equals

$$A_t^5 = \beta_t^5 \hat{Q}_t^5 - a_t^5 = \beta_t^5(R_t^3 + Q_t^5 - Q_t^3) - a_t^5 \tag{7.129}$$

From equation 7.126,

$$A_t^5 = \beta_t^5[(1 - \delta_t^4\beta_t^4)\sum_{s=1}^{2}(b_{t-1}^s - b_t^s) - \delta_t^4(\beta_t^4 Q_t^4 - a_t^4) + b_{t-1}^3 - b_t^3 + Q_t^5] - a_t^5 \tag{7.130}$$

Hence chance-constrained equation 7.115 for site 5 has as its deterministic equivalent

$$\beta_t^5[(1 - \delta_t^4\beta_t^4)\sum_{s=1}^{2}(b_{t-1}^1 - b_t^2) + \delta_t^4 a_t^4 + b_{t-1}^3 - b_t^3] - a_t^5 - A_t^{\min,5} \geq \beta_t^5(\beta_t^4\delta_t^4 Q_t^4 - Q_t^5)^{(P_{3t}^5)} \tag{7.131}$$

This completes the development of the deterministic equivalents of chance-constrained equations 7.113 to 7.115. These equations for the various sites within the basin shown in Figure 7.10, merely serve to illustrate how multi-site chance-constrained models can be developed. This example illustrates just one of many possible types of chance constraints and linear operating rules that could be applied in any specific planning situation.

7.5 CONCLUDING REMARKS

This chapter introduces three types of stochastic screening models for multipurpose reservoir operation and river basin planning. Hydrologic uncertainty has been included in each of the models in different ways. Development of these models was motivated by the desire to estimate reservoir capacities and operating policies, and water-use allocations, that meet different objectives and requirements. These objectives and requirements dictate, in part, the type of modeling approach most appropriate. Each approach has its advantages as well as its limitations.

The stochastic design models are large, especially for multireservoir planning problems, but they have provided relatively accurate estimates of design parameters. They are not very helpful in defining reservoir operating policies, however [3].

The chance-constrained models are small; they define explicit operating policies, but their structure tends to lead to conservative estimates of design variables. The possible range of probabilities that can be assigned to each chance constraint is also limiting [19].

The yield models, although larger than the chance-constrained models, are much smaller than the stochastic design models. They have resulted in relatively good estimates of both design and operating policy variables.

These three model types differ in their data requirements and in the type of information provided in the solution. Hence one's planning situation may influence the decision as to which model type or combinations of model types are most appropriate. It should be clear that modeling is very much an art. Considerable judgment is required in model development as well as in parameter estimation. Certainly, the use of stochastic planning models, such as those outlined in this chapter, both for reservoir operation and for river basin project design and operation, are relatively crude compared to the real prototype. Yet they are generally a better screening tool than the deterministic models introduced in Chapter 5. More details on the applications, as well as various modifications of these models, are discussed in the chapter's references.

APPENDIX 7A

STOCHASTIC LINEAR PROGRAMMING OPERATING MODEL

Section 7.2 discusses how one could formulate a linear programming model to determine the optimal operating policy for a single reservoir. This appendix formalizes the proposed model. In this model, the discrete storage volumes S_{kt} and releases R_{kilt} associated with each index k, i, and l in each period t are known. Hence it is possible to evaluate any measure of system performance that would result from any feasible combination of k, i, and l in any period t. Let this value be denoted as B_{kilt}. Assume that the reservoir is to be operated so as to maximize expected system performance. Since expected performance is the sum of all possible performance values B_{kilt} times their probabilities PR_{kilt}, the objective of the operating policy model will be

$$\text{maximize} \sum_k \sum_i \sum_l \sum_t B_{kilt} \text{PR}_{kilt} \tag{7A.1}$$

The unknown variables are, of course, the joint probabilities PR_{kilt}. Their values are influenced by the reservoir operating policy and by the random inflows. Hence, when solving for the optimal values of the unknown joint probabilities, certain relationships must be maintained. These relationships form the constraint set of the model.

The first condition is that the probability $\text{PS}_{l,t+1}$ of an initial volume index of l in period $t + 1$ must equal the probability of a final volume with the same index l in period t. This stems from the fact that the initial volume in period $t + 1$ is equivalent to the final volume in the previous period t. Hence their probabilities must be the same.

$$\text{PS}_{l,t+1} = \sum_m \sum_j \text{PR}_{l,j,m,t+1} = \sum_k \sum_i \text{PR}_{kilt} \qquad \forall l, t \tag{7A.2}$$

In the equation above, the subscript m denotes the final volume index in period $t + 1$, or equivalently the initial volume index in period $t + 2$.

Next, the joint probability of a final volume $S_{l,t+1}$ in period t followed by an inflow $Q_{j,t+1}$ in period $t + 1$ must equal the joint probability of that initial volume and that inflow in period $t + 1$. Denoting P_{ij}^t as the known conditional or transition probability of an inflow $Q_{j,t+1}$ in period $t + 1$ given an inflow Q_{it} in period t, this second condition can be written as

$$\sum_k \sum_i \text{PR}_{kilt} P_{ij}^t = \sum_m \text{PR}_{l,j,m,t+1} \qquad \forall l, j, t \tag{7A.3}$$

Clearly, this second condition specified by equation 7A.3 includes, and is more restrictive than, the first condition specified by equation 7A.2. Hence equation 7A.2 is redundant and need not be included in the model. The right-hand term equals the joint probability of an initial volume $S_{l,t+1}$ and inflow $Q_{j,t+1}$ in period $t + 1$. The left-hand term defines the joint probability of a final volume l in period t equal to $S_{l,t+1}$ followed by an inflow $Q_{j,t+1}$. These joint probabilities must be equal.

The third and final condition simply requires that the sum of all joint probabilities PR_{kilt} equals one in each period t.

$$\sum_k \sum_i \sum_l \text{PR}_{kilt} = 1 \qquad \forall t \tag{7A.4}$$

Equations 7A.3 (less one for each t) and 7A.4 together with objective function 7A.1 define the linear reservoir operating policy model. Note that constraint equations 7A.3 and 7A.4 are of the same form as the steady-state Markov chain equations 3.74 and 3.81a but instead of only one state index for streamflows, there are three state indices for initial volumes, inflows, and final volumes. Steady-state streamflow probabilities P_i in equations 3.74 and 3.81a (now denoted as PQ_{it}) are replaced by steady-state joint probabilities PR_{kilt}; both are derived from the known streamflow transition probabilities

P_{ij}^t. The difference between equations 3.74 and 3.81a and 7A.3 and 7A.4 is that equations 7A.4 and 7A.3 contain many more variables than equations. Hence there exist many possible solutions. The problem is one of finding the optimal solution, rather than simply solving a set of simultaneous equations similar to equations 3.74 and 3.81a for a single unique solution.

The steady-state probabilities of reservoir storage volumes PS_{kt} and inflows PQ_{it} can be derived from the steady-state joint probabilities PR_{kilt} simply by summing over the appropriate subscripts.

$$\text{PS}_{kt} = \sum_i \sum_l \text{PR}_{kilt} \qquad \forall k, t \tag{7A.5}$$

$$\text{PQ}_{it} = \sum_l \sum_k \text{PR}_{kilt} \qquad \forall i, t \tag{7A.6}$$

Note that the following two sets of conditions should hold.

$$\text{PS}_{l,t+1} = \sum_k \sum_i \text{PR}_{kilt} = \sum_j \sum_m \text{PR}_{l,j,m,t+1} \qquad \forall l, t \tag{7A.7}$$

$$\sum_l \text{PS}_{l,t+1} = 1 \qquad \forall t \tag{7A.8}$$

and

$$\text{PQ}_{j,t+1} = \sum_i \text{PQ}_{it} P_{ij}^t = \sum_l \sum_m \text{PR}_{l,j,m,t+1} \qquad \forall j, t \tag{7A.9}$$

$$\sum_i \text{PQ}_{it} = 1 \qquad \forall t \tag{7A.10}$$

All these conditions are met by a solution to equations 7A.3 and 7A.4 and hence need not be included in the set of model constraints. Equations 7A.5 and 7A.6 may be included if the steady-state probabilities PS_{kt} and PQ_{it} of the storage volume and inflow intervals are desired in the model output. They are definitional constraints only. Using the first and middle terms of equation 7A.9 and equation 7A.10, the probabilities PQ_{it} can be determined simply from the known inflow transition probabilities P_{ij}^t and are not affected by the other model constraints or objective functions.

Equations 7A.1, 7A.3, and 7A.4 represent a potentially large number of equations and an even larger number of variables, for this simplified single-reservoir operating problem. Hence it is usually more efficient to solve directly for the operating policy (i.e., for the optimal final storage volume interval l given an initial volume S_{kt} and inflow Q_{it} in each period t) using stochastic dynamic programming. Once the optimal policy is known, it is possible to modify equations 7A.3 and 7A.4, so that the number of linearly independent equations equals the number of variables, and then solve these equations for the unique set of joint probabilities. This is demonstrated following the development of a stochastic dynamic programming model for determining sequential operating policies in Section 7.2.2.

Figure 7A.1 illustrates the stochastic linear programming model applied to the example problem in Section 7.2.4. The solution of this model is presented in Table 7.6.

Minimize

$$100PR_{1111} + 441PR_{1121} + 121PR_{1221} + 101PR_{2111} + 244PR_{2121} + 100PR_{2211} + 104PR_{2221} + 200PR_{1112} + 500PR_{1122} + 100PR_{1212} + 200PR_{1222} + 121PR_{2122} + PR_{2222}$$

subject to

$$PR_{1112} + PR_{1122} - 0.7PR_{1111} - 0.2PR_{1211} - 0.7PR_{2111} - 0.2PR_{2211} = 0$$

$$PR_{1212} + PR_{1222} - 0.3PR_{1111} - 0.8PR_{1211} - 0.3PR_{2111} - 0.8PR_{2211} = 0$$

$$PR_{2112} + PR_{2122} - 0.7PR_{1121} - 0.2PR_{1221} - 0.7PR_{2121} - 0.2PR_{2221} = 0$$

$$PR_{1111} + PR_{1121} - 0.6PR_{1112} - 0.0PR_{1212} - 0.6PR_{2112} - 0.0PR_{2212} = 0$$

$$PR_{1211} + PR_{1221} - 0.4PR_{1112} - 1PR_{1212} - 0.4PR_{2112} - 1PR_{2212} = 0$$

$$PR_{2111} + PR_{2121} - 0.6PR_{1122} - 0.0PR_{1222} - 0.6PR_{2122} - 0.0PR_{2222} = 0$$

$$PR_{1111} + PR_{1121} + PR_{1211} + PR_{1221} + PR_{2111} + PR_{2121} + PR_{2211} + PR_{2221} = 1$$

$$PR_{1112} + PR_{1122} + PR_{1212} + PR_{1222} + PR_{2112} + PR_{2122} + PR_{2212} + PR_{2222} = 1$$

$$PR_{kilt} \geq 0 \qquad \forall k, i, l, t$$

FIGURE 7A.1. Example stochastic linear programming model for reservoir operation problem defined in Section 7.2.4.

APPENDIX 7B

INPUT AND OUTPUT OF EXAMPLE MIXED-INTEGER RESERVOIR DESIGN PROBLEM

TABLE 7B.1. Input Data for Solving Mixed-Integer Problem Defined in Figure 7.3 Using IBM's MPSX/370 (V1M3) Computer Program (Format same as discussed in Chapter 2, Appendix 2A)

(a) *Program Statements*

```
PROGRAM
INITIALZ
MOVE(XDATA,'EXAMPLE')
MOVE(XPBNAME,'PBFILE')
CONVERT
BCDOUT
TITLE('MIXED INTEGER PROG. EXAMPLE')
SETUP('BOUND','UBS')      (minimization problem)
MOVE(XOBJ,'ROBJ')
MOVE(XRHS,'RIGHT')
CRASH
```

TABLE 7B.1 (Continued)

```
PRIMAL
XMXFNLOG=1      (0 = short output; 1 = long output)
OPTIMIX         (initiates integer solution procedure)
EXIT
PEND
```

(b) *Row Types and Names*

```
ROWS
 N    ROBJ
 E    ROW1
 E    ROW2
 L    ROW3
 E    ROW4
 L    ROW5
 L    ROW6
 L    ROW7
 E    ROW8
 E    ROW9
 E    ROW10
 E    ROW11
 E    ROW12
 E    ROW13
 E    ROW14
 E    ROW15
 E    ROW16
 E    ROW17
 E    ROW18
 E    ROW19
 E    ROW20
 E    ROW21
 E    ROW22
 E    ROW23
 E    ROW24
 E    ROW25
 E    ROW26
 E    ROW27
 E    ROW28
 E    ROW29
 L    ROW30
 L    ROW31
```

TABLE 7B.1 (Continued)

(c) *Column (Variable) Names and Coefficients*

COLUMNS

Column	Row	Coefficient	Row	Coefficient
TS0	ROBJ	50.000	ROW1	1.000
TS1	ROBJ	−250.000	ROW1	1.000
TS1	ROW2	30.000		
TS2	ROBJ	−290.000	ROW1	1.000
TS2	ROW2	50.000		
TS	ROW2	−1.000	ROW18	−1.000
TS	ROW19	−1.000	ROW20	−1.000
TS	ROW21	−1.000		
TR1	ROBJ	−80.000	ROW3	1.000
TR1	ROW4	10.000		
TR2	ROBJ	−140.000	ROW3	1.000
TR2	ROW4	20.000		
TR3	ROBJ	−220.000	ROW3	1.000
TR3	ROW4	60.000		
TR	ROW4	−1.000	ROW10	−1.000
TR	ROW11	−1.000	ROW12	−1.000
TR	ROW13	−1.000	ROW14	−1.000
TR	ROW15	−1.000	ROW16	−1.000
TR	ROW17	−1.000		
K1	ROBJ	5.000	ROW5	1.000
K1	ROW8	1.000		
K2	ROBJ	3.000	ROW6	1.000
K2	ROW8	1.000		
K3	ROBJ	6.000	ROW7	1.000
K3	ROW8	1.000		
K	ROW8	−1.000	ROW30	−1.000
K	ROW31	−1.000		
Z	ROW9	−1.000	ROW22	1.000
Z	ROW23	1.000	ROW24	1.000
Z	ROW25	1.000	ROW26	.500
Z	ROW27	.500	ROW28	.500
Z	ROW29	.500	ROW30	10.000
Z	ROW31	10.000		
R111	ROW10	1.000	ROW22	1.000
R112	ROW14	1.000	ROW26	1.000
R121	ROW11	1.000	ROW24	1.000
R122	ROW15	1.000	ROW28	1.000
R211	ROW12	1.000	ROW23	1.000
R212	ROW16	1.000	ROW27	1.000
R221	ROW13	1.000	ROW25	1.000
R222	ROW17	1.000	ROW29	1.000
D111	ROBJ	1.710	ROW10	1.000
D112	ROBJ	1.200	ROW14	1.000

TABLE 7B.1 (Continued)

D121	ROBJ		1.660	ROW11		1.000
D122	ROBJ		.510	ROW15		1.000
D211	ROW12		1.000			
D212	ROBJ		1.660	ROW16		1.000
D221	ROBJ		6.330	ROW13		1.000
D222	ROBJ		6.630	ROW17		1.000
E111	ROW10	–	1.000			
E112	ROW14	–	1.000			
E121	ROW11	–	1.000			
E122	ROW15	–	1.000			
E211	ROW12	–	1.000			
E212	ROW16	–	1.000			
E221	ROW13	–	1.000			
E222	ROW17	–	1.000			
S11	ROW18		1.000	ROW22	–	.980
S11	ROW24	–	.980	ROW26		1.010
S11	ROW27		1.010	ROW28		1.010
S11	ROW31		1.000			
S12	ROW20		1.000	ROW22		1.020
S12	ROW26	–	.990	ROW28	–	.990
S21	ROW19		1.000	ROW23	–	.980
S21	ROW25	–	.980	ROW29		1.010
S21	ROW30		1.000			
S22	ROW21		1.000	ROW23		1.020
S22	ROW24		1.020	ROW25		1.020
S22	ROW27	–	.990	ROW29	–	.990
D11	ROBJ		5.055	ROW18		1.000
D12	ROBJ		2.565	ROW20		1.000
D21	ROBJ		9.945	ROW19		1.000
D22	ROBJ		12.435	ROW21		1.000
E11	ROW18	–	1.000			
E12	ROW20	–	1.000			
E21	ROW19	–	1.000			
E22	ROW21	–	1.000			
FIRST	'MARKER'			'INTORG'		
Z1	ROBJ		10.000	ROW5	–	20.000
Z1	ROW9		1.000			
Z2	ROBJ		50.000	ROW6		40.000
Z2	ROW7	–	10.000	ROW9		1.000
LAST	'MARKER'			'INTEND'		

(Note marker cards separating integer variables; words FIRST and LAST are arbitrary.)

TABLE 7B.1 (Continued)

(d) *Right-Hand-Side Nonzero Values*

```
RHS
    RIGHT     ROW1          1.000     ROW3           1.000
    RIGHT     ROW22        10.000     ROW23         10.000
    RIGHT     ROW24        20.000     ROW25         20.000
    RIGHT     ROW26        30.000     ROW27         30.000
    RIGHT     ROW28        40.000     ROW29         40.000
```

(e) *Bounds Section and Final ENDATA Card*

```
BOUNDS
 UP UBS     Z              1.000
 UP UBS     E111          10.000
 UP UBS     E112          10.000
 UP UBS     E121          10.000
 UP UBS     E122          10.000
 UP UBS     E211          10.000
 UP UBS     E212          10.000
 UP UBS     E221          10.000
 UP UBS     E222          10.000
 UP UBS     D11            5.000
 UP UBS     D12            5.000
 UP UBS     D21            5.000
 UP UBS     D22            5.000
 UP UBS     E11            7.000
 UP UBS     E12            7.000
 UP UBS     E21            7.000
 UP UBS     E22            7.000
 UP UBS     Z1             1.000
 UP UBS     Z2             1.000
ENDATA
```

TABLE 7B.2. Solution Output for Mixed-Integer Problem

(a) *Maximization Objective Function Value:* 175.965 = expected annual net benefits. (Minimization objective value = −175.965.)

(b) *Row Values, Limits, and Their Dual Variables* (DUAL ACTIVITY is the change in minimization objective function value per unit decrease in right-hand side of constraint.)

ROW	AT	ACTIVITY	SLACK ACTIVITY	LOWER LIMIT	UPPER LIMIT	DUAL ACTIVITY
ROBJ	BS	-175.965	175.965	NONE	NONE	1.000
ROW1	EQ	1.000	.	1.000	1.000	66.382
ROW2	EQ	.	.	.	.	6.121
ROW3	UL	1.000	.	NONE	1.000	100.000
ROW4	EQ	.	.	.	.	2.000
ROW5	UL	.	.	NONE	.	1.000
ROW6	UL	.	.	NONE	.	3.000
ROW7	BS	4.422	4.422	NONE	.	.
ROW8	EQ	.	.	.	.	-6.000
ROW9	EQ	.	.	.	.	70.000
ROW10	EQ	.	.	.	.	-1.710
ROW11	EQ	.	.	.	.	.
ROW12	EQ	.	.	.	.	.
ROW13	EQ	.	.	.	.	2.320
ROW14	EQ	.	.	.	.	.
ROW15	EQ	.	.	.	.	.
ROW16	EQ	.	.	.	.	.
ROW17	EQ	.	.	.	.	2.030
ROW18	EQ	.	.	.	.	.
ROW19	EQ	.	.	.	.	.
ROW20	EQ	.	.	.	.	-1.744
ROW21	EQ	.	.	.	.	-4.376
ROW22	EQ	10.000	.	10.000	10.000	1.710
ROW23	EQ	10.000	.	10.000	10.000	.
ROW24	EQ	20.000	.	20.000	20.000	.
ROW25	EQ	20.000	.	20.000	20.000	2.320
ROW26	EQ	30.000	.	30.000	30.000	.
ROW27	EQ	30.000	.	30.000	30.000	.
ROW28	EQ	40.000	.	40.000	40.000	.
ROW29	EQ	40.000	.	40.000	40.000	-2.030
ROW30	UL	.	.	NONE	.	4.324
ROW31	UL	.	.	NONE	.	1.676

TABLE 7B.2 (Continued)

(c) *Column (Variable) Values, Limits and Reduced Costs* (REDUCED COST is the change in minimization objective function value per unit increase in variable at lower limit.)

COLUMNS	AT	ACTIVITY	INPUT COST	LOWER LIMIT	UPPER LIMIT	REDUCED COST
TS0	LL	.	50.000	0	NONE	116.382
TS1	BS	1.000	−250.000	0	NONE	.
TS2	LL	.	−290.000	0	NONE	82.412
TS	BS	30.000	.	0	NONE	.
TR1	LL	.	− 80.000	0	NONE	40.000
TR2	BS	.918	−140.000	0	NONE	.
TR3	BS	.082	−220.000	0	NONE	.
TR	BS	23.266	.	0	NONE	.
K1	BS	.	5.000	0	NONE	.
K2	BS	40.000	3.000	0	NONE	.
K3	BS	5.578	6.000	0	NONE	.
K	BS	45.578	.	0	NONE	.
Z	UL	1.000	.	0	1.000	−6.985
R111	BS	13.266	.	0	NONE	.
R112	BS	23.266	.	0	NONE	.
R121	BS	23.266	.	0	NONE	.
R122	BS	33.266	.	0	NONE	.
R211	BS	13.266	.	0	NONE	.
R212	BS	23.266	.	0	NONE	.
R221	BS	23.266	.	0	NONE	.
R222	BS	33.266	.	0	NONE	.
D111	BS	10.000	1.710	0	NONE	.
D112	LL	.	1.200	0	NONE	1.200
D121	LL	.	1.660	0	NONE	1.660
D122	LL	.	.510	0	NONE	.510
D211	BS	20.000	.	0	NONE	.
D212	LL	.	1.660	0	NONE	1.660
D221	LL	.	6.330	0	NONE	4.010
D222	LL	.	6.630	0	NONE	8.660
E111	LL	.	.	0	10.000	1.710
E112	BS	.	.	0	10.000	.
E121	BS	.	.	0	10.000	.
E122	BS	10.000	.	0	10.000	.
E211	UL	10.000	.	0	10.000	.
E212	BS	.	.	0	10.000	.
E221	UL	.	.	0	10.000	2.320
E222	UL	10.000	.	0	10.000	−2.030
S11	BS	35.578	.	0	NONE	.
S12	BS	30.000	.	0	NONE	.
S21	BS	35.578	.	0	NONE	.
S22	BS	30.000	.	0	NONE	.
D11	LL	.	5.055	0	5.000	5.055

TABLE 7B.2 (Continued)

COLUMNS	AT	ACTIVITY	INPUT COST	LOWER LIMIT	UPPER LIMIT	REDUCED COST
D12	LL	.	2.565	0	5.000	.821
D21	LL	.	9.945	0	5.000	9.945
D22	LL	.	12.435	0	5.000	8.059
E11	BS	5.578	.	0	7.000	.
E12	LL	.	.	0	7.000	1.744
E21	BS	5.578	.	0	7.000	.
E22	LL	.	.	0	7.000	4.376
Z1	IV	.	10.000	0	1.000	60.000
Z2	IV	1.000	50.000	0	1.000	.

(d) Summary of Solution

Maximum annual expected net benefits: 176
Reservoir capacity: 45.6
Reservoir release target: 23.3
Reservoir storage target: 30.0
Reservoir storage volumes
$t = 1$: 35.6
$t = 2$: 30.0

(Note that the two storage volumes in each period have taken on the same value. In models having more discrete volumes and inflows, this generally does not occur.)

Sequential Reservoir Operating Policy:

Period 1

Storage Volume	INFLOW *10*	INFLOW *20*
35.6	13.3	23.3
	Releases	

Period 2

Storage Volume	INFLOW *30*	INFLOW *40*
30.0	23.3	33.3
	Releases	

EXERCISES

7-1. (a) Compare the method used in Figure 7.3 for modeling the concave portion of the cost function $C(K)$ in Figure 7.2 with method (3) of Figure 2.14.

(b) Will this alternative method work for the maximization of a piecewise-linear convex function as well as for the minimization of a concave function?

(c) Which of the two methods do you prefer, and why?

7-2. Instead of defining a final volume subscript l and m for computing joint probabilities PR_{kilt} in Appendix 7A, assume that subscripts d and e were used to denote different reservoir release volumes. How would the linear programming model equations 7A.1, 7A.3, and 7A.4 be altered to include d and e in place of l and m? How would the dynamic programming recursion equation 7.5 be altered? Finally, how would equations 7.7 and 7.8 differ?

7-3. Given joint probabilities PR_{kit} found from equations 7.7 and 7.8, how would one derive the probability distribution of reservoir releases?

7-4. Solve the example dynamic programming problem for reservoir operation defined by equations 7.12, 7.13, and 7.14, and the data presented in Table 7.1, but using the unconditional probabilities of streamflows rather than the transition probabilities presented in Table 7.2. Compare the annual expected sum of squared deviations and the optimal operating policy. Which is the better policy or estimate of squared deviations?

7-5. Given inflows to an effluent storage lagoon that can be described by a simple first-order Markov chain in each of T periods t, and an operating policy that defines the lagoon discharge as a function of the initial volume and inflow, indicate how you would estimate the probability distribution of lagoon storage volumes.

7-6. Using the data presented in Figure 7.2, except for the subscript relationships, and the model defined by equations 7.17 to 7.25, find the optimal values of the unknown variables when the subscript policy is

(a)

Period $t = 1$ (l values)

k \ i	1	1
1	1	1
2	1	2

Period $t = 2$ (l values)

k \ i	1	2
1	1	2
2	2	2

(b)

Period $t = 1$ (l values)

k \ i	1	2
1	1	1
2	2	2

Period $t = 2$ (l values)

k \ i	1	2
1	1	2
2	2	2

(c)

Period $t = 1$ (l values)

k \ i	1	2
1	1	1
2	2	2

Period $t = 2$ (l values)

k \ i	1	2
1	1	1
2	2	2

7-7. (a) Using the inflows and their transition probabilities defined for the two periods in Table 7.2, find the reservoir operating policy that minimizes the expected sum of squared deviations from a storage volume target of 40 and a release target of 20, where the minimum allowable storage volume is 30 and the maximum allowable volume (reservoir capacity) is 50 in each period t. Find the probabilities

PR_{kilt}, PS_{kt}, PQ_{it}, and also the probabilities for the releases. For this problem, let k or $l = 3, 4, 5$ and $S_k = 10k$.

(b) Compare solutions obtained for (a) above from a dynamic programming model and from a stochastic linear programming model discussed in Appendix 7A.

7-8. (a) In the example reservoir operation problem in Chapter 2, whose solution is given in Table 2.3, replace the deterministic inflows in the three seasons by three possible flows Q_{it}, $i = 1, 2,$ and 3, in each period t, whose unconditional probabilities PQ_{it} are $\frac{1}{3}$ each. The three equally likely flows Q_{it} in each period t are given in the following table:

Value of Flow Q_{it}

t \ i	*1*	*2*	*3*
1	0	10	20
2	40	50	60
3	10	20	30

Note that the mean flows in each of the three periods remains the same as those used in the Chapter 2 example. Solve for the optimal operating policy that minimizes the expected sum of squared deviations from the desired storage and release targets. Show that this minimum annual expected sum is 418, as compared to 275 if only mean flows are considered.

(b) Next assume that the inflows are serially correlated and for all periods t, each flow transition probability $P^t_{ij} = \Pr[Q_{j,t+1} \mid Q_{it}]$ is as specified in the accompanying table. Note again that the mean flows in each period remain the same, namely 10, 50, and 20 for periods $t = 1, 2,$ and 3, respectively. Solve for the optimal reservoir operating policy that again minimizes the expected sum of squared deviations from the desired storage and release targets. Show that in this case the minimum annual expected sum of squared deviations increases to 475.

Transition Probabilities

i \ j	*1*	*2*	*3*
1	0.67	0.33	0
2	0.33	0.34	0.33
3	0	0.33	0.67

(c) Write a computer flowchart and program for solving both parts (a) and (b), using dynamic programming.

7-9. (a) Using the inflow data in Table 7.7, develop and solve a yield model for estimating the storage capacity of a single reservoir required to produce a yield of 1.5 that is 90% reliable in both of the two within-year periods t, and an additional yield of 1.0 that is 70% reliable in period $t = 2$.

(b) Construct a reservoir operating rule that defines reservoir release zones for these yields.

(c) Using the operating rule, simulate the 18 periods of inflow data in Table 7.7 to evaluate the adequacy of the reservoir capacity and storage zones for delivering the required yields and their reliabilities. (Note that in this simulation of the historical record the 90% reliable yield should be satisfied in all 18 periods, and the incremental 70% reliable yield should fail only two times in the 9 years.)

(d) Compare the estimated reservoir capacity with that which is needed using the sequent peak procedure, equation 5.8.

7-10. (a) Solve the following chance-constrained problem for a single reservoir of active capacity K, initial storage volumes S_t, and releases R_t in each of four periods t.

$$\text{minimize } K$$

subject to

$$\Pr[0 \leq S_t \leq K] \geq 0.99 \qquad \forall t$$
$$\Pr[R_t \leq R_t^{\max}] \geq 0.90 \qquad \forall t$$
$$\Pr[R_t \geq R_t^{\min}] \geq 0.80 \qquad \forall t$$

Use the linear decision rules defined by equations 7.74 and 7.75, with all $\lambda_t = 1$ and with all $\lambda_t = 0$, to derive deterministic equivalents of each chance constraint, and compare the two model solutions. The following table lists the values $q_t^{(\alpha)}$ of the random inflows that are exceeded with probability $(1 - \alpha)$, and the values of $R_t^{\max}$ and $R_t^{\min}$.

t	$q_t^{(0.0)}$	$q_t^{(0.1)}$	$q_t^{(0.2)}$	$q_t^{(0.8)}$	$q_t^{(0.9)}$	$q_t^{(0.99)}$	$R_t^{\max}$	$R_t^{\min}$
1	0	10	28	60	80	100	75	20
2	0	2	10	40	50	70	70	20
3	0	5	15	30	60	80	70	20
4	0	8	30	65	90	110	80	20

If either solution is infeasible change $R_t^{\min}$ to achieve feasibility.

(b) How would the model be modified if a flood storage capacity of

40 were required throughout period $t = 3$? What would be the value of the capacity K?

(c) How would the same deterministic equivalent model be modified to include the maximization of an unknown release target T^R such that the probability of a deficit release D_t^R being no greater than 10 is no less than 0.90 in each period t.

$$\Pr[D_t^R \leq 10] \geq 0.90 \qquad \forall t$$

(d) Assume that the reservoir capacity K is known along with each $R_t^{\max}$ and $R_t^{\min}$. Also assume that the distribution functions $F_{Q_t}(q) = \Pr[Q_t \leq q]$ are known for the random inflows Q_t in each period t. Develop the deterministic equivalent model (corresponding to all $\lambda_t = 1$) of the following chance-constrained problem:

$$\text{maximize } P$$

subject to

$$\Pr[S_t \leq K] \geq 0.99$$

$$\Pr[R_t \leq R_t^{\max}] \geq P$$

$$\Pr[R_t \geq R_t^{\min}] \geq P$$

Note that $P = F_{Q_t}(q_t^{(P)})$ and $F_{Q_t}^{-1}(P) = q_t^{(P)}$.

7-11. Using the data presented in Exercise 7-10(a), indicate how the sequent peak algorithm (equation 5.8) can be used to determine the active storage volume K needed to satisfy the two chance constraints for reservoir releases specified in Exercise 7-10(a). Solve for the reservoir capacity K using this algorithm. (*Hint:* The algorithm must be applied to each release constraint separately. The constraint requiring the largest K is the binding constraint.) Compare this solution to that obtained for all $\lambda_t = 1$, and for all $\lambda_t = 0$, in Exercise 7-10(a).

7-12. One possible modification of the yield model of Section 7.3.2 would permit the solution algorithm to determine the appropriate failure years associated with any desired reliability instead of having to choose these years prior to model solution. This modification can provide an estimate of the extent of yield failure in each failure year and include the economic consequences of failures in the objective function. It can also serve as a means of estimating the optimal reliability with respect to economic benefits and losses. Letting F_y be the unknown yield reduction in a possible failure year y, then in place of $\alpha_{py} Y_p$ in the over-year continuity constraint, the term $(Y_p - F_y)$ can be used. What additional constraints are needed to ensure (1) that the average shortage does not exceed $(1 - \alpha_{py})Y_p$ or (2) that at most there are f failure years and none of the shortages exceed $(1 - \alpha_{py})Y_p$.

7-13. Develop a stochastic linear programming model and a chance-constrained model for estimating the active storage volume required within a single reservoir to obtain the maximum *constant* yield y each month that has approximately a 90% chance of being equaled or exceeded in every month of any year. Also develop a yield model for estimating the active storage volume required to deliver a constant yield y each month which is 90% reliable in any year. Assume that there exist 49 years of monthly streamflow records for a particular site at which a reservoir could be constructed, and from these records, the probability distributions can be obtained for the monthly flows at the reservoir site and at the downstream demand site. Be sure to define all variables clearly and describe how all model parameters (and probabilities) would be obtained.

Once each model is developed, indicate the relative advantages and limitations of each modeling approach. Indicate which in your opinion would be the better screening model to use for this problem, and why.

REFERENCES

Stochastic Models

1. GABLINGER, M., and D. P. LOUCKS, Markov Models for Flow Regulation, *Journal of the Hydraulics Division, ASCE*, Vol. 96, No. HY1, 1970.
2. HOUCK, M. H., and J. L. COHON, Sequential Explicitly Stochastic Linear Programming Models: A Proposed Method for Design and Management of Multipurpose Reservoir Systems, *Water Resources Research*, Vol. 14, No. 2, 1978.
3. JACOBY, H. D., and D. P. LOUCKS, The Combined Use of Optimization and Simulation Models in River Basin Planning, *Water Resources Research*, Vol. 8, No. 6, 1972.
4. LOUCKS, D. P., Computer Models for Reservoir Regulation, *Journal of the Sanitary Engineering Division, ASCE*, Vol. 94, No. SA4, 1968.
5. LOUCKS, D. P., and L. M. FALKSON, A Comparison of Some Dynamic, Linear and Policy Iteration Methods for Reservoir Operation, *Water Resources Bulletin*, Vol. 6, No. 3, 1970.
6. LOUCKS, D. P., and H. D. JACOBY, Flow Regulation for Water Quality Management, in *Models for Managing Regional Water Quality*, R. Dorfman, H. D. Jacoby, and H. A. Thomas, Jr. (eds.), Harvard University Press, Cambridge, Mass., 1972.
7. ROEFS, T. G., and L. D. BODIN, Multireservoir Operation Studies, *Water Resources Research*, Vol. 6, No. 2, 1970.
8. SU, S. Y., and R. A. DEININGER, Modeling the Regulation of Lake Superior under Uncertainty of Future Water Supplies, *Water Resources Research*, Vol. 10, No. 1, 1974.
9. THOMAS, H. A., JR., and P. WATERMEYER, Mathematical Models: A Stochastic Sequential Approach, in *Design of Water Resource Systems*, A. Maass, M. M. Hufschmidt, R. Dorfman, H. A. Thomas, Jr., S. A. Marglin, and G. M. Fair (eds.), Harvard University Press, Cambridge, Mass., 1962.

Chance-Constrained Models

10. ASKEW, A. J., Optimum Reservoir Rules and the Imposition of a Reliability Constraint, *Water Resources Research*, Vol. 10, No. 1, 1974.

11. ASKEW, A. J., Chance-constrained Dynamic Programming and the Optimization of Water Resource Systems, *Water Resources Research*, Vol. 10, No. 6, 1974.

12. EASTMAN, J., and C. S. REVELLE, Linear Decision Rule in Reservoir Management and Design. 3. Direct Capacity Determination and Intraseasonal Constraints, *Water Resources Research*, Vol. 9, No. 1, 1973.

13. EISEL, L. M., Comments on "The Linear Decision Rule in Reservoir Management and Design," by C. ReVelle, E. Joeres, and W. Kirby, *Water Resources Research*, Vol. 6, No. 4, 1970.

14. GUNDELACH, J., and C. S. REVELLE, Linear Decision Rule in Reservoir Management and Design. 5. A General Algorithm, *Water Resources Research*, Vol. 11, No. 2, 1975.

15. JOERES, E. F., J. C. LIEBMAN, and C. S. REVELLE, Operating Rules for Joint Operation of Raw Water Sources, *Water Resources Research*, Vol. 7, No. 2, 1971.

16. LANE, M., Designing Reservoir Control Policies with Chance Constrained Programming, Proceedings of the International Symposium on *Uncertainties in Hydrological Water Resources Systems*, Department of Hydrology, University of Arizona, Tucson, Arizona, Vol. 3, 1972.

17. LECLERC, G., and D. H. MARKS, Determination of the Discharge Policy for Existing Reservoir Networks under Differing Objectives, *Water Resources Research*, Vol. 9, No. 5, 1973.

18. LOUCKS, D. P., Some Comments on Linear Decision Rules and Chance Constraints, *Water Resources Research*, Vol. 6, No. 2, 1970.

19. LOUCKS, D. P., and P. DORFMAN, An Evaluation of Some Linear Decision Rules in Chance Constrained Models for Reservoir Planning and Operation, *Water Resources Research*, Vol. 11, No. 6, 1975.

20. NAYAK, S. C., and S. R. ARORA, Optimal Capacities for a Multireservoir System Using the Linear Decision Rule, *Water Resources Research*, Vol. 7, No. 3, 1971.

21. NAYAK, S. C., and S. R. ARORA, Linear Decision Rule: A Note on Control Volume Being Constant, *Water Resources Research*, Vol. 10, No. 4, 1974.

22. REVELLE, C., and W. KIRBY, Linear Decision Rule in Reservoir Management and Design. 2. Performance Optimization, *Water Resources Research*, Vol. 6, No. 4, 1970.

23. REVELLE, C., E. JOERES, and W. KIRBY, The Linear Decision Rule in Reservoir Management and Design. 1. Development of the Stochastic Model, *Water Resources Research*, Vol. 5, No. 4, 1969.

24. REVELLE, C. and J. GUNDELACH, Linear Decision Rule in Reservoir Management and Design. 4. A Rule That Minimizes Output Variance *Water Resources Research*, Vol. 11, No. 2, 1973.

25. ROSSMAN, L. A., Reliability-Constrained Dynamic Programming and Ran-

domized Release Rules in Reservoir Management, *Water Resources Research*, Vol. 13, No. 2, 1977.

26. Sniedovich, M., Reliability-Constrained Reservoir Control Problems, *Water Resources Research*, Vol. 15, No. 6, 1979.

Yield Models

27. Loucks, D. P., Stochastic Models for Reservoir Design, in *Stochastic Approaches to Water Resources*, Vol. 2, H. W. Shen (ed./publ.), Fort Collins, Colo., 1976.
28. Loucks, D. P., Surface Water Quantity Management, in *Systems Approach to Water Management*, A. K. Biswas (ed.), McGraw-Hill Book Company, New York, 1976.
29. Viessman, W., Jr., G. L. Lewis, I. Yomtovian, and N. J. Viessman, A Screening Model for Water Resources Planning, *Water Resources Bulletin*, Vol. 11, No. 2, 1975.

Additional References

30. Butcher, W. S., Stochastic Dynamic Programming for Optimum Reservoir Operation, *Water Resources Bulletin*, Vol. 7, No. 1, 1971.
31. Hall, W. A., and J. A. Dracup, *Water Resources Systems Engineering*, McGraw-Hill Book Company, New York, 1970.
32. Young, G. K., Jr., Finding Reservoir Operating Rules, *Journal of the Hydraulics Division, ASCE*, Vol. 93, No. HY6, 1967.

CHAPTER 8

Irrigation Planning and Operation

8.1 INTRODUCTION

In Chapters 5 and 7 on multipurpose river basin planning, irrigation was considered simply as one of many uses competing for water. What distinguished irrigation from most other uses was that the benefits derived from the allocation of water throughout each growing season were available only at the end of the growing season (i.e., at the time of harvest). The unknown decision variable was the total target allocation. The benefits were expressed as a function of that unknown target allocation, assuming that the allocation was properly distributed, as specified, over the growing season.

In this chapter irrigation planning and operation will be examined in greater detail. This will involve the consideration of other resource inputs, including land, seed, fertilizer, machinery, and labor. It will also involve the development of methods for estimating which crops should be grown on which sites within an irrigation area, the quantity of each crop produced, and, in situations where the quantity produced will affect the unit selling price, the price at which the crops can be sold. Not only is this information required in order to derive the benefit functions needed for the multipurpose river basin models previously discussed, but it is also useful to those responsible for planning and managing irrigation projects.

The maximization of net income from an irrigation project is just one possible objective. Irrigation development objectives may be numerous and conflicting and the optimal irrigation development plans may differ depending on the relative importance given to each objective. A crop production program that provides the greatest net income is not likely to create the highest employment of labor, nor may it produce the most food or foreign exchange from agricultural exports. For developing regions these objectives may be more important than maximization of net income. Each development objective draws upon the same available resources, but not equally. It is not possible to maximize simultaneously net income, employment, food production, and agricultural exports. Trade-offs are possible, and indeed necessary, when formulating a plan that achieves an appropriate compromise among various irrigation development objectives.

In this chapter only one objective will be considered, the maximization of net income. In an actual application of irrigation planning, other objectives should be included in the planning or operating policy study as discussed in Chapter 4.

8.2 AN IRRIGATION PLANNING MODEL

Consider an irrigation district containing various soil types in various regions or areas of the district. If there is a variety of crops that can be grown in the irrigation district, one must decide which crops should be grown on which soil types in which areas. The crop yield that can be expected will depend, in part, on (1) the physical characteristics of the irrigation area, (2) the crop type, (3) the amounts and timing of water and fertilizer applications, (4) the available labor and machinery, and on (5) how effectively these resources are used (i.e., the farm management practices).

In order to estimate how much of each resource input should be allocated to each area of an irrigation district, and to decide the type and quantity of each crop that should be planted and where each crop should be planted, it is convenient to divide the whole irrigation district into compartments or subareas. Each designated subarea should exhibit similar physical characteristics that together will be called a soil type. In other words, each subarea should be confined to a single, specific soil type and to a relatively local and homogeneous region of the district. The latter restriction is necessary to ensure that the costs of transporting any resource input to the subarea and of operating within the subarea are essentially uniform throughout the designated subarea.

A hypothetical irrigation district shown in Figure 8.1 will be used to illustrate the development of an irrigation planning model. This district, the Helga River Valley Irrigation District (site 4 in Figure 5.10), can be subdivided into areas as illustrated in Figure 8.1. In this example assume that it

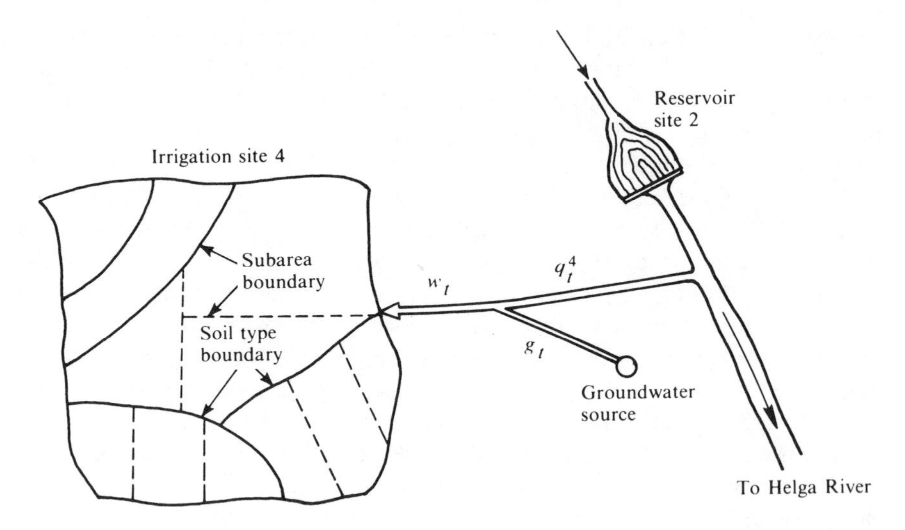

FIGURE 8.1. Helga River irrigation district and its 11 subareas.

is possible to use groundwater to supplement surface water supplies, or as a source of water in times of a surface water shortage.

The possibility of using groundwater as well as surface water will permit the development of what is termed a conjunctive-use model. In this model it will be assumed that the groundwater table is below the streambed, and hence groundwater withdrawals will not significantly affect the surface-water flow.

8.2.1 Resource Inputs

An irrigation planning model can provide estimates of the quantities of all resource inputs (labor, fertilizer, pesticides, seed, machinery, capital or credit, etc.) and their costs; the location, quantity, and types of crops to plant; and, if appropriate, their market price that together maximize net income. The relationships among these resource inputs and crop yields are defined by the constraints of the model. This section will begin the development of the constraints of a simple irrigation planning model to illustrate how the relationships among all resource inputs and the final crop yields can be estimated.

First several indices or subscripts must be defined. Let the subscript j represent a particular subarea having a homogeneous soil type within the irrigation district. If the area of the subarea is denoted as A_j, then $\sum_j A_j$ is the total area of the irrigation district.

Next, let the subscript f distinguish among various farm types. Although it is unlikely that any two farms will be exactly alike, for planning purposes, existing or potential farms can be grouped into various types characterized by their size, their level of technology, their productivity or efficiency, and their financial resources.

Finally, let the index c represent a particular crop, such as maize (corn), cotton, rice, sugar cane, soybean, or whatever is appropriate for the particular subarea and farm type. If any particular crop can be grown under different management practices by a specific farm type f, then each crop–management combination is denoted by a different index c. For example, if maize can be grown under intensive or extensive irrigation, these are considered as two different crops requiring different quantities of resource inputs, including water, and have different yields per unit area, even though the price per unit yield may be the same.

Having defined these three indices, it is now possible to ask how many hectares in subarea j on farm type f should be planted with crop c. These are unknown decision variables, and will be denoted as X_{jfc}. If farms already exist in the irrigation district, then within each subarea j the areas AF_{jf} of each farm type f are known. In this case the total area planted in any crop c by any particular farm type f cannot exceed the area operated by each farm type in each subarea j,

$$\sum_{c} X_{jfc} \leq \mathrm{AF}_{jf} \qquad \forall\, j, f \tag{8.1}$$

where, of course, $\sum_f \mathrm{AF}_{jf} \leq A_j$. In situations where new farms must be developed, and hence the farm type is unknown and to be determined, each farm type area AF_{jf} is unknown, and instead of equation 8.1, constraint 8.2 is required.

$$\sum_{f} \sum_{c} X_{jfc} \leq A_j \qquad \forall\, j \tag{8.2}$$

The particular crops c considered in each subarea j and farm type f may vary for different subareas and farm types, but will include only those crops that are suitable for that particular soil type. If the irrigation district is a single-crop district or if there is only one crop that is suitable for one or more subareas, the subscript c may be dropped from the variables associated with those subareas. In such cases, it is a matter of deciding only how much land area is to be irrigated in each subarea. In this example, crop rotation (i.e., growing different crops on the same subarea in different growing seasons) will not be considered. Crop rotation could be considered if the model included multiple growing seasons.

Given specific resource inputs and the use of these inputs by each farm type, the expected or target yield of each crop c per unit area in subarea j must be estimated from crop production functions or from experience. Denote this yield per hectare as Y_{jfc}. Then the total yield of crop c on farm type f in subarea j is simply the product $Y_{jfc} \cdot X_{jfc}$. The gross income I_f of all farms in each farm type is simply the sum of the total yield times the price P_c per unit yield of each crop c.

$$I_f = \sum_j \sum_c P_c Y_{jfc} X_{jfc} \qquad \forall f \tag{8.3}$$

Both income and costs will be discussed in greater detail shortly.

To obtain these yields, specific resources must be available. Each hectare of crop c will require various quantities of seed, fertilizer and pesticides, labor, machinery, and water. Of course, these resource inputs will vary somewhat in practice, but for irrigation planning (as opposed to operation), it is convenient to assume that they are fixed and known. This assumption implies that shortages in one or more resource inputs will limit the hectares of crop production rather than the yields per hectare. The particular quantities of resources needed per hectare will often depend on the soil type, the farm type, and the crop itself. This information must be developed before an irrigation planning model can be constructed. The required resource inputs are defined by the following constraints.

Water. In each period t of the growing season, let the total quantity of water required per hectare of crop c in each subarea j of each farm type f be designated as W_{jfct}. Thus the total ground and surface water required, $Q_{jft}^{\mathrm{GW}} + Q_{jft}^{\mathrm{SW}}$, by each farm type f in each subarea j in each period t equals

$$Q_{jft}^{\mathrm{GW}} + Q_{jft}^{\mathrm{SW}} = \sum_c W_{jfct} X_{jfc} \qquad \forall j, f, t \tag{8.4}$$

This water can be obtained from surface-water allocations q_t^4 from site 4 (Figure 8.1), or from groundwater pumping g_t, if conjunctive operation of both surface and groundwater supplies is appropriate or acceptable. In this case the total quantity of water w_t required in each period t of the growing season is the sum of the groundwater (Q_{jft}^{GW}) and surface-water (Q_{jft}^{SW}) allocations

$$\sum_j \sum_f Q_{jft}^{\mathrm{GW}} = g_t \qquad \forall t \tag{8.5a}$$

$$\sum_j \sum_f Q_{jft}^{\mathrm{SW}} = q_t^4 \qquad \forall t \tag{8.5b}$$

$$g_t + q_t^4 = w_t \qquad \forall t \tag{8.5c}$$

To deliver this quantity of water to each subarea j in each period t, the channel capacities q, g, and w required to transport water to the irrigation district and the channel capacities q_j within the irrigation district must at least equal the maximum of the appropriate flows q_t^4, g_t, w_t, and $\sum_f Q_{jft}^{\mathrm{GW}} + \sum_f Q_{jft}^{\mathrm{SW}}$, respectively.

$$q \geq q_t^4 \qquad \forall t \tag{8.6}$$

$$g \geq g_t \qquad \forall t \tag{8.7}$$

$$w \geq w_t \qquad \forall t \tag{8.8}$$

$$q_j \geq \sum_f (Q_{jft}^{\mathrm{GW}} + Q_{jft}^{\mathrm{SW}}) \qquad \forall t \tag{8.9}$$

There may also be limitations on the maximum amounts of surface and groundwater available in each period t. These restrictions, if any, can be included as constraints or can be incorporated into the cost functions, as will be discussed later.

Fertilizer and Pesticides. The total quantity F_f^k of fertilizer or pesticide type k required in the growing season by each farm type f will equal the quantity required per hectare F_{jfc}^k times the hectares in that category X_{jfc} summed over each subarea and crop type.

$$F_f^k = \sum_j \sum_c F_{jkc}^k X_{jfc} \qquad \forall f, k \tag{8.10}$$

Labor. The total person-hours of labor L_{ft}^k of skill type k required by each farm type f in each period t of the growing season will equal the sum over all subareas and crop types of all person-hour requirements per hectare L_{jfct}^k times the total respective land areas X_{jfc}.

$$L_{ft}^k = \sum_j \sum_c L_{jfct}^k X_{jfc} \qquad \forall f, k, t \tag{8.11}$$

There may also be restrictions on the total person-hours of labor available for one or more skill types k.

Equipment. The total hours E_{ft}^k of equipment type k required by each farm type f in each period t of the growing season depends on the given hourly requirements E_{jfct}^k per hectare and on the unknown hectares X_{jfc}.

$$E_{ft}^k = \sum_j \sum_c E_{jfct}^k X_{jfc} \qquad \forall f, k, t \tag{8.12}$$

The number of units U_f^k of equipment type k required for irrigation activities by each farm type f in the irrigation district may be of interest if the equipment is to be purchased by each farm. This number can be estimated from a knowledge of the average number of hours H_{kt} each unit of type k equipment will be used per period and the total hours of required use E_{ft}^k by each farm type f.

$$U_f^k \geq \frac{E_{ft}^k}{H_{kt}} \qquad \forall f, k, t \tag{8.13}$$

If the equipment is to be rented from an organization serving the entire district, the total number of units of equipment of each type k required is simply the sum $\sum_f U_f^k$. Clearly, the value of U_f^k should be an integer since a fraction of a particular type of machine or tool is not very useful (except perhaps for spare parts). However, for preliminary estimates of resource requirements and costs, it seems unnecessary at this stage of irrigation planning to resort to mixed-integer models to ensure that each U_f^k is an integer.

Seed. The quantity of seed or planting stock S_{fc} of crop c required by each farm type f can be estimated from a knowledge of the seed requirements per hectare s_{jfc} and the total hectares of crop c planted by farm type f.

$$S_{fc} = \sum_j s_{jfc} X_{jfc} \qquad \forall f, c \tag{8.14}$$

A lack of any resource input may restrict the values of these unknown areas X_{jfc}. If any of the required resource inputs, such as water, labor, equipment, seed, fertilizer, and so on, are limited, or if their use is restricted (as pesticide use might be for environmental reasons), then upper bounds should be placed on the quantities of those resources that are limited or restricted. The constraints are similar to the bounds that define the maximum land area available (equation 8.1 or 8.2) or that will be used to define the maximum capital that may be borrowed (equation 8.25).

8.2.2 Crop Diversification

Farmers have a variety of reasons for not planting too much of their land in a single crop. On-farm needs for additional crops and the possibility of a crop or market failure are two reasons why more than one type of crop is often planted on each farm. Without performing a complex analysis for estimating the extent to which risk aversion may dictate crop diversification at the expense of expected net income, constraints ensuring at least some prespecified crop diversification can be added to the model. This requires the selection of a set C_f of crops c' ($c' \in C_f$) for each farm type f whose planted area $\sum_j X_{jfc'}$ must be at least some fraction $\alpha_{fc'}$ of the total planted area. Constraints that ensure these conditions are

$$\sum_j X_{jfc'} \geq \alpha_{fc'} \sum_j \sum_c X_{jfc} \qquad \forall f, c' \in C_f \tag{8.15}$$

These crop diversification constraints can also be expressed in terms of crop yields. Letting $\beta_{fc'}$ be the minimum acceptable crop yield of crop type c' for farm type f, then

$$\sum_j Y_{jfc'} X_{jfc'} \geq \beta_{fc'} \qquad \forall f, c' \in C_f \tag{8.16}$$

8.2.3 Annual Costs

The total cost of each resource for each farm type can now be estimated. These resources include land and possibly drainage works in addition to the above-mentioned water, fertilizer, pesticides, equipment, labor, and seed or planting stock.

Land and Drainage Works. If the irrigated land is to be drained, and/or is rented, taxed, or otherwise incurs a cost C^D_{jf} per unit area that is dependent on its location and soil type j, and perhaps also the farm type f, then the total land cost TC^D_f for farm type f is

$$TC^D_f = \sum_j \sum_c C^D_{jf} X_{jfc} \qquad \forall f \tag{8.17}$$

Especially for drainage works, there also may be fixed costs that apply. These fixed costs would then have to be added to the variable costs defined by the right-hand side of equation 8.17.

Water. The cost of water TC_f^W for each farm type f may include an allocated portion δ_f of the fixed costs $C_g^W(g)$, $C_q^W(q)$, and $C_w^W(w)$ associated with the channel capacities g, q, and w, respectively, and groundwater pumps. The cost will also include the fraction γ_{jf} of the cost $C_j(q_j)$ of channel capacity q_j to farm type f. In addition, there are variable costs associated with each water allocation Q_{jft}. These variable costs include the cost to obtain and transport water to each subarea j in each period t. Denoting the variable costs per unit quantity of groundwater as C_{jt}^{GW}, and those of surface water as C_{jt}^{SW}, the total cost of water for each farm type f is

$$\begin{aligned} TC_f^W = \sum_j \{\delta_f[C_g^W(g) + C_q^W(q) + C_w^W(w)] + \gamma_{jf}C_j(q_j) \\ + \sum_t (C_{jt}^{GW}Q_{jft}^{GW} + C_{jt}^{SW}Q_{jft}^{SW})\} \end{aligned} \tag{8.18}$$

If some cost functions are nonlinear, they need to be made piecewise linear for inclusion in any linear optimization model.

Fertilizer and Pesticides. The cost of fertilizer TC_f^F and/or pesticides for each farm type f will depend on the unit cost C_k^F of each type k of fertilizer or pesticide. Recalling that F_f^k is the quantity of type k required by farm type f, the total cost is

$$TC_f^F = \sum_k C_k^F F_f^k \qquad \forall f \tag{8.19}$$

Labor. The cost of labor TC_f^L for each farm type f may include wages per unit of time worked or fixed salaries per individual, or a combination of both, depending on the type of job and skill level k required. Assume that labor of type k is paid an hourly wage rate of C_k^L. Recalling that L_{ft}^k are the required person-hours of the kth type of labor for farm type f in period t, the total labor cost for each farm type f is

$$TC_f^L = \sum_t \sum_k C_k^L L_{ft}^k \qquad \forall f \tag{8.20}$$

Equipment. The cost of equipment TC_f^E for each farm type f will depend on the fixed annual costs $FC_k^E(U_f^k)$ of each unit of equipment U_f^k and on the variable hourly operating costs VC_k^E multiplied by the hours E_{ft}^k of type k equipment usage by each farm type f in each period t.

$$TC_f^E = \sum_k [FC_k^E(U_f^k) + \sum_t VC_k^E E_{ft}^k] \qquad \forall f \tag{8.21}$$

If each type of equipment is rented at a uniform hourly rate HC_k^E by each farm, the total equipment cost for each farm type is

$$TC_f^E = \sum_t \sum_k HC_k^E E_{jt}^k \qquad \forall f \tag{8.22}$$

Seed. The cost of seed or planting stock TC_f^S for each farm type f will depend on the unit cost C_c^S of each type c of seed. Recalling that S_{fc} is the quantity of type c seed or stock required by farm type f, the total cost is simply

$$TC_f^S = \sum_c C_c^S S_{fc} \qquad \forall f \tag{8.23}$$

Capital. It may be necessary to borrow money at the beginning of the growing season in order to have sufficient cash to be able to pay those costs which occur prior to the receipt of income from the sale of crops. If this is the case, it is appropriate to include a budget constraint limiting the total cost TC_f to be no greater than the available capital M_f^A plus the borrowed capital M_f^B.

$$TC_f \leq M_f^A + M_f^B \qquad \forall f \tag{8.24}$$

The capital that can be borrowed by any farm type may itself be limited to some amount $M_f^{B,\max}$.

$$M_f^B \leq M_f^{B,\max} \qquad \forall f \tag{8.25}$$

Assuming an interest rate of r_f for borrowed capital, over the period borrowed, which might vary for different farm types f, the cost of borrowed capital will be $r_f M_f^B$.

Total Cost. The total cost TC_f for each farm type f is the sum of all the individual annual costs of all resource inputs just defined, plus the cost of any borrowed capital.

$$TC_f = TC_f^D + TC_f^W + TC_f^F + TC_f^L + TC_f^E + TC_f^S + r_f M_f^B \qquad \forall f \tag{8.26}$$

8.2.4 Annual Net Income

The annual net income for each farm type f will equal the total annual gross income I_f (equation 8.3) less the total cost TC_f (equation 8.26). Assuming an objective of maximizing total net income from the irrigation district, the objective can be written

$$\text{maximize} \sum_f (I_f - TC_f) \tag{8.27}$$

Equally important is individual farm net income. The average net income to each farm within each type f is simply the net income for the farm type $I_f - TC_f$, divided by the number of farms within the type f. If this number is not already established or known, then it will depend, in part, on the area $\sum_c X_{jfc}$ within each subarea j allocated to farm type f. Since different farm types use different mixes of resource inputs, solutions that favor labor-intensive as opposed to capital-intensive agriculture, for example, can be obtained by the assignment of appropriate relative weights or shadow prices to each component of the objective function 8.27 (as discussed in more detail in Chapter 4).

Objective function 8.27 subject to constraints and definitions 8.1 to 8.26, as appropriate, constitute an irrigation planning model that can be solved by linear optimization techniques (as discussed in Chapter 2). The solution of the foregoing model will provide estimates of each X_{jfc}, the hectares of subarea j of crop c managed by a farm of type f.

For large projects that contribute a major portion of the total annual production of one or more crops, the possible effect of project yields on the market price and demand must be considered. In addition, for projects funded by public monies, the maximization of net income may not be the most appropriate objective. These issues are discussed below.

8.2.5 Annual Net Benefits and Varying Demands

It has been assumed so far that there is a fixed price P_c per unit quantity of crop c produced. For competitive markets in which the total quantity of each crop produced by the irrigation district does not significantly increase the total quantity sold in the market, or for crops whose prices are fixed by the government, these fixed prices will apply (Figure 8.2b). In other situations, the quantity of the crop produced may affect the price at which it can be sold. In these cases, the constant unit price P_c in equation 8.3 should be replaced by the price function $P_c(Y_c)$, where Y_c is the total crop yield

$$Y_c = \sum_j \sum_f Y_{jfc} X_{jfc} \qquad \forall c \tag{8.28}$$

Given a decreasing unit price as the total yield increases, the total revenue functions $R_c(Y_c)$,

$$R_c(Y_c) = P_c(Y_c) \cdot Y_c \qquad \forall c \tag{8.29}$$

will be concave, as illustrated in Figure 8.2b.

In situations where public funds are used for irrigation development and the unit prices are functions of the yields as defined by inverse demand functions $P_c(Y_c)$, the maximization of total net income or revenue may not be an appropriate objective. This is because it is not a measure of social net benefits. As discussed in Chapter 4, social benefits are measured by the area under the price–quantity function $P_c(Y_c)$. Hence where appropriate, the maximization of total net benefits $B_c(Y_c) - \sum_f \mathrm{TC}_f$ should be used as an objective:

$$\text{maximize} \sum_c \int_0^{Y_c} P_c(y)\, dy - \sum_j \mathrm{TC}_f \tag{8.30}$$

The difference between the total benefits $B_c(Y_c)$ and total revenue $R_c(Y_c)$ is illustrated in Figure 8.2. For the simple case in which the unit price is a linear

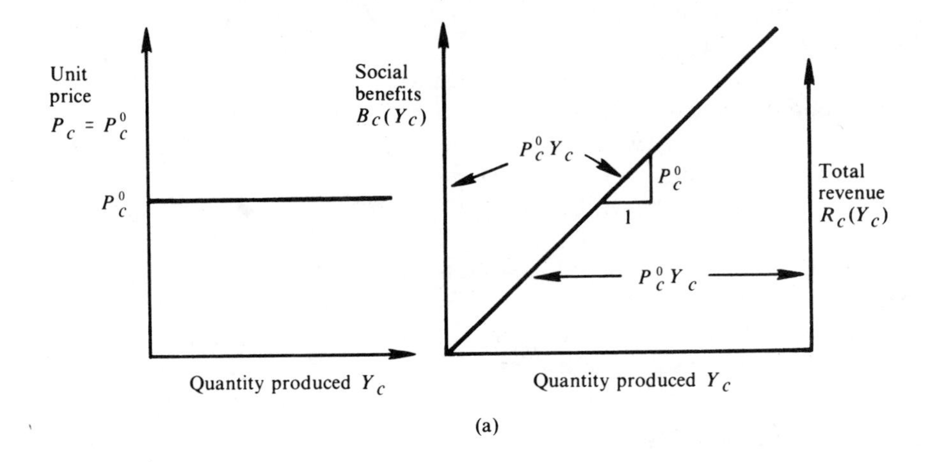

(a)

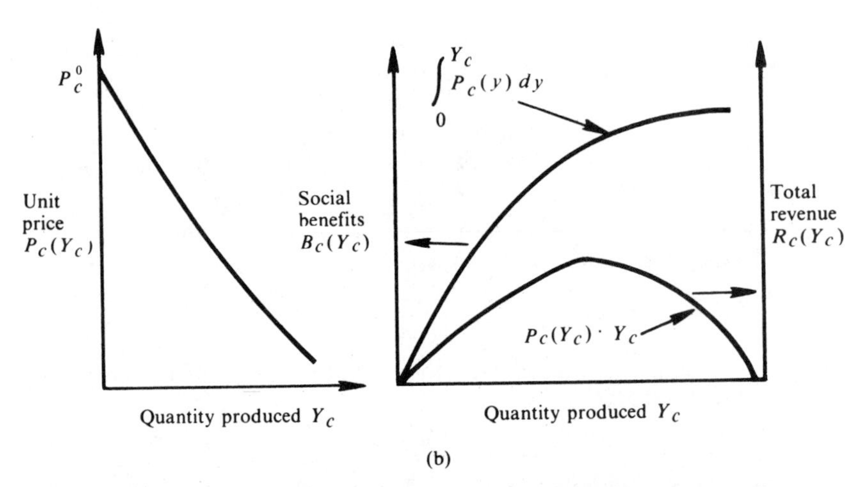

(b)

FIGURE 8.2. (a) Demand function, social benefits, and total revenue where price is independent of quantity produced Y_c. (b) Demand function, social benefits, and total revenue where price decreases with quantity sold.

function of the quantity Y_c supplied $[P_c(Y_c) = P_c^0 - b_c Y_c,\ b_c > 0]$, the total benefits associated with a crop yield of Y_c would equal

$$\int_0^{Y_c} (P_c^0 - b_c y) y\, dy = P_c^0 Y_c - 0.5 b_c Y_c^2 \qquad \forall c \tag{8.31}$$

This concave function, or its piecewise-linear approximation, defines the total benefits and would replace the total revenue term $\sum_f I_f$ in objective function 8.27. The net revenue or producer's income defined by each term $I_f - \mathrm{TC}_f$ in equation 8.27 is still of interest and may be evaluated after the solution of the model using the unit price $P_c(Y_c) = (P_c^0 - b_c Y_c)$ for P_c in equation 8.3.

8.2.6 Annual Net Benefit Function for Irrigation Surface-Water Allocations

To define a net benefit function $B^4(T^4)$ associated with the total irrigation surface-water allocation target T^4 at site 4 in the Helga River valley (Figure 8.1), it is first necessary to estimate the maximum net benefits that can be obtained from irrigation, assuming no surface-water allocations q_t^4 from the Helga River. These would be the net benefits from the irrigation district that could be obtained from the use of groundwater g_t or water from sources other than the Helga River. The model just developed for irrigation planning, with all q_t^4 equal to zero, could be used to estimate the maximum possible annual net benefits obtainable without surface water from the Helga River. Let these benefits be denoted as NB_0^4.

Next, the irrigation planning model can again be solved for various values of the surface-water target allocation T^4. This target quantity is allocated among the periods within the growing season subject to the constraint that

$$\sum_t q_t^4 \leq T^4 \tag{8.32}$$

The difference between the resulting net benefits and NB_0^4 (when T^4 is zero) is the net benefits $B^4(T^4)$ associated with that particular value of T^4. Having these estimates of the incremental benefits for a variety of target allocations T^4 permits the estimation of the benefit function $B^4(T^4)$ itself.

Having estimated the optimal value of each q_t^4 for each of a variety of fixed T^4, the fraction f_t^4 of the total allocation required in each period t can be determined for each discrete value of T^4 as

$$f_t^4 = \frac{q_t^4}{T^4} \tag{8.33}$$

Since these fractions f_t^4 may differ for different values of the total allocation T^4, the actual values of f_t^4 (used in equation 5.44 of the river basin planning model to define and evaluate basin-wide water resources development and allocation policies) may have to be based on some assumed allocation target T^4. Once the model is solved, if the computed T^4 differs from the T^4 assumed, the fractions can be adjusted and the river basin model resolved.

8.2.7 An Example Irrigation Planning Problem

To summarize the basic elements of the irrigation planning model just discussed, consider a simple two-period, two-soil type, two-farm type, three-crop irrigation planning problem. This example can easily be extended to more periods within an irrigation season, more subareas, more farm types, and additional crops. To save space and present a more compact summary, this simple planning problem should suffice. Unit prices P_c of each crop c are constant; hence the maximization of total net benefits is the same as the maximization of total income less total costs.

Before defining the objective, the constraints will be developed. These constraints will include the resource requirements and the definitions of yields Y_{fc} and total costs TC_f for each crop c and farm type f.

Land. Assume that the irrigation district contains 1000 hectares and is divided into 700 hectares of soil type $j = 1$ and 300 hectares of soil type $j = 2$. Four farms of farm type $f = 1$ average about 100 hectares each. The two farms of farm type $f = 2$ average 300 hectares each. Figure 8.3 illustrates the various soil types (subareas) j and farms of each type f. Half of each farm of type $f = 2$ is within each soil type. Assume that on soil type $j = 1$, crops $c = 1$ and 2 can be grown. On soil type $j = 2$, crops 1 and 3 can be grown. Crop 2 cannot survive on soil type 2, and similarly crop 3 cannot grow well on soil type 1. Hence equation 8.1 limiting the land areas X_{jfc} in each soil type or subarea j and by farm type f becomes

for $j = 1$ and $f = 1$:

$$X_{111} + X_{112} \leq 400$$

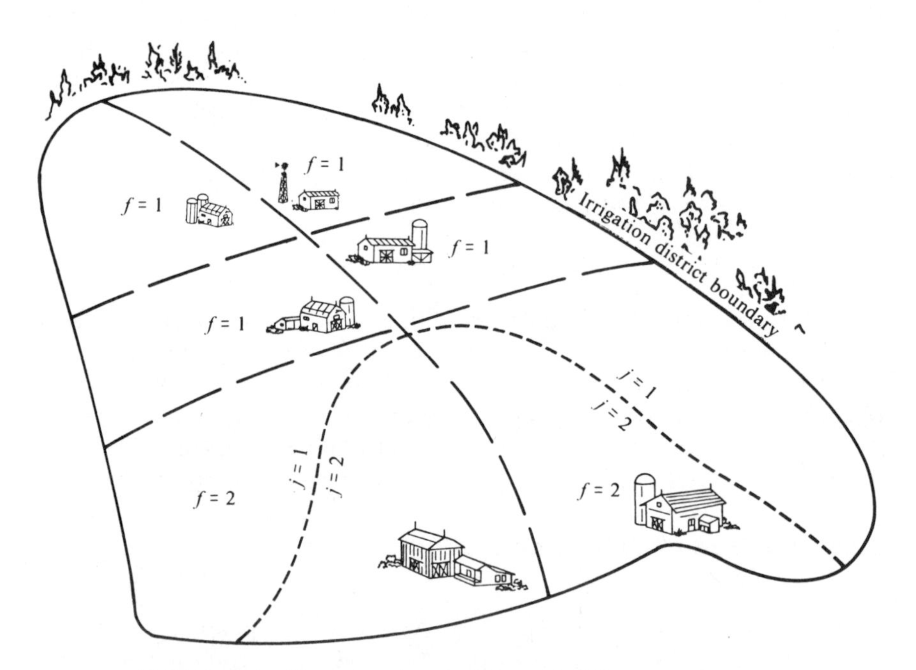

FIGURE 8.3. Map showing location of different soil types $j = 1, 2$ and various farms (two types $f = 1, 2$) in irrigation district.

for $j = 1$ and $f = 2$:

$$X_{121} + X_{122} \leq 300$$

and for $j = 2$ and $f = 2$:

$$X_{221} + X_{223} \leq 300 \tag{8.1E}$$

Water. Table 8.1 specifies the water requirements W_{jfct} per hectare for each soil type, for each crop type, and for each farm type in each of the two

TABLE 8.1. Water Requirements per Hectare

Soil Type,	*Farm Type,*	*Crop Type,*	WATER REQUIREMENTS	
j	*f*	*c*	W_{jfc1}	W_{jfc2}
1	1	1	20	10
1	1	2	35	17
1	2	1	30	15
1	2	2	45	20
2	2	1	10	5
2	2	3	15	8

periods of the growing season. These water requirement coefficients are needed to define the water allocations Q_{jft}^{GW} and Q_{jft}^{SW} from ground- and surface-water sources.

$$\begin{aligned}
Q_{11t}^{GW} + Q_{11t}^{SW} &= \sum_{c=1}^{2} W_{11ct}X_{11c} = \begin{cases} 20X_{111} + 35X_{112} & t = 1 \\ 10X_{111} + 17X_{112} & t = 2 \end{cases} \\
Q_{12t}^{GW} + Q_{12t}^{SW} &= \sum_{c=1}^{2} W_{12ct}X_{12c} = \begin{cases} 30X_{121} + 45X_{122} & t = 1 \\ 15X_{121} + 20X_{122} & t = 2 \end{cases} \\
Q_{22t}^{GW} + Q_{22t}^{SW} &= \sum_{c=1,3} W_{22ct}X_{22c} = \begin{cases} 10X_{221} + 15X_{223} & t = 1 \\ 5X_{221} + 8X_{223} & t = 2 \end{cases}
\end{aligned} \tag{8.4E}$$

Assume that the surface-water source is limited to a total amount of water T.

$$Q_{111}^{SW} + Q_{112}^{SW} + Q_{121}^{SW} + Q_{122}^{SW} + Q_{221}^{SW} + Q_{222}^{SW} \leq T \tag{8.32E}$$

To simplify the problem, the channel capacity constraint equations of the type 8.5 to 8.9 will be ignored, assuming that the channel capacities are fixed and nonrestrictive. Such capacity constraints are easily added if required.

Fertilizer and Pesticides. There are two types of fertilizer and pesticides required for each crop. The requirements per hectare F_{jfc}^{k} are specified in Table 8.2. Using the information in Table 8.2, the total fertilizer and pesticide requirements F_{f}^{k} of each type k for each farm type f are

TABLE 8.2. Fertilizer and Pesticide Requirements per Hectare

Soil Type, j	Farm Type, f	Crop Type, c	REQUIREMENTS F^1_{jfc}	F^2_{jfc}
1	1	1	100	20
1	1	2	75	30
1	2	1	120	45
1	2	2	90	35
2	2	1	150	30
2	2	3	20	70

$$F^1_1 = \sum_{c=1}^{2} F^1_{11c}X_{11c} = 100X_{111} + 75X_{112}$$

$$F^1_2 = \sum_{j=1}^{2}\sum_{c=1}^{3} F^1_{j2c}X_{j2c} = 120X_{121} + 90X_{122} + 150X_{221} + 20X_{223}$$

$$F^2_1 = \sum_{c=1}^{2} F^2_{11c}X_{11c} = 20X_{111} + 30X_{112} \tag{8.10E}$$

$$F^2_2 = \sum_{j=1}^{2}\sum_{c=1}^{3} F^2_{j2c}X_{j2c} = 45X_{121} + 35X_{122} + 30X_{221} + 70X_{223}$$

Labor and Equipment. Only one type of labor and one type of equipment is included. But, now consider the possibility of substituting a portion of labor for equipment, or vice versa. Let h_f be the person-hours of labor time that are equivalent to 1 unit-hour of equipment time, for each farm type f. The constraint defining the person-hours of labor time L_{ft} and machine-hours of equipment time E_{ft} required by each farm type f is

$$h_f E_{ft} + L_{ft} = \sum_j \sum_c L_{jfct}X_{jfc} \qquad \forall f, t \tag{8.34}$$

or, equivalently,

$$E_{ft} + \left(\frac{1}{h_f}\right)L_{ft} = \sum_j \sum_c E_{jfct}X_{jfc} \qquad \forall f, t \tag{8.34a}$$

The left-hand side of equation 8.34 represents the equivalent person-hours of labor required by farm type f in period t. Equation 8.34a defines this requirement in terms of equivalent machine-hours.

Complete substitution is usually not possible. At least a portion γ^L_{ft} of the equivalent labor-hours must be in labor; hence

$$L_{ft} \geq \gamma^L_{ft}(L_{ft} + h_f E_{ft}) \tag{8.35}$$

Similarly, if γ^E_{ft} is the proportion of the equivalent person-hours of labor (using equation 8.34) that must be in machine time, then

$$E_{ft} \geq \gamma^E_{ft}(L_{ft} + h_f E_{ft}) \tag{8.36}$$

Clearly, γ^E_{ft} in equation 8.36 must be no greater than $1/h_f$, since h_f person-hours of labor are equivalent to 1 machine-hour.

TABLE 8.3. Labor and Equipment Requirements per Hectare

Soil Type, j	*Farm Type,* f	*Crop Type,* c	EQUIVALENT LABOR REQUIREMENTS		PROPORTIONS FOR ALL t	
			L_{jfc1}	L_{jfc2}	γ_{ft}^L	γ_{ft}^E
1	1	1	90	120	0.4	0.05
1	1	2	50	90		
1	2	1	70	80	0.3	0.02
1	2	2	40	50		
2	2	1	80	100		
2	2	3	30	40		

Table 8.3 defines the hourly labor requirements per hectare L_{jfct}, and the minimum required proportions γ_{ft}^L and γ_{ft}^E of equivalent labor time that must be specifically in labor time and in machine time. These proportions are assumed to be the same in each of the two periods t. Assuming that 10 person-hours of labor is equivalent to one machine-hour (i.e., that $h_f = 10$ for each farm type f), constraints 8.34 to 8.36 can be written as

Total labor and equipment requirements:

$$
\begin{aligned}
&10E_{11} + L_{11} = 90X_{111} + 50X_{112} && f = 1, t = 1 \\
&10E_{12} + L_{12} = 120X_{111} + 90X_{112} && f = 1, t = 2 \\
&10E_{21} + L_{21} = 70X_{121} + 40X_{122} + 80X_{221} + 30X_{223} && f = 2, t = 1 \\
&10E_{22} + L_{22} = 80X_{121} + 50X_{122} + 100X_{221} + 40X_{223} && f = 2, t = 2
\end{aligned}
\tag{8.34E}
$$

Minimum labor requirements:

$$
\begin{aligned}
&(1 - 0.4)L_{1t} - (10)(0.4)E_{1t} \geq 0 && f = 1, t = 1, 2 \\
&(1 - 0.3)L_{2t} - (10)(0.3)E_{2t} \geq 0 && f = 2, t = 1, 2
\end{aligned}
\tag{8.35E}
$$

Minimum equipment requirements:

$$
\begin{aligned}
&[1 - 0.05(10)]E_{1t} - 0.05L_{1t} \geq 0 && f = 1, t = 1, 2 \\
&[1 - 0.02(10)]E_{2t} - 0.02L_{2t} \geq 0 && f = 2, t = 1, 2
\end{aligned}
\tag{8.36E}
$$

Seed. For simplicity, assume that the seed or planting stock requirements per hectare are 10 units for each soil type, farm type, and crop type. Hence the total seed requirements S_{fc} for each farm type f and crop c are

$$
\begin{aligned}
&S_{1c} = 10X_{11c} && c = 1, 2 \\
&S_{21} = 10X_{121} + 10X_{221} \\
&S_{22} = 10X_{122} \\
&S_{23} = 10X_{223}
\end{aligned}
\tag{8.14E}
$$

For risk adversion and for on-farm needs, assume that at least 20% of each farm area planted will be allocated to each crop. This requirement is enforced by the constraints.

$$\left.\begin{aligned} X_{111} &\geq 0.2(X_{111} + X_{112}) \\ X_{112} &\geq 0.2(X_{111} + X_{112}) \end{aligned}\right\} f = 1$$

$$\left.\begin{aligned} X_{121} + X_{221} &\geq 0.2(X_{121} + X_{122} + X_{221} + X_{223}) \\ X_{122} &\geq 0.2(X_{121} + X_{122} + X_{221} + X_{223}) \\ X_{223} &\geq 0.2(X_{121} + X_{122} + X_{221} + X_{223}) \end{aligned}\right\} f = 2 \qquad (8.15E)$$

where $X_{111} + X_{112}$ is the total acreage planted on farms of type 1 and $X_{121} + X_{122} + X_{221} + X_{223}$ is the total acreage planted on farms of type 2.

All annual costs are expressed as a cost per unit of resource input. Table 8.4 specifies these costs, which are then included in equations 8.17 to 8.23.

TABLE 8.4. Cost per Unit of Resource Input and Available Capital

Annual Resource Input		*Cost Units per Unit Resource Input*
Drainage equipment	Soil type 1	$C^D_{1f} = 120(f = 1); 150(f = 2)$
	Soil type 2	$C^D_{22} = 330$
Water (groundwater source)		$C^{GW}_{jt} = 10 \ \forall j, t$
Fertilizer/pesticides	Type 1	$C^F_1 = 0.05$
	Type 2	$C^F_2 = 0.10$
Labor	Farm type 1	$C^L_1 = 5$
	Farm type 2	$C^L_2 = 8$
Equipment		$HC^E = 65$
Seed	Crop 1	$C^S_1 = 0.01$
	Crop 2	$C^S_2 = 0.20$
	Crop 3	$C^S_3 = 1.00$
Capital available	Farm type 1	$M^A_1 = 300{,}000$
	Farm type 2	$M^A_2 = 400{,}000$
Borrowed capital limit	Farm type 1	$M^{B,\max}_1 = 200{,}000$
	Farm type 2	$M^{B,\max}_2 = 300{,}000$

Hence for each farm type f:

1. Total land drainage cost:

$$\begin{aligned} TC^D_1 &= 120X_{111} + 120X_{112} \\ TC^D_2 &= 150X_{121} + 150X_{122} + 330X_{221} + 330X_{223} \end{aligned} \qquad (8.17E)$$

2. Total water cost:

$$\begin{aligned} TC^W_1 &= 10(Q^{GW}_{111} + Q^{GW}_{112}) \\ TC^W_2 &= 10(Q^{GW}_{121} + Q^{GW}_{122} + Q^{GW}_{221} + Q^{GW}_{223}) \end{aligned} \qquad (8.18E)$$

3. Total fertilizer and pesticide cost:

$$TC_1^F = 0.05F_1^1 + 0.10F_1^2$$
$$TC_2^F = 0.05F_2^1 + 0.10F_2^2 \tag{8.19E}$$

4. Total labor cost:

$$TC_1^L = 5(L_{11} + L_{12})$$
$$TC_2^L = 8(L_{21} + L_{22}) \tag{8.20E}$$

5. Total equipment costs:

$$TC_1^E = 65(E_{11} + E_{12})$$
$$TC_2^E = 65(E_{21} + E_{22}) \tag{8.22E}$$

6. Total seed cost:

$$TC_1^S = 0.01S_{11} + 0.20S_{12}$$
$$TC_2^S = 0.01S_{21} + 0.20S_{22} + 1.00S_{23} \tag{8.23E}$$

7. Total farm cost: Assuming interest rates of 0.12 and 0.10 for borrowed capital for farm types 1 and 2 respectively, the total cost for each farm type is

$$TC_1 = TC_1^D + TC_1^W + TC_1^F + TC_1^L + TC_1^E + TC_1^S + 0.12M_1^B$$
$$TC_2 = TC_2^D + TC_2^W + TC_2^F + TC_2^L + TC_2^E + TC_2^S + 0.10M_2^B \tag{8.26E}$$

where

$$TC_1 \leq 300{,}000 + M_1^B$$
$$TC_2 \leq 400{,}000 + M_2^B \tag{8.24E}$$
$$M_1^B \leq 200{,}000$$
$$M_2^B \leq 300{,}000 \tag{8.25E}$$

Table 8.5 contains the yields Y_{jfc} per hectare times the unit price per unit

TABLE 8.5. Income per Hectare for Various Soil, Farm, and Crop Types

Soil Type, j	*Farm Type,* f	*Crop Type,* c	*Income/ha,* P_cY_{jfc}
1	1	1	1500
1	1	2	1600
1	2	1	1600
1	2	2	1800
2	2	1	900
2	2	3	1000

yield. The sum of these incomes per hectare times the appropriate hectares determines the total farm income I_f.

$$\begin{aligned} I_1 &= 1500X_{111} + 1600X_{112} \\ I_2 &= 1600X_{121} + 1800X_{122} + 900X_{221} + 1000X_{223} \end{aligned} \tag{8.3E}$$

Total net income NI_f to each farm type f is

$$\begin{aligned} NI_1 &= I_1 - TC_1 \\ NI_2 &= I_2 - TC_2 \end{aligned} \tag{8.27E}$$

As an example of one of many possible objective functions, the total net income can be maximized:

$$\text{maximize } NI_1 + NI_2 \tag{8.27aE}$$

This objective function plus the 58 constraint equations (denoted by the letter E after the equation number) listed in this example section define an irrigation planning model. The model can be solved using linear programming. The model includes 56 unknown variables, including the surface-water allocation T (whose value is fixed in a separate bounds section for ease of programming multiple solutions, corresponding to multiple values of T, in a single computer run). Some of the constraints are merely definitions included for bookkeeping purposes only.

The coefficient matrix for this problem is illustrated in Figure 8.4. The variables are written along the top of the matrix, and the constraint equations are as indicated at the left of the matrix. The letters in the matrix signify ranges of coefficient values, as defined in Table 2A.3.

The model was solved for various values of the annual surface-water allocation T to determine the net-benefit function NB(T) for irrigation at this site. The results are summarized in Table 8.6. The values for each farm of type f were obtained by dividing the farm-type values in the solution by the number of farms of that type.

The values of the net-benefit function NB(T) equal the increase in the total net benefits (objective function 8.27aE) compared to those obtained when $T = 0$. Note that the slope of the net-benefit function NB(T) at each given value of T equals the value of the dual variable asociated with equation 8.32E restricting the surface-water allocation to be no greater than T.

The particular surface and groundwater allocations in each period t shown in Table 8.6 are not unique. In this example problem the optimal solution is the same with any combination of ground- and surface-water allocations as long as the total allocations in each period t are maintained. Hence the values of f_t (the fraction of the total surface water allocation T in period t as defined by equation 8.33) listed in Table 8.6 are not uniquely determined. When there is no groundwater allocation (i.e., when $T \geq 40$), the values of f_t are $27.2/(27.2 + 13.0)$ and $13.0(27.2 + 13.0)$ or approximately 0.68 and

0.32, respectively. These fractions of the total surface-water allocation allocated in each period t can also apply for all values of T greater than 0, as seen in Table 8.6, without changing the values of the other decision variables or objective function. However, if the cost of alternative water supplies were to vary in different periods t, each fraction f_t may not be the same for all values of available surface water T.

From the solutions in Table 8.6, note the effect surface-water availability has on individual farm operation. The types of crops planted do not vary, but the area planted increases in most cases, sometimes even doubling as surface-water availability increases. Farm income for both farm types is considerably increased, even though labor, seed, and equipment costs are higher because of the increasing size of the irrigated area. At the same time that the cost of water decreases, groundwater usage decreases. Interestingly, the optimal amount of borrowed capital, and its cost, shift from farm type 2 only, when there is no surface water available, to farm type 1 only, when there is ample surface water available.

Whether or not any of the solutions in Table 8.6 might be considered acceptable, or truly optimal, by anyone is problematical. The solutions to models such as these are influenced strongly by the objective function, and often by the diversification constraints. It would be appropriate at this stage to alter some of the assumptions made in these and other portions of the model to examine the sensitivity of the farm planting and resource management plans to other than the change in surface-water availability. Whatever the final outcome with regard to the actual objective function and constraints used, models such as the one just discussed provide an excellent means of ensuring the consistency of any proposed irrigation plan. They provide a systematic means of estimating the farm income and expense budget, the quantities of the essential resources required to develop and operate an irrigation district, and for estimating the extent to which one or more objectives are likely to be satisfied, and the additional resources, if any, which will make the biggest impact on these objectives.

8.3 AN IRRIGATION OPERATION MODEL

Once each crop has been planted in any subarea of the irrigation district, it is possible to estimate the water allocation policy that together with existing soil moisture conditions and plant growth potential will maximize the expected yields or economic returns. Definition of this water allocation policy will involve the development of a discrete stochastic dynamic programming model.

The decision variables of the irrigation operating model will, in this example, be limited to the allocations of water in every period within the growing

Model variables

No.	Type	X 1 1 1	X 1 1 2	X 1 2 1	X 1 2 2	X 2 2 1	X 2 2 3	QG W 1 1 1	QG W 1 1 2	QG W 1 2 1	QG W 1 2 2	QG W 2 2 1	QG W 2 2 2	QS W 1 1 1	QS W 1 1 2	QS W 1 2 1	QS W 1 2 2	QS W 2 2 1	QS W 2 2 2	T	F 1 1	F 1 2	F 2 1	F 2 2	E 1 1	E 1 2	E 2 1	E 2 2	L 1 1	L 1 2	L 2 1	L 2 2	S 1 1	S 1 2	S 2 1	S 2 2	S 2 3	TC D 1	TC D 2	TC W 1	TC W 2	TC F 1	TC F 2	TC L 1	TC L 2	TC E 1	TC E 2	TC S 1	TC S 2	T C 1	T C 2	M B 1	ıM B 2	I 1	I 2	N I 1	N I 2	R H S
	N																																																							1	1	
(8.1E)	L	1	1																																																							C
	L			1	1																																																			C
	L					1	1																																																	C
(8.4E)	E	-B	-B					1						1																																										
	E	-A	-B						1						1																																									
	E			-B	-B					1						1																																								
	E			-B	-B						1						1																																							
	E					-A	-B					1						1																																						
	E					-A	-A						1						1																																					
(8.32E)	L													1	1	1	1	1	1	-1																																						
(8.10E)	E	-B	-B																		1																																					
	E			-C	-B	-C	-B															1																																				
	E	-B	-B																				1																																			
	E			-B	-B	-B	-B																	1																																		
(8.34E)	E	-B	-B																						A				1																													
	E	-C	-B																							A				1																											
	E			-B	-B	-B	-B																				A				1																										
	E			-B	-B	-B	-B																					A				1																									
(8.35E)	G																								-A				T																													
	G																									-A				T																											
	G																										-A				T																										
	G																											-A				T																									
(8.36E)	G																								T				-U																													
	G																									T				-U																											
	G																										T				-U																										
	G																											T				-U																									
(8.14E)	E	-A																															1																									
	E		-A																															1																								
	E			-A																															1																							
	E				-A																															1																						
	E						-A																														1																					

Model variables

								Q	Q	Q	Q	Q	Q	Q	Q	Q	Q	Q	Q																																											
								G	G	G	G	G	G	S	S	S	S	S	S																																									
		X	X	X	X	X	X	W	W	W	W	W	W	W	W	W	W	W	W																			T	T	T	T	T	T	T	T	T	T	T	T									
Constraint		1	1	1	1	2	2	1	1	1	1	2	2	1	1	1	1	2	2		F	F	F	F	E	E	E	E	L	L	L	L	S	S	S	S	S	C	C	C	C	C	C	C	C	C	C	C	C	T	T	M	M			N	N	R
		1	1	2	2	2	2	1	1	2	2	2	2	1	1	2	2	2	2		1	1	2	2	1	1	2	2	1	1	2	2	1	1	2	2	2	D	D	W	W	F	F	L	L	E	E	S	S	C	C	B	B	I	I	I	I	H
No.	Type	1	2	1	2	1	3	1	2	1	2	1	2	1	2	1	2	1	2	T	1	2	1	2	1	2	1	2	1	2	1	2	1	2	1	2	3	1	2	1	2	1	2	1	2	1	2	1	2	1	2	1	2	1	2	1	2	S
	G	T	−T																																																							
	G	−T	T																																																							
(8.15E)	G			T	−T	T	−T																																																	
	G			−T	T	−T	−T																																																	
	G			−T	−T	−T	T																																																	
(8.17E)	E	−C	−C																																	1																				
	E			−C	−C	−C	−C																														1																			
(8.18E)	E							−A	−A																													1																		
										−A	−A	−A	−A																										1																	
(8.19E)	E																			−U		−U																		1																
																					−U		−U																		1															
(8.20E)																												−A	−A														1													
	E																													−A	−A														1													
(8.22E)	E																							−B	−B																					1												
	E																									−B	−B																				1											
(8.23E)	E																															−V	−T																1										
	E																																	−V	−T	−1												1									
(8.26E)	E																																				1		1		1		1		1		1		−1		T						
	E																																					1		1		1		1		1		1		−1		U					
(8.24E)	L																																																1		−1						F
	L																																																	1		−1					F
(8.25E)	L																																																		1						F
	L																																																			1					F
(8.3E)	E	D	D																																																	−1				
	E			D	D	C	C																																														−1			
(8.27E)	E																																																1				−1		1		
	E																																																	1				−1		1	

FIGURE 8.4. Coefficient matrix for example irrigation problem; letters indicate ranges of numbers as defined in Tables 2A.3 of Appendix 2A, Chapter 2.

TABLE 8.6. Summary of Solutions to Irrigation Planning Model for Various Amounts of Surface Water Available, T

		TOTAL SURFACE WATER AVAILABLE TO IRRIGATION DISTRICT															
		0		*5*		*10*		*15*		*20*		*30*		*40*		*45*	
		PERIOD t IN GROWING SEASON															
WATER ALLOCATIONS (× 1000)		*1*	*2*	*1*	*2*	*1*	*2*	*1*	*2*	*1*	*2*	*1*	*2*	*1*	*2*	*1*	*2*
Ground water		19.9	9.4	15.5	8.8	16.2	3.1	10.9	3.9	7.1	4.5	3.2	0.6	0	0	0	0
Surface water		—	—	4.4	0.6	3.8	6.2	9.4	5.6	14.4	5.6	19.8	10.2	27.1	12.9	27.2	13.0
Total		29.3		29.3		29.3		29.8		31.6		33.8		40.0		40.2	
		FARM TYPE f															
TOTAL HECTARES OF:		*1*	*2*	*1*	*2*	*1*	*2*	*1*	*2*	*1*	*2*	*1*	*2*	*1*	*2*	*1*	*2*
Soil Type	*Crop Type*																
1	1	40.3	75.0	40.3	75.0	40.3	75.0	42.9	75.0	49.6	75.0	59.1	75.0	80.0	88.0	80.0	89.5
1	2	161.1	225.0	161.1	225.0	161.1	225.0	171.6	225.0	198.5	225.0	236.7	225.0	320.0	212.0	320.0	210.5
2	1	—	0.0	—	0.0	—	0.0	—	0.0	—	0.0	—	0.0	—	0.0	—	0.0
2	3	—	75.0	—	75.0	—	75.0	—	75.0	—	75.0	—	75.0	—	140.0	—	147.3
Total hectares		201.4	375.0	201.4	375.0	201.4	375.0	214.5	375.0	248.1	375.0	295.8	375.0	400.0	440.0	400.0	447.3

	TOTAL SURFACE WATER AVAILABLE TO IRRIGATION DISTRICT															
	0		*5*		*10*		*15*		*20*		*30*		*40*		*45*	
	FARM TYPE f															
INDIVIDUAL FARM COSTS (× 1000)	*1*	*2*	*1*	*2*	*1*	*2*	*1*	*2*	*1*	*2*	*1*	*2*	*1*	*2*	*1*	*2*
Land	6	35	6	35	6	35	6	35	7	35	9	35	12	46	12	46
Water	24	99	24	74	24	29	21	34	12	34	0	19	0	0	0	0
Fert./pest.	0.3	1.6	0.3	1.6	0.3	1.6	0.4	1.6	0.4	1.6	0.5	1.6	0.7	1.8	0.7	1.9
Labor	19	53	19	53	19	53	21	54	26	54	28	54	38	61	38	62
Equipment	25	76	25	76	25	76	27	76	31	76	37	76	50	87	50	89
Seed	0.1	0.6	0.1	0.6	0.1	0.6	0.1	0.6	0.1	0.6	0.1	0.6	0.2	0.9	0.2	1.0
Interest	0	11	0	7	0	3	0	0	0	0	0	0	14	0	14	0
Individual net farm income (× 1000)	4.5	29	4.5	57	4.5	84	9.8	100	26	100	42	115	53	135	53	135
Total net benefits	76.5		130.8		185.2		239.0		292.0		396.3		481.2		481.4	
NB(T)	0		54.3		108.7		162.5		215.5		319.8		404.7		404.9	
$\frac{\partial \text{(total net benefits)}^a}{\partial T}$	10.87		10.87		10.87		10.60		10.60		10.00		1.80		0	

[a] Value of dual variable of equation 8.32E (at $T = 35{,}000$, the value is 8.99).

season. Other resource inputs could also be considered, if warranted, but for simplicity they will be omitted here. The model will be stochastic in that both the future available water supply and the future effective rainfall are not known with certainty. The policy defines each period's water allocation as a function of the available water in storage, current streamflow levels, and of the quantity of water in the root zone of the plants.

The fraction of the maximum possible plant growth in any period t in the growing season depends on the percentage of available soil moisture in the plant root zone, a zone that increases in each period t. Figure 8.5 illustrates the general functional relationship between the fraction of maximum plant growth $\mathrm{PG}_t(M_t)$ in period t and the percentage soil moisture content M_t in the root zone in that period. Obviously, the general relationship shown in Figure 8.5 will vary depending on the period t within the growing season and on the crop and soil type.

The average percentage soil moisture content in each period t is

$$M_t = \frac{100}{2}\left(\frac{\mathrm{SZ}_t}{h_t} + \frac{\min\left[h_t, \max\left(\mathrm{SZ}_t + e_t w_t + R_t - \mathrm{ET}_t, 0\right)\right]}{h_t}\right) \qquad (8.37)$$

where

SZ_t = available soil moisture in the plant root zone at the beginning of period t (mm)

h_t = active storage capacity of the root zone at the beginning of period t (mm)

e_t = surface irrigation efficiency

w_t = quantity of irrigation water applied in period t (mm)

R_t = effective rainfall (rainfall that enters the ground) in period t (mm)

ET_t = evapotranspiration potential during period t (mm)

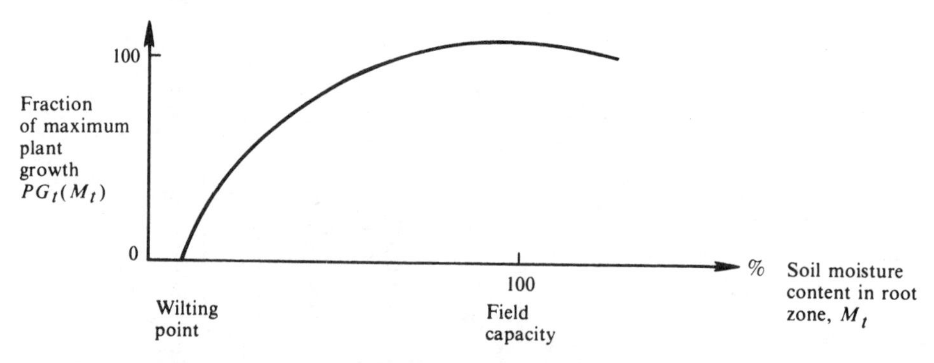

FIGURE 8.5. General relationship between percentage soil moisture and plant growth.

The objective of the water allocation policy will be to maximize the total expected crop yield. The total crop yield will equal the product of each growth fraction $PG_t(M_t)$ times the maximum obtainable crop yield Y given the other resource inputs. For example, if in a three-period growing season the fraction of maximum plant growth in those periods were 0.90, 0.70, and 1.00, the total crop yield at the end of the growing season (period T) would equal $(0.9) \cdot (0.7) \cdot (1.0)\,Y$, or $0.63\,Y$. Thus the objective is to

$$\text{maximize } E\Big[\prod_{t=1}^{T} PG_t(M_t)\Big]\,Y \tag{8.38}$$

Constraints limiting the value of function 8.38 might include restrictions on the allocation of water w_t in each period, which would affect the percentage of soil moisture content M_t (as defined by equation 8.37).

The above model may be formulated as a dynamic programming problem. The stages in the model will be the periods of the growing season. The states of the system will include:

V_t = quantity or volume of water available (from surface- or groundwater storage) for irrigation use in period t and all remaining periods of the growing season (m^3)

Q_t = additional water during period t that enters surface storage and is available for use in period t and all remaining periods (m^3)

SZ_t = soil moisture content in the root zone at the beginning of period t (mm)

Ultimately, each of these state variables will need to be discretized for model solution. However, for this presentation, only the streamflow Q_{it} and rainfall R_{it} will be discretized. For convenience, the vector $\mathbf{S}_{it}$ will denote the values $\{V_t, Q_{it}, SZ_t\}$ in period t.

The quantity of water V_t in storage and the additional quantity or streamflow Q_t that becomes available in period t could be combined into a single state variable, except for the fact that the value of Q_t affects the probability distribution of the streamflow Q_{t+1} in the following period $t + 1$. Depending on the distance between the source of streamflow and the irrigation area and on the characteristics of the region, Q_t can be either statistically independent or dependent on the effective rainfall R_t. In this example the effective rainfall R_t is taken to be a deterministic function $R_t(Q_t)$ of the streamflow Q_t. This function is the inverse of a rainfall–runoff model in which the streamflow is estimated from the rainfall. In this case, the conditional or transition probability of particular values of both $Q_{j,t+1}$ and $R_{j,t+1}$, given values Q_{it} and R_{it}, will be denoted as P^t_{ij}.

$$P^t_{ij} = \Pr[Q_{t+1} = Q_{j,t+1} \text{ and } R_{t+1} = R_{j,t+1} \mid Q_t = Q_{it}] \qquad \forall i, j, t \tag{8.39}$$

For given values of the state variables $\mathbf{S}_{it}$, and w_t and R_t, let the incremental yield $\mathrm{PG}_t(M_t)$ be denoted as

$$\mathrm{PG}_t(M_t) = \mathrm{PY}_t(\mathbf{S}_{it}, w_t, R_t) \tag{8.40}$$

where M_t is defined by equation 8.37, R_t is the effective rainfall, and w_t is the quantity of water allocated in period t.

If there is only one period (period T) remaining in the irrigation season, the optimal water allocation w_T is obtained by maximizing the yield given the current state $\mathbf{S}_{iT}$.

$$F_1(\mathbf{S}_{iT}) = \underset{\substack{w_t \\ 0 \le Aw_T \le V_T + Q_{iT} \\ R_{iT} = R_T(Q_{iT})}}{\text{maximum}} \ \mathrm{PY}_t(\mathbf{S}_{iT}, w_T, R_{iT})Y \tag{8.41}$$

The function $F_1(\mathbf{S}_{iT})$ is the maximum yield obtained, given $\mathbf{S}_{iT}$ with only one period remaining in the irrigation season. The streamflow Q_{iT} in state $\mathbf{S}_{iT}$ defines the effective rainfall R_{iT}. Denoting A (ha/10 $= 10^3$ m^2) as the total irrigation area, the water volume allocation Aw_t cannot exceed the quantity available $V_T + Q_{iT}$.

When only two periods remain before the end of the growing season, allocations Aw_t and Aw_{t+1} should be made that maximize the total yield. The maximum yield obtained given state $\mathbf{S}_{it}$ in period $t = T - 1$ is

$$F_2(\mathbf{S}_{it}) = \underset{\substack{w_t \\ 0 \le Aw_t \le V_t + Q_{it} \\ R_{it} = R_t(Q_{it})}}{\text{maximum}} \{\mathrm{PY}_t(\mathbf{S}_{it}, w_t, R_{it})E[F_1(\mathbf{S}_{t+1})]\} \tag{8.42}$$

where the state vector $\mathbf{S}_{t+1}$ can be any of the possible state vectors $\mathbf{S}_{j,t+1}$ which include streamflows $Q_{j,t+1}$ and

$$V_{t+1} = V_t + Q_{it} - Aw_t \tag{8.43}$$

$$\mathrm{SZ}_{t+1} = \min(h_t, \mathrm{SZ}_t + e_t w_t + R_{it} - \mathrm{ET}_t) \tag{8.44}$$

Equations 8.43 and 8.44 are termed state transition functions from period $t = T - 1$ to period $t + 1 = T$.

The expected value $E[F_1(\mathbf{S}_{t+1})]$ of the maximum yields $F_1(\mathbf{S}_{t+1})$, given state $\mathbf{S}_{it}$, is

$$E[F_1(\mathbf{S}_{t+1})] = \sum_j P^t_{ij} F_1(\mathbf{S}_{j,t+1}) \tag{8.45}$$

where state vector $\mathbf{S}_{j,t+1}$ includes streamflow $Q_{j,t+1}$.

With $n \ge 2$ periods to go, the general recursion equation is

$$F_n(\mathbf{S}_{it}) = \underset{\substack{w_t \\ 0 \le Aw_t \le V_t + Q_{it} \\ R_{it} = R_t(Q_{it})}}{\text{maximum}} \{\mathrm{PY}_t(\mathbf{S}_{it}, w_t, R_{it})[\sum_j P^t_{ij} F_{n-1}(\mathbf{S}_{j,t+1})]\} \tag{8.46}$$

Beginning with $n = 1$, the solution of this sequence of recursive equations defines a sequential water allocation policy w_t for each period t that is dependent on the storage volume V_t, the streamflow Q_t, and the soil moisture content SZ_t. Such sequential policies are similar to those defined for reservoir operation in Chapters 2 and 7. Sequential operating policies can be illustrated in matrix form, as shown in Figure 8.6.

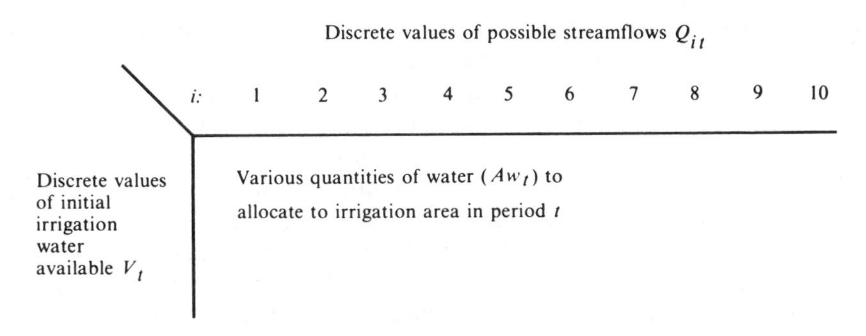

FIGURE 8.6. Allocation policy for period t given initial soil moisture content SZ_t. A sequential policy is given by such tables for each value of SZ_t for each given period t.

Depending on the number of discrete values allowed for each state variable, this three-state-variable sequential operating policy may require a substantial time to derive using dynamic programming. Hence it may not be very practical to do this unless relatively high capacity/high speed computer facilities are available. If water is not limiting, or if there is no reservoir storage to consider, a two-state-variable stochastic model may be constructed and would be easier to solve. In addition, there are other approaches that have been proposed that the reader may wish to consider. Many of these are contained in the references of this chapter.

8.4 CONCLUSION

The irrigation planning and operating models outlined in this chapter are intended to illustrate an approach, rather than to serve as actual models for analyzing alternative irrigation plans and policies in any particular situation. Only the basic structure of irrigation planning and operating models has been presented. Many modifications and extensions may be required before such models can be used for analyzing a particular planning or operating problem. For example, it may be necessary to consider more accurately (1) hydrologic risk; (2) the possibility of storing harvested crops when demand is low and selling them later when demand is high; (3) the uncertainty of future benefits and costs and the changing limitations of various resources used or required for irrigation farming and how these costs and resource limitations may differ for various groups of irrigation farmers; (4) the possible restrictions in salinity, pesticide or nutrient concentrations in the irrigation runoff or return flow; and (5) many other important conditions that might apply in specific planning and operating situations.

The irrigation planning and operating models discussed in this chapter were oriented toward water resources planning. For agricultural planning these models could be extended and become a part of agricultural sector models. These sector models would encompass several irrigation districts or regions and distinguish possible differences in irrigation intensities, in the structure of the labor market, and constraints on labor migration and availability, in marketing options, in the potential for livestocks or perennial crops in some regions but not others, and so on. One of the more successful applications of agricultural sector modeling is discussed by Duloy and Norton [8]. The references at the end of the chapter provide additional modeling suggestions and data pertaining to irrigation planning and operation.

EXERCISES

8-1. In Indonesia there exists a wet season followed by a dry season each year. In one area of Indonesia all the farmers within an irrigation district plant and grow rice during the wet season. This crop brings the farmer the greatest income per hectare; thus they would all prefer to continue growing rice during the dry season. However, there is insufficient water during the dry season for irrigating all 5000 hectares of available irrigable land for rice production. Hence some hectares must be allocated to a second crop. Assume an available irrigation water supply of 32×10^6 m^3 at the beginning of each dry season, and a minimum requirement of 7000 m^3/ha for rice and 1800 m^3/ha for the second crop.

(a) What proportion of the 5000 hectares should the irrigation district manager allocate for rice during the dry season each year, provided that all available hectares must be given sufficient water for rice or the second crop?

(b) Suppose that crop production functions are available for the two crops, indicating the increase in yield per hectare per m^3 of additional water, up to 10,000 m^3/ha for rice and 2000 m^3/ha for the second crop. Develop a model in which the water allocation per hectare, as well as the hectares allocated to each crop, is to be determined, assuming a specified price or return per unit of yield of each crop. Under what conditions would the solution of this model be the same as in part (a)?

8-2. Construct the long-run benefit function NB(T) for irrigation water for the example problem solved in this chapter, the solution of which is summarized in Table 8.6. Show that the dual variables of the constraint on the available surface water supply, T, are indeed the slopes of the benefit function NB(T) at the various values of T given in the table.

8-3. How would the example problem in the text be modified to examine the following alternative objectives?
(a) Maximize total employment.
(b) Maximize total crop production.
(c) (a) or (b), but at minimum capital expense.
(d) Income redistribution among farm types.

8-4. Along the Nile River in Egypt, irrigation farming is practiced for the production of cotton, maize, rice, sorghum, full and short berseem for animal production, wheat, barley, horsebeans, and winter and summer tomatoes. Cattle and buffalo are also produced, and together with the crops they require labor, water, fertilizer, and land area (feddans). Farm types or management practices are fairly uniform, and hence in any analysis of irrigation policies in this region this distinction need not be made. Given the accompanying data develop a model for determining the tons of crops and numbers of animals to be grown that will maximize (a) net economic benefits based on Egyptian prices, and (b) net economic benefits based on international prices. Identify all variables used in the model.

Known parameters:

C_i = miscellaneous costs of land preparation per feddan

p_i^E = Egyptian price per 1000 tons of crop i

p_i^I = international price per 1000 tons of crop i

v = value of meat and dairly production per animal

g = annual labor cost per worker

f^P = cost of P fertilizer per ton

f^N = cost of N fertilizer per ton

y_i = yield of crop i, tons/feddan

α = feddans serviced per animal

β = tons straw equivalent per ton of berseem carryover from winter to summer

r^w = berseem requirements per animal in winter

s^{wh} = straw yield from wheat, tons per feddan

s^{ba} = straw yield from barley, tons per feddan

r^s = straw requirements per animal in summer

μ_i^N = N fertilizer required per feddan of crop i

μ_i^P = P fertilizer required per feddan of crop i

l_{im} = labor requirements per feddan in month m, man-days

w_{im} = water requirements per feddan in month m, 1000 m^3

h^i_m = land requirements per month, fraction (1 = full month)

Required constraints. (assume known resource limitations for labor, water, and land):

(a) Summer and winter fodder (berseem) requirements for the animals.
(b) Monthly labor limitations.
(c) Monthly water limitations.
(d) Land availability each month.
(e) Minimum number of animals required for cultivation.
(f) Lower bounds on cotton area (assume this is known).
(g) Upper bounds on summer and winter tomatoes (assume these are known).

Other possible constraints:

(a) Crop balances.
(b) Fertilizer balances.
(c) Labor balance.
(d) Land balance.

8-5. In Algeria there are two distinct cropping intensitites, depending on the availability of water. Consider a single crop that can be grown under intensive rotation or extensive rotation on a total of A hectares. Assume that the annual water requirements for the intensive rotation policy are 16,000 m^3 per hectare, and for the extensive rotation policy they are 4000 m^3 per hectare. The annual net production returns are 4000 and 2000 dinars, respectively. If the total water available is 320,000 m^3, show that as the available land area A increases, the rotation policy that maximizes total net income changes from one that is totally intensive to one that is increasingly extensive.

Would the same conclusions hold if instead of fixed net incomes of 4000 and 2000 dinars per hectare of intensive and extensive rotation, the net income depended on the quantity of crop produced? Assuming that intensive rotation produces twice the crop yield of extensive rotation, and that the net income per unit of crop yield Y is defined by the simple linear function $5 - 0.05Y$, develop and solve a linear programming model to determine the optimal rotation policies if A equals 20, 50, and 80. Need this net income or price function be linear to be included in a linear programming model?

8-6. In the example problem whose solution is summarized in Table 8.6, show that the objective function value will not be affected by a reallocation of the surface and groundwater allocations so that 68% of the surface water allocation T is in period 1 and 32% of the surface water allocation T is in period 2 for all solutions when $T > 0$. Hence in the river basin model the function f_t (in equation 8.33 or 5.44) can be 0.68 and 0.32, respectively, for all values of T in this example problem.

REFERENCES

1. CUMMINGS, R. G., and J. W. MCFARLAND, Groundwater Management and Salinity Control, *Water Resources Research*, Vol. 10, No. 5, 1974, pp. 909–915.
2. DELUCIA, R. J., Operating Policies for Irrigation Systems under Stochastic Regimes, Harvard Water Program, Harvard University, Cambridge, Mass., 1969.
3. DUDLEY, N. J., Irrigation Planning. 4. Optimal Interseasonal Water Allocation, *Water Resources Research*, Vol. 8, No. 3, 1972, pp. 586–594.
4. DUDLEY, N. J., D. T. HOWELL, and W. F. MUSGRAVE, Optimal Intraseasonal Irrigation Water Allocation, *Water Resources Research*, Vol. 7, No. 4, 1971, pp. 770–788.
5. DUDLEY, N. J., D. T. HOWELL, and W. F. MUSGRAVE, Irrigation Planning. 2. Choosing Optimal Acreages within an Irrigation Season, *Water Resources Research*, Vol. 7, No. 5, 1971, pp. 1051–1063.
6. DUDLEY, N. J., D. T. HOWELL, and W. F. MUSGRAVE, Irrigation Planning. 3. The Best Size of Irrigation Area for a Reservoir, *Water Resources Research*, Vol. 8, No. 1, 1972, pp. 7–17.
7. DUDLEY, N. J., and O. R. BURT, Stochastic Reservoir Management and System Design for Irrigation, *Water Resources Research*, Vol. 9, No. 3, 1973, pp. 507–522.
8. DULOY, J. H., and R. D. NORTON, CHAC, A Programming Model of Mexican Agriculture, Chap. IV.1 in *Multi-level Planning: Case Studies in Mexico*, L. M. Goreau, and A. S. Manne (eds.), North-Holland Publishing Co., Inc., Amsterdam, 1973.
9. HALL, W. A., and J. A. DRACUP, *Water Resources Systems Engineering*, McGraw-Hill Book Company, New York, 1970.
10. HEDGES, T. R., Water Supplies and Cost in Relation to Farm Resource Use Decisions and Profits on Sacramento Valley Farms, Giannini Foundation Research Report 320, University of California, Davis, Calif., March 1974.
11. HELWEG, O. J., and J. W. LABADIE, Linked Models for Managing River Basin Salt Balance, *Water Resources Research*, Vol. 13, No. 2, 1977, pp. 329–336.
12. HEXEM, R. W., and E. O. HEADY, *Water Production Functions for Irrigated Agriculture*, Iowa State University Press, Ames, Iowa, 1978.
13. MAASS, A., and R. L. ANDERSON, ... *and the Desert Shall Rejoice: Conflict, Growth, and Justice in Arid Environments*, The MIT Press, Cambridge, Mass., 1978.
14. MATANGA, G. B., and M. A. MARINO, Irrigation Planning, 1, Cropping Pattern; and 2, Water Allocation for Leaching and Irrigation Purposes, *Water Resources Research*, Vol. 15, No. 3, 1979, pp. 672–683.

15. ROGERS, P., and D. V. SMITH, The Integrated Use of Ground and Surface Water in Irrigation Project Planning, *American Journal of Agricultural Economics*, Vol. 52, No. 1, 1970, pp. 13–24.

16. ROGERS, P., and D. V. SMITH, An Algorithm for Irrigation Project Planning, *ICID Bulletin*, International Commission of Irrigation and Drainage, Vol. 46, 1970.

17. SCHERER, C. R., Water Allocation and Pricing for Control of Irrigation-Related Salinity in a River Basin, *Water Resources Research*, Vol. 13, No. 2, 1977, pp. 225–238.

18. STEWART, J. I., R. M. HAGAN, and W. O. PRUITT, Functions to Predict Optimal Irrigation Programs, *Journal of the Irrigation and Drainage Division, ASCE*, Vol. 100, No. IR2, 1974, pp. 179–199.

19. YARON, D., and A. OLIAN, Application of Dynamic Programming in Markov Chains to the Evaluation of Water Quality in Irrigation, *American Journal of Agricultural Economics*, Vol. 55, 1973, pp. 467–471.

PART IV

Water Quality Management

So far this book has focused on various methods for analyzing water quantity management alternatives. This is only one aspect of water resources planning. Water quality prediction and management is another important component of water resource systems planning. This is especially true in regions where the discharge of natural or man-made waste products or residuals into water bodies may affect public health or reduce the economic benefit derived from the use of those water bodies.

The following two chapters are devoted to surface-water quality planning. The first introduces models that have been used to predict, and hence simulate, a variety of water quality constituents in different types of water bodies. The second reviews some models that have been used to define and evaluate various water quality management alternatives and their costs. Both chapters are concerned with surface-water quality.

It is indeed artificial to assume that these topics are independent. Clearly, water quantity changes affect water quality, and surface-water quality often affects groundwater quality, and vice versa. While water quality cannot be separated from water quantity, and surface-water quality cannot be considered independently of groundwater quality and quantity, analyses of water quantity and quality problems are often carried out assuming a fixed set of boundary conditions. This is done to simplify the analyses and sometimes to conform to institutional policies and responsibilities that typically do not include both quantity and quality management. Still, it is essential to recognize that these two aspects of water resources planning are integrated in nature. If nothing else, one should perform sufficient sensitivity analyses to understand the impacts of varying boundary conditions.

It would be a lengthy undertaking to outline even each major type of mathematical model currently proposed, and sometimes used, for predicting and managing water quality in various water bodies. Hence the following two chapters only introduce this subject by illustrating the development of a few of the more commonly used types of water quality models for rivers, estuaries, and lakes. Readers interested in additional detail may refer to journals, books, and reports devoted solely to this subject, as well as to annotated bibliographies on water quality modeling and management, such as those published each June in the *Journal of the Water Pollution Control Federation*, and by the U.S. Department of the Interior (Water Resources Scientific Information Center, Office of Water Research and Technology).

CHAPTER 9

Water Quality Prediction and Simulation

9.1 INTRODUCTION

Achieving regional water quality goals, especially in the more developed areas of the world, often involves substantial capital investments and changes in public attitudes concerning resource management. Economic impacts include the cost of facilities designed to reduce the discharge of contaminants into natural waters or to improve the quality of waste receiving waters, and limitations on economic activities and economic development in a particular region or river basin. Those responsible for the formulation and adoption of water quality plans or management policies must have a means of estimating and evaluating the temporal and spatial economic, environmental, and ecologic impacts of these plans and policies. This need has stimulated the development and application of a wide range of mathematical modeling techniques for predicting the impacts of alternative pollution control plans.

There are many different types of water quality models. For any specific situation, the appropriate model and the required data depend on the purpose of the study. Long-range regional water quality planning does not require the detail that is appropriate, for example, when evaluating a single proposed industrial waste outfall or discharge site. There is no best single water quality model for all water bodies and for all planning situations. An important decision that must be made early in the planning process is the selection of the modeling method or methods appropriate for a planning exercise given the limits of available time and money. In some cases models should be relatively simple; in other cases they may have to be more complex.

Most predictive water quality models in use today apply to water bodies receiving wastewater from point sources. As the quantities of wastes discharged from point sources are reduced, nonpoint or distributed sources of pollutants from agricultural and urban runoff become increasingly important. Models are needed to help predict nonpoint-source waste inputs to surface waters. The outputs of these nonpoint-source wastewater generation models provide the inputs to receiving water quality models. For example, urban storm water management models often contain models for runoff, sewer routing, and receiving water quality prediction.

This chapter is a review of a number of the more typical predictive water quality models developed for and applied to waste receiving water bodies. Models used to predict nonpoint wastewater loadings are also reviewed. These models will range from the relatively simple to the relatively complex, yet each has proven to be effective in certain planning situations. The inclusion of various water quality management alternatives, and their costs, within these predictive models is reviewed in Chapter 10.

9.2 TYPES OF WATER QUALITY MODELS

It is useful to distinguish between certain types of models and to discuss their characteristics. This will provide an opportunity to define a few of the terms and to illustrate some of the concepts used by those who develop and apply water quality models. From a manager's viewpoint, models can be classified based on (1) their applicability to various hydrologic systems, (2) the particular aspects of the hydrologic system that are simulated, and (3) the method of model solution or analysis.

Many of the water quality models in use today are extensions of two simple equations proposed by Streeter and Phelps in 1925 [51] for predicting the biochemical oxygen demand (BOD) of various biodegradable constituents, and the resulting dissolved oxygen (DO) concentration in rivers [53]. Often used with these BOD–DO models are other fairly simple first-order exponential decay, dilution, and sedimentation models for other nonconservative and conservative substances.

More complex nonlinear multiconstituent water quality models have also been proposed and applied to predict the physical, chemical, and biological interactions of many constituents and organisms found in natural water bodies [4, 5, 16, 50]. These multiconstituent water quality simulation models generally require more data and computer time, but they also can provide a more detailed and comprehensive description of the quantity and quality of water resulting from various water and land management policies.

Water quality models can be used to evaluate steady-state conditions, for which the values of the water quality and quantity variables do not change with time. They can also be used to evaluate dynamic or time-varying conditions. The latter types of models permit an evaluation of transient phenomena such as nonpoint storm runoff and accidental spills of pollutants. Steady-state models are usually simpler and require less computational effort than dynamic or transient models, and are more relevant to long-term planning than to short-term management and control.

Assumptions pertaining to the mixing of pollutants in water bodies dictate the spatial dimensionality of the model. Although all real physical systems are three-dimensional, sufficient accuracy may be obtained in many river systems by modeling only one or two dimensions. One-dimensional models of river systems assume complete mixing in the vertical and lateral directions. One-dimensional lake models usually assume complete mixing in all but the vertical direction. Two-dimensional models may assume either lateral mixing, as in stratified estuaries or lakes, or vertical mixing, as in relatively shallow and wide rivers.

Undoubtedly, the most data-demanding model type is the stochastic or probabilistic model, as compared to its deterministic counterpart. Most deterministic models yield estimates of mean values of various quality constituents, whereas probabilisitc models explicitly take into account the randomness or uncertainty of various physical, biological, or chemical processes. Validation of stochastic models is especially difficult due to the quantity of prototype data necessary to compare probability distributions of variables rather than just their expected or mean values. This introduction to water quality models will be confined to a review of one-dimensional, steady-state flow, deterministic models. These relatively simple models are used for both long- and short-term planning, management, and control of rivers, estuaries, and lakes during periods when complete mixing in the other two dimensions, and constant flows over time, exist.

9.3 COMPUTATIONAL METHODS

Model solution techniques play a significant role in model development. Solution methods range from hand computation and the use of nomographs to computer-aided optimization and simulation procedures. If the water quality management problem is simple enough to be solved manually or with the aid of a nomograph or programmable calculator, this is by far the most inexpensive and preferable method to use.

Simple models are superior to complex ones if they indeed provide the information needed. For more complex problems, model solutions may require thousands of computations that would be too expensive and too prone to errors if attempted without the aid of a computer.

The basic argument for using computers is that computers can yield results for relatively complex simulation and optimization models quickly, accurately, and cheaply. The largest cost component of computer modeling is often model development and calibration prior to verification and simulation.

Most water quality models designed for computer solution are simulation models. Simulation models indicate the values of various water quality variables given hydrologic and waste parameters and the type and extent of measures designed to reduce waste discharges or to otherwise control their effects on the receiving water body. For each set of such assumptions, a simulation run is required. For many situations the number of reasonable water quality management alternatives is sufficiently large to preclude a simulation of each alternative. In cases where the time and/or cost would prohibit trial-and-error simulation, optimization models have been developed and applied as a means of substantially reducing the number of management alternatives that need be simulated.

Optimization models are usually more limited than simulation models, especially with respect to the number of water quality parameters included in the model. But they explicitly include variables defining the range of management alternatives and objectives being considered. If used properly, optimization models, such as those discussed in Chapter 10, can assist in identifying management alternatives that best satisfy management objectives. Examining these particular management alternatives with a verified simulation model provides a more accurate basis for comparing various economic and environmental impacts.

A substantial portion of the literature on water quality simulation modeling is devoted to solution procedures. Many computer models include the solution procedure in the program. The solution procedure often involves the use of finite differences, finite elements, quasi-linerization, and other methods [2, 11, 27, 59]. Even when using a well-documented computer program, an understanding of the limitations of the solution procedures is important when defining river reach lengths, lake volume segment heights, and other parameters that may affect the stability and accuracy of model solutions. This chapter will introduce some of the more easily solved and applied water quality prediction and simulation models, and the assumptions upon which these models are based.

9.4 MODEL DEVELOPMENT, CALIBRATION, AND VERIFICATION

A water quality model is a set of mathematical expressions defining the physical, biological, and chemical processes that are assumed to take place in a water body. Most water quality models consist of equations based on conservation of mass and/or energy. Given the particular water quality parameters of interest and the important physical, biological, and chemical processes that affect these quality parameters, a mass balance is developed that takes into account three phenomena: the inputs of constituents to the water body from outside the system, the transport of constituents through the water body, and the reactions within the water body that either increase or decrease constituent concentrations or masses.

The inputs of pollutant constituents usually come from natural processes and wastewater discharges from municipal, industrial, or agricultural activities in the form of point sources or nonpoint runoff. The physical and biochemical characteristics of the point waste sources are much better understood, and measurable, than those of the nonpoint sources. The complex, often random, time-variable nature of both point and nonpoint sources is often disregarded because of insufficient data and modeling complexity, with a consequent reduction in model output reliability.

The transport of constituents, by dispersion and/or advection, is dependent on the hydrologic and hydrodynamic characteristics of the water body. Advective transport dominates in flowing rivers. In contrast, dispersion is the predominant transport phenomenon in estuaries subjected to tidal action. As one would expect, variations in surface-water runoff and groundwater inflow have more of an effect on freshwater river quality than on estuarine river quality. Lake-water quality prediction is complicated by the influence of random wind directions and velocities that often affect surface mixing, currents, and stratification.

Transport phenomena of freshwater and estuarine river systems are better understood than those of lakes and oceans. Hence water quality models of river systems tend to be more reliable than those of lakes and oceans. In addition, steady-state assumptions, often made to reduce model complexity and data requirements, are usually more realistic for rivers than for lakes and oceans. While steady-state assumptions may be more realistic for rivers, such assumptions often preclude complete evaluation of the effects of time-varying inputs of constituents to river systems.

Biological, chemical, and physical reactions among constituents are also better understood in river systems than in lakes or oceans. This is especially true for wastes affecting the biochemical oxygen demand and dissolved oxygen concentrations in rivers, and for wastes affecting the concentration of bacteria, various forms of carbon and nitrogen, and chemical compounds commonly contained in industrial effluents. Knowledge of reactions involving heavy metals and many complex synthetic compounds and toxic materials is relatively limited. The interactive nonlinear and time-varying reactions of nutrients associated with eutrophication are sufficiently understood to permit some reasonable modeling of these constituents. But these fairly sophisticated and complex models require considerable data and time to develop, calibrate, and verify. Models of eutrophication are often of greater relevance to lake quality management than to river quality management.

The development of water quality models is both a science and an art. Although most models are developed according to some basic modeling procedures, each model reflects the creativity of its developer. A model reflects what the modeler sees as the important components and relationships between these components of the prototype. The modeler must decide just how much detail to include in the model. This in turn is a function of the purpose of the model (i.e., its intended use) and of how quickly the model is needed for planning and decision making.

Once the general outline and purpose of the model is defined, model development proceeds from conceptualization (sometimes just in words), to mathematical definition or representation, to modification if necessary for computational reasons (e.g., the incorporation of piecewise-linear functions, finite differences, or finite elements) to calibration, to verfication and sensitivity testing, to documentation, and finally to use in defining and evaluating possible policy alternatives. This process is illustrated in Figure 9.1.

Model calibration is performed using one or more observed data sets of both inputs and outputs. The model parameters and indeed the model itself are adjusted or modified so as to produce an output that is as close to the actual observed water quality as is possible. The evaluation of model parameters, usually a subjective trial-and-error procedure, can be aided by least-squares and quasilinearization methods, and is called model or parameter identification [2, 43, 50, 52].

Model verification requires an independent set of input and output data to test the calibrated model. The verification data must be independent of that used to calibrate the model. A model is verified if the model's predictions, for a range of conditions other than those used to calibrate the model, compares favorably with observed field data. Here again the criteria for deciding whether or not model output and observed field data are essentially the same, for the same input conditions, are largely subjective. What constitutes a satisfactory comparison depends on the nature of the problem, the type of model developed and its purpose, and the extent and reliability of available input and output data. The fact that most models cannot accurately predict what actually happens does not detract from their value. Even rela-

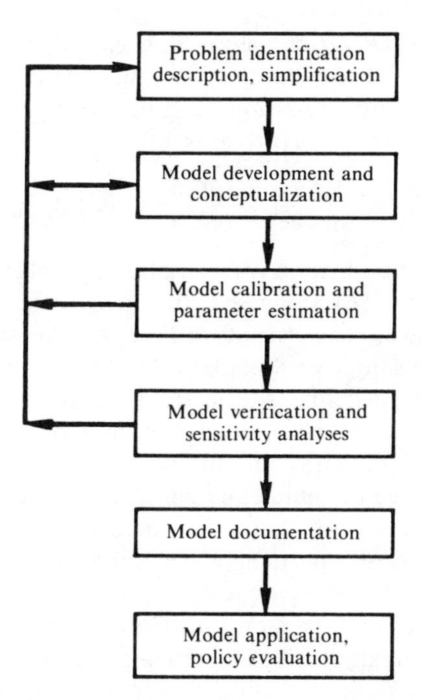

FIGURE 9.1. Steps in the development and use of water quality models.

tively simple models can help planners understand the prototype and estimate the relative change in water quality associated with given changes in the inputs that may result from alternative wastewater management policies.

9.5 STEADY-STATE MODELS OF RIVER AND ESTUARINE SYSTEMS

Several models are developed in this section to illustrate the basic approaches to water quality modeling of river and estuarine systems. Some of the main mathematical expressions or components of these predictive water quality models will also apply to other types of water bodies.

First, consider a one-dimensional river, extending from a small stream to the tidal river and finally to the saline estuary. Complete mixing is assumed in the vertical and lateral directions. The concentration (ML^{-3}) of various constituents C_i, $i = 1, 2, \ldots$, is a function of the rate of inputs and outputs (sources and sinks) of the constituents, of the dispersion and advection of the constituents, and of the various physical, chemical, biological, and possibly radiological reactions that affect the constituent concentrations.

The partial differential equation defining the concentration of any constituent C in a one-dimensional river is

$$\frac{\partial C}{\partial t} = \frac{1}{A}\frac{\partial}{\partial X}\left(EA\frac{\partial C}{\partial X} - UAC\right) \pm \sum_k S_k \tag{9.1}$$

in which C is the concentration (ML^{-3}) of a particular constituent, t is time (T), X is the distance (L) along the stream, E is the dispersion coefficient (L^2T^{-1}), U is net downstream velocity (LT^{-1}), A is the stream's cross-sectional area (L^2) and S_k is a source or sink ($ML^{-3}T^{-1}$) of the constituent C. Equation 9.1 states that at a particular site in the river system, the change in concentration with respect to time, $\partial C/\partial t$, depends on the change in the constituent flux in the X direction due to dispersion $EA(\partial C/\partial X)$ and advection UAC plus any sources or minus any sinks S_k. These source or sink terms include the various reactions that either increase or decrease the concentration of constitutent C at a particular site in the river system.

The expression within the parentheses in equation 9.1 is termed the total flux (MT^{-1}). Flux due to dispersion $EA(\partial C/\partial X)$ is assumed to be proportional to the concentration gradient. Constituents are transferred by dispersion from zones of higher concentration to zones to lower concentration. The coefficient of dispersion E depends, in part, on the amplitude and frequency of the tide, if any, as well as upon the turbulence of the river waters. It is common practice to include in E all the effects on the distribution of C that cannot otherwise be accounted for by advective transport. The term UAC is the advective flux caused by the movement of water containing the constituent at a rate U across an area A. Equation 9.1 is derived in Appendix 9A.

The relative importance of the dispersion or advection terms depends on how well defined the velocity field is within which the pollutant is being distributed. A good description of the velocity U in a spatial and temporal sense minimizes the significance of the dispersion term and may justify its elimination altogether on the grounds that it may be reduced to the level of acceptable error in computation. On the other hand, a crude description of fluid flow, such as averaging across irregular cross sections or approximating transients by steady flows, may lead to a dominance of the dispersion term.

Many of the reactions affecting the decrease or increase of constituent concentrations are often represented by first-order kinetics that assume the reactions rates are proportional to the constituent concentration. While higher-order kinetics may be more correct in certain situations, predictions of constituent concentrations based on first-order kinetics have often been found to be acceptable for natural aquatic systems.

9.5.1 Steady-State Models of a Single Constituent

With a constant streamflow, constant cross-sectional area, a constant dispersion coefficient, and first-order kinetics, equation 9.1 for steady-state conditions in one-dimensional reaches of streams, rivers, and estuaries becomes

$$0 = \frac{\partial C}{\partial t} = E\frac{\partial^2 C}{\partial X^2} - U\frac{\partial C}{\partial X} - KC \tag{9.2}$$

where K is a reaction or decay rate coefficient (T^{-1}). The net downstream velocity U is averaged over the tidal cycle in estuaries. This steady-state equation may apply to many flow conditions in river systems, including low-flow conditions often found in late summer or early fall. It is at this time when high temperatures and low velocities are coupled with low quantities of dilution water that the most severe water quality conditions often occur.

Equation 9.2 can be integrated in this simple case where river parameters are constant. Consider a long section of a river or estuary in which K, E, A, and U are constant and such that the concentration of the constituent C approaches 0 both upstream ($X < 0$) and downstream ($X > 0$) of the waste discharge site ($X = 0$). These are called *boundary conditions*. Then the pollutant concentration at any point X resulting from a discharge of the constituent at a constant rate W_0 (MT^{-1}) at the point $X = 0$ is (see Appendix 9B for derivation)

$$C(X) = \begin{cases} \dfrac{W_0}{Qm}\exp\left[\dfrac{U}{2E}(1+m)X\right] & X < 0 \\ \dfrac{W_0}{Qm}\exp\left[\dfrac{U}{2E}(1-m)X\right] & X \geq 0 \end{cases} \tag{9.3}$$

where

$$m = \sqrt{1 + \frac{4KE}{U^2}} = \frac{\sqrt{U^2 + 4KE}}{U} \tag{9.4}$$

Note that $m \geq 1$ and the velocity U and dispersion coefficient E are positive so that the exponents in equation 9.3 are always negative. This means that $C(X)$ goes to zero for large X as required by the boundary conditions. Equation 9.3 assumes that there are no sources or sinks of the constituent, other than the natural decay $-KC$ and the constant discharge at $X = 0$.

For conditions of complete mixing at the point of discharge, the initial constituent concentration due to the discharge of W_0 at $X = 0$ is

$$C(0) = \frac{W_0}{Qm} \tag{9.3a}$$

In freshwater rivers not under a tidal influence, the dispersion coefficient E is often small and hence m, defined by equation 9.4, is essentially 1. In these situations

$$C(0) = \frac{W_0}{Q} \tag{9.3b}$$

If E is small and dispersion is negligible, then equation 9.3 takes on a simpler form. In particular, for $X < 0$, $C(X)$ equals zero, while for $X > 0$

$$C(X) = \frac{W_0}{Q} \exp\left(-\frac{KX}{U}\right) \tag{9.5}$$

This represents an exponential decay over time as the constituent travels downstream from a single point of discharge at $X = 0$.

As rivers approach the sea, the dispersion coefficient E increases and the net downstream velocity U decreases. Because the flow Q equals the cross-sectional area A times the velocity U, $Q = AU$, and m can be expressed as $[(U^2 + 4KE)^{1/2}]/U$, then as U approaches 0, the term Qm in equation 9.3 approaches $A\sqrt{4KE}$. Likewise the exponent

$$\frac{UX}{2E}(1 \pm m) = \frac{UX}{2E}\left[\frac{U \pm (U^2 + 4KE)^{1/2}}{U}\right]$$

approaches $\pm X\sqrt{K/E}$. Thus for small U equation 9.3 becomes

$$C(X) = \begin{cases} \dfrac{W_0}{2A\sqrt{KE}} \exp\left(+X\sqrt{K/E}\right) & X < 0 \\ \dfrac{W_0}{2A\sqrt{KE}} \exp\left(-X\sqrt{K/E}\right) & X \geq 0 \end{cases} \tag{9.6}$$

Here dispersion is much more important than advective transport and the concentration profile is symmetric about the point of discharge at $X = 0$. Figure 9.2 illustrates the concentration profiles specified by equations 9.3, 9.5, and 9.6.

Equation 9.3, 9.5, or 9.6 can be used as the basis for a one-dimensional steady-state biochemical oxygen demand (BOD) water quality model for a river with many BOD sources. Assume that the model parameters K, E, A, and U are constant over a long homogeneous region of the river and that the concentration $C(X)$ approaches zero for large X. Let the index or subscript i represent a particular site along the river at location X_i at which a quantity of carbonaceous BOD_i^C is discharged at a constant rate (MT^{-1}). The waste from each source i is allowed to have its own reaction rate K_i^C.

For this simple model, the BOD concentration (ML^{-3}) at any site j in the homogeneous river having velocity U, cross-sectional area A, and flow $Q = UA$ as a result of the discharge at site i equals

$$B_{ij}^C = \text{BOD}_i^C b_{ij}^C \tag{9.7}$$

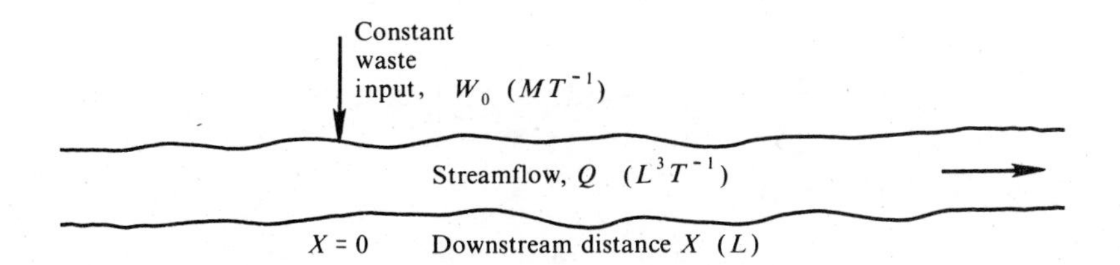

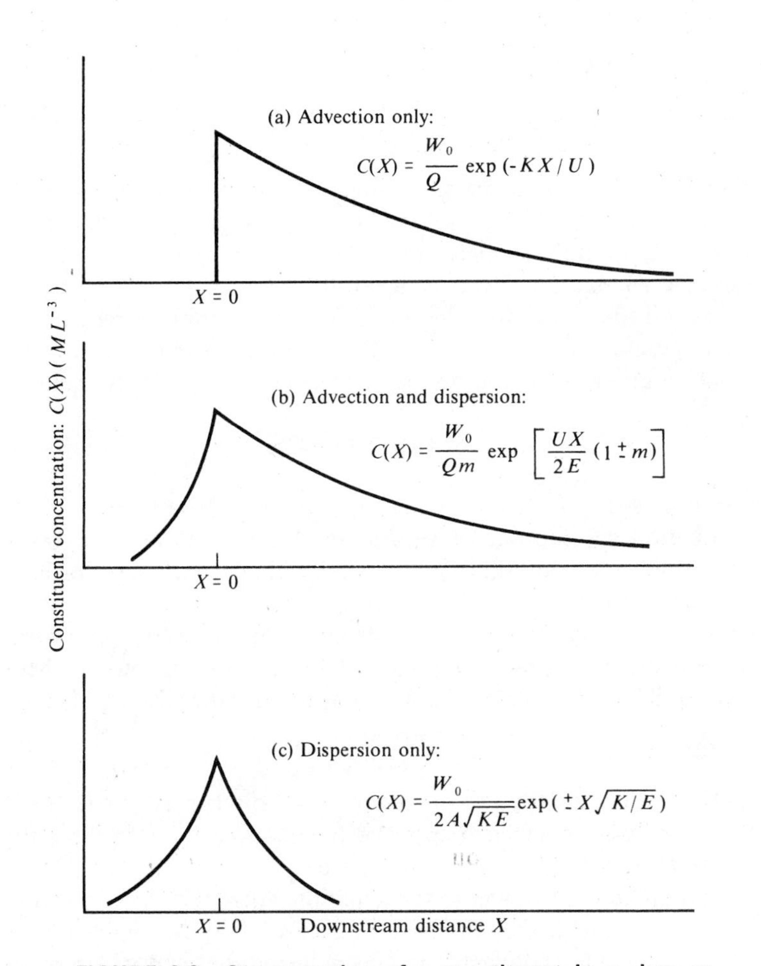

FIGURE 9.2. Concentration of a constituent in a river or estuary following discharge from a single point source.

where (1) for $E = 0$

$$b_{ij}^C = \begin{cases} 0 & X_j < X_i \\ \dfrac{1}{Q} \exp\left[-\dfrac{K_i^C(X_j - X_i)}{U}\right] & X_j \geq X_i \end{cases} \tag{9.8a}$$

(2) for $E > 0$ and $U > 0$

$$b_{ij}^C = \begin{cases} \dfrac{1}{Qm_i^C} \exp\left[\dfrac{U(X_j - X_i)(1 + m_i^C)}{2E}\right] & X_j < X_i \\ \dfrac{1}{Qm_i^C} \exp\left[\dfrac{U(X_j - X_i)(1 - m_i^C)}{2E}\right] & X_j \geq X_i \end{cases} \tag{9.8b}$$

and (3) for $E > 0$ but $U \simeq 0$

$$b_{ij}^C = \frac{1}{2A\sqrt{K_i^C E}} \exp\left[-|X_j - X_i|\sqrt{K_i^C/E}\right] \tag{9.8c}$$

with m_i^C calculated using equation 9.4 substituting the constant K_i^C for K. In equation 9.8a for $X_j \geq X_i$, $(X_j - X_i)/U$ is the average travel time between the source site i and site j at which the BOD concentration might be measured.

The total carbonaceous BOD concentration at site j resulting from the discharges at all sites i is simply the sum of the concentrations resulting from each discharge, plus the carbonaceous BOD concentration $\hat{B}_j^C$ at j resulting from BOD^C sources, including nonpoint sources, not explicitly included in the model.

$$B_j^C = \sum_i (BOD_i^C b_{ij}^C) + \hat{B}_j^C \tag{9.9}$$

The b_{ij}^C parameters (TL^{-3}) are defined in equations 9.8. Because each b_{ij}^C term in equation 9.9 approaches zero as the location X_j of site j becomes distant from the X_i's, the term in parentheses will approach zero for large X as is required by the boundary conditions.

As formulated, equation 9.9 only accounts for the BOD concentration resulting from organic carbon compounds. Often the nitrogenous biochemical oxygen demand is also modeled. In that case, one has the additional equation

$$B_j^N = \sum_i (BOD_i^N b_{ij}^N) + \hat{B}_j^N \tag{9.10}$$

where BOD_i^N is the quantity of nitrogenous BOD discharged at site i (MT^{-1}) and b_{ij}^N is the transfer coefficient specified in equations 9.8 using the appropriate decay rate coefficients K_i^N for each site.

Before expanding this model to the situation where the model parameters K_i^C, K_i^N, K_a, E, A, and U are not constant over a relatively long distance, the water quality model for coupled-reactions is introduced and its simple solution presented. Then this more general model is extended to the situation where the model parameters have different values in different river sections.

9.5.2 Coupled-Reaction Models

Some constituent concentrations are functions of two or more simultaneous reactions. Dissolved oxygen and BOD concentrations illustrate such coupled reactions. The differential equation for the dissolved oxygen deficit concentration D (i.e., the difference between the water's saturated dissolved oxygen concentration and the actual concentration) in steady-state conditions when the parameters K^C, K^N, K_a, E, A and U are constant is

$$0 = E\frac{d^2D}{dX^2} - U\frac{dD}{dX} + K^C B_X^C + K^N B_X^N - K_a D \quad (9.11)$$

$$0 = \begin{bmatrix}\text{dispersion}\end{bmatrix} - \begin{bmatrix}\text{advective} \\ \text{transport}\end{bmatrix} + \begin{bmatrix}\text{oxidation} \\ \text{of BOD}^C\end{bmatrix} + \begin{bmatrix}\text{oxidation} \\ \text{of BOD}^N\end{bmatrix} - \begin{bmatrix}\text{reaeration} \\ \text{across water} \\ \text{surface}\end{bmatrix}$$

where K_a is the reaeration rate constant (T^{-1}) and B_X^C and B_X^N are the biochemical oxygen demand concentrations at X, defined by the respective terms in the right-hand side of equations 9.9 and 9.10. Integration of 9.11 yields the dissolved oxygen deficit concentration D_{ij} (ML^{-3}) at any site j resulting from the discharge of carbonaceous and nitrogenous BOD at site i in a homogeneous river section with constant parameters. The solution can be written using the transfer coefficients d_{ij} as

$$D_{ij} = (\text{BOD}_i^C d_{ij}^C + \text{BOD}_i^N d_{ij}^N) \quad (9.12)$$

where (1) for $E = 0$

$$d_{ij} = \begin{cases} 0 & X_j < X_i \\ \dfrac{K_i}{Q(K_a - K_i)}\{\exp[-K_i(X_j - X_i)/U] - \exp[-K_a(X_j - X_i)/U]\} & X_j \geq X_i \end{cases} \quad (9.13a)$$

(2) for $E > 0$ and $U > 0$

$$d_{ij} = \frac{K_i}{Q(K_a - K_i)}\left\{\frac{1}{m_i}\exp\left[\frac{U(X_j - X_i)(1 \pm m_i)}{2E}\right] - \frac{1}{m^a}\exp\left[\frac{U(X_j - X_i)(1 \pm m^a)}{2E}\right]\right\}$$

using $(1 + m)$ when $X_j < X_i$
and $(1 - m)$ when $X_j \geq X_i$ (9.13b)

and (3) for $E > 0$ and $U \simeq 0$

$$d_{ij} = \frac{1}{2A(K_a - K_i)}\left\{\frac{1}{\sqrt{K_i E}}\exp[-|X_j - X_i|\sqrt{K_i/E}] - \frac{1}{\sqrt{K_a E}}\exp[-|X_j - X_i|\sqrt{K_a/E}]\right\} \quad (9.13c)$$

The coefficients d_{ij}^C and d_{ij}^N (TL^{-3}) in equation 9.12 are obtained by substituting K_i^C and K_i^N, respectively, into equations 9.13 and also 9.4 to obtain m_i^C and m_i^N; the quantity m^a is obtained by substituting K_a for K in the definition of m, equation 9.4. The parameters d_{ij}^C and d_{ij}^N in equation 9.12 specify the dissolved oxygen deficit concentration that occurs at site j per unit of BOD discharge (MT^{-1}) at site i.

The dissolved oxygen deficit concentration in a river is affected by the dissolved oxygen deficit in discharges to the river as well as the biochemical oxygen demand exerted by substances carried by those discharges. The steady-state differential equation predicting the dissolved oxygen deficit concentration D caused by oxygen deficit discharges may be written

$$0 = E\frac{d^2D}{dX^2} - U\frac{dD}{dX} - K_aD \tag{9.14}$$

This is equation 9.11 without the extra terms added by the oxidation of organic substances. Without these terms, equation 9.11 reduces to equation 9.14 which is the same as equation 9.2. For a river reach in which K_a, E, A, and U are constant and where $D(X)$ goes to zero for large X, the dissolved oxygen deficit concentration at site j resulting from an oxygen deficit DOD_i in the discharge at site i (MT^{-1}) is simply $\text{DOD}_i b_{ij}^a$ where b_{ij}^a is given by equation 9.8 with K_a substituted for K_i^C and m^a for m^C.

The total dissolved oxygen deficit concentration D_j at site j is the sum of the deficits caused by BOD discharges BOD_i^C and BOD_i^N at each site i in addition to the oxygen deficit DOD_i in those discharges and the deficit concentration $\hat{D}_j$ at site j that results from BOD and oxygen deficit sources not explicitly included in the model.

$$D_j = \sum_i [\text{BOD}_i^C d_{ij}^C + \text{BOD}_i^N d_{ij}^N + \text{DOD}_i b_{ij}^a] + \hat{D}_j \tag{9.15}$$

In nontidal $(E = 0)$ freshwater systems, equation 9.15 applied to only one waste discharge site i in a homogeneous river reach defines the classical oxygen sag curve, shown in Figure 9.3.

9.5.3 Multi-Reach Models

The water quality models developed in Sections 9.5.1 and 9.5.2 apply to a single homogeneous river reach in which the model parameters K, E, A, and U are constant and in which the constituent concentration $C(X)$ approaches zero for large X. These models are extended in this section to rivers which are described by any number of distinct reaches. Of the three cases in equation 9.8, the simplest to extend to the multi-reach river system is the flowing river in which dispersion is negligible so that one may let $E = 0$. In this case, all constituents move downstream at the river velocity.

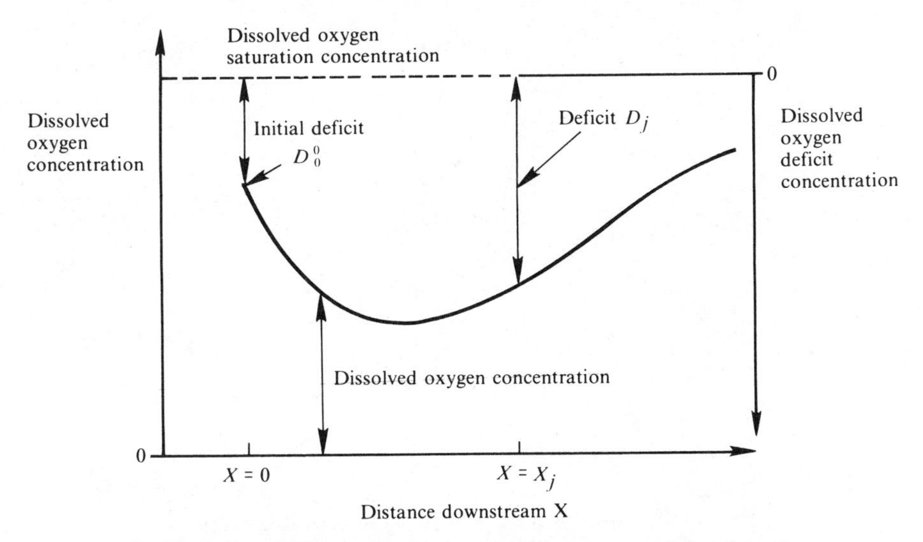

FIGURE 9.3. Dissolved oxygen "sag" curve resulting from single point discharge of BOD and initial oxygen deficit concentration at $X = 0$.

Water Quality in Flowing Rivers. To illustrate this simple case ($E = 0$), consider the river system drawn in Figure 9.4. Let sites 1 to 5 define five river reaches where the reaeration rate K_i^a, river flow Q_i, river cross-sectional area A_i, and velocity U_i are assumed to be constant in each reach and where a reach is denoted by the index of the site defining the top of the reach. Suppose one wants to determine the dissolved oxygen deficit concentration D_5 at site 5 which results from upstream discharges and from the initial BOD and DOD concentrations in the river water at sites 0 and 1. For this simple case where $E_i = 0$, equations 9.9, 9.10, and 9.15 can still be used to estimate the BOD and DOD concentrations at site 5, B_5 and D_5, if the b_{ij}'s and d_{ij}'s are appropriately defined when sites i and j are not in the same homogeneous river reach. Here the use of different deoxygenation rate constants K_i^C and K_i^N for each waste source is changed to the use of different constants in each reach. Figure 9.4 illustrates the appropriate technique (when $E_i = 0$) for calculating any b_{ij} or d_{ij} when sites i and j are not in the same homogeneous reach. In the figure, b_{25} is calculated by determining the impact of a discharge BOD_2 (MT^{-1}) at site 2 on the BOD concentration at various sites downstream.

For a discharge rate BOD_2 at site 2, Figure 9.4 indicates the resulting BOD concentration at site 3, denoted B_{23}, is $BOD_2 b_{23}$. Here b_{23} is calculated using equation 9.8a with $K = K_2^C$ or K_2^N as appropriate and where the subscript 2 now indicates the reach. Given this concentration in the stream at site 3, the total constituent flux from reach 2 into reach 3 is $Q_2 B_{23}$. Thus the BOD

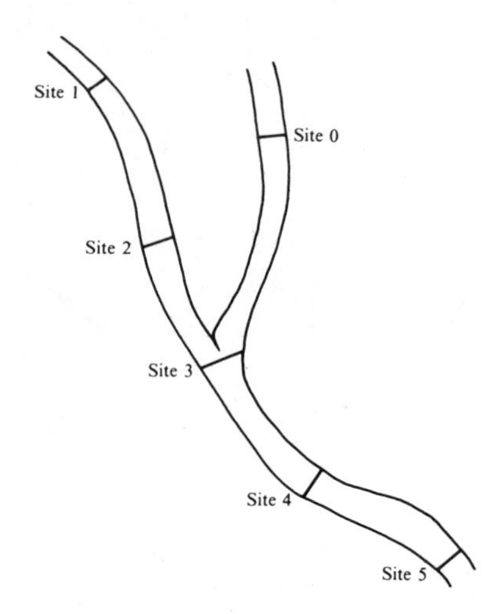

Definitions

B_{ij} = BOD concentration at site j resulting from a BOD discharge at site i (ML^{-3})

D_{ij} = DO deficit concentration at site j resulting from a BOD discharge at site i (ML^{-3})

Q_j = streamflow in the reach below site j (L^3T^{-1})

Calculation of b_{25}	Calculation of d_{25}
$\text{BOD}_2 b_{23} = B_{23}$	$\text{BOD}_2 d_{23} = D_{23}$
$(Q_2 B_{23})b_{34} = B_{24}$	$(Q_2 B_{23})d_{34} + (Q_2 D_{23})b^a_{34} = D_{24}$
$(Q_3 B_{24})b_{45} = B_{25}$	$(Q_3 B_{24})d_{45} + (Q_3 D_{24})b^a_{45} = D_{25}$
$B_{25}/\text{BOD}_2 = b_{25}$	$D_{25}/\text{BOD}_2 = d_{25}$

FIGURE 9.4. Simple multi-reach river system and illustration of the calculation of transfer coefficients b_{ij} and d_{ij} for sites in different river reaches as explained in the text.

concentration at site 4 resulting from the discharge at site 2, denoted B_{24}, equals $(Q_2B_{23})b_{34}$. Continuing this procedure, B_{25} is obtained and one can calculate $b_{25} = B_{25}/\text{BOD}_2$. The calculations are simplest if one lets BOD_2 equal unity. The procedure works in the same way for calculating b^a_{ij} for the transfer of dissolved oxygen deficits as it does for calculating either b^C_{ij} for BOD^C or b^N_{ij} for BOD^N.

In the example,

$$b_{25} = Q_2 b_{23}\, Q_3 b_{34}\, b_{45} \tag{9.16}$$

Substituting equation 9.8a for b_{ij} into this expression for b_{25} one obtains

$$b_{25} = \frac{1}{Q_4} \exp\left[-\sum_{i=2}^{4} \frac{K_i(X_{i+1} - X_i)}{U_i}\right] \tag{9.17}$$

where Q_4 is the flow at site 5. If K_i is independent of i, then the exponent in equation 9.17 is just $-K \cdot$ [average travel time between sites 2 and 5]. In general, if K^C_i, K^N_i or K^a_i is independent of i, then

$$b_{ij} = \frac{1}{Q_{j-1}} \exp\left[-K\tau_{ij}\right] \tag{9.18}$$

where τ_{ij} is the average travel time between i and j for $X_j > X_i$ and Q_{j-1} is the flow just above site j.

Calculation of the DOD concentration at site 5 resulting just from a BOD discharge rate BOD_2 at site 2 proceeds in a similar fashion except now the DO deficit and BOD concentrations caused by the initial discharge must both be taken into account. The discharge rate BOD_2 results in a DOD concentration $D_{23} = (BOD_2)d_{23}$ at site 3 and also a residual BOD concentration $B_{23} = BOD_2 b_{23}$. The combination of the two mass loadings $Q_2 B_{23}$ and $Q_2 D_{23}$ on the water in reach 3 results in a DOD concentration $D_{24} = (Q_2 B_{23})d_{34} + (Q_2 D_{23})b^a_{34}$ at site 4. Finally

$$d_{25} = (Q_2 b_{23})(Q_3 b_{34})d_{45} + [(Q_2 b_{23})d_{34} + (Q_2 d_{23})b^a_{34}]Q_3 b^a_{45}$$

The complexity of this last expression points out the advantage of the simple recursive calculation procedure shown in Figure 9.4. However, in the special case where K^a_i and K^C_i or K^N_i are independent of the reach index i,

$$d_{ij} = \frac{K}{Q_{j-1}(K_a - K)}\left[\exp(-K\tau_{ij}) - \exp(-K_a\tau_{ij})\right] \tag{9.19}$$

where τ_{ij} is the travel time between sites i and j ($\tau_{ij} \geq 0$), K is the appropriate reach-independent BOD decay coefficient, and K_a is the reach-independent oxygen reaeration coefficient.

Equations 9.9, 9.10, and 9.12 can now be extended to provide a multi-reach river water quality model when E is negligible or zero. Consider the simple multi-reach river system diagrammed in Figure 9.4. For simplicity, one can incorporate into the discharges BOD^S_0 and BOD^S_1 ($S = C$ or N) and DOD_0 and DOD_1 (MT^{-1}), the initial BOD and DOD loadings at sites 0 and 1 due to other than perfectly clean and aerated water entering the river system at these sites. Thus sites 0 and 1 can be thought of as the sources of the river flow and the constituent concentrations at those upstream sites. With this convention, the B^S concentration ($S = C$ or N) at any site in the river due to the initial concentration at site $i = 0$ and/or 1 and all discharges at sites i can be written

$$B^S_j = \sum_{\substack{\text{all sites } i \\ \text{upstream of} \\ \text{site } j}} (BOD^S_i b^S_{ij}) \tag{9.20}$$

for $S = C$ or N. Likewise, for the DO deficit concentration at any site j due to the initial deficit, oxygen deficit discharges, biochemical oxygen demand, and reaeration

$$D_j = \sum_{\substack{\text{all sites } i \\ \text{upstream of} \\ \text{site } j}} (BOD^C_i d^C_{ij} + BOD^N_i d^N_{ij} + DOD_i b^a_{ij}) \tag{9.21}$$

These equations have a particularly simple form if in addition to the condition $E = 0$, the rate constants K_i^C, K_i^N and K_i^a are independent of the reach index i so that equations 9.18 and 9.19 can be substituted for the b_{ij}'s and d_{ij}'s.

Modeling Rivers with Dispersion. When the dispersion of constituents along the river's length is an important component of a constituent's flux, $EAdC/dX - QC$, then the simple model developed in equations 9.17 to 9.21 does not apply. When dispersion is important, a constituent's concentration at any site is affected by the concentrations both upstream and downstream of that site. As a result, it is generally necessary to solve a set of simultaneous equations to determine the constituent concentrations throughout a multi-reach river system.

There are two basic approaches to solving the general steady-state differential equation for a constituent's concentration

$$0 = \frac{1}{A}\frac{d}{dX}\left[EA\frac{dC}{dX} - UAC\right] - KC \pm \sum_k S_k \tag{9.22}$$

Continuous Solution. Consider first the continuous solution approach. Assume that the river system can be described by a number of reaches where in each reach r, the model parameters K_r, E_r, A_r, and U_r are relatively constant. Then equations 9.9, 9.10, and 9.15 can be used to construct a multi-reach model. Let

$$B_r^S(X) = \sum_{\substack{\text{sites } i \\ \text{within reach } r}} (\text{BOD}_i^S b_{iX}^S)$$

and

$$D_r(X) = \sum_{\substack{\text{sites } i \\ \text{within reach } r}} [\text{BOD}_i^C d_{iX}^C + \text{BOD}_i^N d_{iX}^N + \text{DOD}_i b_{iX}^a] \tag{9.23}$$

Here $B_r^S(X)$ ($S = C$ or N) and $D_r(X)$ specify the BOD and DOD concentrations within each reach that one might expect due to the activities within that reach ignoring the effect of the reach's upstream and downstream boundaries and activities beyond those boundaries.

Figure 9.5 contains a linear multi-reach river system which will serve to illustrate this modeling approach. One assumes that the concentrations of BOD^C, BOD^N, and DOD at the most upstream point x_0 and downstream point x_n of the modeled river system are specified. Other boundary conditions are also possible. However, given concentrations at x_0 and x_n, the complete multi-reach water quality model solution can be obtained by joining together the individual reach functions $B_r^C(X)$, $B_r^N(X)$ and $D_r(X)$ so as to insure the continuity of each constituent concentration and flux, $EAdC/dX - QC$, at the boundaries of each reach. Continuity is required to avoid infinite concentration gradients which are impossible with a nonzero dispersion

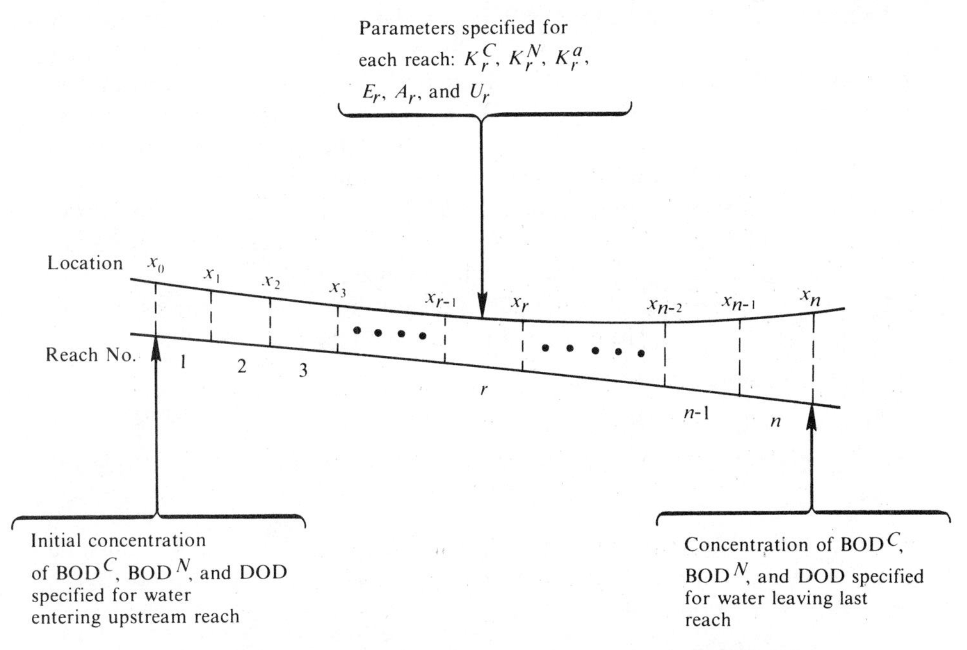

FIGURE 9.5. Simple multi-reach estuary having different parameter values in each reach.

coefficient. Requiring equal constituent fluxes is equivalent to requiring the conservation of constituent mass.

To solve equation 9.22 with its boundary conditions one can introduce boundary functions $BB_r^s(X)$ for BOD^S and $DB_r(X)$ for DOD which are solutions to the corresponding homogeneous differential equation ($S_k = 0$) as discussed in Appendix 9B. For each reach r,

$$BB_r^C(X) = BA_{1,r}^C b_{r,X}^C + BA_{2,r}^C b_{r-1,X}^C \tag{9.24a}$$

$$BB_r^N(X) = BA_{1,r}^N b_{r,X}^N + BA_{2,r}^N b_{r-1,X} \tag{9.24b}$$

$$\begin{aligned} DB_r(X) = {} & (BA_{1,r}^C d_{r,X}^C + BA_{2,r}^C d_{r-1,X}^C) \\ & + (BA_{1,r}^N d_{r,X}^N + BA_{2,r}^N d_{r-1,X}^N) \\ & + (DA_{1,r} b_{r,X}^a + DA_{2,r} b_{r-1,X}^a) \end{aligned} \tag{9.24c}$$

Here the constants $BA_{1,r}^C$, $BA_{2,r}^C$, $BA_{1r,}^N$, $BA_{2,r}^N$, $DA_{1,r}$, and $DA_{2,r}$ are selected to reflect the effect of the boundary conditions of each reach r on the concentration at any point in each reach. These boundary conditions account for the partial reflection of constituents at a reach's boundary that might otherwise disperse across that boundary and the impact of sources of BOD and DO outside of reach r on the concentation of those constituents in reach r. Note that $d_{r,X}^C$, $d_{r,X}^N$, and $b_{r,X}^a$ are the transfer functions given in equations

9.8b and c and 9.13b and c for the BOD and DO concentrations at a point X resulting from a discharge at x_r where X and x_r are both in the same homogeneous reach with parameters K_r^C, K_r^N, or K_r^a; E_r; A_r; and U_r. Unfortunately the expression for $\mathrm{DB}_r(X)$, equation 9.24c, is rather complicated because the BOD boundary conditions also affect the DO deficit concentrations within each reach. The BOD and DO concentrations at any point X in reach r will be the sum of these boundary functions $BB_r^S(X)$ and $DB_r(X)$ and the respective BOD and DO concentrations $B_r^S(X)$ and $D_r(X)$ caused by discharges directly into the waters within the reach (equation 9.23).

For dissolved oxygen, the boundary conditions requiring the continuity of concentration and of flux at each reach boundary are

Continuity of Concentration

$$
\begin{aligned}
&D_0(x_0) = D_1(x_0) + DB_1(x_0) \\
&D_r(x_r) + DB_r(x_r) = D_{r+1}(x_r) + DB_{r+1}(x_r) \qquad \text{for } r = 1, \ldots, n-1 \\
&D_n(x_n) + DB_n(x_n) = D_{n+1}(x_n)
\end{aligned}
\tag{9.25}
$$

where the boundary concentrations $D_0(x_0)$ and $D_{n+1}(x_n)$ at x_0 and x_n must be specified prior to model solution. Note that DB_r is a linear function of $DA_{1,r}$ and $DA_{2,r}$ (equation 9.24c).

Continuity of Flux

$$
\begin{aligned}
&\left\{E_r A_r \frac{d[D_r(x_r) + DB_r(x_r)]}{dX} - Q_r[D_r(x_r) + DB_r(x_r)]\right\} \\
&\quad = \left\{E_{r+1} A_{r+1} \frac{d[D_{r+1}(x_r) + DB_{r+1}(x_r)]}{dX}\right. \\
&\qquad \left. - Q_{r+1}[D_{r+1}(x_r) + DB_{r+1}(x_r)]\right\} \\
&\qquad \text{for } r = 1, 2, 3, \ldots, n-1
\end{aligned}
\tag{9.26}
$$

There are $n + 1$ concentration continuity and $n - 1$ flux continuity constraints for each constituent providing $2n$ equations for each constituent. These equations turn out to be linear but rather involved. Their solution for each constituent determines the values of BA_{kr}^C, BA_{kr}^N and DA_{kr} $(k = 1, 2)$ needed to specify the BOD and DO concentrations in every reach.

Finite Difference Approximations. An alternative approach to solving the differential equations for water quality constituents, when the dispersion coefficient E is other than zero, is to develop a finite difference or finite section approximation to the differential equation. Figure 9.6 illustrates the possible division of a river or estuary into sections. Consider the mass balance for dissolved oxygen in some particular section k. Let the DO deficit concentration at the center of the section be D_k while the corresponding BOD^C and BOD^N concentrations are B_k^C and B_k^N. Thus when approximating the mass balance for DO in section k, the total rate of reaeration is approximately

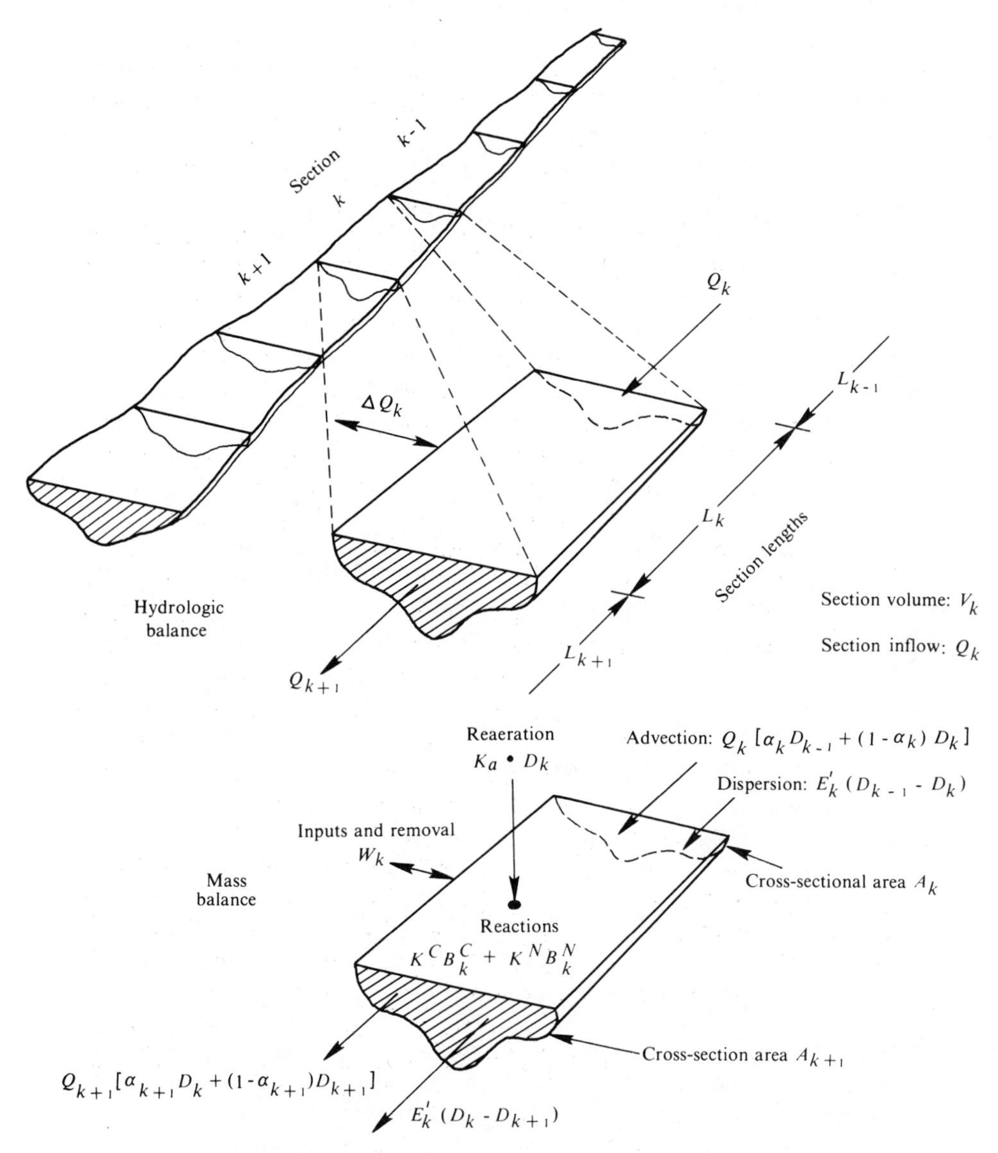

FIGURE 9.6. Division of river or estuary into sections for finite difference approximation.

$V_k(K_a D_k)$ where V_k is the volume of the section. The rate of oxygen consumption for biochemical oxidation is approximately $V_k(K_k^C B_k^C + K_k^N B_k^N)$.

The difficulty arises in approximating the rates of dispersion and advection into and out of each section. If D_k is the concentration at the center of each section k, then the average concentration at the boundary separating section $k - 1$ and k can be approximated as

$$D_{k-1,k} \simeq \frac{L_k D_{k-1} + L_{k-1} D_k}{L_k + L_{k-1}} = \alpha_k D_{k-1} + (1 - \alpha_k) D_k$$

where L_k is the length of reach or section k. Thus the advection into section k should be approximately

$$Q_k D_{k-1,k} = Q_k[\alpha_k D_{k-1} + (1 - \alpha_k) D_k] \tag{9.27}$$

The dispersion into section k across its common boundary with section $k - 1$ depends on the concentration gradient between the sections $(D_{k-1} - D_k)/[(L_k + L_{k-1})/2]$, the cross-sectional area of the boundary A_k, and the dispersion coefficient, E_k. Thus, the net flux across A_k into section k is approximately

$$E_k A_k \frac{(D_{k-1} - D_k)}{(L_{k-1} + L_k)/2} = E'_k(D_{k-1} - D_k)$$

where E'_k is the bulk dispersion (or exchange) coefficient. Combining these terms for fluxes into and out of section k at both boundaries and reactions within the section, continuity of mass requires that

$$\begin{aligned} 0 = {} & E'_k(D_{k-1} - D_k) - E'_{k+1}(D_k - D_{k+1}) + Q_k[\alpha_k D_{k-1} \\ & + (1 - \alpha_k) D_k] - Q_{k+1}[\alpha_{k+1} D_k + (1 - \alpha_{k+1}) D_{k+1}] \\ & + V_k K_k^C B_k^C + V_k K_k^N B_k^C - V_k K_k^a D_k + W_k \end{aligned} \tag{9.28}$$

where W_k is the rate of addition or removal of the DO deficit into section k because of discharges, withdrawals, or artificial aeration. (MT^{-1}).

Rearranging equation 9.28, one obtains for each section k,

$$\begin{aligned} & -(E'_k + \alpha_k Q_k) D_{k-1} + [(E'_k + E'_{k+1}) - Q_k(1 - \alpha_k) \\ & + Q_{k+1}\alpha_{k+1} + V_k K_k^a] D_k - (E'_{k+1} - (1 - \alpha_k) Q_{k+1}) D_{k+1} \\ & = V_k K_k^C B_k^C + V_k K_k^N B_k^N + W_k \end{aligned} \tag{9.29}$$

This system of linear equations with unknowns D_k can be solved when the boundary conditions are specified for the first and last section and after similar sets of equations are first solved to determine B_k^C and B_k^N.

Because of the approximations involved in developing equation 9.29 and the instabilities that can arise in the solution, the segment lengths L_k must generally be rather small to obtain good results. As a contrast, the continuous solution procedure can be used with very long reaches if the streamflow model parameters are reasonably constant. However, the continuous solution procedure is generally considered to be conceptually more difficult than the finite difference technique.

The parameters b_{ij} and d_{ij} and equation 9.29 apply only for aerobic conditions. If the dissolved oxygen deficit concentration results in anaerobic conditions at any point in the river system, the deoxygenation rate constants K are no longer valid, and hence neither are the parameters that are functions of these rate constants.

Modifications of these equations, identification of their parameters, and methods of incorporating these equations into models of actual river systems are discussed in more detail elsewhere [9, 24, 31, 35, 43, 53]. Applications of these steady-state equations to actual river systems have been extensive. One application of these equations to the multiple-reach Black River in the northeastern United States is illustrated in Figure 9.7. Although these relatively simple models do not capture the detail of time-varying multispecies aquatic ecosystem models now being developed and used, they require less data. Hence, they are useful in situations where limited time and budget preclude the application of more detailed models.

9.5.4 Detailed Nitrogen Models

In river systems where nitrogen is a major constituent, the assumption of a constant reaction rate K_i^N for BOD^N may not be satisfactory. The reaction rate may vary along the river system due to the biochemical processes which effect the oxidation of the various chemical forms of nitrogen. A more detailed nitrogen model together with data on the input of each form of nitrogen and various reaction rates will enable the prediction of the separate components of the nitrogen cycle, illustrated in Figures 9.8a and 9.8b.

O'Connor et al. [37] proposed sequential reaction models to account for each nitrogen component. The following four nitrogen concentration components (ML^{-3}) can be considered:

$$N_1 = \text{organic nitrogen,}$$
$$N_2 = \text{ammonia nitrogen, } NH_3\text{-N}$$
$$N_3 = \text{nitrite nitrogen, } NO_2\text{-N}$$
$$N_4 = \text{nitrate nitrogen, } NO_3\text{-N}$$

Here K_i^N is the first-order decay rate including settling and conversion of nitrogen form i (T^{-1}), $K_{i,i+1}^N$ represents the forward reaction rate coefficient (T^{-1}) and $W_i(s)$ is the discharge rate of form i at site s (MT^{-1}). Recalling that E is the dispersion coefficient (L^2T^{-1}), and U is the net downstream velocity

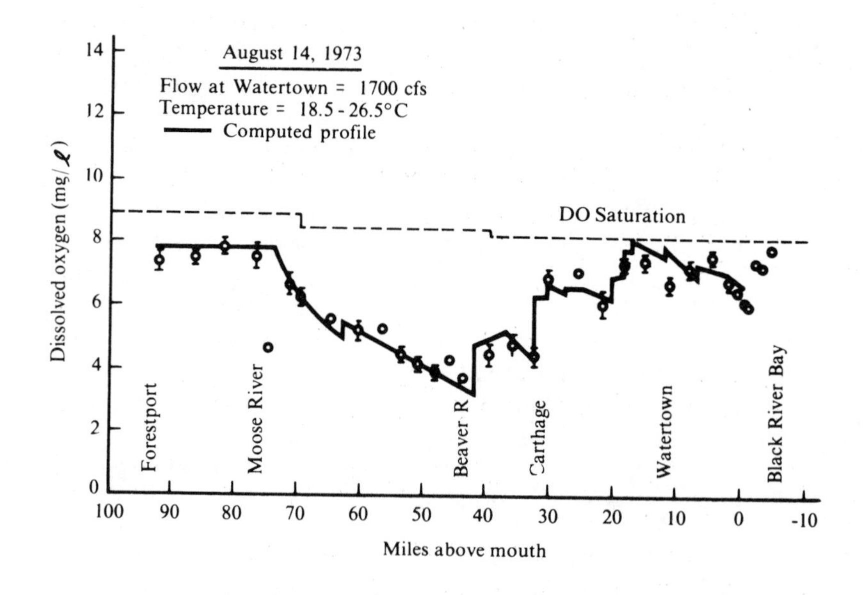

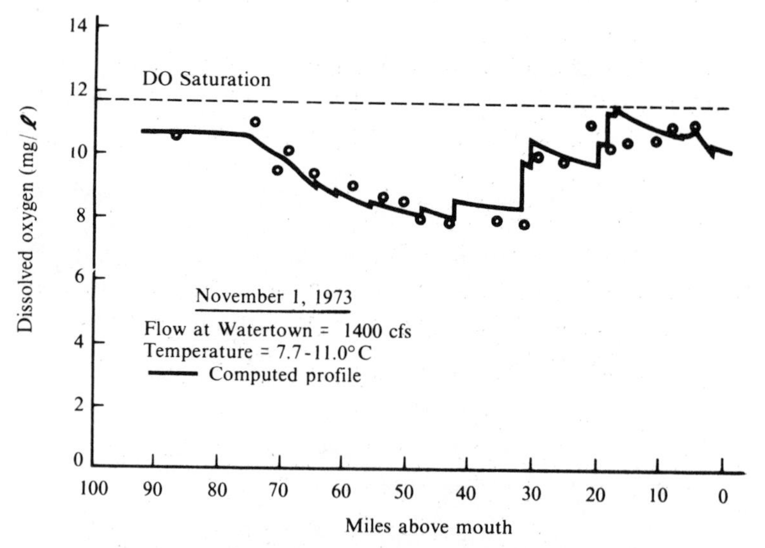

FIGURE 9.7. Predicted dissolved oxygen profiles and measured data for the Black River in New York. (Hydroscience, Inc., "Water Pollution Investigation: Black River of New York," U.S. EPA Report, No. EPA-905/9-74-009, December 1974.)

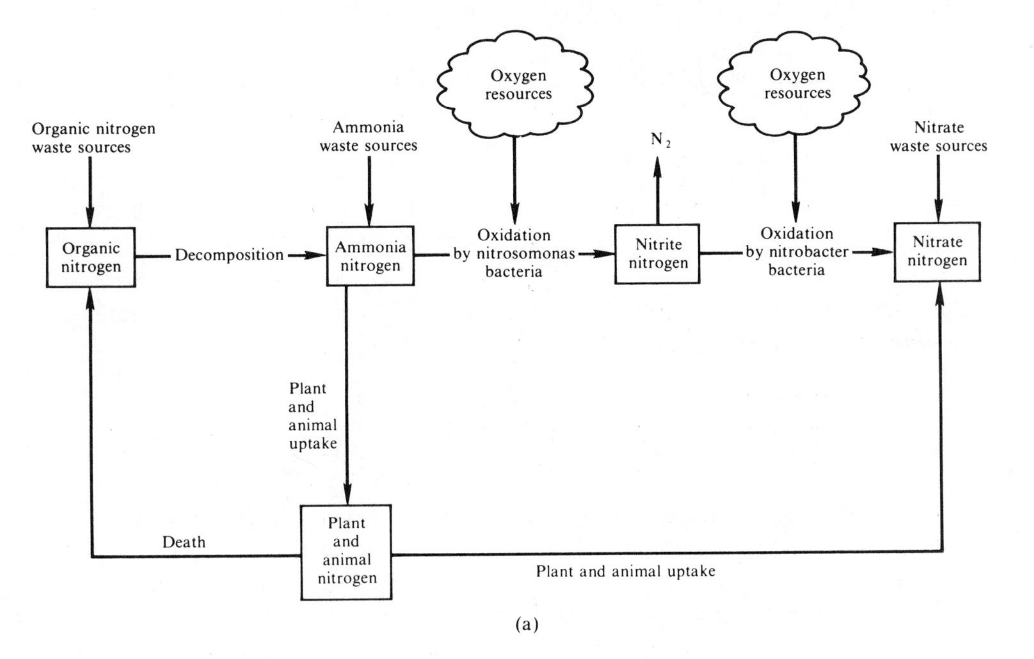

(a)

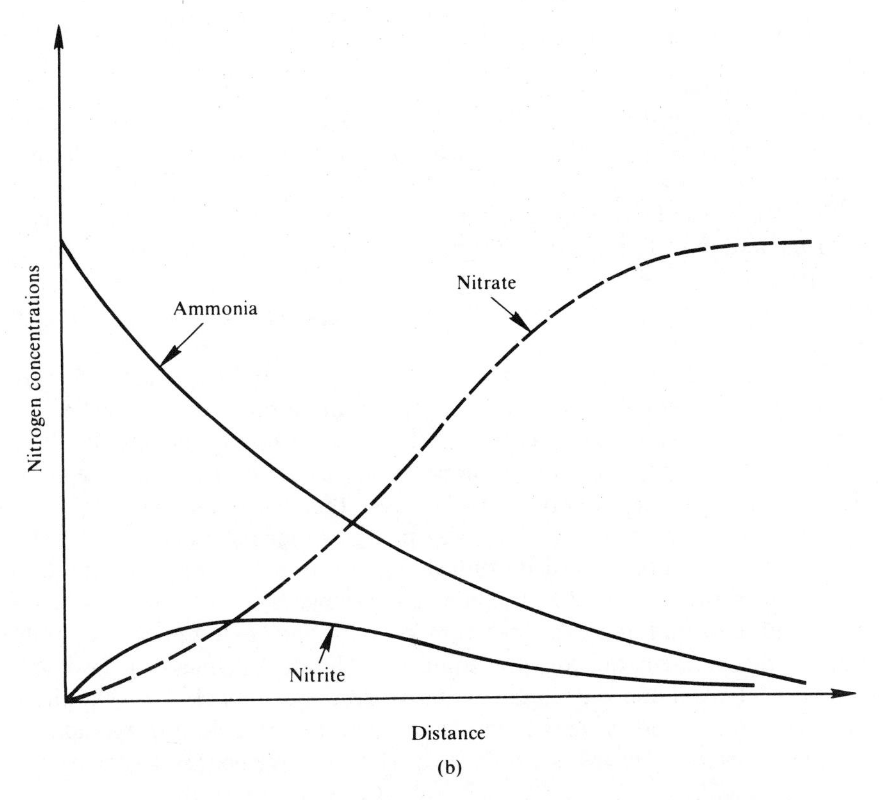

(b)

FIGURE 9.8. (a) Major features of the nitrogen cycle in aerobic waters. (b) Sequential reactions in nitrification with increasing distance downstream from point course of ammonia waste.

(LT^{-1}), the solution of the following four equations permits prediction of the concentration of each form of nitrogen in steady-state river systems

$$
\begin{aligned}
0 &= E\frac{d^2N_1}{dX^2} - U\frac{dN_1}{dX} - K_1^N N_1 \\
0 &= E\frac{d^2N_i}{dX^2} - U\frac{dN_i}{dX} - K_i^N N_i + K_{i-1,i}^N N_{i-1} \qquad i = 2, 3, 4
\end{aligned}
\tag{9.30}
$$

provided the boundary conditions just above X_s^- and just below X_s^+ each discharge are satisfied:

$$N_i(X_s^+) = N_i(X_s^-) \qquad i = 1, 2, 3, 4$$

$$\left[EA\frac{dN_i}{dX}\bigg|_{X_s^+} - QN_i(X_s^+)\right] - \left[EA\frac{dN_i}{dX}\bigg|_{X_s^-} - QN_i(X_s^-)\right] = W_i(s)$$

A decrease in the dissolved oxygen concentration in such river systems is caused by ammonia oxidation ($3.43K_{23}^n N_3$) and nitrate oxidation ($1.14K_{34}^n N_3$). The dissolved oxygen deficit concentration D^N resulting from the oxidation of these two nitrogen forms can be predicted by the equation

$$0 = E\frac{d^2D^N}{dX^2} - U\frac{dD^N}{dX} - K_aD^N + 3.43K_{23}^N N_2 + 1.14K_{34}^N N_3 \tag{9.31}$$

in which K_a is the reaeration rate constant (T^{-1}) and N_2 and N_3 are obtained from the solution of equations 9.30. Figure 9.9 illustrates the application of a finite difference approximation of this nitrogen model to the Delaware estuary in the northeastern United States.

This model can be further expanded to include the utilization of ammonia-nitrogen and nitrate-nitrogen by phytoplankton. Algae, in turn, produce organic nitrogen, thereby completing a very simplified version of the nitrogen cycle. Ecological models which incorporate algal growth are discussed in Section 9.5.6.

The models discussed in this section and Sections 9.5.1 through 9.5.3 are all first-order kinetic steady-state models. Although they vary in complexity and in their data requirements, they are relatively simple models. In these simple models, flows and temperatures do not vary with time and the complex nonlinear kinetic interactions between the microorganisms and the constituents are approximated by linear or first-order reactions.

Although for more detailed studies, nonlinear non-steady-state models may be required, many problems can be approached by first assuming steady-state conditions and first-order reaction kinetics. Such assumptions certainly simplify the mathematics and the solution techniques. Unless the problem context and available data warrant more complex models [16, 50], these relatively simple models can be, and have been, used to help understand and analyze a variety of river system water quality management problems.

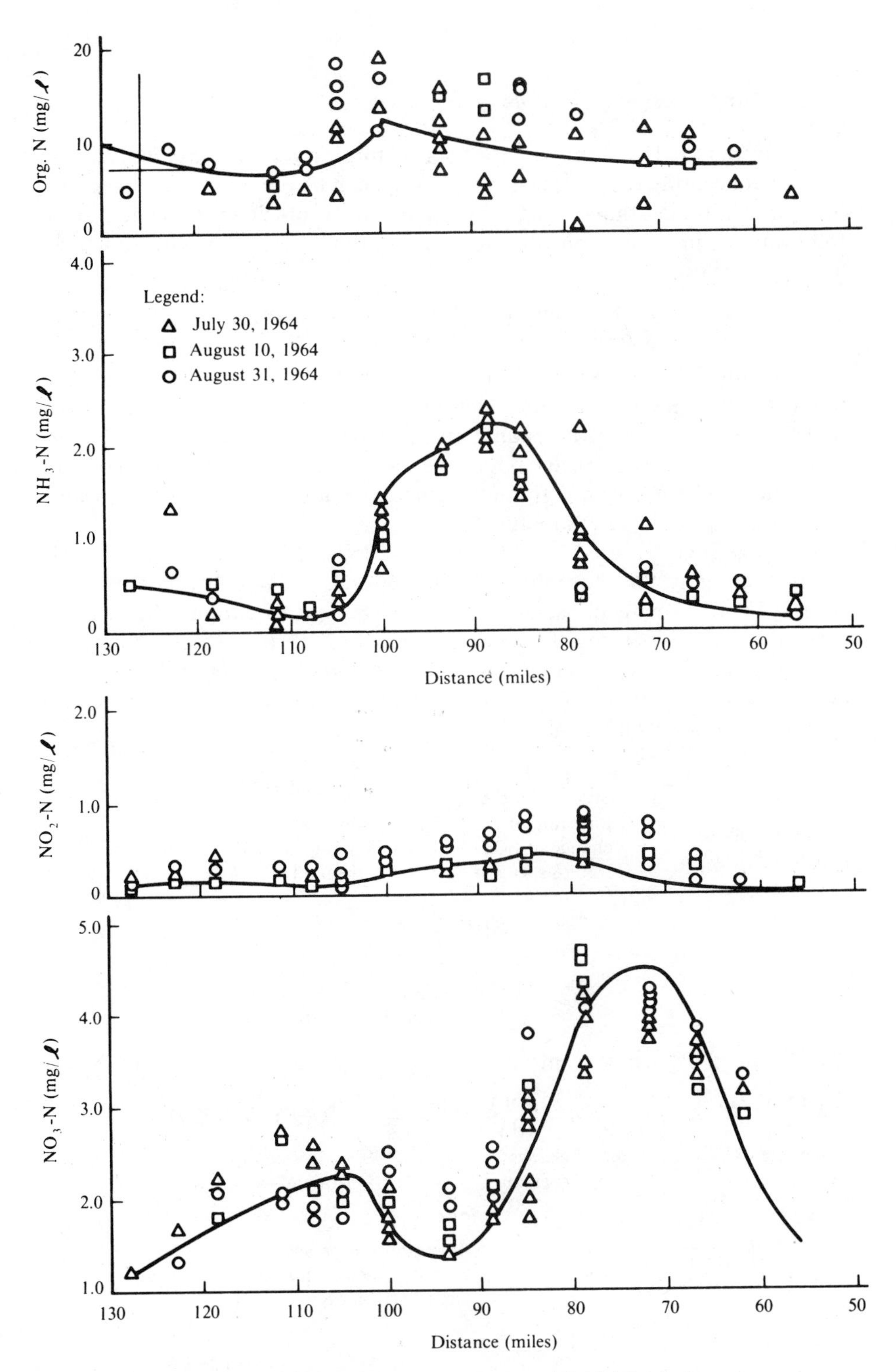

FIGURE 9.9. Observed versus computed (solid line) nitrogen profile of the Delaware River, August 1964. (Redrawn from O'Connor et al. [37].)

9.5.5 Time-Varying Models

When required, a representation of natural river systems with time-varying concentrations and steady-state flows can be obtained from integrating, or approximating (using finite differences or finite elements), the following form of the one-dimensional continuity or transport equation (equation 9.1):

$$\frac{\partial C}{\partial t} = \frac{1}{A}\frac{\partial}{\partial X}\left(EA\frac{\partial C}{\partial X}\right) - \frac{1}{A}\frac{\partial}{\partial X}(QC) \pm \sum_k S_k(U, X, t) \tag{9.32}$$

where again C is the constitutent (degradable or nondegradable) concentration (ML^{-3}), E the dispersion coefficient (L^2T^{-1}), A the cross-sectional area (L^2) at location X, X the longitudinal river location (L), U the net downstream freshwater velocity (LT^{-1}), and $\sum_k S_k$ the sources and sinks ($ML^{-3}T^{-1}$) of the constituent. The problem is to find a general solution for the concentration of C at all locations X at any time t.

Investigations reported by O'Connor [34, 35] and Tracor Associates [55] are among many that have applied equation 9.32 or modifications of it to river systems with steady-state flows. For example, equation 9.32 can be used to predict biochemical oxygen demand $B(X, t)$ and dissolved oxygen concentrations $DO(X, t)$ in river systems [34, 53]. Equations for carbonaceous (BOD^C) and for nitrogeneous (BOD^N) biochemical oxygen demand prediction have the following form:

$$\frac{\partial B(X,t)}{\partial t} = \frac{1}{A}\frac{\partial}{\partial X}\left[EA\frac{\partial B(X,t)}{\partial X}\right] - \frac{1}{A}\frac{\partial}{\partial X}\left[Q(X)\cdot B(X,t)\right]$$

[time rate of change in BOD] = [time rate of change in BOD from dispersion] − [time rate of change in BOD from advection]

$$- K[B(X,t)] + \frac{\sigma_{BOD}}{A} \tag{9.33}$$

− [time rate of decrease due to decay] + [time rate of addition from distributed runoff]

The dissolved oxygen concentration satisfies the equation:

$$\frac{\partial DO(X,t)}{\partial t} = \frac{1}{A}\frac{\partial}{\partial X}\left[EA\frac{\partial DO(X,t)}{\partial X}\right] - \frac{1}{A}\frac{\partial}{\partial X}\left[Q(X)\cdot DO(X,t)\right]$$

[time rate of DO change] = [rate of change from dispersion] − [rate of change from advection]

$$+ K_a[DO_{SAT}(X,t) - DO(X,t)] - K^C B^C(X,t) - K^N B^N(X,t) \tag{9.34}$$

+ [rate of addition from reaeration] − [time rate of decrease caused by deoxygenation from BOD decay]

$$+ \quad P(X, t) \quad - \quad R(X, t) \quad - \quad S_B(X, t) \quad + \quad \frac{\sigma_{\text{DO}}}{A}$$

$$+ \begin{bmatrix}\text{photosynthesis} \\ \text{addition rate}\end{bmatrix} - \begin{bmatrix}\text{respiration} \\ \text{demand rate}\end{bmatrix} - \begin{bmatrix}\text{benthal} \\ \text{demand rate}\end{bmatrix} + \begin{bmatrix}\text{distributed runoff} \\ \text{addition rate}\end{bmatrix}$$

The continuous solution approach can be used to solve equations 9.33 and 9.34 in the simple case of flowing streams where the dispersion is negligible compared to advective transport. In that case, assuming K^C, K^N, K_a, A, and U are constant throughout the reach, equations 9.33 and 9.34 can be readily integrated to determine the time-varying longitudinal distribution of BOD and DO. One also needs the BOD concentrations $B^C(0, t)$ and $B^N(0, t)$ and also the dissolved oxygen deficit concentration $D(0, t)$ at the upstream end of the reach at $X = 0$ as a function of time t. Given these inflow conditions, the solution to equation 9.34 for the DOD concentration within the reach is

$$D(X, t) = D(0, t - X/U)e_a \tag{9.35a}$$

$$+ B^C(0, t - X/U)\left[\frac{K^C(e_C - e_a)}{K_a - K^C}\right] \tag{9.35b}$$

$$+ B^N(0, t - X/U)\left[\frac{K^N(e_N - e_a)}{K_a - K^N}\right] \tag{9.35c}$$

$$+ \frac{\sigma^C_{\text{BOD}}}{A}\left[\frac{1 - e_a}{K_a} - \frac{e_C - e_a}{K_a - K^C}\right] \tag{9.35d}$$

$$+ \frac{\sigma_{\text{DO}}}{A}\left[\frac{1 - e_a}{K_a}\right] \tag{9.35e}$$

$$+ \frac{1}{U}\int_{x=0}^{x=X} \exp\left[-K_a(X - x)/U\right]\left\{R\left(x, t - \frac{X - x}{U}\right)\right. \tag{9.35f}$$

$$\left. - P\left(x, t - \frac{X - x}{U}\right) + S^B\left(x, t - \frac{X - x}{U}\right)\right\} dx$$

where

K_a, K^C, K^N = reaeration, BOD^C decay and BOD^N decay rate constants (T^{-1})

$e_a = \exp(-K_a X/U)$

$e_C = \exp(-K^C X/U)$

$e_N = \exp(-K^N X/U)$

σ^C_{BOD} = uniformly distributed carbonaceous BOD input in runoff along river $(ML^{-1}T^{-1})$

σ_{DO} = uniformly distributed DO deficit in runoff along river $(ML^{-1}T^{-1})$

U = river velocity (LT^{-1})

$P(X, t)$ = time varying oxygen production rate from photosynthesis ($ML^{-3}T^{-1}$)

$R(X, t)$ = time varying oxygen consumption rate from respiration ($ML^{-3}T^{-1}$)

$S_B(X, t)$ = time varying oxygen consumption rate from benthal demand ($ML^{-3}T^{-1}$)

Portions of equation 9.35 represent different effects. Part (a) is the contribution to the downstream dissolved oxygen deficit from the initial deficit concentration $D(0, t)$ at $X = 0$. Parts (b) and (c) are the dissolved oxygen deficit which results from the oxidation of the BOD^C and BOD^N concentrations in the water at $X = 0$. Part (d) represents the deficit contribution from uniformly distributed carbonaceous BOD sources such as from runoff, when the quantity of runoff is small relative to the streamflow. A similar expression could apply to nitrogenous BOD also. Part (e) is the net deficit resulting from respiration, photosynthesis, and benthal oxygen demands.

The integral in 9.35e can often be evaluated analytically if the functions R, P, or S_B take on simple forms. For example, $S_B(x, t)$ is often assumed to be a constant demand $\bar{S}_B$ ($ML^{-3}T^{-1}$) independent of time. In this case,

$$\frac{1}{U}\int_{x=0}^{x=X} \exp\left[-K_a(X - x)/U\right]\bar{S}_B\, dx = \frac{\bar{S}_B(1 - e_a)}{K_a} \qquad (9.35g)$$

Sometimes $R(X, t) - P(X, t)$ is approximated by a Fourier series. Then the integral in 9.35e can again be evaluated analytically. For example, if one lets

$$R(X, t) - P(X, t) = RP \sin(\omega t)$$

then

$$\frac{1}{U}\int_{0}^{X} \exp\left[-K_a(X - x)/U\right]RP \sin \omega\left(t - \frac{X - x}{U}\right) dx$$

$$= RP(K_a^2 + \omega^2)^{-1}\{K_a \sin \omega t + \omega \cos \omega t \qquad (9.35h)$$

$$- e^{-KX/U}\left[K_a \sin \omega\left(t - \frac{X}{U}\right) + \omega \cos \omega\left(t - \frac{X}{U}\right)\right]\}$$

where for t in seconds and one R-P cycle per day the angular frequency ω would equal (2π radians/day)/[(60 seconds/hr)(24 hr/day)] and the arguments of the trigonometric sine and cosine functions are in radians.

Because each part of equation 9.35 is linear, those parts that are applicable for a particular river system can be added together to account for all contributions to the oxygen deficit. Figure 9.10 illustrates the functional form of several components of equation 9.35.

Using equation 9.35, the DO deficit concentration may be estimated at any distance X downstream from a point source of BOD, at any time of day

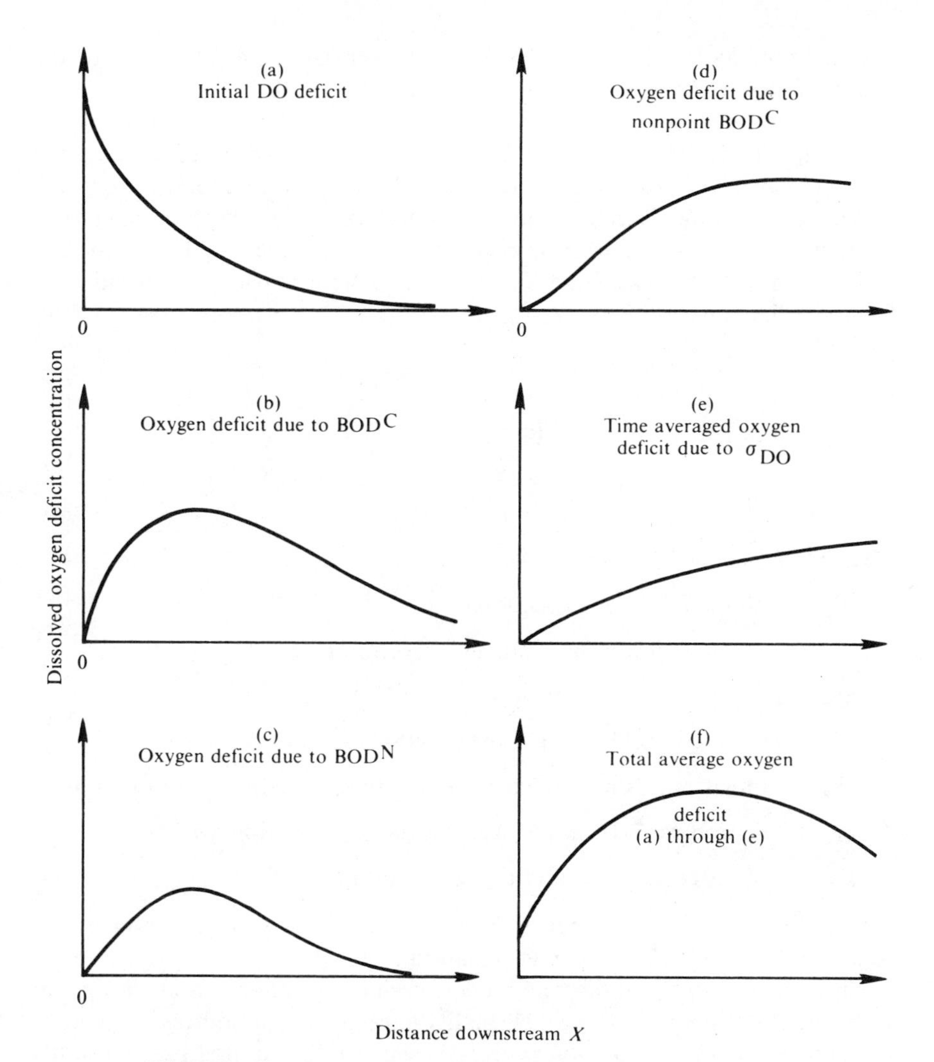

FIGURE 9.10. Components of dissolved oxygen model, equation 9.26.

t. By fixing a constant value of X and varying t over a 24-hour period, the behavior of the deficit concentration at a single site can be estimated. By fixing the value of t and varying X along the length of the river segment, a profile of the deficit versus distance may be determined (similar to Figure 9.9) for a particular time of day.

If the time variation in inputs BOD(X, t) is substantial, equation 9.35 for nondispersive streams may not be a very good predictor simply because dispersion may become important. But in any case, the time parameter t of the time-varying wastewater inputs BOD(0, t) must be considered together with the time of flow X/U in order to achieve a reasonable prediction of the effects of time-varying waste inputs at some location $X > 0$.

Equation 9.35 illustrates the continuous solution approach to water quality modeling. The finite-section solution approach involves dividing the river system into finite sections, developing a continuity equation for each section, and solving all section equations simultaneously at each point in time. In a finite-difference approach, each section is assumed to be completely mixed. The mass-balance equation for the kth section, using the notation of Section 9.5.3 equation 9.29, is

$$
\begin{aligned}
V_k \frac{dC_k}{dt} = {} & E'_k(C_{k-1} - C_k) - E'_{k+1}(C_k - C_{k-1}) \\
& + Q_k[\alpha_k C_{k-1} + (1 - \alpha_k)C_k] \\
& - Q_{k+1}[\alpha_{k+1} C_k + (1 - \alpha_{k+1})C_{k+1}] \\
& - V_k K_k C_k + W_k
\end{aligned}
\tag{9.36}
$$

where:

V_k = volume of water in section k (L^3)

C_k = concentration of constituent in section k (ML^{-3})

Q_k = flow into section k from $k - 1$ (L^3T^{-1})

E'_k = exchange coefficient between sections k and $k - 1$ (L^3T^{-1})

α_k = dimensionless mixing coefficient between sections k and $k - 1$

K_k = reaction-rate constant for constituent for section k (T^{-1})

W_k = rate of constituent addition to section k (MT^{-1})

Figure 9.6, with some change in variables, illustrates the components of equation 9.36, which is similar to equation 9.29.

These continuous and discrete models permit an estimation of the spatial and temporal distribution of water quality parameters and indices. The problem of predicting water quality involves measuring the required rate constants for a particular river system, developing the model in a form that can be solved, calibrating and verifying the model and then interpreting the results of various solutions corresponding to alternative management plans.

9.5.6 Multi-constituent River Ecosystem Models

Over the past decade there has been an increasing emphasis on the effects of various substances, especially nutrients, on aquatic ecosystems, and in particular on the production and populations of bacteria, protozoans, phytoplankton, zooplankton, fish, and other organisms [5]. QUAL II, developed for the U.S. Environmental Protection Agency [59], is a good example of an operational aquatic ecosystem simulation model. QUAL II numerically integrates equation 9.32 for a variety of water quality constituents, including conservative substances; algal biomass; ammonia, nitrite, and nitrate nitrogen; phosphorus; carbonaceous biochemical oxygen demand; dissolved oxygen; coliforms and radionuclides. The following paragraphs define the mathematical relationships that describe the individual reactions and interactions among these constituents. The dispersion and advection terms are not shown in the following equations, but are included in the actual model, as summarized in Table 9.1 at the end of this discussion.

Chlorophyll *a*, C_a, is considered to be proportional to the concentration of phytoplanktonic algal biomass A:

$$C_a = \alpha_0 \mathrm{A} \tag{9.37}$$

The time-varying ($d\mathrm{A}/dt$) growth and production of algal biomass A is dependent on their growth rate μ_A (T^{-1}), the specific-loss rate ρ_A (T^{-1}), the settling rate σ_1 (LT^{-1}), and the average stream depth D_a (L), all at a particular location X in the river. Although not in the QUAL II model, the concentration of algae is also a function of grazing G_Z by higher trophic levels such as zooplankton Z. Hence

$$\frac{d\mathrm{A}}{dt} = \mu_\mathrm{A}\mathrm{A} - \rho_\mathrm{A}\mathrm{A} - \frac{\sigma_1 \mathrm{A}}{D_a} - G_Z \tag{9.38}$$

The algal growth rate is dependent on temperature and the availability of nutrients (nitrogen, carbon, phosphorus) and light. The standard Michaelis-Menten formulation, illustrated in Figure 9.11, defines the specific

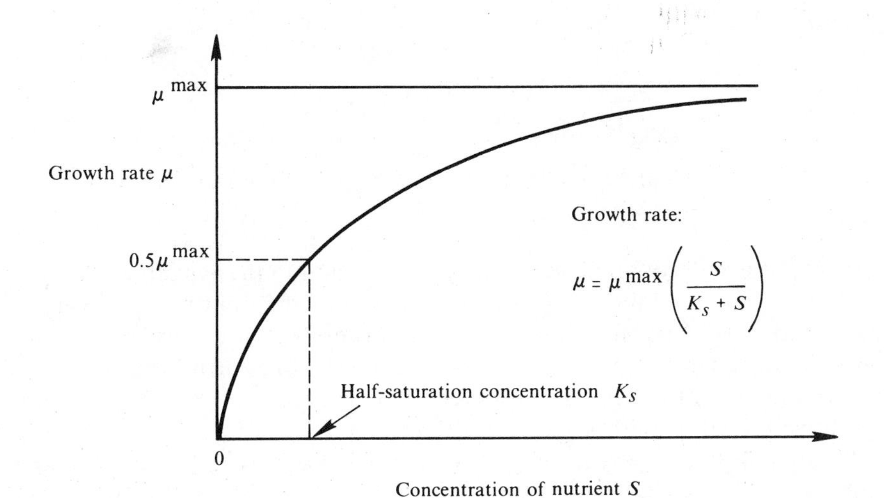

FIGURE 9.11. Michaelis-Menten kinetics expression for microbial growth.

growth rate μ at a given site in a river system as a function of a nutrient's concentration. This Michaelis-Menten formulation asserts that the growth rate μ is an increasing function of the essential nutrient concentration S:

$$\mu = \mu^{\max}\frac{S}{K_s + S} \tag{9.39}$$

The parameter K_s is the value of S when μ equals half of the maximum growth rate $\mu^{\max}$. It is called the half-saturation concentration. Note that at low values of nutrient concentrations, μ is proportional to S. At high values of S, μ approaches the limiting saturation value $\mu^{\max}$. In the QUAL II model the algal growth rate μ_A may be limited by nitrate-nitrogen NO_3, phosphrous P, carbon C, light L, or temperature. Hence

$$\mu_A = (\mu_{A,20}^{\max})(\theta^{T-20})\left(\frac{NO_3}{K_{NO_3} + NO_3}\right)\left(\frac{P}{K_P + P}\right)\left(\frac{C}{K_C + C}\right) \times \frac{1}{\lambda D_a}\ln\left(\frac{K_L + L}{K_L + Le^{-\lambda D_a}}\right) \tag{9.40}$$

where

μ_A = specific growth rate of algal biomass (T^{-1})

$\mu_{A,20}^{\max}$ = maximum growth rate of algal biomass at 20°C (T^{-1})

θ = temperature coefficient ranging from 1.02 to 1.06

T = actual water temperature in °C

NO_3, P, C = nitrate–nitrogen, orthophosphate–phosphorus, and carbon concentrations (ML^{-3})

K_C, K_{NO_3}, K_P = temperature dependent half-saturation concentrations for algal growth (ML^{-3})

L = light intensity (langleys/T)

K_L = half-saturation coefficient for light (langleys/T)

λ = light extinction coefficient in the river (L^{-1})

D_a = average stream depth (L).

Equation 9.40 couples algal biomass production to the available supply of nutrients, light, and temperature; hence algae and chlorophyll *a* will vary in time and space in response to the concentrations of elements needed for growth. In equation 9.40, if any of the critical growth element concentrations are zero, the algal growth rate μ_A is zero. This expression is based on a multiplicative growth hypothesis, as contrasted to a threshold growth hypothesis. The latter approach includes only the smallest valued Michaelis-Menten term for the most limiting nutrient, or light, that constrains algal growth. This threshold hypothesis is expressed in equation 9.41.

$$\mu_A = (\mu_{A,20}^{max}\theta^{T-20})\left(\frac{NO_3}{NO_3 + K_{NO_3}}\right) \quad \text{if } NO_3 \text{ limiting}$$

$$\mu_A = (\mu_{A,20}^{max}\theta^{T-20})\left(\frac{P}{P + K_P}\right) \quad \text{if P limiting}$$

$$\mu_A = (\mu_{A,20}^{max}\theta^{T-20})\left(\frac{C}{C + K_C}\right) \quad \text{if C limiting} \tag{9.41}$$

$$\mu_A = (\mu_{A,20}^{max}\theta^{T-20})\left[\frac{1}{\lambda D_a}\,\ell n\left(\frac{K_L + L}{K_L + Le^{-\lambda D_a}}\right)\right] \quad \text{if } L \text{ limiting}$$

The biomass specific-loss rate is also temperature dependent:

$$\rho_A = \rho_{20}\theta^{T-20} \tag{9.42}$$

where again T is the temperature in °C, θ is a constant, and ρ_{20} is the specific loss rate (T^{-1}) at 20°C. All rate constants that are temperature dependent are defined by equations having the form of equation 9.42.

The growth rate of zooplankton μ_Z could be defined by

$$\mu_Z = (\mu_{Z,20}^{max})(\theta^{T-20})\left(\frac{A}{K_A + A}\right) \tag{9.43}$$

where K_A is the temperature-dependent half-saturation-rate constant for algal biomass (ML^{-3}). The growth rate of zooplankton μ_Z (T^{-1}) times the zooplankton concentration Z (ML^{-3}) times a conversion coefficient $F_{A,Z}$ (MM^{-1}), which indicates the mass of algal biomass required per unit mass of zooplankton, estimates the loss in algal biomass due to zooplankton grazing G_Z. The reliability of current models for predicting trophic levels higher than phytoplankton (such as zooplankton and fish) is relatively poor [29], thus the omission of these higher trophic levels from operational models such as QUAL II.

The nitrogen cycle in QUAL II is described by differential equations governing the transformation of nitrogen from one form to another. For ammonia-nitrogen NH_3,

$$\frac{dNH_3}{dt} = \alpha_1\rho_A A - \beta_1 NH_3 + \frac{\sigma_3}{A_X} \tag{9.44}$$

where

NH_3 = concentration of ammonia-nitrogen (ML^{-3})

α_1 = fraction of non living algal biomass resolubilized as ammonia-nitrogen by bacteria

ρ_A = temperature-dependent specific-loss rate (T^{-1})

A = algal biomass concentration (ML^{-3})

β_1 = temperature-dependent rate of biological oxidation of NH_3 (T^{-1})

σ_3 = benthos source rate for NH_3 ($MT^{-1}L^{-1}$)

A_X = average stream cross-sectional area at location X (L^2)

For nitrite-nitrogen NO_2,

$$\frac{dNO_2}{dt} = \beta_1 NH_3 - \beta_2 NO_2 \tag{9.45}$$

where

NO_2 = concentration of nitrite-nitrogen (ML^{-3})

β_1 = rate of oxidation of NH_3 to NO_2 (T^{-1})

β_2 = rate of oxidation of NO_2 to NO_3 (T^{-1})

For nitrate-nitrogen NO_3,

$$\frac{dNO_3}{dt} = \beta_2 NO_2 - \alpha_1 \mu_A A \tag{9.46}$$

in which the parameters and variables are as defined after equations 9.44 and 9.45. Equations 9.44 to 9.46 describe the nitrogen cycle. In QUAL II equations 9.44 and 9.46 express the assumption that the fraction of non-living algal biomass resolubilized as NH_3-N equals the fraction of biomass that is NO_3-N.

The phosphorus cycle is modeled in a less detailed manner than the nitrogen cycle. Only the interaction of phosphorus and algae, plus a sink term, are considered. Thus the differential equation describing the time-varying concentration of orthophosphate-phosphorus P (ML^{-3}) is written as

$$\frac{dP}{dt} = \alpha_2 A(\rho_A - \mu_A) + \frac{\sigma_2}{A_X} \tag{9.47}$$

where

α_2 = fraction of algal biomass that is phosphorus

σ_2 = benthos source rate for phosphorus ($MT^{-1}L^{-1}$)

Carbonaceous biochemical oxygen demand BOD^C (ML^{-3}), is formulated as a first-order reaction

$$\frac{dBOD^C}{dt} = -K^C BOD^C - K_s BOD^C \tag{9.48}$$

where

K^C = temperature-dependent deoxygenation or decay rate constant of carbonaceous BOD (T^{-1})

K_s = rate of loss of carbonaceous BOD due to settling (T^{-1})

The oxygen uptake by benthic deposits is assumed to be dependent only on the cross-sectional area A_X at location X, and equals σ_b/A_X where σ_b is the constant benthic source rate ($MT^{-1}L^{-1}$).

The differential equation that describes the rate of change in dissolved oxygen is

$$\frac{d}{dt}\mathrm{DO}(X) = K_a[\mathrm{DO}_{\mathrm{SAT}}(X) - \mathrm{DO}(X)] + (\alpha_3\mu_{\mathrm{A}} - \alpha_4\mu_{\mathrm{A}})\mathrm{A} - K^C\mathrm{BOD}^C - \sigma_b/A_X - \alpha_5\beta_1\mathrm{NH}_3 - \alpha_6\beta_2\mathrm{NO}_2 \tag{9.49}$$

where, in addition to the terms defined above,

$\mathrm{DO}_{\mathrm{SAT}}(X)$ = temperature-dependent dissolved oxygen saturation at location X (ML^{-3})

$\mathrm{DO}(X)$ = dissolved oxygen concentration at location X (ML^{-3})

K_a = temperature-dependent reaeration rate constant (T^{-1})

α_3 = rate of oxygen production through photosynthesis per unit of algal biomass (MM^{-1})

α_4 = rate of oxygen uptake from respiration per unit of algal biomass (MM^{-1})

α_5 = rate of oxygen uptake per unit of oxidation of ammonia nitrogen (MM^{-1})

α_6 = rate of oxygen uptake per unit of nitrite-nitrogen oxidation (MM^{-1})

Other water quality indicators are coliform and radionuclide concentrations. For the most probable number (MPN) of coliforms, the rate of change with respect to time equals

$$\frac{d\mathrm{F}}{dt} = -K_d\mathrm{F} \tag{9.50}$$

where F is the coliform MPN and K_d is the die-off rate (T^{-1}). The differential equation describing the rate of change in radionuclides R is

$$\frac{d\mathrm{R}}{dt} = -\gamma_r\mathrm{R} - \gamma_a\mathrm{R} \tag{9.51}$$

where

γ_r = radioactive decay rate (T^{-1})

γ_a = radioactive absorption rate to settled particles and other loss from water column (T^{-1})

The final constituent that must be included is heat which affects the reaction rates of most other constituents in addition to being a possible pollutant itself. Temperature modeling of one-dimensional streams involves the same continuity and transport equation, 9.1 or 9.32. The constituent term C in those equations can be defined as $\rho c T$, the amount of heat per unit volume of water, where T is the water temperature (°C), ρ is the density of water (ML^{-3}), and c is the specific heat of water ($HM^{-1}{}^{\circ}C^{-1}$).

The heat source or sink term, S_k, in equation 9.1 or 9.32, can be defined as S_H in units of heat (calories or Btu's) per unit volume per unit time ($HL^{-3}T^{-1}$). This source or sink term accounts for internal heat generation and all heat transferred across the system boundaries (i.e., across the air-water interface and across the wetted interface). The latter, in the absence of groundwater flow, is often insignificant compared to the surface heat interchange.

It is often convenient to represent the interfacial heat transfer rate as a flux, H_N, per unit area having units of ($HL^{-2}T^{-1}$). For a stream reach of length ΔX and mean surface width of W, the total rate of heat transfer across the water surface is $H_N(\Delta X)W$. Assuming a mean cross-sectional area of A and a uniform temperature distribution throughout the volume $A(\Delta X)$, S_H will equal H_N/D, where D is the hydraulic depth, A/W, of the reach. Thus equation 9.32 for temperature prediction over time t and distance X can be written

$$\frac{\partial T}{\partial t} = \frac{1}{A}\frac{\partial}{\partial X}\left(EA\frac{\partial T}{\partial X}\right) - \frac{1}{A}\frac{\partial}{\partial X}(TQ) \pm \frac{H_N}{\rho c D} \tag{9.52}$$

[time rate of change in temperature] = [rate of change from dispersion] − [rate of change from advection] ± [rate of change from sources/sinks]

The net heat flux, H_N, can be expressed as $\lambda(T_E - T)$, where T_E is the equilibrium water surface temperature at which the net rate of heat flux through the air-water interface is zero. The parameter λ is the net heat flux per degree of temperature difference. This parameter and the equilibrium temperature T_E are dependent on a number of atmospheric and water quality conditions, and hence are not constants.

Table 9.1 summarizes the complete set of equations that are solved by QUAL II except the temperature relationships of the type defined by equation 9.52. These equations include the effects of dispersion, advection, constituent reactions, and interactions up through the phytoplankton trophic level. They also include a source term S_C (MT^{-1}) that is assumed uniform over the length ΔX of the river section at location X.

If the left-hand sides of all the equations listed in Table 9.1 are set to 0, this becomes a static model. Both static and time-varying models defined by these equations can be solved by numerical procedures that permits a rigorous treatment of the coupling effects among all constituents and the factors that characterize the aquatic environment.

Table 9.2 lists the input parameters defined above, the range of values of those parameters, whether or not they vary with location and/or temperature, and their relative reliability [59].

TABLE 9.1. **Summary of Differential Equations Solved by QUAL II Simulation Model**

Conservative mineral (C)	$\frac{\partial \mathrm{C}}{\partial t} = \frac{\partial\left(A_X E \frac{\partial \mathrm{C}}{\partial X}\right)}{A_X \partial X} - \frac{\partial(A_X U \mathrm{C})}{A_X \partial X} + \frac{S_\mathrm{C}}{A_X \Delta X}$
Algae (A)	$\frac{\partial \mathrm{A}}{\partial t} = \frac{\partial\left(A_X E \frac{\partial \mathrm{A}}{\partial X}\right)}{A_X \partial X} - \frac{\partial(A_X U \mathrm{A})}{A_X \partial X} + \frac{S_\mathrm{A}}{A_X \Delta X} + \left(\mu_\mathrm{A} - \rho_\mathrm{A} - \frac{\sigma_1}{D_a}\right)\mathrm{A}$
Ammonia-nitrogen (NH_3)	$\frac{\partial \mathrm{NH_3}}{\partial t} = \frac{\partial\left(A_X E \frac{\partial \mathrm{NH_3}}{\partial X}\right)}{A_X \partial X} - \frac{\partial(A_X U \mathrm{NH_3})}{A_X \partial X} + \frac{S_{\mathrm{NH_3}}}{A_X \Delta X} + \left(\alpha_1 \rho_\mathrm{A} \mathrm{A} - \beta_1 \mathrm{NH_3} + \frac{\sigma_3}{A_X}\right)$
Nitrite-nitrogen (NO_2)	$\frac{\partial \mathrm{NO_2}}{\partial t} = \frac{\partial\left(A_X E \frac{\partial \mathrm{NO_2}}{\partial X}\right)}{A_X \partial X} - \frac{\partial(A_X U \mathrm{NO_2})}{A_X \partial X} + \frac{S_{\mathrm{NO_2}}}{A_X \Delta X} + (\beta_1 \mathrm{NH_3} - \beta_2 \mathrm{NO_2})$
Nitrate-nitrogen (NO_3)	$\frac{\partial \mathrm{NO_3}}{\partial t} = \frac{\partial\left(A_X E \frac{\partial \mathrm{NO_3}}{\partial X}\right)}{A_X \partial X} - \frac{\partial(A_X U \mathrm{NO_3})}{A_X \partial X} + \frac{S_{\mathrm{NO_3}}}{A_X \Delta X} + (\beta_2 \mathrm{NO_2} - \alpha_1 \mu_\mathrm{A} \mathrm{A})$
Phosphate-phosphorus (P)	$\frac{\partial \mathrm{P}}{\partial t} = \frac{\partial\left(A_X E \frac{\partial \mathrm{P}}{\partial X}\right)}{A_X \partial X} - \frac{\partial(A_X U \mathrm{P})}{A_X \partial X} + \frac{S_\mathrm{P}}{A_X \Delta X} + \left(\alpha_2(\rho_\mathrm{A} - \mu_\mathrm{A})\mathrm{A} - \frac{\sigma_2}{A_X}\right)$
Biochemical oxygen demand (BOD^C)	$\frac{\partial \mathrm{BOD}^C}{\partial t} = \frac{\partial\left(A_X E \frac{\partial \mathrm{BOD}^C}{\partial X}\right)}{A_X \partial X} - \frac{\partial(A_X U \mathrm{BOD}^C)}{A_X \partial X} + \frac{S_{\mathrm{BOD}^C}}{A_X \Delta X} - (K^C + K_s)\mathrm{BOD}^C$
Dissolved oxygen (DO)	$\frac{\partial \mathrm{DO}}{\partial t} = \frac{\partial\left(A_X E \frac{\partial \mathrm{DO}}{\partial X}\right)}{A_X \partial X} - \frac{\partial(A_X U \mathrm{DO})}{A_X \partial X} + \frac{S_\mathrm{DO}}{A_X \Delta X}$
	$+ \left[K_a(\mathrm{DO_{SAT}} - \mathrm{DO}) + (\alpha_3 \mu_\mathrm{A} - \alpha_4 \rho_\mathrm{A})\mathrm{A} - K^C \mathrm{BOD}^C - \frac{\sigma_b}{A_X} - \alpha_5 \beta_1 \mathrm{NH_3} - \alpha_6 \beta_2 \mathrm{NO_2}\right]$
Coliform (F)	$\frac{\partial \mathrm{F}}{\partial t} = \frac{\partial\left(A_X E \frac{\partial \mathrm{F}}{\partial X}\right)}{A_X \partial X} - \frac{\partial(A_X U \mathrm{F})}{A_X \partial X} + \frac{S_\mathrm{F}}{A_X \Delta X} - K_d \mathrm{F}$
Radioactive material (R)	$\frac{\partial \mathrm{R}}{\partial t} = \frac{\partial\left(A_X E \frac{\partial \mathrm{R}}{\partial X}\right)}{A_X \partial X} - \frac{\partial(A_X U \mathrm{R})}{A_X \partial X} + \frac{S_\mathrm{R}}{A_X \Delta X} - \gamma_r \mathrm{R} - \gamma_a \mathrm{R}$

Source: Water Resources Engineers, Inc., Computer Program Documentation for the Stream Quality Model QUAL II, Prepared for the U.S. Environmental Protection Agency, Washington, D.C., 1973.

TABLE 9.2. Input Parameters for QUAL II

Names in Equations	*Description*	*Units*	*Range of Values*	*Variable by Reach?*	*Temperature-Dependent?*	*Reliability*
α_0	Ratio of chlorophyll *a* to algal biomass	$\frac{\mu g\ Chl\text{-}a}{mg\ A}$	50–100	Yes	No	Fair
α_1	Fraction of algae biomass which is N	$\frac{mg\ N}{mg\ N}$	0.08–0.09	No	No	Good
α_2	Fraction of algae biomass which is P	$\frac{mg\ P}{mg\ A}$	0.012–0.015	No	No	Good
α_3	O_2 production per unit of algae growth	$\frac{mg\ O}{mg\ A}$	1.4–1.8	No	No	Good
α_4	O_2 uptake per unit of algae loss	$\frac{mg\ O}{mg\ A}$	1.6–2.3	No	No	Fair
α_5	O_2 uptake per unit of NH_3 oxidation	$\frac{mg\ O}{mg\ N}$	3.0–4.0	No	No	Good
α_6	O_2 uptake per unit of NO_2 oxidation	$\frac{mg\ O}{mg\ N}$	1.0–1.14	No	No	Good
$\mu^{max}_{A,20}$	Maximum specific growth rate of algae, 20°C	$\frac{1}{day}$	1.0–3.0	No	Yes	Good
ρ_A	Algae specific loss rate	$\frac{1}{day}$	0.05–0.5	No	Yes	Fair
β_1	Rate constant for biological oxidation of $NH_3 \longrightarrow NO_2$	$\frac{1}{day}$	υ.1–0.5	Yes	Yes	Fair
β_2	Rate constant for biological oxidation of $NO_2 \longrightarrow NO_3$	$\frac{1}{day}$	0.5–2.0	Yes	Yes	Fair

σ_1	Local settling rate for algae	$\frac{m}{day}$	0.2–2.0	Yes	No	Fair
σ_2	Benthos source rate for phosphorus	$\frac{mg\ P}{day\text{-}m}$	[a]	Yes	No	Poor
σ_3	Benthos source rate for NH_3	$\frac{mg\ N}{day\text{-}m}$	[a]	Yes	No	Poor
σ_b	Benthos source rate for BOD	$\frac{mg}{day\text{-}m}$	[a]	Yes	No	Poor
K^C	Carbonaceous BOD decay rate	$\frac{1}{day}$	0.1–2.0	Yes	Yes	Poor
K_a	Reaeration rate	$\frac{1}{day}$	0.0–100	Yes	Yes	Good
K_s	Carbonaceous BOD sink rate	$\frac{1}{day}$	[a]	Yes	No	Poor
K_d	Coliform die-off rate	$\frac{1}{day}$	0.5–4.0	Yes	Yes	Fair
γ_r	Radionuclide decay rate	$\frac{1}{day}$	[a]	No	No	Poor
γ_a	Radionuclide absorption rate	$\frac{1}{day}$	[a]	No	No	Poor
K_{NO_3}	Nitrate–nitrogen half-saturation constant for algae growth	$\frac{mg}{\ell}$	0.2–0.4	No	No	Fair to good
K_P	Phosphorus half-saturation constant for algae growth	$\frac{mg}{\ell}$	0.03–0.05	No	No	Fair to good
K_L	Light half-saturation constant for algae growth	$\frac{langleys}{day}$	260	No	No	Good

[a]Highly variable.

Source: Water Resources Engineers, Inc., Computer Program Documentation for the Stream Quality Model QUAL II, Prepared for the U.S. Environmental Protection Agency, Washington, D.C., 1973.

9.6 WATER QUALITY MODELING OF LAKES AND RESERVOIRS

The prediction of water quality in surface-water impoundments is based on mass-balance relationships which are similar to those used to predict water quality in streams and estuaries. There are also significant differences in the problems of predicting the water quality of lakes or reservoirs compared to those of river systems. These differences will be the focus of the discussion in this section of the chapter.

Natural surface-water impoundments that have a more-or-less constant volume throughout the year are often modeled using a simple input-output approach. For well-mixed lakes, and input-output approach seems to be adequate. However, for reservoirs in which the foregoing conditions do not apply, more complex models have been developed to predict thermal gradients, density stratification, and the impact that various reservoir designs and operating rules may have on these and other physical, chemical, and biological quality characteristics of the impounded water.

9.6.1 Input-Output Models for Well-Mixed, Constant-Volume Lakes

For lakes that are completely mixed, the change in the mass of any constituent equals the mass input less the mass output less losses due to decay or sedimentation, if any. Assuming a constant input rate N (MT^{-1}) of a constituent or nutrient with net decay and sedimentation rate K (T^{-1}) a mass balance can be written for any interval of time ΔT. Letting V equal the constant volume (L^3), Q the constant inflow and discharge (L^3T^{-1}), and C the concentration of the nutrient (ML^{-3}),

$$\Delta C = \frac{1}{V}[N\,\Delta T - QC\,\Delta T - KCV\,\Delta T] \tag{9.53}$$

Dividing equation 9.53 by ΔT and letting ΔT approach 0, one obtains the differential equation

$$\frac{dC}{dt} = \frac{N}{V} - \frac{QC}{V} - KC \tag{9.54}$$

Integration of equation 9.54 yields a predictive expression for the concentration $C(t)$ of the nutrient at any time t. Assuming that the initial concentration is zero [i.e., $C(0) = 0$],

$$C(t) = \frac{N}{Q + KV}\left\{1 - \exp\left[-t\left(\frac{Q}{V} + K\right)\right]\right\} \tag{9.55}$$

The equilibrium concentration C^* can be determined by letting t approach infinity in equation 9.55 or setting $dC/dt = 0$ in equation 9.54. In either case

$$C^* = \frac{N}{Q + KV} \tag{9.56}$$

Solving for the time t_α required to reach a given fraction α of the equilibrium concentration [i.e., $C(t)/C^* = \alpha$] results in

$$t_\alpha = \frac{-V}{Q + KV} \ln (1 - \alpha) \tag{9.57}$$

Similar equations can be developed to estimate the concentrations and times associated with a decrease in a pollutant concentration. For the perfectly mixed lake having an initial nutrient concentration $C(0) = C_0$ (at time $t = 0$), no nutrient input ($N = 0$), a constant volume V, discharge Q, and net decay and sedimentation coefficient K, the change in concentration with respect to time, from equation 9.54, equals

$$\frac{dC}{dt} = -C\left(\frac{Q}{V} + K\right) \tag{9.58}$$

Integrating equation 9.58, the concentration $C(t)$ at any time t is

$$C(t) = C_0 e^{-t(Q/V+K)} \tag{9.59}$$

In this case one can solve for the time t_α required for the nutrient concentration to reach a fraction $1 - \alpha$ of the initial concentration C_0 [i.e., $C(t)/C_0 = 1 - \alpha$]. The result is equation 9.57.

Vollenwieder and Dillon [57, 58], Larsen and Mercier [26], and others use this input-output model, or extensions, to predict phosphorus concentrations in large lakes while O'Connor and Mueller [36] use it to predict chloride concentrations.

9.6.2 Stratified Impoundments

The input-output model in Section 9.6.1 assumes that the lake is completely mixed throughout its entire volume. Many lakes and other surface-water impoundments become stratified during particular times of the year so that temperature gradients with depth effectively prevent mixing. In particular in the summer, lakes often exhibit two zones: an upper volume of warm water called the *epilimnion* and a lower colder volume called the *hypolimnion*. Each of the two are fairly well mixed while their different densities prevent complete mixing between the two. The transition zone is called the *thermocline*.

Snodgrass and O'Melia [48] have developed a model that considers the epilimnion and hypolimnion as two distinct lake volumes during summer stratification, with some interchange across the thermocline. They also developed a separate winter mixing model. Their summer model consists of four interdependent linear differential mass-balance equations for orthophosphate (OP) and particulate phosphorus (PP) concentrations (ML^{-3}) in each of the two zones. These are illustrated in Figure 9.2.

During the summer, the epilimnion is assumed to be well mixed. Discharges Q_j (L^3T^{-1}) into the lake at sites j have concentrations OP_j and PP_j of the two forms of phosphorus and are assumed to enter the epilimnion. These discharges are balanced by the lakes outflow Q_e which is also from the epilimnion. The time rate of change of the concentrations of the two forms of phosphorus in the lake's epilimnion of constant volume V_e are

$$V_e \frac{d\mathrm{OP}_e}{dt} = \sum_j Q_j \mathrm{OP}_j - Q_e \mathrm{OP}_e - K^s_{\mathrm{OP}} V_e \mathrm{OP}_e + \frac{vA}{D_T}(\mathrm{OP}_h - \mathrm{OP}_e) \tag{9.60}$$

$$\begin{bmatrix}\text{OP}_e\text{ mass}\\\text{rate of}\\\text{change}\end{bmatrix} = \begin{bmatrix}\text{mass}\\\text{inflow}\end{bmatrix} - \begin{bmatrix}\text{mass}\\\text{outflow}\end{bmatrix} - \begin{bmatrix}\text{net}\\\text{production}\\\text{of PP}_e\text{ from}\\\text{OP}_e\end{bmatrix} + \begin{bmatrix}\text{net vertical}\\\text{transfer}\\\text{across}\\\text{thermocline}\end{bmatrix}$$

$$V_e \frac{d\mathrm{PP}_e}{dt} = \sum_j Q_j \mathrm{PP}_j - Q_e \mathrm{PP}_e + K^s_{\mathrm{OP}} V_e \mathrm{OP}_e$$

$$\begin{bmatrix}\text{PP}_e\\\text{mass}\\\text{rate of}\\\text{change}\end{bmatrix} = \begin{bmatrix}\text{mass}\\\text{inflow}\end{bmatrix} - \begin{bmatrix}\text{mass}\\\text{outflow}\end{bmatrix} + \begin{bmatrix}\text{net}\\\text{production}\\\text{of PP}_e\text{ from}\\\text{OP}_e\end{bmatrix}$$

$$- g_e A(\mathrm{PP}_e) + \frac{vA}{D_T}(\mathrm{PP}_h - \mathrm{PP}_e) \tag{9.61}$$

$$- \begin{bmatrix}\text{settling}\end{bmatrix} + \begin{bmatrix}\text{net vertical}\\\text{transfer}\\\text{across}\\\text{thermocline}\end{bmatrix}$$

where

K^s_{OP} = rate coefficient for production of PP from OP in summer epilimnion (T^{-1})

v = vertical exchange coefficient that includes effects of molecular and turbulent diffusion, erosion of hypolimnion, and similar fluid processes on the transfer of materials across the thermocline (L^2T^{-1})

A = interfacial area of thermocline (L^2)

g_e = effective settling velocity for PP_e (LT^{-1})

D_T = thickness of thermocline (L)

For the well-mixed summer hypolimnion (denoted by the subscript h) of volume V_h,

$$V_h \frac{d\mathrm{OP}_h}{dt} = K^s_{\mathrm{PO}} V_h \mathrm{PP}_h + \frac{vA}{D_T}(\mathrm{OP}_e - \mathrm{OP}_h) \tag{9.62}$$

$$\begin{bmatrix}\text{OP}_h\text{ mass}\\\text{rate of}\\\text{change}\end{bmatrix} = \begin{bmatrix}\text{net conversion}\\\text{of PP}_h\\\text{to OP}_h\end{bmatrix} + \begin{bmatrix}\text{net vertical}\\\text{transfer across}\\\text{thermocline}\end{bmatrix}$$

$$
\begin{aligned}
V_h \frac{d\mathrm{PP}_h}{dt} \quad &= \quad g_e A(\mathrm{PP}_e) \quad - \quad g_h S(\mathrm{PP}_h) \\
\begin{bmatrix} \mathrm{PP}_h \text{ mass} \\ \text{rate of} \\ \text{change} \end{bmatrix} &= \begin{bmatrix} \text{net settling} \\ \text{of } \mathrm{PP}_e \text{ across} \\ \text{thermocline} \end{bmatrix} - \begin{bmatrix} \text{net settling} \\ \text{of } \mathrm{PP}_h \text{ on} \\ \text{sediment interface} \end{bmatrix} \\
&\quad - \quad K_{\mathrm{PO}}^s V_h \mathrm{PP}_h \quad + \quad \frac{vA}{D_T}(\mathrm{PP}_e - \mathrm{PP}_h) \\
&\quad - \begin{bmatrix} \text{net conversion} \\ \text{of } \mathrm{PP}_h \\ \text{to } \mathrm{PO}_h \end{bmatrix} + \begin{bmatrix} \text{net vertical} \\ \text{transfer} \\ \text{across thermocline} \end{bmatrix}
\end{aligned}
\tag{9.63}
$$

where

K_{PO}^s = rate coefficient for conversion of PP_h to OP_h in summer hypolimnion (T^{-1})

S = interfacial area of sediment-water interface (L^2)

g_h = effective settling velocity for PP_h (LT^{-1}) in hypolimnion

These equations describe the dynamics of the concentrations of the two phosphorus species in the lake during the summer stratified period.

The winter model assumes that during the fall overturn, the lake becomes completely mixed. Thus for the winter model, the initial phosphorus concentrations throughout the lake are

$$\mathrm{OP} = \frac{V_e \mathrm{OP}_e + V_h \mathrm{OP}_h}{V_e + V_h} \tag{9.64}$$

$$\mathrm{PP} = \frac{V_e \mathrm{PP}_e + V_h \mathrm{PP}_h}{V_e + V_h} \tag{9.65}$$

During the winter, the lake is assumed to remain well mixed but production of PP from OP is assumed to only take place in the upper euphotic zone of the lake where there is sufficient light for algal growth. Letting $V = V_e + V_h$ equal the total volume of the lake, and V_u equal the volume of the euphotic zone, the differential equations describing orthophosphate and particulate phosphorus concentrations in the lake are

$$
\begin{aligned}
V \frac{d\mathrm{OP}}{dt} \quad &= \sum_j Q_j \mathrm{OP}_j - Q(\mathrm{OP}) \\
\begin{bmatrix} \text{OP mass} \\ \text{rate of} \\ \text{change} \end{bmatrix} &= \begin{bmatrix} \text{mass} \\ \text{inflow} \end{bmatrix} - \begin{bmatrix} \text{mass} \\ \text{outflow} \end{bmatrix} \\
&\quad - \quad K_{\mathrm{OP}}^w V_u(\mathrm{OP}) \quad + \quad K_{\mathrm{PO}}^w V(\mathrm{PP}) \\
&\quad - \begin{bmatrix} \text{net production of} \\ \text{PP from OP} \\ \text{in euphotic zone} \end{bmatrix} + \begin{bmatrix} \text{net conversion of} \\ \text{PP to OP} \\ \text{throughout lake} \end{bmatrix}
\end{aligned}
\tag{9.66}
$$

$$
\begin{aligned}
V\frac{d\text{PP}}{dt} &= \sum_j Q_j\text{PP}_j - Q(\text{PP}) + K^w_{\text{OP}}V_u(\text{OP}) \\
\begin{bmatrix}\text{PP mass}\\ \text{rate of}\\ \text{change}\end{bmatrix} &= \begin{bmatrix}\text{mass}\\ \text{inflow}\end{bmatrix} - \begin{bmatrix}\text{mass}\\ \text{outflow}\end{bmatrix} + \begin{bmatrix}\text{net production of}\\ \text{PP from OP}\\ \text{in euphotic zone}\end{bmatrix} \\
&\quad - K^w_{\text{OP}}V(\text{PP}) - g_hS(\text{PP}) \\
&\quad - \begin{bmatrix}\text{net conversion of}\\ \text{PP to OP}\\ \text{throughout lake}\end{bmatrix} - \begin{bmatrix}\text{net settling of}\\ \text{PP on}\\ \text{lake bottom}\end{bmatrix}
\end{aligned}
\tag{9.67}
$$

At the beginning of spring stratification the concentrations OP_e, OP_h and PP_e, PP_h should be set equal to the concentrations OP and PP, respectively, representing the concentrations at the end of the winter period. Equations 9.60 to 9.67 are illustrated in Figure 9.12.

The summer phosphorus model assumes a knowledge of the volumes of the epilimnion and the hypolmnion during the summer season. Models have been developed to predict the temperature gradients in stratified impoundments and hence to provide estimates of the location of the thermocline at any time associated with different reservoir designs and operating policies. The simulation of deep thermally stratified lakes and reservoirs has been accomplished by the application of one-dimensional advection-diffusion (or

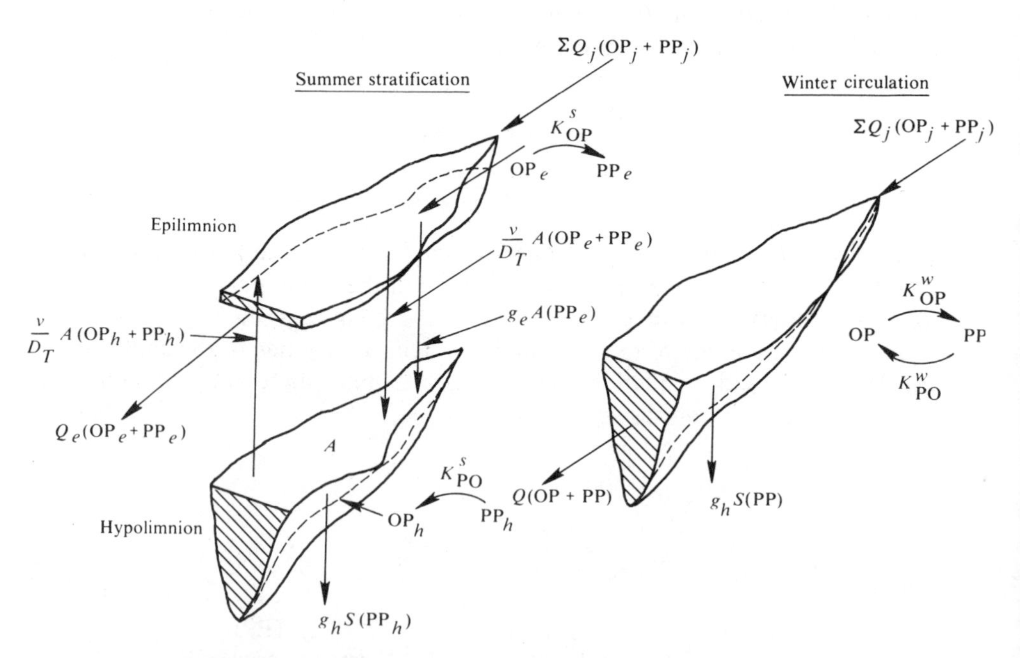

FIGURE 9.12. Schematic of Snodgrass–O'Melia model, Equations 9.60 to 9.67 for predicting orthophosphate (OP) and particulate phosphorus (PP) concentrations in lakes.

dispersion) and heat-energy conservation equations. These types of models are exemplified by those developed by Harleman and his students at MIT [20, 21], and by Orlob and his associates [60]. These models have been extended to include more dimensions and a variety of water quality and ecological variables and parameters [1, 5, 6, 7].

9.7 RELIABILITY OF RECEIVING WATER QUALITY SIMULATION MODELS

The water quality models outlined in the preceding sections of this chapter provide only an introduction to the variety of types of steady-state models used to simulate or predict water quality in water bodies. The relative reliabilities of various constituent concentration predictions provided by current water quality models are listed in Table 9.3. Table 9.3 also summarizes the major impacts of various constituents in natural river systems.

9.8 NONPOINT-SOURCE POLLUTION MODELS

So far this review of water quality prediction and simulation models was based on the assumption that the inputs of various potential pollutants were known. These inputs are not always easily estimated, especially if from nonpoint sources. This section reviews some models developed for estimating nonpoint-source discharges. Major nonpoint-source pollutants include carbonaceous and nitrogeneous BOD, pathogens, nutrients, sediment, pesticides, and metals.

Point-source discharges from identifiable outlets of wastewater collection and treatment systems can be measured and hence are more easily estimated and included in water quality simulation models. Steady-state models are often adequate for evaluation of water quality impacts from these discharges.

Nonpoint-source discharges are associated with stormwater runoff events and can be measured only indirectly by upstream-downstream mass flux differences. Without continuous monitoring and comprehensive description of instream sources and sinks, such measurements are seldom reliable. Given these difficulties, nonpoint-source loadings to receiving waters are often estimated using stormwater runoff models. Effects on receiving waters may subsequently be determined by nonsteady-state models.

Nonpoint-source models range from very simple mathematical relationships, used without calibration to estimate the approximate magnitude of nonpoint-source loadings, to complex simulation models for evaluating alter-

TABLE 9.3. Quality Impacts and Current State of Modeling

Models of	*Where Problems Arise*	TYPE OF WATER BODY		
		Streams	*Estuaries*	*Lakes and Reservoirs*
Transport, steady state		Good	Fair	—
Transport, dynamic		Fair	Poor	Fair
Conservative substances	Water supplies	Fair	Fair	Poor
Suspended solids	Water supplies, recreation	Poor	Poor	Poor
Bacteria, protozoans	Water supplies, recreation	Fair	Poor	Poor
BOD-dissolved oxygen	Ecosystem impacts	Good	Good	Fair
Simple chemicals and metals	Water supply, ecosystem impacts	Fair	Fair	Poor
Synthetic chemicals and complex metals	Water supply, ecosystem impacts	Poor	Poor	Unsatisfactory
Nutrients	Ecosystem impacts, recreation	Fair	Fair	Fair
Eutrophication (algae)	Water supply, recreation, ecosystems	Fair	Poor	Poor
Zooplankton and fish	Recreation	Unsatisfactory	Unsatisfactory	Unsatisfactory
Temperature	Ecosystem impacts	Fair	Fair	Fair
Virus	Water supply, recreation	Unsatisfactory	Unsatisfactory	Unsatisfactory
Floating substances	Recreation	Unsatisfactory	Unsatisfactory	Unsatisfactory
Color and turbidity	Recreation, water supply	Unsatisfactory	Unsatisfactory	Unsatisfactory

native nonpoint-source management practices. The latter often contain parameters which are not physically measurable and can be estimated only by calibration. The calibrated models must also be validated. Both steps require a comparison between model output and observed nonpoint-source loadings, and since these loadings are difficult to measure, comparisons cannot be precise.

Characteristics of nonpoint-source models can be seen in Figure 9.13. Within a source area or catchment such as an urban parking lot or farmer's field, precipitation comes in contact with waste material and other sources of pollution. The runoff leaves the catchment carrying some of this material. Some fraction of this runoff may reach a stream or other surface-water body. Within urban areas, pollutant transport to receiving water bodies is usually through well-defined systems of combined or separate storm sewers. Transport of agricultural pollution is poorly understood, but can occur via overland flow, ephemeral surface drainage channels, and tile drainage.

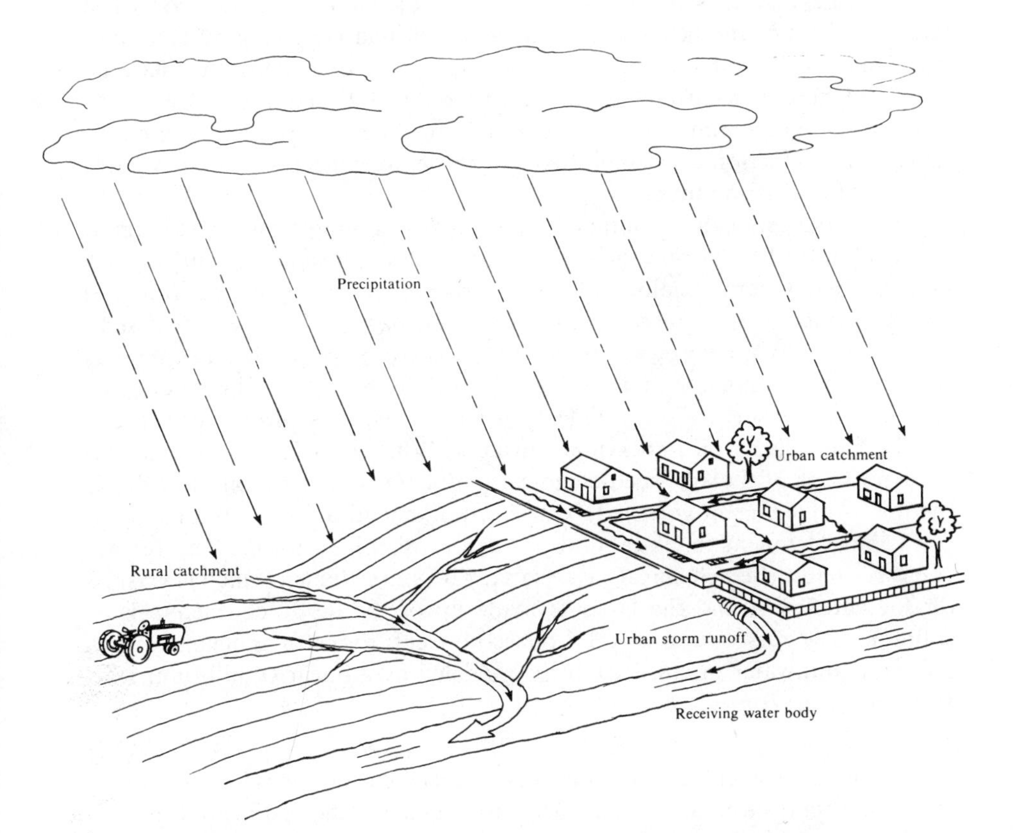

FIGURE 9.13. Components of nonpoint-source pollution.

In many ways, the state of the art in nonpoint-source modeling is primitive. Most progress has been made in the modeling of source area or catchment runoff quantity and quality. The remainder of this chapter is devoted to a discussion of some features of such models. Pollutant transport models are not included in this discussion. In the case of urban runoff, transport can be described by the basic hydraulic flow routing methods used for sewer design [19]. Transport models for agricultural runoff are generally limited to empirical "delivery ratios" which are multiplied by catchment pollutant runoff to determine the amounts of pollutants that actually reach a receiving water body [32].

9.8.1 Characteristics of Nonpoint-Source Runoff Models

A detailed discussion of any of the available nonpoint-source simulation models is beyond the scope of this chapter. These models typically consist of large numbers of mathematical relationships in relatively lengthy computer programs. Rather than focus on any one of these models, the material presented here concentrates on certain fundamental methods and assumptions that are common to nonpoint-source models. The discussion is essentially an introduction to nonpoint-source modeling concepts and not a guide to the use of existing models.

These concepts can be outlined by means of a simple model for runoff losses of adsorbed chemicals such as pesticides, phosphates, and certain metals. These potential pollutants are adsorbed by sediment that accumulates on the surface of the source area (field, street, parking lot, etc.). This sediment is typically a heterogeneous mixture, consisting of particulates such as soil, dust, and organic matter. On agricultural catchments, sediment consists of the erodible surface soil, while in urban areas, it is often the dirt and debris that collects on impervious surfaces. When rain falls on the source area, portions of the adsorbed chemicals will enter into solution (desorb). If the rain is of sufficient volume to cause runoff, chemicals will be transported from the catchment in both solid-phase and dissolved forms. The former consists of adsorbed chemicals associated with the sediment which is washed off the catchment and the latter is made up of the desorbed chemicals in solution. The dissolved and solid-phase concentrations of a chemical in a sediment and water mixture are often modeled by a linear equilibrium relationship of the form

$$s = kd \tag{9.68}$$

where d is the concentration of the chemical in solution (ppm or mg/ℓ) and s is the concentration of the chemical adsorbed on the sediment (ppm or mg/kg). The partition coefficient, k, can be measured in laboratory adsorption experiments and will in general be a function of both sediment and chemical properties.

Equation 9.68 is the basis for a simple model that can be used to estimate the losses of a chemical from a source area due to a rainfall-runoff event. Assume that at the time of the rainstorm the concentration of the chemical in the sediment on the catchment is s_0 (mg/kg) and the storm results in a runoff volume of Q (m^3) and a sediment loss in runoff of X (kg). The total chemical loss, P, can be partitioned into solid-phase (adsorbed) and dissolved constituents, S and D (mg), respectively:

$$P = S + D = sX + d(1000Q) \tag{9.69}$$

Assuming that a significant portion of the catchment's accumulated sediment is uniformly washed off by the storm, P can be estimated as s_0X. Equations 9.68 and 9.69 can then be solved for the concentrations s and d.

The solid-phase and dissolved chemical losses are

$$S = \left[\frac{ks_0X}{kX + 1000Q}\right]X \tag{9.70}$$

and

$$D = \left[\frac{s_0X}{kX + 1000Q}\right]1000Q \tag{9.71}$$

As an example, consider a storm on a 2-ha catchment which produces 1 cm of runoff ($Q = 200$ m^3) and a sediment loss of $X = 2000$ kg. Table 9.4 shows solid-phase and dissolved losses for chemicals with different adsorption characteristics. These losses are based on an initial chemical concentration of $s_0 = 50$ mg/kg. In any runoff event, the ratio of solid-phase to

TABLE 9.4. Solid-Phase and Dissolved Chemical Losses in Runoff for an Example of Runoff Event

Partition Coefficient k	LOSSES IN RUNOFF (mg)	
	Solid-Phase S	*Dissolved* D
1	990	99,010
5	4,760	95,240
25	20,000	80,000
100	50,000	50,000
500	83,330	16,670

dissolved losses, S/D, is

$$\frac{S}{D} = \frac{k}{1000} \cdot \frac{X}{Q} \tag{9.72}$$

where X/Q is the sediment concentration in runoff. Table 9.5 shows the

TABLE 9.5. Ratio of Solid-Phase to Dissolved Chemical Losses in Runoff

Partition Coefficient k	*S/D*, RATIO OF SOLID-PHASE TO DISSOLVED CHEMICAL LOSSES		
	X/Q: *1*	*10*	*50 (kg/m³)*
1	0.001	0.01	0.05
10	0.01	0.1	0.5
100	0.1	1.0	5.0
1000	1.0	10.0	50.0

ratio S/D for the ranges of partition coefficients and sediment concentrations typically encountered in runoff studies.

This simple model demonstrates the factors that must be included in a realistic nonpoint-source model. The starting points for any model are methods to predict catchment runoff volumes and sediment losses. The relative importance of these two predictions will depend on the nature of the pollutant being modeled. As indicated in Table 9.5, weakly adsorbed chemicals are lost mainly in the dissolved form and hence sediment losses are relatively unimportant. For strongly adsorbed pollutants, sediment losses must be estimated with care, but there is little incentive for accurate runoff volume predictions unless they affect estimated sediment losses. A key requirement for modeling of a pollutant is knowledge of its affinity for sediment, and hence a determination of whether its losses are associated with sediment losses, runoff volumes, or both. A final requirement is a method for estimating the amount of the chemical that is actually available for runoff loss ($s_0 X$ in the simple model). The degree of model complexity is often dictated by the accuracy that is desired for these estimates. Often, the pollutant quantities are based on empirical data extrapolated from a limited number of measurements in actual catchments. In the case of agricultural catchment models, pollutant availability is sometimes predicted with soil chemistry models.

9.8.2 Urban Models

The key parameter in urban runoff models is the fraction of a catchment's area which is impervious (e.g., street, parking lot, and roof surfaces). Runoff rates from such areas are much greater than those from pervious surfaces. Also, waste materials common to urban areas, such as dust and other atmospheric fallout, litter, animal wastes, leaves, and such, tend to collect in the drainage gutters of impervious surfaces. In general, the nonpoint-source load from an urban catchment is directly related to its impervious surface area.

Runoff Volumes. When rain R_t (cm) falls on an urban catchment on day t, a certain fraction, $1 - C$, will infiltrate the catchment's surface. The remaining portion of the rain, $C \cdot R_t$, will remain on the surface. However, not all of this water will leave the catchment as runoff, since portions will be detained in depression storages, including mud puddles, potholes, and small ponds. Denoting the available depression storage as DS_t (cm), then Q_t (cm), the total runoff from the storm on day t, is

$$Q_t = C \cdot R_t - DS_t, \text{ for } C \cdot R_t > DS_t \tag{9.73}$$

As shown in Figure 9.14, depression storage can be viewed as a catchment storage reservoir with capacity DS^*. Preceding storms may have filled a portion of the storage, so that DS_t, the available storage for the current day's storm, is less than the capacity, DS^*. In general, DS_t is equal to the previous day's available storage, DS_{t-1}, minus the water collected from runoff $C \cdot R_{t-1}$ plus loss L_{t-1} due to evaporation or seepage. Thus

$$DS_t = \text{Max}\,(DS_{t-1} - C \cdot R_{t-1} + L_{t-1};\, 0) \tag{9.74}$$

The losses L_t cannot exceed $DS^* - DS_t$ or some constant $L_t^{\max}$, however estimated. Alternatively one can assume that depressions dry up after 1 day if no rain occurs. In this case

$$DS_t = \begin{cases} \text{Max}\,(DS_{t-1} - C \cdot R_{t-1};\, 0), & R_{t-1} > 0 \\ DS^*, & R_{t-1} = 0 \end{cases} \tag{9.75}$$

Equations 9.73 and 9.75 constitute an urban runoff model based on daily rainfall data, runoff coefficient C, and depression storage capacity DS^*. These two parameters can be estimated from I, the fraction of the catchment's surface that is impervious. If c_i and c_p are the fractions of rain that run off the impervious and pervious surfaces, respectively, then the average runoff coefficient for the catchment is

$$C = c_i I + c_p(1 - I) \tag{9.76}$$

Similarly, if impervious and pervious depression storage capacities are d_i and d_p (cm), respectively, the total depression storage capacity can be estimated as

$$DS^* = d_i I + d_p(1 - I) \tag{9.77}$$

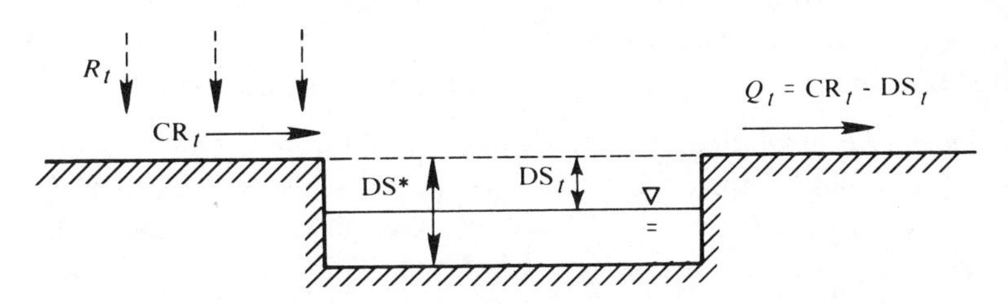

FIGURE 9.14. Conceptual model of urban runoff process.

The values $c_i = 0.90$, $c_p = 0.15$, $d_i = 0.15$, and $d_p = 0.60$ have been used in Level I of the U.S. Environmental Protection Agency's Stormwater Management Model (SWMM) [17].

The general form of the runoff equation 9.73 has been confirmed in field studies. For example, Miller and Viessman [33] examined rainfall and runoff data from four small urban watersheds in the United States and fitted regression equations of the form $Q_t = aR_t - b$ to the data. Their results are compared in Table 9.6 with predictions based on equation 9.73 with $DS_t = DS^*$ using the SWMM–Level I coefficients.

TABLE 9.6. Comparison of Different Predictive Equations for Urban Runoff

			RUNOFF PREDICTIONS (cm) FOR TWO STORMS			
			$R_t = 2$ cm		$R_t = 4$ cm	
Watershed	*Impervious Fraction, I*	*Regression Equation* [33]	*Regression*	*Eq. 9.73*	*Regression*	*Eq. 9.73*
Oakdale	0.55	$Q_t = 0.48R_t - 0.11$	0.85	0.77	1.81	1.89
Northwood	0.68	$Q_t = 0.56R_t - 0.26$	0.86	1.03	1.98	2.35
Gray Haven	0.52	$Q_t = 0.46R_t - 0.14$	0.78	0.71	1.70	1.79
Boneyard	0.44	$Q_t = 0.26R_t - 0.05$	0.47	0.56	0.99	1.52

Impervious fractions are best estimated directly from aerial photographs or land-use maps. When these are not available, regression equations based on population density are sometimes used. The equation used in SWMM–Level I can be approximated as

$$I = 0.069\, PD_d^{0.48} \tag{9.78}$$

In this equation I is the impervious fraction of the developed portion of the catchment and PD_d is the population density (persons/ha) in the developed area. For example, if a 100-ha catchment with a population of 1200 people is 60% developed then $PD_d = 1200/60 = 20$ persons per hectare and $I = 0.29$. Thus, within the 60 developed hectares, there are approximately 17.4 ha of impervious surfaces. In estimating runoff from the entire catchment the impervious surfaces within the undeveloped area are either assumed to be negligible or must be determined from street maps or aerial photographs.

Sediment Accumulation. Daily sediment buildup in urban areas is based on average values obtained from field surveys. These values are often reported as unit loadings per length of curb, since street gutters are major collecting surfaces for dirt and debris. Sediment buildup, Y, in kg/day is then given by

$$Y = yL \tag{9.79}$$

where L is the total curb length (km) within the catchment and y is the daily sediment accumulation per unit curb length (kg/km-day). Estimated values of y have been summarized as functions of land use and geographic regions [32]. Similar values, which are used in SWMM, are given in Table 9.7. Curb length L can be estimated as twice the total street length in the catchment.

TABLE 9.7. Data for Sediment and Pollutant Buildup on Urban Surfaces Used in SWMM

		POLLUTANT CONCENTRATION IN SEDIMENT, C (mg/kg)			
Land Use	*Sediment, y (kg/km-day)*	*Biochemical Oxygen Demand*	*Chemical Oxygen Demand*	*Nitrogen*	*Phosphate*
Single-family residential	10	5000	40,000	480	50
Multifamily residential	35	3600	40,000	610	50
Commercial	50	7700	39,000	410	70
Industrial	70	3000	40,000	430	30
Undeveloped or parkland	20	5000	20,000	50	10

Source: W. C. Huber et al., Stormwater Management Model User's Manual—Version II, Rept. EPA-670/2-75-017, U.S. Environmental Protection Agency, Cincinnati, Ohio, 1975.

Alternatively, a rough approximation can be obtained from a regression on gross population density, PD (persons/ha) [32].

$$L = A[0.31 - 0.27(0.93)^{PD}] \tag{9.80}$$

where A is the catchment area (ha).

Pollutant Availability. Urban nonpoint-source pollutants are usually assumed to be adsorbed to sediment, and hence to accumulate in proportion to the sediment buildup. If the concentration of pollutant i in the sediment is s_i (mg/kg), then the daily accumulation of pollutant i in kg/day is

$$Y_i = 10^{-6} s_i y L \tag{9.81}$$

Equation 9.81 is sometimes referred to as a pollutant *loading function*. Average concentrations for several pollutants are given in Table 9.7.

In applying equation 9.81 to an urban catchment, concentrations and sediment accumulations are weighted by land use. Thus, if a_j is the fraction of the catchment area in land use j and s_{ij} and y_j are the pollutant concentrations and daily sediment accumulation for this use, then

$$Y_i = 10^{-6} L \sum_j a_j s_{ij} y_j \tag{9.82}$$

Example of an Urban Nonpoint-Source Simulation Model. Equations 9.73 to 9.82 can be combined to form a daily simulation model which can be used to estimate pollutant losses in runoff and to evaluate the effects of street cleaning on the losses. If Y_{it} is the accumulation of pollutant i (kg) at the beginning of day t, then

$$Y_{i,t+1} = Y_{it} + Y_i - P_{it} - W_{it} \tag{9.83}$$

where Y_i is given by equation 9.82, and P_{it} and W_{it} are the pollutant removal by runoff and cleaning, respectively (kg), on day t. If streets are cleaned with cleaning efficiency e every h days after $t = 0$, then

$$W_{it} = \begin{cases} e(Y_{it} + Y_i), & t = h, 2h, 3h, \ldots \\ 0, & t \neq h, 2h, 3h, \ldots \end{cases} \tag{9.84}$$

When runoff occurs, a fraction w_t of the accumulated sediment and pollutant will be washed from the surface. The fraction depends on runoff volume, and can be modeled by a threshold runoff Q_0 (cm) above which complete washoff occurs and below which washoff is assumed linear. Thus

$$P_{it} = w_t(Y_{it} + Y_i) \tag{9.85}$$

and

$$w_t = \begin{cases} 1, & Q_t > Q_0 \\ Q_t/Q_0, & Q_t \leq Q_0 \end{cases} \tag{9.86}$$

where Q_t is the runoff (cm) on day t as given by equation 9.73. The complete model thus consists of equations 9.73, 9.75, 9.82, and 9.83 to 9.86. Hydrologic input data are daily rainfall amounts R_t and model output are daily runoff quantity Q_t and quality P_{it}.

The effects of the pollutant on receiving waters and the pollutant removals achieved by off-site control alternatives, such as storage and treatment, will depend on the distribution of solid-phase and dissolved pollutant constituents in the runoff. This partitioning is similar to that outlined in Section 9.8.1.

9.8.3 Agricultural Models

The major waste materials found in agricultural runoff are nutrients, pesticides, and sediment. Most agricultural nonpoint-source models are based on models of soil erosion and/or runoff which were developed for other purposes (e.g., maintenance of soil productivity, flood control, or drainage). Since eroded soil (as sediment) and runoff water are the transporting media for nutrient and other potential pollutants from cropped fields, these models form a logical basis for agricultural nonpoint-source models.

Runoff. A commonly used runoff model is the simple U.S. Soil Conservation Service's curve number equation for predicting the runoff Q_{ijt} (cm) from crop j on soil i during day t.

$$Q_{ijt} = \begin{cases} \dfrac{(R_t - 0.2S_{ijt})^2}{R_t + 0.8S_{ijt}} & \text{if } R_t \geq 0.2S_{ijt} \\ 0 & \text{otherwise} \end{cases} \tag{9.87}$$

In this equation, R_t is the rainfall during day t (cm) and S_{ijt} is a detention parameter (cm) which is a function of soil hydrologic group and condition, crop, and supporting practice as well as antecedent 5-day rainfall. Rainfall events are considered to be of 1-day duration. For example, a prolonged rainstorm lasting for 3 days would be treated as three separate events. The detention parameter S_{ijt} is calculated from "curve numbers" that have been determined and tabulated for most U.S. soils and a variety of crops [38, 49].

Equation 9.87 has the advantage of simplicity and wide applicability. However, it does not predict the runoff hydrograph. It is not clear how important this defect is, but two available runoff models, the Stanford watershed model [28] and the USDA hydrologic model [18], can be used for hydrograph prediction. These models are parametric; that is, they contain parameters that can only be determined using observed runoff for calibration.

Sediment Loss. Most models used to predict erosion (and by inference sediment losses) from croplands are based on the universal soil loss equation [63]. This equation is of the form

$$X_{ij} = E(K_i LS_i CF_j P_i) \tag{9.88}$$

where X_{ij} is the average annual soil loss from field i with crop j (tons/ha), E the rainfall erosivity index (EK_i has units of tons/ha), K_i a soil erodibility factor, LS_i a topographic factor which is determined from field slope and length of slope, CF_j a cover factor for crop j, and P_i a supporting practice (contouring, terracing, etc.) factor. The last three factors are dimensionless. Since the parameters for equation 9.88 have been determined and are tabulated in several readily available sources [46, 63], estimation of average annual erosion is straightforward.

Nonpoint-source loadings are dynamic stormwater events, so that average annual soil loss estimates are of limited value in determining effects of loadings on receiving waters. For example, the average annual sediment input to a river might be low, but infrequent peak loads could still decimate aquatic life. The universal soil loss equation has been modified to provide event-based predictions. One such modification is that proposed by Williams [61] for predicting X_{ijt}, the soil loss (tons/ha) on day t associated with runoff quantity V_{ijt} (m^3) and peak runoff q_{ijt} (m^3/s).

$$X_{ijt} = \frac{11.8}{A_{ij}}(V_{ijt}q_{ijt})^{0.56}LS_iCF_{jt}P_i \tag{9.89}$$

In equation 9.89, A_{ij} is the area of the catchment (ha) and CF_{it} is the cover factor for day t.

Pollutants in Agricultural Runoff. Most nonpoint-source pollutants will have both dissolved and adsorbed forms, and hence will leave croplands both in solution with the runoff and adsorbed to sediment particles.

The simplest models for estimating catchment or "edge of field" pollutant losses multiply runoff and erosion quantities by average or typical pollutant concentrations. For example, dissolved losses in runoff are

$$L_{ijt} = 0.1C_j^rQ_{ijt}A_{ij} \tag{9.90}$$

where L_{ijt} is pollutant loading from crop j on soil (or field) i during day t (kg), C_j^r the pollutant concentration in runoff (mg/ℓ) from crop j, and Q_{ijt} and A_{ij} were defined previously. Typical concentrations of dissolved nitrogen and phosphorus in agricultural runoff are given in Table 9.8.

TABLE 9.8. Average Concentrations of Dissolved Nitrogen and Phosphorus in Agricultural Runoff

CROP	DISSOLVED NITROGEN (mg/ℓ) RUNOFF FROM:		DISSOLVED PHOSPHORUS (mg/ℓ) IN RUNOFF FROM:	
	Rain	*Snowmelt*	*Rain*	*Snowmelt*
Fallow	2	4	0.1	0.2
Pasture	2	3	0.3	0.3
Hay	1	3	0.2	0.2
Small grains	3	1	0.3	0.3
Corn	2	3	0.2	0.3

Source: D. A. Haith and L. J. Tubbs, Modelling Nutrient Export in Rainfall and Snowmelt Runoff, in R. C. Loehr et al., *Best Management Practices for Agriculture and Silviculture*, Ann Arbor Science Press, Ann Arbor, Mich., 1979.

Pollutant losses in eroded soil can be determined from

$$L_{ijt} = 10^{-3}e_iC_i^sX_{ijt}A_{ij} \tag{9.91}$$

where C_i^s is the pollutant concentration (mg/kg) in the soil and L_{ijt}, X_{ijt}, and A_{ij} were defined previously. The term e_i is a dimensionless enrichment ratio which is generally greater than 1, reflecting the fact that pollutant concentrations in eroded soil usually exceed those of the *in situ* soil. Typical values of enrichment ratios are from 2 to 4 for nitrogen and 1.5 to 2.5 for phosphorus [32]. In estimating nutrient losses, nitrogen and phosphorus concentrations (C_i^s in equation 9.91) can sometimes be determined from soil surveys.

Pesticide and heavy-metals concentrations must generally be based on localized soil samples.

Equations 9.87 and 9.89 through 9.91 can be combined into a simple simulation model for estimating losses of nonpoint-source pollutants from an agricultural catchment area. The model uses a daily time increment for computation and requires weather, soil, and crop practice parameters as inputs. The principal drawback to the model is the assumption of uniform concentrations which are unique functions of crop (C_j^r) or soil type (C_i^s). A more realistic simulation model would be based on pollutant and moisture balances in the soil. This type of model has been developed to estimate nutrient losses from cropland [56]. In addition, simulation models based on the Stanford and USDA runoff models are available [8, 12].

9.9 CONCLUDING REMARKS

This chapter has been an introduction to water quality simulation models of rivers, estuaries, and surface-water impoundments, and their associated nonpoint-source waste loadings. The modeling procedures or approaches have ranged from fairly simple single dimension steady-state approximations to ones that are considerably more involved. Just as with every modeling approach discussed in this book, the best water quality simulation model is the simplest one that will adequately predict the water quality impacts within a particular water body associated with a particular water quality management policy.

The simulation models outlined in this chapter, or their more complex extensions, are relatively crude approximations of the interactions among various constituents that occur in water bodies. Yet in spite of their current limitations, they are the only reasonable means available for predicting surface water quality. The state of the art in water quality modeling and the understanding of the physical, chemical, and biological processes that affect water quality are improving rapidly. Readers interested in pursuing this area of modeling activity are encouraged to study in greater detail alternative water quality modeling approaches and solution procedures, many of which are cited in the references.

Having finished this introduction to water quality simulation models, it is appropriate to focus next on the development and use of water quality optimization models. These models can assist in defining and evaluating the economic and environmental impacts of alternative water quality management policies. Such models can serve to screen out many of the less desirable policy alternatives prior to a more detailed analysis using simulation models such as those introduced in this chapter.

APPENDIX 9A

DERIVATION OF GOVERNING EQUATIONS

Most of Section 9.5 pertains to various special cases of the general partial differential equation for predicting the concentrations of water quality constituents in a river or estuary. This appendix presents a simple derivation of that general differential equation and illustrates the processes which give rise to the various terms in the equation. The general partial differential equation for the concentration of a substance over time and space $C(X, t)$ can be obtained by considering the mass balance for the substance in a small slice of the river shown in Figure 9A.1. The following processes are considered:

1. Advective transport of the constituent due to river flow Q (L^3/T).
2. Dispersion of the constituent due to turbulence over one or more tidal cycles.
3. Loss of the constituent from the water column as a result of decay or settling.
4. Distributed sources of the constituent along the sides or bottom of the stream resulting from release from benthal deposits, seepage or runoff into the stream.
5. Sources or sinks of the constituent within the water column such as the oxidation of organic carbon or nitrogen compounds.

The river is assumed to be well mixed in the vertical and horizontal directions so that only longitudinal and temporal variations in concentrations are of interest. Consider the change in mass of the constituent within the slice of the river illustrated in Figure 9A.1 in a short period of time Δt. The center of the slice is at x with its upper and lower boundaries at $x - \delta/2$ and $x + \delta/2$. Thus the volume of the slice is essentially $\delta A(x)$ where δ is the length of the slice and $A(x)$ is the cross-sectional area (L^2) of the river at x. The total change in mass of the constituent within the time interval t to $t + \Delta t$ will be

$$\delta A(x)[C(x, t + \Delta t) - C(x, t)] \tag{9A.1}$$

with units $(L)(L^2)[(ML^{-3})] = (M)$. This change in mass must equal the net change in mass within the slice due to transport of the constituent in and out of the slice and due to losses and additions of mass to the water within the slice. The five processes listed above are discussed in turn.

Advective transport. The river flow $Q(x)$ often depends on x, the point at which the flow is measured. The flow in the river may vary with distance due to seepage of water into or out of the river channel, evaporation, distributed withdrawals from or distributed runoff into the channel. The flux of

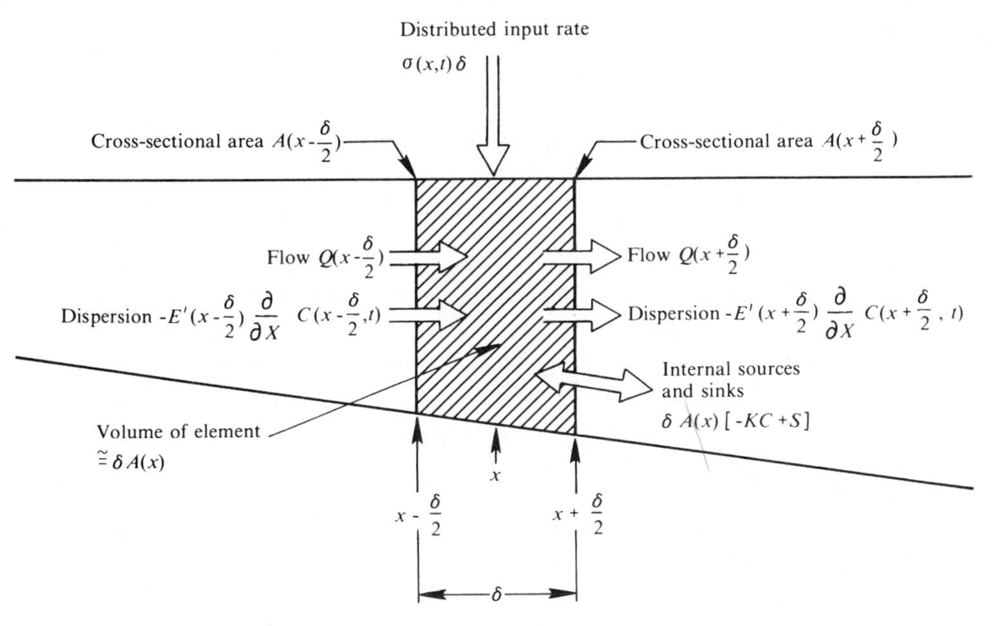

FIGURE 9A.1. Illustration of a small slice of river used to derive partial differential equation for a constituent's concentration.

constituent mass into the slice due to advective transport is $Q(x - \frac{\delta}{2})\ C(x - \frac{\delta}{2},\ t)$ while the flux out of the slice is $Q(x + \frac{\delta}{2})\ C(x + \frac{\delta}{2},\ t)$. Thus the net addition of mass to the slice in a short time Δt due to advective transport is

$$\left[Q\left(x - \frac{\delta}{2}\right)C\left(x - \frac{\delta}{2},\ t\right) - Q\left(x + \frac{\delta}{2}\right)C\left(x + \frac{\delta}{2},\ t\right)\right]\Delta t \qquad (9A.2)$$

with units $[(L^3/T)(M/L^3)](T) = (M)$.

Dispersion. The rate at which the constituent disperses into the slice of the river depends on the gradient or slope of the concentration profile at the volume's upstream and downstream boundaries across which the dispersion occurs. At the upstream boundary, the mass flux into the slice due to dispersion is

$$-E\left(x - \frac{\delta}{2}\right)A\left(x - \frac{\delta}{2}\right)\frac{\partial C\left(x - \frac{\delta}{2},\ t\right)}{\partial X} = -E'\left(x - \frac{\delta}{2}\right)\frac{\partial C\left(x - \frac{\delta}{2},\ t\right)}{\partial X}$$

where $E(x)$ is the dispersion coefficient on a unit area basis (L^2T^{-1}) and $E'(x)$ is a bulk dispersion or exchange coefficient (L^4T^{-1}) equal to $E(x)A(x)$. The minus sign appears because dispersion occurs in the direction of decreasing concentration. The total change in mass within the section due to dispersion across the slice's upper and lower cross-sectional areas in a short period of time Δt is

$$\left[-E'\left(x - \frac{\delta}{2}\right)\frac{\partial}{\partial X}C\left(x - \frac{\delta}{2}, t\right) + E'\left(x - \frac{\delta}{2}\right)\frac{\partial}{\partial X}C\left(x + \frac{\delta}{2}, t\right)\right]\Delta t \tag{9A.3}$$

with units $[(L^4T^{-1})(L^{-1})(ML^{-3})](T) = (M)$.

Decay. The total loss of the constituent mass due to decay, such as reaeration for DOD or decay of BOD, or due to settling, is generally assumed to be proportional to the mass of the constituent within the box, which for small δ is essentially $\delta A(x)C(x, t)$. Thus for decay rate constant K, the loss of mass in a short time Δt is

$$K[\delta A(x)]C(x, t)\Delta t \tag{9A.4}$$

with units $(T^{-1})[(L)(L^2)](ML^{-3})(T) = (M)$.

Distributed Sources. Often direct runoff into a stream or benthal deposits act like a source per unit length $\sigma(x, t)$ of the constituent $(ML^{-1}T^{-1})$. Thus the increase in mass of the constituent within the river section during a time interval Δt is

$$\sigma(x, t)\delta\Delta t \tag{9A.5}$$

with units $(ML^{-1}T^{-1})(L)(T) = (M)$. There are other approachs to describe distributed nonpoint sources. Let C_r be the constituent concentration in runoff into the stream, where $\partial Q/\partial X$ is the rate of addition of runoff to the stream per unit channel length. Then the constituent mass added to the small slice of the stream in a period of time Δt is

$$C_r \frac{\partial Q}{\partial X}\,\delta\,\Delta t \tag{9A.6}$$

with units $(ML^{-3})(L^3T^{-1})(L^{-1})(L)(T) = (M)$. One may use either equation 9A.5 or 9A.6 to describe distributed constituent sources.

Other Sources and Sinks. Finally, various biochemical and physical processes can act as sources or sinks $S(x, t)$ that are expressed on a unit volume basis $(ML^{-3}T^{-1})$. Thus the associated change in the mass within the segment in a time interval Δt equals the volume of the slice times S times Δt

$$[\delta A(x)]S(x, t)\Delta t \tag{9A.7}$$

with units $[(L)(L^2)](ML^{-3}T^{-1})(T) = (M)$.

Synthesis. One can equate the total change in the constituent's mass in a time interval Δt, expression 9A.1, with the individual transport gains and losses, terms 9A.2 and 9A.3, along with any sources and sinks, 9A.4 to 9A.7, to obtain

$$\delta A(x)[C(x, t + \Delta t) - C(x, t)]$$

$$= \left[-E'\left(x - \frac{\delta}{2}\right) \frac{\partial}{\partial X} C\left(x - \frac{\delta}{2}, t\right) + E'\left(x + \frac{\delta}{2}, t\right) \frac{\partial}{\partial X} C\left(x + \frac{\delta}{2}, t\right)\right] \Delta t$$
$$+ \left[Q\left(x - \frac{\delta}{2}\right) C\left(x - \frac{\delta}{2}, t\right) - Q\left(x + \frac{\delta}{2}\right) C\left(x + \frac{\delta}{2}, t\right)\right] \Delta t$$
$$- K[\delta A(x)] C(x, t) \Delta t + \sigma(x, t) \delta \Delta t + C_r \frac{\partial Q}{\partial X} \delta \Delta t + [\delta A(x)] S(x, t) \Delta t \tag{9A.8}$$

This can be converted from a difference equation to a differential equation. Dividing both sides by Δt and δ one obtains

$$A(x)\left[\frac{C(x, t + \Delta t) - C(x, t)}{\Delta t}\right]$$
$$= \left[\frac{E'\left(x + \frac{\delta}{2}, t\right) \frac{\partial}{\partial X} C\left(x + \frac{\delta}{2}, t\right) - E'\left(x - \frac{\delta}{2}, t\right) \frac{\partial}{\partial X} C\left(x - \frac{\delta}{2}, t\right)}{\delta}\right]$$
$$- \left[\frac{Q\left(x + \frac{\delta}{2}\right) C\left(x + \frac{\delta}{2}, t\right) - Q\left(x - \frac{\delta}{2}\right) C\left(x - \frac{\delta}{2}, t\right)}{\delta}\right]$$
$$- K A(x) C(x, t) + \sigma(x, t) + C_r \frac{\partial Q}{\partial X} + A(x) S(x, t) \tag{9A.9}$$

Letting the time interval Δt and the longitudinal length δ of the slice approach zero, the quantities in brackets become partial derivatives:

$$A(X) \frac{\partial}{\partial t} C(X, t) = \frac{\partial}{\partial X}\left[E'(X) \frac{\partial}{\partial X} C(X, t)\right] - \frac{\partial}{\partial X}[Q(X) C(X, t)]$$
$$- K A(X) C(X, t) + \sigma(X, t) + C_r \frac{\partial Q}{\partial X} + A(X) S(X, t) \tag{9A.10}$$

Dividing equation 9A.10 by $A(X)$ and substituting $E(X)A(X)$ for the bulk dispersion coefficient $E'(X)$, one obtains the more common expression

$$\frac{\partial}{\partial t} C(X, t) = \frac{1}{A} \frac{\partial}{\partial X}\left[EA \frac{\partial}{\partial X} C\right] - \frac{1}{A} \frac{\partial}{\partial X}[QC]$$
$$- KC + \frac{\sigma}{A} + \frac{C_r}{A} \frac{\partial Q}{\partial X} + S(X, t) \tag{9A.11}$$

Here the units on both sides of the equation are time rates of change of concentration ($ML^{-3}T^{-1}$). Note that the first term involves the derivative of the product $EA\partial C/\partial X$. The spatial variation of the dispersion coefficient and cross-sectional area affect the rate of dispersion as do variations in the concentration gradient $\partial C/\partial X$. It is also interesting to note that because σ is the distributed source rate per river length, it is divided by the cross-sectional area A in equation 9A.11 to obtain the source rate on a unit volume basis.

This completes the basic derivation of the general partial differential equation for a constituent's concentration in a river or estuary. A number of solutions for special cases of this equation are given in the text. The next section reviews the derivation of the steady-state solution of equation 9A.11 for a river reach with constant cross-sectional area A, flow Q and dispersion coefficient E.

APPENDIX 9B

SOLUTION OF STEADY-STATE GOVERNING EQUATIONS

The steady-state or time invariant solution to equation 9A.11 corresponds to the situation where $\partial C/\partial t = 0$. In the special case when the model parameters E, A, Q, and K are independent of X, analytic expressions for the constituents concentration can be derived. Many of these are given in Sections 9.5.1 and 9.5.2. Omitting the term $C_r \, \partial Q/\partial X$ for simplicity, the steady-state constant-coefficient differential equation for a water quality constituent's concentration becomes

$$0 = E\frac{d^2C}{dX^2} - U\frac{dC}{dX} - KC + \frac{\sigma}{A} + S \tag{9B.1}$$

where the velocity U can replace the flow rate Q because $Q = UA$. To derive the general solution to this equation, consider first the *homogeneous differential equation* obtained by omitting the terms $\sigma/A + S$, that do not depend on C.

$$0 = E\frac{d^2C}{dX^2} - U\frac{dC}{dX} - KC \tag{9B.2}$$

A solution to this homogeneous differential equation is of the form

$$C(X) = C_\lambda \exp(\lambda X) \tag{9B.3}$$

for some constants λ and C_λ. Substituting this expression for $C(X)$ into equation 9B.2 one finds that

$$C_\lambda(E\lambda^2 - Q\lambda - K)\exp(\lambda X) = 0 \tag{9B.4}$$

This equation is satisfied for all X and any C_λ if λ satisfies

$$E\lambda^2 - U\lambda - K = 0 \tag{9B.5}$$

or

$$\lambda = \frac{U \pm \sqrt{U^2 + 4EK}}{2E} = \frac{U}{2E}(1 \pm m) \tag{9B.6}$$

where

$$m = \sqrt{1 + \frac{4EK}{U^2}} \geq 1$$

Note that the dimensionless parameter m is never less than one so that $1 + m > 0$ and $1 - m \leq 0$ for any positive values of E, K and U.

Because the homogeneous differential equation 9B.2 is linear in the function $C(X)$, the sum of any two solutions is also a solution. Thus the two solutions found in equations 9B.4 and 9B.5 can be added to obtain the general solution to the homogeneous differential equation.

$$C(X) = C_1 \exp\left[\frac{UX}{2E}(1 + m)\right] + C_2 \exp\left[\frac{UX}{2E}(1 - m)\right] \qquad (9B.7)$$

where C_1 and C_2 are arbitrary constants. If one requires a solution such that $C(X)$ approaches zero as X becomes arbitrarily large ($X \longrightarrow +\infty$), then C_1 must be zero. Likewise, if $C(X)$ approaches zero as X becomes very negative ($X \longrightarrow -\infty$), C_2 must be zero. These are possible *boundary conditions.*

In Section 9.5.2, equation 9B.2 was used to describe the concentration of a water quality constituent in a river with a single point discharge W_0 (MT^{-1}) at point $X = 0$. Equation 9B.7 then gives the general equation for the constituent's concentration in the river above ($X < 0$) and below ($X > 0$) the discharge. However, these two solutions must join at $X = 0$ so that the constituent concentration is continuous at that point; infinite concentration gradients are impossible in a system with dispersion. In addition, the total mass flux away from $X = 0$ must equal W_0. Let the solution for $X > 0$ be

$$C(X) = C_1^+ \exp\left[\frac{UX}{2E}(1 + m)\right] + C_2^+ \exp\left[\frac{UX}{2E}(1 - m)\right] \qquad (9B.8a)$$

while for $X < 0$

$$C(X) = C_1^- \exp\left[\frac{UX}{2E}(1 + m)\right] + C_2^- \exp\left[\frac{UX}{2E}(1 - m)\right] \qquad (9B.8b)$$

In Section 9.5.2 it was assumed that $C(X)$ approached zero as $X \longrightarrow +\infty$. This means that C_1^+ must be zero. Requiring that $C(X)$ approach zero as $X \longrightarrow -\infty$ means that C_2^- must also be zero. The condition that the constituent concentration is a continuous function of X at $X = 0$ implies that

$$C(0) = C_1^- = C_2^+$$

The condition which determines the actual values of C_2^+ and C_1^- is that the total constituent flux away from the point $X = 0$ must equal the mass input rate W_0. Letting $X = 0^+$ and $X = 0^-$ correspond to points just downstream and upstream of the discharge at $X = 0$, the conservation of the constituents flux at $X = 0$ requires that

$$\left[-EA\frac{dC}{dX} + QC\right]_{X=0^+} - \left[-EA\frac{dC}{dX} + QC\right]_{X=0^-} = W_0 \qquad (9B.9)$$

Substituting the appropriate expressions for $C(X)$ for $X \geq 0$ and $X \leq 0$ and evaluating the functions at $X = 0$ yields

$$C_2^+\left\{-EA\left[\frac{U}{2E}(1-m)\right]+Q\right\}-C_1^-\left\{-EA\left[\frac{U}{2E}(1+m)\right]+Q\right\}=W_0$$

or

$$W_0=C(0)\left\{\frac{AU}{2}[-(1-m)+(1+m)]+Q-Q\right\}=C(0)(Qm)$$

so that

$$C(0)=C_2^+=C_1^-=W_0/(Qm)$$

Thus the final solution $C(X)$ which approaches zero for large $|X|$ and which satisfies the boundary conditions at $X=0$ is

$$C(X)=\begin{cases}\dfrac{W_0}{Qm}\exp\left[\dfrac{UX}{2E}(1+m)\right] & \text{for } X<0\\ \dfrac{W_0}{Qm}\exp\left[\dfrac{UX}{2E}(1-m)\right] & \text{for } X\geq 0\end{cases} \qquad (9B.10)$$

which is equation 9.3. Other solutions result for other boundary conditions.

If σ or S are not zero, the general solution to equation 9B.1 is obtained by adding to equation 9B.7 additional terms to account for the impact of these constituent sources on the constituent's concentration. In the case where the functions $\sigma(X)$ and $S(X)$ are constant and do not depend on X, a simple solution to equation 9B.1 which does not vary with X is

$$C(X)=\frac{1}{K}\left[\frac{\sigma}{A}+S\right] \qquad (9B.11)$$

The general solution to equation 9B.1 is then the sum of this *particular solution* and the general solution, equation 9B.7, to the homogenious differential equation.

Another situation that arises in Section 9.5.2 is where $C(X)$ corresponds to the dissolved oxygen deficit concentration $D(X)$ and S corresponds to the oxygen used in the oxidation of organic compounds. The resulting equation is

$$0=E\frac{d^2D}{dX^2}-Q\frac{dD}{dX}-K_aD+\frac{K^CW_0}{Qm^C}\exp\left[-\frac{UX}{2E}(1\pm m^C)\right] \qquad (9B.12)$$

with $1+m$ for $X<0$ and $1-m$ for $X>0$. A particular solution that accounts for the impact of the last term in equation 9B.12 on the dissolved oxygen deficit concentration D or $D(X)$, is of the form

$$D(X)=D_p\exp\left[\frac{UX}{2E}(1\pm m^C)\right] \qquad (9B.13)$$

for some constant D_p and with $(1+m^C)$ for $X<0$ and $(1-m^C)$ for $X>0$. Substituting equation 9B.13 into equation 9B.12 yields

$$D_p\left\{E\left[\frac{U}{2E}(1\pm m^C)\right]^2-Q\left[\frac{U}{2E}(1\pm m^C)\right]-K_a\right\}+\frac{K^CW_0}{Qm^C}=0 \qquad (9B.14)$$

Note that by the derivation of m^C

$$E\left[\frac{U}{2E}(1 \pm m^C)\right]^2 - Q\left[\frac{U}{2E}(1 \pm m^C)\right] - K^C = 0 \tag{9B.15}$$

Hence equation 9B.14 reduces to

$$D_p(K^C - K_a) = -\frac{K^C W_0}{Qm^C}$$

so that

$$D_p = \frac{K^C}{K_a - K^C}\left(\frac{W_0}{Qm^C}\right) \tag{9B.16}$$

Now the general solution to equation 9B.12 is

$$D(X) = D_p \exp\left[\frac{UX}{2E}(1 + m^C)\right] + D_1^- \exp\left[\frac{UX}{2E}(1 + m^a)\right] + D_2^- \exp\left[\frac{UX}{2E}(1 - m^a)\right] \qquad \text{for } X < 0 \tag{9B.17a}$$

and

$$D(X) = D_p \exp\left[\frac{UX}{2E}(1 - m^C)\right] + D_1^+ \exp\left[\frac{UX}{2E}(1 + m^a)\right] + D_2^+ \exp\left[\frac{UX}{2E}(1 - m^a)\right] \qquad \text{for } X \geq 0 \tag{9B.17b}$$

The constants D_1^+, D_2^+, D_1^- and D_2^- are determined by a problem's boundary conditions. If $D(X)$ approaches zero for large $|X|$, then $D_2^- = 0$ and $D_1^+ = 0$. Then requiring continuity of the concentration at $X = 0$ means that $D_2^+ = D_1^-$.

Finally, to model only the impact of the BOD discharge at $X = 0$, the dissolved oxygen deficit flux at $X = 0$ must be continuous. That is, here it is assumed that the discharge at $X = 0$ contributes to the DO deficit concentration only by the oxidation of the discharged wastes. For the DOD flux to be continuous at $X = 0$, one must have

$$\left[EA\frac{d}{dx}D(X) - QD(X)\right]_{X=0^+} = \left[EA\frac{d}{dx}D(X) - QD(X)\right]_{X=0^-} \tag{9B.18}$$

Because $D(X)$ is continuous at $X = 0$ and the flow Q, dispersion E, and area A, are assumed to be constant, equation 9B.18 reduces to

$$\left.\frac{d}{dx}D(X)\right|_{X=0^+} = \left.\frac{d}{dx}D(X)\right|_{X=0^-} \tag{9B.19}$$

or

$$D_p\frac{U}{2E}(1 - m^C) + D_2^+\frac{U}{2E}(1 - m^a) = D_p\frac{U}{2E}(1 + m^C) + D_1^-\frac{U}{2E}(1 + m^a)$$

Denoting $D_2^- = D_1^+$ as D_0, one obtains

$$D_0 = -\frac{m^C}{m^a}D_p \tag{9B.20}$$

Putting all these results together, the solution to 9B.12 with the boundary conditions that $D(X)$ approach zero for large $|X|$ is

$$D(X) = \left(\frac{K^C}{K_a - K^C}\right)\left(\frac{W_0}{Q}\right)\left\{\frac{1}{m^C}\exp\left[\frac{UX}{2E}(1 \pm m^C)\right] - \frac{1}{m^a}\exp\left[\frac{UX}{2E}(1 \pm m^a)\right]\right\} \tag{9B.21}$$

which is equations 9.12 and 9.13b in the text.

If $S(x)$ is a very complicated expression, then a simple analytic solution to equation 9B.1 may not exist and numerical procedures become advantageous.

EXERCISES

9-1. The commons version of the Streeter–Phelps equations for predicting biochemical oxygen demand BOD and dissolved oxygen deficit D concentrations are based on the following two differential equations:

(a) $\dfrac{d(\text{BOD})}{d\tau} = -K_d(\text{BOD})$

(b) $\dfrac{dD}{d\tau} = K_d(\text{BOD}) - K_a D$

where K_d is the deoxygenation rate constant (T^{-1}), K_a is the reaeration-rate constant (T^{-1}), and τ is the time of flow along a uniform reach of stream in which dispersion is not significant. Show the relationship of the integrated forms of (a) and (b) to equations 9.5, 9.12, and 9.13a.

9-2. Based on the integrated differential equations in Exercise 9-1:

(a) Derive the equation for the distance X_c downstream from a single point source of BOD that for a given streamflow will have the lowest dissolved oxygen concentration.

(b) Determine the relative sensitivity of the deoxygenation-rate constant K_d and the reaeration-rate constant K_a on the critical distance X_c and on the critical deficit D_c. For initial conditions, assume that the reach has a velocity of 2 m/s (172.8 km/day), a K_d of 0.30 per day, and a K_a of 0.4 per day. Assume that the DO saturation concentration is 8 mg/ℓ, the initial deficit is 1.0 mg/ℓ, and the BOD concentration at the beginning of the reach (including that discharged into the reach at that point) is 15 mg/ℓ.

9-3. (a) Compute the coefficients b_{ij}, d_{ij}, and b^a_{ij} defined by equations 9.8 and 9.13 for the mid- and endpoints of a three-reach stream, assuming the following data. Let i and j take on the values of

1, 2, . . . , 7, representing the beginning, mid-, and endpoints of the three reaches, in which the flow is from reach 1 to reach 3. Additions or withdrawals of flows are made at the endpoints of the reach. Assume $E = 0$ in all reaches.

Parameter	*Reach 1*	*Reach 2*	*Reach 3*
K_d (days^{-1})	0.3	0.27	0.25
K_a (days^{-1})	0.4	0.45	0.65
Total time of flow (days)	0.8	2.00	1.20
Streamflow (10^6 m^3/day)	2.0	2.3	2.1
Velocity (km/day)	3.5	3.8	3.6

(b) Compute the BOD and DO profile associated with the following conditions, based on these values of b_{ij}, d_{ij}, and b^a_{ij} found for part (a) above.

	VALUES		
Parameter	*Reach 1*	*Reach 2*	*Reach 3*
DO saturation concentration (mg/ℓ)	8.0	7.5	7.0
Initial DO deficit (mg/ℓ)	1.0	—	—
Initial BOD concentration (mg/ℓ)	3.0	—	—
BOD discharge into beginning of reach (kg/day)	15,700	7800	0
DO deficit of waste discharge at beginning of reach (kg/day)	120	140	0

9-4. To account for settling of BOD, in proportion to the BOD concentration, and for a constant rate of BOD addition R due to runoff and scour, and oxygen production ($A > 0$) or reduction ($A < 0$) due to plants and benthal deposits, the following differential equations have been proposed:

$$\frac{d\text{BOD}}{d\tau} = -(K_d + K_s)\,\text{BOD} + R \qquad (1)$$

$$\frac{dD}{d\tau} = K_d\text{BOD} - K_aD - A \qquad (2)$$

where K_s is a settling-rate constant (T^{-1}) and τ is the time of flow. Integrating these two equations results in the following deficit equation:

$$D_\tau = \frac{K_d}{K_a - (K_d + K_s)}\left[\left(\text{BOD}_0 - \frac{R}{K_d + K_s}\right)\{\exp\,[-(K_d + K_s)\tau] - \exp\,(-K_a\tau)\}\right] + \frac{K_d}{K_a}\left\{\left(\frac{R}{K_d + K_s} - \frac{A}{K_d}\right)[1 - \exp\,(-K_a\tau)]\right\} + D_0\exp\,(-K_a\tau) \qquad (3)$$

where BOD_0 and D_0 are the BOD and DO deficit concentrations at $\tau = 0$.

(a) Compare this equation with that found in Exercise 9-1 if K_s, R, and A are 0 and with equation 9.35.

(b) Integrate equation (1) to predict the BOD_τ at any flow time τ.

9-5. Using Figure 9.6 and equation 9.29 as a guide, develop finite difference equations for predicting the steady-state nitrogen component and DO deficit concentrations D in a multi-section one-dimensional estuary. Define every variable or parameter used.

9-6. Using Michaelis-Menten kinetics illustrated in Figure 9.11 develop equations for

(a) predicting the time rate of change of a nutrient concentration N (dN/dt) as a function of the concentration of bacterial biomass B;

(b) predicting the time rate of change in the bacterial biomass B (dB/dt) as a function of its maximum growith rate $\mu_B^{\max}$, temperature T, B, N, and the specific-loss rate of bacteria ρ_B; and

(c) predicting the time rate of change in dissolved oxygen deficit (dD/dt) also as a function of N, B, ρ_B, and the reaeration-rate constant K_a (T^{-1}).

How would these three equations be altered by the inclusion of protozoans P that feed on bacteria, and in turn require oxygen? Also write the differential equation for the time rate of change in the concentration of protozoans P (dP/dt).

9-7. Equation 9.52 for predicting stream temperature is in Eulerian coordinates. The actual behavior of the stream temperature is more easily demonstrated if Lagrangian coordinates (i.e., time of flow τ rather than distance X) are used. Assuming insignificant dispersion, the "time-of-flow" rate of temperature change of a water parcel as it moves downstream is

$$\frac{dT}{d\tau} = \frac{\lambda}{\rho c D}(T_E - T)$$

(a) Assuming that λ, D, and T_E are constant over an interval of time of flow $\tau_2 - \tau_1$, integrate the equation above to derive the temperature at location X_2 $[X_1 + U(\tau_2 - \tau_1)]$ given an initial temperature T_1 at location X_1.

(b) Develop a model for predicting the temperature at a point in a nondispersive stream downstream from multiple point sources (discharges) of heat.

9-8. Consider three well-mixed bodies of water that have the following constant volumes and freshwater inflows:

Water Body	*Volume* (m^3)	*Flow* (m^3/s)	*Displacement Time*
1	3×10^{12}	3×10^3	31.7 years
2	3×10^8	3×10^2	11.6 days
3	3×10^4	3	2.8 hours

The first body is representative of the Great Lakes in North America, the second is characteristic in size to the upper New York harbor with the summer flow of the Hudson River, and the third is typical of a small bay or cove. Compute the time required to achieve 99% of the equilibrium concentration, and that concentration, of a substance having an initial concentration of 0 (at time = 0) and an input of N (MT^{-1}) for each of the three water bodies. Assume that the decay-rate constant K is 0, 0.01, 0.05, 0.25, 1.0, and 5.0 days^{-1} and compare the results.

9-9. Use the simple model for adsorbed chemical losses in runoff (equations 9.68 to 9.71) to estimate the losses of several chemicals from a 10-ha urban watershed in the northeastern United States during August 1976. Recorded precipitation was as follows:

Day	1	6	7	8	9	10	13	14	15	26	29
R_t (*cm*)	1.8	0.7	2.6	2.9	0.1	0.3	2.9	0.1	1.4	3.7	0.8

Solids (sediment) build up on the watershed at the rate of 50 kg/ha-day, and chemical concentrations in the solids are 100 mg/kg. Assume that each runoff event washes the watershed surface clean. Assume also that there was a previous runoff event on July 31, so that there is no initial sediment buildup on August 1. Runoff volumes should be computed using equation 9.73. The watershed is 30% impervious. The following results should be given for each storm:

(a) Runoff in cm and m^3.

(b) Sediment loss (kg).

(c) Chemical loss (g), in dissolved and solid-phase form for chemicals with three different adsorption coefficients, $k = 5, 100, 1000$.

9-10. There exists a modest-sized urban subdivision of 100 ha containing 2000 people. Land uses are 60% single-family residential, 10% commerical, and 30% undeveloped. An evaluation of the effects of street cleaning practices on nutrient losses in runoff is required for this catchment.

This evaluation is to be based on the simulation model presented in

	PRECIPITATION (cm)						
Day	*A*	*M*	*J*	*J*	*A*	*S*	*O*
1			0.6		1.4	0.7	1.9
2	1.1	0.4				0.5	
3						0.1	0.1
4	0.1	1.5					
5	0.1	0.9		1.4	1.9	0.3	
6	0.1		1.4	1.1	1.0		0.5
7			0.1	0.7	0.7		
8	0.1	0.1		0.1			0.4
9		0.6	1.6				1.5
10					0.1		
11				0.2			
12				0.2	0.2		0.2
13	0.1					1.5	
14			0.2		0.5	3.5	0.8
15							1.0
16				0.8		4.3	2.8
17			0.7	0.1	0.5	0.8	1.9
18		0.5	0.4			0.8	0.1
19			0.4	0.3		0.4	0.9
20			0.7			2.3	
21				0.1		0.3	
22	0.1				0.4		
23	2.0						
24	3.2	0.1			0.2	4.7	
25	0.1		0.6	3.0		2.8	
26						1.6	
27							
28			0.3			0.1	
29		0.2	1.1		0.6		
30		0.1		0.3	0.2		
31							

Section 9.8.2, the data in Table 9.7, and on the 7-month precipitation record given on page 498. Present the results of the simulations as 7-month PO_4 and N losses as functions of street-cleaning interval and efficiency (i.e., show these losses for ranges of intervals and efficiencies). Assume a runoff threshold for washoff of $Q_0 = 0.5$ cm.

REFERENCES

1. BACA, R. G., M. W. LORENZEN, R. D. MUDD, and L. V. KIMMELL, A Generalized Water Quality Model for Eutrophic Lakes and Reservoirs, Report to EPA 211B01601, Pacific Northwest Laboratory, Battelle Memorial Institute, Rev. November 1974. 112 p.
2. BELLMAN, R. E., and R. E. KALABA, *Quasilinearization and Nonlinear Boundary-Value Problems*, American Elsevier Publishing Co., Inc., New York, 1965.
3. BISWAS, A. K., Systems Approach to Water Management, and Mathematical Modeling and Water Resources Decision-Making, Chaps. 1 and 11 in *Systems Approach to Water Management*, A. K. Biswas (ed.), McGraw-Hill Book Company, New York, 1976.
4. CANALE, R. P., *Modeling Biochemical Processes in Aquatic Ecosystems*, Ann Arbor Science Publishers, Inc., Ann Arbor, Mich., 1976.
5. CHEN, C. W., and G. T. ORLOB, Ecological Simulation of Aquatic Environments, Chap. 12 in *Systems Analysis and Simulation in Ecology*, Vol. 3, B. C. Patten (ed.), Academic Press, Inc., New York, 1975.
6. CHEN, C. W., M. LORENZEN, and D. J. SMITH, A Comparative Water Quality-Ecologic Model for Lake Ontario, Report to the Great Lakes Environmental Research Laboratory, NOAA, by Tetra Tech, Inc., Pasadena, California, October 1975. 202 p.
7. DITORO, D. M., D. J. O'CONNOR, R. J. THOMANN, and J. L. MANCINI, Phytoplankton–Zooplankton–Nutrient Interaction Model for Western Lake Erie, Chap. 11 in *Systems Analysis and Simulation in Ecology*, Vol. 3, B. C. Patten (ed.), Academic Press, Inc., New York, 1975, pp. 423–474.
8. DONIGIAN, A. S., JR., and N. H. CRAWFORD, Modeling Pesticides and Nutrients on Agricultural Lands, Report EPA-600/2-76-043, U.S. Environmental Protection Agency, Athens, Ga., 1976.
9. DORFMAN, R., H. D. JACOBY, and H. A. THOMAS, JR., *Models for Managing Regional Water Quality*, Harvard University Press, Cambridge, Mass., 1972.
10. DORNBUSH, J. N., J. R. ANDERSON, and L. L. HARMS, Quantification of Pollutants in Agricultural Runoff, Report EPA 660/2-74-005, U.S. Environmental Protection Agency, Washington, D.C., 1974.
11. DOUGLAS, J., JR., and T. DUPONT, Galerkin Methods for Parabolic Equations, *Journal of Numerical Analysis*, *SIAM*, Vol. 7, December 1970, pp. 575–626.
12. FRERE, M. H., C. A. ONSTAD, and H. N. HOLTAN, ACTMO, An Agricultural Chemical Transport Model, Report ARS-H-3, U.S. Department of Agriculture, Agricultural Research Service, Hyattsville, Md., 1975.
13. GRIMSRUD, G. P., E. J. FINNEMORE, and H. J. OWEN, Evaluation of Water

Quality Models; A Management Guide for Planners, Report EPA-600/5-76-004, U.S. Environmental Protection Agency, Washington, D.C., July 1976.

14. HAITH, D. A., and J. V. DOUGHERTY, Nonpoint Source Pollution from Agricultural Runoff, *Journal of the Environmental Engineering Division, ASCE*, Vol. 102, No. EE2, 1976.

15. HAITH, D. A., and L. J. TUBBS, Modelling Nutrient Export in Rainfall and Snowmelt Runoff, in *Best Management Practices for Agriculture and Silviculture*, R. C. Loehr et al. (eds.), Ann Arbor Science Publishers, Inc., Ann Arbor, Mich., 1979.

16. HARLEMAN, D. R. F., A Comparison of Water Quality Models of the Aerobic Nitrogen Cycle, Research Memorandum RM-78-34, International Institute for Applied Systems Analysis, Laxenburg, Austria, July 1978.

17. HEANEY, J. P., W. C. HUBER, and S. J. NIX, Stormwater Management Model Level I—Preliminary Screening Procedures, Report EPA-600/2-76-275, U.S. Environmental Protection Agency, Cincinnati, Ohio, 1976.

18. HOLTON, H. N., G. J. STILTNER, W. H. HENSON, and N. C. LOPEZ, USDAHL-74 Revised Model of Watershed Hydrology, Technical Bulletin 1518, U.S. Department of Agriculture, Agricultural Research Service, Washington, D.C., 1975.

19. HUBER, W. C., et al., Stormwater Management Model User's Manual-Version II, Report EPA-670/2-75-017, U.S. Environmental Protection Agency, Cincinnati, Ohio, 1975.

20. HUBER, W. C., and D. R. F. HARLEMAN, Laboratory and Analytical Studies of the Thermal Stratification of Reservoirs, MIT Hydrodynamics Laboratory Technical Report 112, October 1968.

21. HUBER, W. C., D. R. F. HARLEMAN, and P. J. RYAN, Temperature Prediction in Stratified Reservoirs, *Journal of the Hydraulics Division, ASCE*, Vol. 98, No. HY4, Paper 8839, 1972, pp. 645–666.

22. Hydroscience, Inc., Simplified Mathematical Modeling of Water Quality, Submitted to the U.S. Environmental Protection Agency, Washington, D.C., March 1971; and "Addendum," May 1972.

23. Hydroscience, Inc., Report on Water Quality Evaluations, Submitted to the National Commission on Water Quality, Washington, D.C., December 1975.

24. KELLY, R. A., The Delaware Estuary, in *Ecological Modeling in a Resource Management Framework*, C. S. Russell (ed.), RFF Working Paper QE-1, Johns Hopkins University Press, Baltimore, Md., July 1975.

25. KELLY, R. A., and W. O. SPOFFORD, JR., Application of an Ecosystem Model to Water Quality Management: The Delaware Estuary, in *Ecosystem Modeling in Theory and Practice*, C. A. S. Hall and J. W. Day, Jr. (eds.), John Wiley & Sons, Inc., York, 1977.

26. LARSEN, D. P., and H. T. MERCIER, Lake Phosphorus Loading Graphs: An Alternative, National Eutrophication Survey Working Paper 174, U.S. Environmental Protection Agency, Cincinnati, Ohio, July 1975.

27. LEE, E. S., *Quasilinearization and Invariant Imbedding*, Academic Press, Inc., New York, 1968.

28. LINSLEY, R. K., Rainfall-Runoff Models, Chap. 2 in *Systems Approach to Water Management*, A. K. Biswas, (ed.), McGraw-Hill Book Company, New York, 1976.

29. LOUCKS, D. P., et al., A Review of the Literature in Water Pollution Control: Systems Analysis, *Journal of the Water Pollution Control Federation*, Vols. 41–47, No. 6, 1970–1978.

30. LOUCKS, D. P. (ed.), A Selected Bibliography on the Analysis of Water Resources Systems, Vols. 3–8, Water Resources Scientific Information Center, Office of Water Research and Technology, USDI, Washington, D.C., 1972–1978.

31. LOUCKS, D. P., Surface Water Quality Management, Chap. 6 in *Systems Approach to Water Management*, A. K. Biswas (ed.), McGraw-Hill Book Company, New York, 1976.

32. MCELROY, A. D., S. Y. CHIU, J. W. NEBGEN, A. ALETI, and F. W. BENNETT, Loading Functions for Assessment of Water Pollution from Nonpoint Sources, Report EPA-600/2-76-151, U.S. Environmental Protection Agency, Washington, D.C., 1976.

33. MILLER, C. R., and W. VIESSMAN, JR., Runoff Volumes from Small Urban Watersheds, *Water Resources Research*, Vol. 8, No. 2, 1972.

34. O'CONNOR, D. J., Estuarine Distribution of Non-Conservative Substances, *Journal of the Sanitary Engineering Division, ASCE*, Vol. 91, 1965, pp. 23–42.

35. O'CONNOR, D. J., and D. M. DITORO, Photosynthesis and Oxygen Balance in Streams, *Journal of the Sanitary Engineering Division, ASCE*, Vol. 96, No. SA2, 1970.

36. O'CONNOR, D. J., and J. A. MUELLER, A Water Quality Model of Chlorides in Great Lakes, *Journal of the Sanitary Engineering Division, ASCE*, Vol. 96, No. 4, 1970.

37. O'CONNOR, D. J., R. V. THOMANN, and D. M. DITORO, Ecologic Models, Chap. 8 in *Systems Approach to Water Management*, A. K. Biswas (ed.), McGraw-Hill Book Company, New York, 1976.

38. OGOSKY, H. O., and V. MOCKUS, Hydrology of Agricultural Lands, Chap. 21 in *Handbook of Applied Hydrology*, V. T. Chow (ed.), McGraw-Hill Book Company, New York, 1964.

39. ONSTAD, C. A., and G. R. FOSTER, Erosion Modeling on a Watershed, *Transactions of the ASAE*, Vol. 18, No. 2, 1975.

40. ORLOB, G. T., Mathematical Modeling of Surface Water Impoundments, Vol. 1, Report prepared for the Office of Water Research and Technology, Resource Management Associates, Lafayette, Calif., June 1977.

41. OVERTON, D. E., and M. E. MEADOWS, *Stormwater Modeling*, Academic Press, Inc., New York, 1976.

42. PAVONI, J. L. (ed.), *Handbook of Water Quality Management Planning*, Van Nostrand Reinhold Company, New York, 1977.

43. RINALDI, S., R. SCONCINI-SESSA, H. STEHFEST, and H. TAMURA, *Modeling and Control of River Quality*, McGraw-Hill Book Company, New York, 1978.

44. RUSSELL, C. S. (ed.), *Ecological Modeling in a Resource Management Framework*, RFF Working Paper QE-1, Johns Hopkins University Press, Baltimore, Md., July 1975.

45. SCAVIA, D., and A. ROBERTSON, *Perspectives on Lake Ecosystem Modeling*, Ann Arbor Science Publishers, Inc., Ann Arbor, Mich., December 1978.

46. SCHWAB, G. O., R. K. FREVERT, T. EDMINISTER, and K. K. BARNES, *Soil and Water Conservation Engineering*, John Wiley & Sons, Inc., New York, 1966.

47. SIMONS, T. J., Development of Three-Dimensional Numerical Models of the Great Lakes, Canada Centre for Inland Waters, Burlington, Ontario, Science Series 12, 1973, 26 pp.

48. SNODGRASS, W. J., and C. R. O'MELIA, Predictive Model for Phosphorus in Lakes, *Environmental Science and Technology*, Vol. 9, No. 10, 1975.

49. SOIL CONSERVATION SERVICE, *National Engineering Handbook*, Sec. 4, Hydrology, U.S. Department of Agriculture, Washington, D.C., 1972.

50. STEHFEST, H., Mathematical Modeling of Self-Purification of Rivers, Professional Paper PP-77-11, International Institute for Applied Systems Analysis, Laxenburg, Austria, October 1977.

51. STREETER, H. W., and E. P. PHELPS, A Study of the Pollution and Natural Purification of the Ohio River, U.S. Public Health Service, Publication Health Bulletin 146, February 1925.

52. TAKAMATSU, T., I. HASHIMOTO, and S. SIOVA, Model Identification of Wet Air Oxidation Process Thermal Decomposition, *Water Research*, Vol. 4, 1970, pp. 33–59.

53. THOMANN, R. V., *Systems Analysis and Water Quality Management*, Environmental Research and Applications, Inc., New York, 1972.

54. TORNO, H. C., Stormwater Management Models, Proceedings of the Research Conference *Urban Runoff Quantity and Quality*, American Society of Civil Engineers, New York, 1975.

55. Tracor Associates, Estuarine Modeling, An Assessment, Capabilities and Limitations for Resource Management and Pollution Control, Prepared for the National Coastal Pollution Research Water Quality Office of the U.S. Environmental Protection Agency, Washington, D.C., February 1971.

56. TUBBS, L. J., and D. A. HAITH, Simulation of Nutrient Losses from Cropland, Paper 77-2502, presented at the 1977 Winter Meeting of the American Society of Agricultural Engineers, Chicago, 1977.

57. VOLLENWEIDER, R. A., Input-Output Models, *Schweizerische Zeitschrift für Hydrologie*, Vol. 37, 1975, pp. 53–84.

58. VOLLENWEIDER, R. A., and P. J. DILLON, The Application of the Phosphorus Loading Concept to Eutrophication Research, Paper prepared for the Associate Committee on Scientific Criteria for Environmental Quality, Burlington, Ontario, June 1974.

59. Water Resources Engineers, Inc., Computer Program Documentation for the Stream Quality Model QUAL II, Prepared for the U.S. Environmental Protection Agency, Washington, D.C., 1973.

60. Water Resources Engineers, Inc., Prediction of Thermal Energy Distribution in Streams and Reservoirs, Report to California Department of Fish and Game, June 1967; rev. August 1968. 90 p.

61. WILLIAMS, J. R., Sediment-Yield Prediction with Universal Equation Using Runoff Energy Factor, in *Present and Prospective Technology for Predicting Sediment Yields and Sources*, Report ARS-S-40, U.S. Department of Agriculture, Washington, D.C., 1975.

62. WILLIAMS, J. R., and H. D. BERNDT, Sediment Yield Computed with Universal Equation, *Journal of the Hydraulics Division, ASCE*, Vol. 98, No. HY12, 1972.

63. WISCHMEIER, W. H., and D. D. SMITH, Predicting Rainfall-Erosion Losses, *Agriculture Handbook 537*, U.S. Government Printing Office, Washington, D.C., 1978.

CHAPTER 10

Water Quality Management Modeling

10.1 INTRODUCTION

Water quality planning involves the identification and evaluation of management alternatives for satisfying various economic and water quality goals. Economic goals are often expressed in terms of cost effectiveness (cost minimization) and an equitable distribution of the cost among those who should pay. Water quality goals are usually expressed in the form of wastewater effluent standards, or water quality standards in waste-receiving water bodies, or both. The effectiveness of any management plan may be measured in terms of how well the plan accomplishes these goals. Water quality management models can assist planners in identifying and evaluating possible management plans and in determining which plans best meet these economic and water quality goals.

Water quality management models are extensions of the simulation or predictive models discussed in Chapter 9. In addition to the predictive equations, management models include as unknowns the design and operating policy variables of each management alternative. Relationships are included that describe the resulting effluent or water quality, and the cost of each management alternative as functions of the design and operating policy variables. Also included in management models are constraints defining the desired effluent and/or water quality standards. The purpose of this chapter is to introduce water quality management models and to discuss some of the approaches used in developing and solving these models.

Many water quality management models are optimization models. As such, the relationships that define economic costs or the resulting effluent or water quality as functions of the unknown design or operating variables must conform to what is required for model solution using one or more particular optimization procedures. If simplifications or modifications are necessary for model solution, then (as emphasized throughout this text) the resulting solution should be checked using more accurate water quality simulation models. Management models are best used for a preliminary evaluation of various alternatives and for identifying what data are important and needed prior to the implementation of a more extensive (and usually more expensive) data collection and simulation study.

10.2 MANAGEMENT ALTERNATIVES FOR WATER QUALITY CONTROL

The first and most obvious method of water quality control is to limit the amount of waste discharged into water bodies. This type of control can take on a number of forms, including:

1. Requirements that each waste producer discharge less waste through, for example, process changes or removal of at least some minimum specified fraction of the waste prior to releasing the remainder onto land or into natural waters. Removal of waste can be accomplished by a variety of physical, biological, and chemical processes.
2. Storing a portion of the treated wastewater effluent which if released into the natural water body would result in a lower-than-desired quality. Ponds or tanks can be used for effluent storage. The quantity and timing of stored effluent discharges to land areas or water bodies should depend in part on the waste assimilative capacity of the receiving land area or water body.
3. Piping waste, either prior to or following some treatment, to areas within or outside the region for additional treatment and/or disposal at land or water sites having greater waste assimilative capacities. This alternative also permits the processing of wastes at larger regional facilities which take advantage of economies of scale in construction and operating costs, as well as of increased operating efficiencies.
4. Instream quality improvement from artificial aeration or flow augmentation. Dissolved oxygen deficit concentrations can be decreased

by the transfer of oxygen into the water by injection of air. Increasing the streamflow in periods of low flows by releasing water from upstream reservoirs may also improve the stream quality by dilution and by changing velocities and temperatures, which in turn affect the reaction rates of various quality constituents.

Each of these means of water quality management will be discussed in greater detail in later sections of this chapter. Prior to this discussion some remarks concerning management objectives and the criteria that are used for evaluating alternative combinations of the various management options are appropriate.

10.3 MANAGEMENT OBJECTIVES AND QUALITY STANDARDS

Water quality management objectives are multiple and conflicting. Those in control of activities that generate wastes would naturally prefer to dispose of their wastes at no cost to themselves and, if possible, to others as well. This policy leads to higher profits, if income is being derived from the waste-making activities, or less taxes if the wastes are derived from human settlements such as cities or other municipalities. However, if the discharge of wastes does result in added costs (i.e., in environmental damages) elsewhere, those who incur these costs and damages would also prefer not to incur them. They can argue that those who discharge wastes into water bodies should pay for the environmental damage caused by that waste or not be allowed to discharge the waste at all. Yet because those who discharge waste are usually not affected by the damage caused by that waste, there is seldom an economic incentive for them to control their discharge. Water pollution is said to be an economic externality (i.e., the activities of some impose costs on others without their consent.). This is the central conflict in water quality management in river basins throughout the world.

Because the private market system fails to charge each polluter an amount equal to the damages resulting from their waste discharge, regulatory action is often required. The types of incentives that water quality regulatory agencies have used to compensate for the failure of individual polluters to consider the damages they impose on others are (1) legislative, including direct regulation, the establishment of effluent or stream-quality standards, licensing, and zoning; (2) legal, including compensation for damages and fines for violation of law; and (3) economic, including effluent charges or taxes, subsidies, accelerated depreciation allowances, and the like. Whatever the methods used, the objective should be to achieve a more efficient and equitable allocation of resources from the standpoint of society as a whole.

One of the difficulties in finding a plan that is both efficient and equitable is the problem of quantifying water quality benefits or damages. This problem is similar to that of attributing a monetary benefit to such things as aesthetics and clean air. There is also the problem of determining equitable distributions of costs and benefits. Thus the selection of the desired water quality and the question of who will pay for it often become political issues. This political aspect is reflected in both the water quality management objectives and in the quality standards. Political systems have clearly demonstrated their sensitivity to these multiobjective aspects of water quality management problems.

Those who develop regional water quality management models usually assume the actual or potential existence of some governmental institution that has the authority to control water quality within its region, either by economic incentives such as effluent charges and/or by legal means such as effluent standards. The main purpose of most regional water quality models is to examine alternatives that will reduce both the abatement costs and damages associated with water pollution.

To begin quantifying the rather general objective for regional water quality management, consider a river basin in which there are several groups s that discharge pollutants into the natural water courses. Included among those individuals or groups of individuals are organizations such as state and federal pollution-control agencies that have interests, financially as well as politically, in the quality of the natural water within the basin. Water quality control alternatives such as wastewater treatment and effluent storage impose costs to private and, because of cost-sharing programs, public or governmental agencies as well. Quality-control alternatives such as flow augmentation and artificial aeration may only add to the cost paid by those public agencies. Regardless of who pays, the cost to each individual or group s can, for the moment, be denoted as a function of the scale of all alternatives used for water quality control.

Denoting by S_i the scale of some waste-reduction alternatives at site i costing $C_i^s(S_i)$ to each group s, a cost-effective objective without regard to cost distribution, or the political influence of each group s, can be written

$$\text{minimize} \sum_i \sum_s C_i^s(S_i) \tag{10.1}$$

Such an objective may or may not result in an acceptable solution. This depends in part on the quality standards imposed. Usually, the minimum allowable quality standards are not intended to represent the desired quality. Environmental protection goals and allowances for future growth and uncertainties often result in targets or desired qualities that are higher than specified minimum quality standards. One way to achieve a quality that comes closer to the targets is through the use of effluent charges or taxes T_i^s levied against each group s discharging wastes at site i.

The fraction $P_i = P_i(S_i)$ of the waste reduced or removed at each site i is, of course, a function of the scale S_i of waste reduction measures employed at each site. If W_i is the quantity of waste generated at site i prior to the implementation of any waste reduction measures, then $W_i(1 - P_i)$ is the quantity of waste that will be discharged after waste treatment. Ideally, any tax on the waste discharged should reflect the external damages attributable to that discharge. The purpose of the tax or subsidy is to provide an economic incentive for reducing the external damages, if any, that result from the discharge of wastes.

Let $C_i^A(S_i)$ represent the annual agency costs of measures S_i taken at site i to improve water quality and $C_i^s(S_i)$ be the individual or private costs (excluding taxes), as before. The goal of the control agency is to establish effluent tax rates T_i^s per unit of waste discharge $W_i(1 - P_i)$ in such a way as to minimize the total cost of water quality control:

$$\text{minimize} \sum_i \left[C_i^A(S_i) + \sum_s C_i^s(S_i)\right] \tag{10.2a}$$

$$\begin{bmatrix}\text{agency}\\ \text{cost}\end{bmatrix} + \begin{bmatrix}\text{other private}\\ \text{and/or public}\\ \text{costs}\end{bmatrix}$$

Sometimes, such a management agency may have to collect sufficient revenue through taxes to cover the costs of their operations. If so,

$$\sum_s \sum_i T_i^s W_i[1 - P_i(S_i)] \geq \sum_i C_i^A(S_i) \tag{10.2b}$$

$$\begin{bmatrix}\text{total effluent}\\ \text{tax income}\end{bmatrix} \geq \begin{bmatrix}\text{total agency}\\ \text{cost}\end{bmatrix}$$

In addition, it must insure that the desired water quality is maintained at all sites j at which standards are established:

$$\mathbf{Q}_j(\mathbf{S}, \mathbf{W}) \geq \mathbf{Q}_j^{\min} \tag{10.2c}$$

$$\begin{bmatrix}\text{quality at site } j\end{bmatrix} \geq \begin{bmatrix}\text{minimum desired}\\ \text{quality at } j\end{bmatrix}$$

The individuals who must pay either the cost of waste reduction or treatment and/or an effluent tax on the waste $W_i(1 - P_i)$ discharged at sites i are, of course, interested in minimizing their total costs:

$$\text{minimize} \quad \sum_i \{C_i^s(S_i) \quad + T_i^s[W_i(1 - P_i)]\} \qquad \forall s \tag{10.3}$$

$$\begin{bmatrix}\text{waste reduction}\\ \text{cost}\end{bmatrix} + \begin{bmatrix}\text{effluent charge}\end{bmatrix}$$

Note that the objectives above do not attempt to quantify the benefits or damages associated with the resulting water quality. Rather, each tries to identify the least cost approach for meeting predefined water quality standards. One of the uses of this model is to estimate the costs incurred by each group

s and the water quality achieved with various proposed combinations of effluent charges and effluent and stream standards.

The problem defined by equations 10.2 and 10.3 is an example of a multilevel, multiobjective planning problem. The controlling agency objective 10.2a is to minimized total cost subject to the minimization of a number of objectives, 10.3, which are not controlled by the agency except through the establishment of the effluent tax rates T_i^s in those objectives. Methods for solving multiobjective problems (i.e., for finding noninferior solutions to 10.3), given the tax rates, are discussed in Chapter 4. However, research has only begun to identify approaches for solving the overall multilevel, multiobjective problem. In the absence of satisfactory solution procedures, the agency level objective by itself (i.e., the minimization of total cost of all measures implemented for water quality control, equation 10.2a) is typically chosen for a preliminary evaluation of water quality management alternatives.

Yet with or without effluent taxes, the agency objective of cost minimization has not been generally accepted. This is in part because there indeed are other objectives at other levels of planning. It is seldom obvious how these multiple water quality management objectives at various decision making levels should be combined to simulate, in a specific situation, what might be the actual response to a decision at the central agency level.

The political process of establishing effluent charges and minimum acceptable qualities, either in the form of effluent or stream-quality standards, involves the participation of each group of interested individuals within the river basin. Some interest groups clearly have more political influence than others. This depends not only on their political skills but also on how strongly they feel about certain issues. To include the effect of this political influence in water quality models, it is sometimes assumed that relative weights can be defined and used in the objective function. Each weight w_s reflects the relative influence that group s might exert compared to all other groups included in the model. Using the proper weights, charges, and standards, a socially or politically acceptable and efficient water quality management policy might result from the use of the following objective:

$$\text{minimize} \sum_s w_s \sum_i [C_i^s(S_i) + T_i^s W_i(1 - P_i)] \tag{10.3a}$$

The difficulty here, of course, is that the relative political weights are unknown, even to the decision makers, until the final decision is made. But as discussed in Chapter 4, by varying the relative weights, an analyst can define some of the efficient alternatives among the many possible alternatives. If the objective is piecewise linear, the number of efficient alternatives defined by this procedure will be smaller than the number of possible efficient alternatives. Efficient alternatives can also be defined by setting upper bounds on all but one of the terms enclosed within the braces { } in equation 10.3 and minimizing the other. Clearly, both the former weighting approach and this latter target approach, and even other more efficient iterative multiobjective approaches, are predictive rather than normative.

If all the weights are assigned values of unity, the objective function 10.3a will represent the minimization of total costs and effluent charges without regard to the distribution of costs borne by the various polluters. As the relative weights or target levels change, so will the alternatives associated with those weights or targets. Relatively high weights or low targets will correspond to those polluters having a relatively strong political position and interest, that may reduce their share of the total cost. Varying the weights or targets permits an examination of the stream quality that is likely to be associated with various cost distributions.

Other means have been used to incorporate considerations of equity into otherwise strictly cost-effective models (e.g., those having objective functions of the type 10.2a). These include constraints specifying equal scales of various alternatives or some function of these scales, such as equal treatment efficiencies or equal costs per capita contributing to the total waste at various sites. If S_i is the scale of quality control alternatives employed at site i, then constraints requiring equal scales of control could be written as

$$S_i = S_k \qquad \forall i \in Z_k \tag{10.4}$$

where Z_k is the set of sites in zone k of the region. Zones within a region may be defined geographically or by other criteria such as industrial activity.

Although numerous analysts have included equity within the constraint set of their models, it could be argued that equity is an objective—one of many that water quality planners consider. The relative weight given to an equity objective depends in part on the economic cost of achieving it, as well as on the administrative and political cost of ignoring it (i.e., the unquantifiable cost associated with implementing a plan calling for a wide range of quality control requirements within a region that minimizes only the total economic costs). Equity within any zone of a river basin can be expressed as an objective by defining and then minimizing the absolute difference between the minimum and maximum scale of water quality control within the zone. Let $S_k^{\min}$ and $S_k^{\max}$ denote variable lower and upper bounds on the scale S_i of a single control alternative within zone k, so that

$$S_k^{\min} \leq S_i \leq S_k^{\max} \qquad \forall i \in Z_k \tag{10.5}$$

A portion of the equity objective can involve minimizing the sum of the weighted differences between $S_k^{\min}$ and $S_k^{\max}$ over all zones k:

$$\text{minimize} \sum_k w_k(S_k^{\max} - S_k^{\min}) \tag{10.6}$$

Constraints on water quality can be defined as (1) effluent standards, restricting the waste released at site i, $W_i(1 - P_i)$, to be no greater than the maximum allowable quantity

$$W_i(1 - P_i) \leq W_i^{max} \tag{10.7}$$

or as (2) receiving water quality standards requiring the quality $Q_j(\mathbf{S})$ at a site or reach j within the water body to be no less than some minimum allowable quality.

$$\mathbf{Q}_j(\mathbf{S}) \geq \mathbf{Q}_j^{min} \tag{10.8}$$

These objectives and constraints will be defined in greater detail below.

10.4 WATER QUALITY CONTROL ALTERNATIVES

In the models that follow, wastewater treatment and/or reduction through process changes (i.e., the removal of some fraction of the total waste load prior to discharging the remainder into the receiving water bodies) will be included as methods of controlling water quality. Models have been developed for assisting in the design of wastewater treatment facilities and in the design of treatment processes within these facilities [7, 8, 14, 20, 21]. Such models are useful for defining the costs of wastewater treatment. Since the design and cost of alternative treatment facilities are fairly well defined [24, 28, 34], the discussion in this section will begin with an examination of what to do with the treated wastewater effluent. One alternative is disposal on land.

10.4.1 Wastewater Disposal on Land

There are two alternatives for disposing of wastewater effluent from a treatment plant. These consist of either discharging the effluent into a receiving water body or onto land for further waste reduction prior to drainage into a water body or groundwater. Wastewater disposal on land, or land application as it is commonly called, may be attractive in certain areas where land is available. Land application permits further removal of nitrogen, phosphorus, organics, pathogens, and heavy metals in the wastewater effluent. Of these constituents, nitrogen is the most mobile in soils, and hence its removal usually controls the application rate of wastewater on land [9, 13]. Nitrogen in the drainage waters from well-aerated soils exists mostly in the form of nitrate-nitrogen, NO_3–N.

Outlined in this section is a simple simulation model for assisting in evaluating irrigation land application alternatives that meet predefined maximum allowable NO_3–N concentrations in the drainage waters. Effluent disposal on land is commonly accomplished using spray irrigation methods. Nitrogen removal in the soil results mostly from plant uptake and subsequent harvest or consumption by grazing.

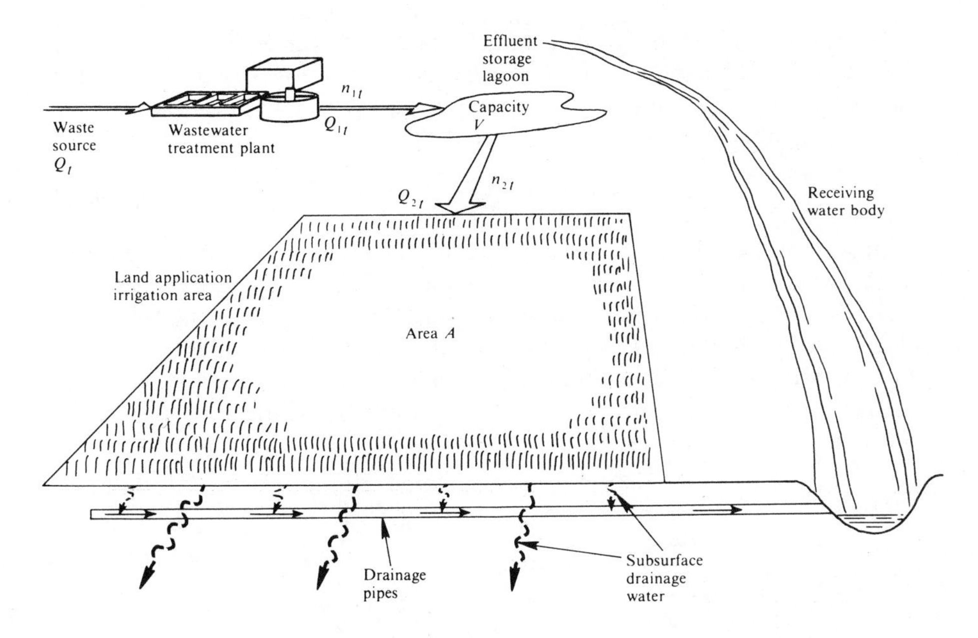

FIGURE 10.1. Components of land application system.

Figure 10.1 illustrates a typical land application system. Of interest is the storage lagoon volume capacity V, the land application area A, the irrigation volumes Q_{2t} in each period t within a year, and the maximum discharge rate Q_2^* during any period of length Δt,

$$Q_2^* = \underset{t}{\text{maximum}} \left(\frac{Q_{2t}}{\Delta t} \right) \tag{10.9}$$

that minimizes the total annual system cost $C(V, A, Q_2^*)$. This cost can then be compared to the additional cost of advanced wastewater treatment required to meet the same effluent standards.

The relationships between the components of the land application system can be defined by a series of mass-balance equations. The wastewater volume mass balance for the storage lagoon equates the final storage volume S_{t+1} to the initial storage volume, S_t, plus the inflow volume, Q_{1t}, less the outflow volume, Q_{2t}, in each period t of the year.

$$S_{t+1} = S_t + Q_{1t} - Q_{2t} \qquad \forall t \tag{10.10}$$

In the equations above, if period t is the last within-year period, then period $t + 1 = 1$. The lagoon storage capacity V equals the maximum of all the storage volumes S_t.

$$V = \underset{t}{\text{maximum}} \, (S_t) \qquad \forall t \tag{10.11}$$

The inflow volumes are known, but the outflow volumes, Q_{2t}, and lagoon capacity, V, are unknown decision variables. Equation 10.10 can be modified to include evaporation losses if desired. Such losses could be based on the average lagoon volume $\bar{S}_t$ in each period t.

$$\bar{S}_t = \frac{S_t + S_{t+1}}{2} \qquad \forall t \tag{10.12}$$

Nitrogen in the lagoon may be removed by ammonia volatilization, denitrification, algal uptake, and settling. Assuming a first-order decay process, the mass of nitrogen (concentration $n_{2,t+1}$ times volume S_{t+1}) in the lagoon at the end of each period t equals the initial lagoon nitrogen mass $n_{2t}S_t$, plus the mass input $n_{1t}Q_{1t}$, less the mass in effluent $Q_{2t}(n_{2t+1} + n_{2t})/2$, less the mass decay $k_t\bar{S}_t(n_{2,t+1} + n_{2,t})/2$, where k_t is the temperature-dependent nitrogen removal rate in period t (T^{-1}).

$$\begin{aligned} n_{2,t+1}S_{t+1} &= n_{2t}S_t + Q_{1t}n_{1t} - \frac{Q_{2t}(n_{2,t+1} + n_{2t})}{2} \\ \begin{bmatrix}\text{mass at end}\\ \text{of period } t\end{bmatrix} &= \begin{bmatrix}\text{mass at}\\ \text{beginning}\\ \text{of period}\\ t\end{bmatrix} + \begin{bmatrix}\text{mass in}\\ \text{influent}\end{bmatrix} - \begin{bmatrix}\text{mass in effluent}\end{bmatrix} \\ &\quad - \frac{k_t\bar{S}_t(n_{2,t+1} + n_{2t})}{2} \\ &\quad - \begin{bmatrix}\text{mass decay in}\\ \text{period } t\end{bmatrix} \end{aligned} \tag{10.13}$$

This completes the equations involving the storage lagoon. What remains to be described is the spray irrigation site.

Assuming that the soil moisture content, expressed as a depth (L), is maintained at field capacity M throughout the year (since otherwise additional lagoon storage volume capacity may be required), the water balance for the irrigated area is defined by equating the irrigation rate (Q_{2t}/A) to the evapotranspiration rate E_t plus the drainage rate d_t less the average precipitation P_t in each period t. Each of these terms is expressed in units of length (L).

$$\frac{Q_{2t}}{A} = E_t + d_t - P_t \qquad \forall t \tag{10.14}$$

To prevent surface runoff, the irrigation rate Q_{2t}/A less loss by evaporation E_t plus the precipitation should not exceed the maximum drainage capacity $d^{\max}$, or

$$d_t \leq d^{\max} \qquad \forall t \tag{10.15}$$

Drainage occurs (i.e., $d_t > 0$) when the application rate exceeds that required to just maintain the soil moisture content at field capacity.

Soil nitrogen relationships can be approximated by separately defining mass balance equations for organic and inorganic nitrogen. Average organic nitrogen levels in the soil must reach an equilibrium value F (in mass per unit area, ML^{-2}) if the waste disposal system is to be operated at a steady state. This value of F may be determined by the native fertility of the soil or if an objective of land application is to build up soil productivity, F will be a desired equilibrium value. Soil organic nitrogen levels may deviate from the equilibrium value during any time period due to mineralization of some fraction m_t of the organic nitrogen or addition of organic nitrogen XO_t (ML^{-2}) from wastewater irrigation during the period. Let O_t denote the deviation of the organic nitrogen levels from the target F at the beginning of period t (ML^{-2}). Then the total organic nitrogen level at the beginning of period $t + 1$ is $F + O_{t+1}$:

$$F + O_{t+1} = F + O_t - m_t(F + O_t) + XO_t$$

or

$$O_{t+1} = (1 - m_t)O_t - m_t F + XO_t \tag{10.16}$$

The nitrogen addition per unit land area is the fraction α of the nitrogen in the lagoon effluent $Q_{2t}(n_{2,t+1} + n_{2t})/2$ that is in the organic form, divided by the total land area A.

$$XO_t = \frac{\alpha Q_{2t}(n_{2,t+1} + n_{2t})}{2A} \qquad \forall t \tag{10.17}$$

The soil inorganic nitrogen content I_{t+1} per unit land area (ML^{-2}) at the end of each period t equals the initial inorganic nitrogen content I_t, plus that fraction m_t of the organic nitrogen that is mineralized to inorganic nitrogen in the period, plus the inorganic nitrogen addition XI_t in the wastewater effluent, less that leached from the soil L_t by drainage and that removed from the soil N_t by plant growth during the period.

$$I_{t+1} = I_t + m_t(O_t + F) + XI_t - L_t - N_t \qquad \forall t \tag{10.18}$$

The additional inorganic nitrogen is that which is contained in the lagoon effluent divided over the whole irrigation area A.

$$XI_t = \frac{(1 - \alpha)Q_{2t}(n_{2,t+1} + n_{2t})}{2A} \qquad \forall t \tag{10.19}$$

In the equations above, soil nitrogen losses from ammonia volatilization, denitrification and surface runoff are assumed to be insignificant. The mineralization fractions m_t will depend on the average soil temperature during the period t. Of course, when the soil is frozen, very little mineralization and drainage d_t take place.

The inorganic nitrogen loss from leaching L_t (ML^{-2}) depends on the average inorganic nitrogen concentration $(I_{t+1} + I_t)/2M$ in the soil water when leaching or drainage occurs (i.e., when $d_t > 0$).

$$L_t = \frac{d_t(I_t + I_{t+1})}{2M} \qquad \forall t \tag{10.20}$$

The uptake of nitrogen N_t by plants will depend on the type of cover crop grown and harvested or consumed as well as on the available inorganic nitrogen in the soil. Assuming that $N_t^{\max}$ is the upper limit of nitrogen uptake by plants (which will depend on the type) and that up to 70% of the soil nitrogen would be available to the irrigated crops, N_t will equal the minimum of these two maximum limits.

$$N_t = \text{minimum}\,\{0.7[m_t(O_t + F) + XI_t + I_t],\, N_t^{\max}\} \qquad \forall t \qquad (10.21)$$

The final model constraint applies to the quality of the drainage water. Letting $n_t^{\max}$ be the maximum allowable nitrate-nitrogen concentration, then

$$\frac{I_t + I_{t+1}}{2M} \leq n_t^{\max} \qquad \forall t \qquad (10.22)$$

Model Solution Procedure. A simulation procedure can be developed to solve this model. Input data include wastewater inflows Q_{1t} to the storage lagoon and their nitrogen concentrations n_{1t}, the nitrate–nitrogen decay rate k_t, the hydrologic parameters P_t and E_t, and the soil data parameters M, d, and m_t. Then a particular set of volume discharges Q_{2t} from the lagoon can be selected whose total equals the sum of the lagoon inflows Q_{1t}. The maximum of all Q_{2t} determines the capacity Q_2^* of the pumps and pipes required from the lagoon to the irrigation area. The outflows Q_{2t}, together with the known inflows Q_{1t}, determine the storage capacity V of the lagoon as found from the simultaneous solutions of equations 10.10 and then equation 10.11. (Since one of equations 10.10 is linearly dependent on the others, each S_t in the solution to equation 10.10 can be adjusted by a constant amount to ensure any desired minimum storage volume for increased detention times and nitrogen removal.)

Having determined each lagoon storage volume S_t and knowing the influent nitrate–nitrogen concentrations n_{1t} permits the simultaneous solutions of equations 10.13, yielding the nitrate–nitrogen concentrations n_{2t} of the lagoon effluent.

Finally, equations 10.14 and 10.15 can be used to compute the minimum irrigation area required given the effluent discharges Q_{2t}.

Equation 10.15 ensures that for any particular land area A, the drainage d_t cannot exceed the drainage capacity d. Knowing the effluent nitrate–nitrogen concentrations n_{2t} permits the solution of equations 10.17 and 10.19 for the nitrogen additions per unit land area XO_t and XI_t. It is then possible to solve equations 10.16.

The simultaneous solutions of equation 10.16 can provide an estimate of deviations in soil organic matter per unit area O_t at the beginning of each period t associated with the particular values of A and Q_{2t}. Having a knowledge of each O_t permits the simultaneous solutions of equations 10.18, 10.20, and 10.21, assuming that the plant uptake of nitrogen N_t equals $0.7[m_t(O_t + F) + XI_t + I_t]$. If, after the solution of these equations for each inorganic nitrogen mass I_t per unit area, any N_t exceeds $N_t^{\max}$, those N_t's are set equal to

$N_t^{\max}$ and the equations must be resolved. This procedure is continued until all equations 10.21 are satisfied.

All that remains is to check to see if the constraint equations 10.22 limiting the nitrate–nitrogen concentration in the drainage water are satisfied. If not, the irrigation area A must be increased if any actual nitrate–nitrogen concentration exceeds the maximum allowable $n_t^{\max}$. The area can be altered if all nitrate–nitrogen concentrations are less than the maximum allowable. Once the minimum area A has been found, the annual costs $C(V, A, Q_2^*)$ can be determined [13].

This simulation procedure can be repeated for different combinations of volume discharges Q_{2t} in an effort to identify the least-cost design. As described above, the simulation procedure involves nothing more complex than the simultaneous solution of several sets of linear equations, some of which may not even be necessary if, say after a period when the soil is frozen, the initial storage volume in the lagoon, and its nitrate–nitrogen concentration, are known. In using the foregoing equations for any particular problem, some unit conversion coefficients may be necessary to maintain the desired units.

10.4.2 Regional Treatment and Transport

An alternative to numerous separate treatment facilities is the transport of partially treated wastewater to one or more regional advanced wastewater treatment facilities for further removal of constituents prior to the discharge of the wastewater into a water body or onto a land area. This usually results in increased treatment efficiencies and reliabilities, and possibly lower total costs due to economies of scale in wastewater treatment. Yet added to the cost of advanced wastewater treatment at any regional facility is the cost of wastewater transport.

Figure 10.2 illustrates a possible situation in which a regional treatment facility might be considered. Each existing treatment plant needs to be upgraded to meet new effluent standards and the increasing volumes of wastewater flow. Assume that the type and concentrations of wastes in the effluent of each plant are approximately the same, and therefore the annual costs $C_i(Q_i^T)$ of increased waste removal capacity at each treatment site i can be defined as a function of the treated wastewater flow Q_i^T at that site.

An alternative to increasing the efficiency of each existing plant is to transport wastewater effluent from one or more existing treatment plants to one or more regional advanced waste treatment facilities that could be located at various existing or new treatment plant sites. The wastewater volume Q_i^T that is to be treated at each site i will be the difference between the total wastewater inflow to that site and the total wastewater outflow. The inflow at any site i is that volume Q_i^0 collected at that site plus that volume trans-

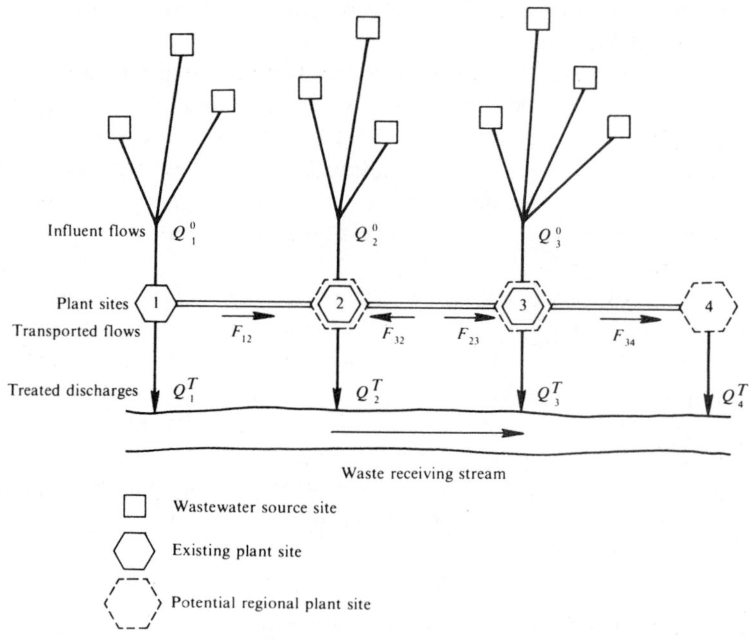

FIGURE 10.2. Existing and potential wastewater treatment sites.

ported to that site, F_{ji}, from adjacent sites j. The outflow from any site i is the volume F_{ij} transported to adjacent sites j. Hence at each site 1, 2, 3, and 4, the total inflow must equal the total outflow plus the volume treated and discharged, if any.

$$Q_1^0 = Q_1^T + F_{12} \tag{10.23}$$

$$Q_2^0 + F_{12} + F_{32} = Q_2^T + F_{23} \tag{10.24}$$

$$Q_3^0 + F_{23} = Q_3^T + F_{32} + F_{34} \tag{10.25}$$

$$F_{34} = Q_4^T \tag{10.26}$$

Because site $i = 1$ is not a potential regional wastewater treatment site, but can be expanded to meet the required effluent standards for the existing influent Q_1^0, the treated flow Q_1^T at site 1 cannot exceed the existing flow Q_1^0. This condition is enforced by equation 10.23.

The annual pipeline construction and pumping cost $C_{ij}(F_{ij})$ associated with the transport of wastewater from one site i to another adjacent site j will be some function of the flow F_{ij}. Because of the difference in elevation between those sites, these costs may depend on the direction of flow. The capacity of each pipe segment between sites i and $i + 1$ must equal at least the flow $F_{i,i+1}$ or $F_{i+1,i}$. If flow in either direction is possible, as between sites 2 and 3, one or both of these flows will undoubtedly be zero, as will its cost, in the model solution. In the example problem as illustrated in Figure

10.2, the capacity of the pipeline between sites $i = 1$ and 2 must equal F_{12}. The pipeline capacity between sites 2 and 3 will equal $F_{23} + F_{32}$ (since one will be zero), and the pipeline capacity between sites 3 and 4 will equal F_{34}.

Assume a cost-minimization objective:

$$\text{minimize} \sum_{i=1}^{4} [C_i(Q_i^T) + \sum_j C_{ij}(F_{ij})] \tag{10.27}$$

The two components in the objective above include annual wastewater treatment, pipeline construction, and pumping costs. The sum over sites j includes only those adjacent to site i.

If the cost functions exhibit fixed costs and economies of scale (i.e., decreasing marginal costs for certain ranges of Q_i^T and F_{ij}), the problem can be formulated as a linear mixed-integer programming model. In this case it is probable that only one regional plant, if any, will be in the solution, even though the model constraints (defined by equations 10.23 to 10.26) allow for more than one regional facility. When deriving the treatment plant cost functions $C_i(Q_i^T)$, the waste concentration in the influent as well as the maximum allowable concentrations in the effluent must be known and considered.

Example Problem. To illustrate a solution of the foregoing model applied to the regional wastewater treatment and transport planning problem defined in Figure 10.2, assume the simplified cost functions and wastewater flow data listed in Table 10.1. To incorporate the cost functions listed in Table 10.1 in

TABLE 10.1. Data for Example Regional Treatment Problem

Site, i	*Wastewater Source Flow,* Q_i^0	*Annual Cost*[a] *of Additional Treatment,* $C_i(Q_i^T)$	*Adjacent Site(s), j*	*Annual Cost*[b] *of Wastewater Transport,* $C_{ij}(F_{ij})$
1	100	$600 + 7(Q_1^T)$	2	$50 + 10(F_{12})$
2	150	$500 + 8(Q_2^T)$	3	$80 + 3(F_{23})$
3	50	$750 + 5(Q_3^T)$	2	$80 + 12(F_{32})$
			4	$100 + 10(F_{34})$
4	—	$450 + 6(Q_4^T)$	—	

[a]Cost is 0 if $Q_i^T = 0$.
[b]Cost is 0 if $F_{ij} = 0$.

the objective function 10.27, two sets of nonnegative integer variables y_i and z_{ij} can be defined. Let

$$y_i = \begin{cases} 1 & \text{if additional treatment at site } i \\ 0 & \text{otherwise} \end{cases}$$

and

$$z_{ij} = \begin{cases} 1 & \text{if wastewater flow is transported from site } i \text{ to an adjacent site } j \\ 0 & \text{otherwise} \end{cases}$$

To ensure these conditions, the following constraints must be added to the model:

$$Q_i^T \leq 300y_i \qquad i = 1, 2, 3, 4 \tag{10.28}$$

$$y_i \leq 1 \qquad i = 1, 2, 3, 4 \tag{10.29}$$

$$\begin{aligned} F_{ij} &\leq 300z_{ij} \\ z_{ij} &\leq 1 \end{aligned} \qquad \begin{cases} i = 1, 2, 3 \\ j = \begin{cases} i+1 & \text{if } i = 1, 2, 3 \\ i-1 & \text{if } i = 2, 3 \end{cases} \end{cases} \qquad \begin{matrix} (10.30) \\ (10.31) \end{matrix}$$

Equations 10.30 and 10.31 apply for all adjacent pairs of i and j for which F_{ij} are defined in equations 10.23 to 10.26. The maximum value of each Q_i^T and F_{ij} is never greater than the total wastewater flow of 300. The inequalities 10.28 through 10.31 force the respective integer variables y_i or z_{ij} to equal 1 if the flows Q_i^T or F_{ij} are greater than zero.

Having defined these integer variables, the objective function 10.27 can be written as:

minimize

1. Additional annual wastewater treatment costs

$$600y_1 + 7Q_1^T + 500y_2 + 8Q_2^T + 750y_3 + 5Q_3^T + 450y_4 + 6Q_4^T$$

plus

2. Annual wastewater transport costs

$$+ 50z_{12} + 10F_{12} + 80z_{23} + 3F_{23} + 80z_{32} + 12F_{32} + 100z_{34} + 10F_{34} \tag{10.32}$$

The foregoing model includes eight integer variables and eight continuous nonnegative variables. Substituting some variables for others, this total number of variables could be reduced if desired. The three wastewater source flows Q_i^0 are known and are specified in Table 10.1. Table 10.2 lists the solu-

TABLE 10.2. Least-Cost Solution to Example Regional Waste Treatment Problem

Site, i	*Additional Treated Wastewater Flow, Q_i^T*	*Annual Treatment Cost, $C_i(Q_i^T)$*	*Adjacent Site, j*	*Wastewater Transport, F_{ij}*	*Annual Transport Cost, $C_{ij}(F_{ij})$*
1	100	1300	2	0	0
2	0	0	3	150	530
3	200	1750	2, 4	0	0
4	0	0	—	—	—
Total annual costs		3050			530
Minimum total annual cost			3580		

tion of this example problem. This solution can be compared to the total annual cost of 4000 if no wastewater transport is permitted in this example problem.

This example problem is a very simplified one that can be solved by complete enumeration of all reasonable alternatives rather than by linear mixed-integer programming procedures. But it serves to illustrate the modeling approach that can be used to solve regional treatment planning problems having a greater number of alternatives and more realistic nonlinear costs or other objective functions.

10.4.3 Multiple-Point-Source Waste Reduction to Meet Water Quality Standards

The models presented so far have assumed the existence of wastewater effluent quality standards, specifying the maximum allowable constituent concentrations in the wastewater effluent that can be discharged into a water body. There may also exist stream quality standards specifying the maximum allowable constituent concentrations within the water body. These maximum allowable concentrations may vary depending on their location within the water body (e.g., along a stream or estuary). Numerous models have been proposed for use in estimating the degree of waste removal at various point source sites along a water body that will meet both effluent and water quality standards. The most common of these models applies to the management of the dissolved oxygen concentrations in streams and rivers.

The oxygen required for the decomposition or assimilation of biodegradable material is expressed as its biochemical oxygen demand, BOD. The oxygen demand of a waste can be separated into two components, the amount required for the assimilation of the carbonaceous waste material BOD^C and the amount required for the assimilation of the nitrogenous waste material BOD^N. This division permits a more accurate description of the oxygen demand at any point in the stream than would the total BOD load, because the rates of deoxygenation associated with these two BOD components differ. Another reason for explicitly considering the nitrogenous component of BOD is that as the percentage of carbonaceous BOD removal increases, the percentage of the nitrogenous component in the remaining wastewater effluent increases [16]. With water quality standards requiring increasingly higher waste removals or treatment efficiencies, the nitrogenous wastes discharged into natural waters become increasingly important for the prediction of dissolved oxygen concentrations.

The depletion of dissolved oxygen by the metabolic processes of waste-consuming organisms, plant respiration, benthal deposits, and the like, is offset by the absorption of oxygen from the atmosphere, from plant photosynthesis, and possibly from other natural and artificial means. Differential equations describing these processes of oxygen depletion and replacement were described in Chapter 9. The solution of these differential equations, subject to the appropriate boundary and initial conditions, represents the temporal and longitudinal distribution of BOD^C, BOD^N, and dissolved oxygen concentrations along a water course. If both natural and wastewater flows are constant, a steady-state condition can be assumed. For those water quality control alternatives that are inflexible with respect to time, it is often reasonable to base the scale of these alternatives on some critical steady-state conditions that can occur at specified locations during certain times of the year.

Denoting each waste source site along a river by the index i and each water quality monitoring site in the river by the index j, the residual concentration of oxygen-demanding waste B_j (ML^{-3}) at any quality site j resulting from the discharge of a mass of BOD_i^C and BOD_i^N per unit time (MT^{-1}) at a source site i can be predicted. Recall from equations 9.9 and 9.10 that this predictive equation can be written as

$$B_j = \{\sum_i (b_{ij}^C BOD_i^C + b_{ij}^N BOD_i^N)\} + \hat{B}_j \tag{10.33}$$

where each parameter b_{ij} (TL^{-3}) is defined by equation 9.8a and $\hat{B}_j$ is the BOD concentration (ML^{-3}) at site j resulting from all sources other than at sites i. Similarly, the dissolved oxygen deficit concentation D_j at any quality site j, based on equation 9.15, can be written as

$$D_j = \{\sum_i (d_{ij}^C BOD_i^C + d_{ij}^N BOD_i^N)\} + \hat{D}_j \tag{10.34}$$

In equation 10.34, the term $\hat{D}_j$ is the dissolved oxygen deficit concentration (ML^{-3}) at site j resulting from all BOD sources other than at sites i.

In the equations above the mass of BOD_i^C and BOD_i^N discharged into the water body at each site i in each period may be an unknown decision variable. Denoting W_i^C and W_i^N (MT^{-1}) as the total mass of carbonaceous and nitrogenous oxygen-demanding waste produced per unit time at site i, and P_i^C as the fraction of the carbonaceous waste removed by wastewater treatment, then

$$BOD_i^C = W_i^C(1 - P_i^C) \tag{10.35}$$

Similarly, letting P_i^N be the fraction of the nitrogenous waste removed by wastewater treatment, then

$$BOD_i^N = W_i^N(1 - P_i^N) \tag{10.36}$$

Equations 10.33 to 10.36 can be combined to form a mathematical model whose solution can identify various combinations of wastewater treatment efficiencies along a river or estuary that will satisfy both effluent and water quality standards.

Consider the planning problem illustrated in Figure 10.3. There exist four point sources of waste and numerous water quality monitoring sites in the river and estuary. The problem is to determine the degree of treatment P_i^C and P_i^N at each site i that satisfies effluent standards BOD_i^{max} at the waste-

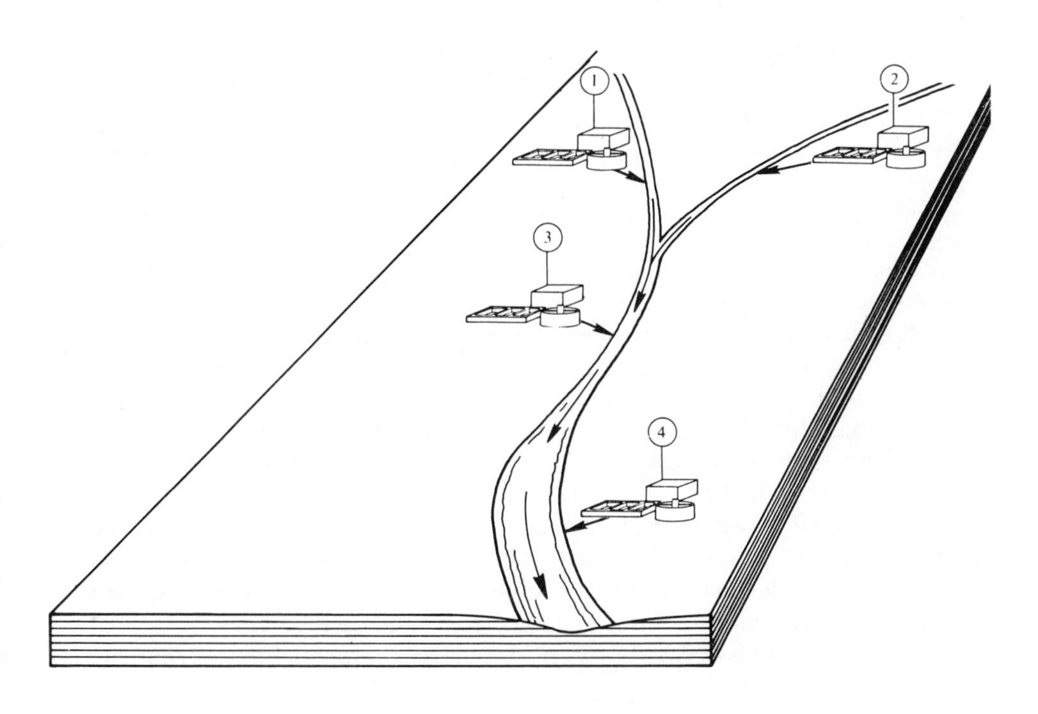

FIGURE 10.3. Wastewater discharge sites along a river system.

water discharge sites i and stream quality standards $B_j^{\max}$ and $DO_j^{\min}$ for both BOD and dissolved oxygen concentrations at various sites j. Since there are usually many alternative combinations of wastewater treatment efficiencies that will meet the standards, the objective of the analysis will be to identify those that minimize the sum of the cost of wastewater treatment $C_i(P_i^C, P_i^N)$ at each site i:

$$\text{minimize} \sum_{i=1}^{4} C_i(P_i^C, P_i^N) \tag{10.37}$$

Constraint equations associated with the BOD effluent standards $BOD_i^{\max}$ (ML^{-3}) at each site i are

$$\frac{1}{Q_i^w}\{W_i^C(1 - P_i^C) + W_i^N(1 - P_i^N)\} \leq BOD_i^{\max} \tag{10.38}$$

where Q_i^w is the wastewater discharge flow at site i.

For any water quality site j in the waste-receiving river–estuary system, the stream quality standards for BOD and DO concentrations can be expressed as

$$\sum_i \{b_{ij}^C W_i^C(1 - P_i^C) + b_{ij}^N W_i^N(1 - P_i^N)\} + \hat{B}_j \leq B_j^{\max} \qquad (10.39)$$

$$\text{DO}_j^S - \sum_i \{d_{ij}^C W_i^C(1 - P_i^C) + d_{ij}^N W_i^N(1 - P_i^N)\} - \hat{D}_j \geq \text{DO}_j^{\min} \qquad (10.40)$$

The term DO_j^S in equation 10.40 is the saturation dissolved oxygen concentration and $\hat{D}_j$ is the uncontrollable deficit concentration at site j.

A sufficient number of quality sites j must be selected to ensure that the maximum BOD or dissolved oxygen deficit concentrations within the entire river section of interest are not greater than the maximum acceptable concentrations. An alternative to the initial selection of numerous quality sites j is to select a few such sites, solve the model, determine where the maximum BOD and oxygen deficit concentrations occur, and if these concentrations are unacceptable, constrain the concentrations at these critical sites and repeat the procedure. For long river systems this interative trial-and-error procedure may be less costly than solving a model that includes a large number of quality sites j. Two or three iterations are usually all that are required.

Assuming the availability of mixed integer programming, each cost function $C_i(P_i^C, P_i^N)$ can be approximated as

$$C_i = \sum_k C_{ik} Z_{ik} \qquad \forall i \qquad (10.41)$$

where each k represents a predefined feasible treatment plant design that costs C_{ik} and removes a known fraction P_{ik}^C of carbonaceous BOD and P_{ik}^N of nitrogenous BOD. Defining 0, 1 variables Z_{ik},

$$P_i^C = \sum_k P_{ik}^C Z_{ik} \qquad \forall i \qquad (10.42)$$

and

$$P_i^N = \sum_k P_{ik}^N Z_{ik} \qquad \forall i \qquad (10.43)$$

where

$$\sum_k Z_{ik} \leq 1 \text{ and } Z_{ik} \text{ integer} \qquad \forall i \qquad (10.44)$$

Often adequate results for preliminary estimates of appropriate removal efficiencies at each site i can be obtained using non-integer linear programming. The unknown variables are the waste removel efficiencies P_i^C and P_i^N of the treatment plants at the waste source sites i and the variables Z_{ik}. The coefficients of the model b_{ij} and d_{ij} must reflect the design flow conditions of the water body between sites i and j. They can be determined by methods discussed in Chapter 9 (Figure 9.4 and equations 9.8 and 9.13, as appropriate).

An alternative to the analytical equations used in the foregoing model to predict BOD and DO concentrations in rivers or estuaries is the finite-section model, Figure 9.6 [32, 33]. The river or estuary is divided into n reaches, each reach denoted by the subscript i. The reaches are defined so that the parameters that affect the change in the dissolved oxygen concentration within that reach are constant. Waste discharges from each source within a specific reach are assumed to have similar characteristics, even though their locations differ. Similarly, the effect of a given waste discharge into a specific reach is assumed to be uniform throughout that reach. Therefore, the lengths and locations of these reaches must be chosen to assure that such approximations and assumptions will result in a reasonably accurate portrayal of the actual constituent concentrations.

To predict both the BOD and dissolved oxygen concentrations in reaches of the river or estuary, a two-stage model can be developed. This first stage consists of one equation for each reach or section. A mass balance of the BOD concentrations (B_i^C or B_i^N) in any reach i requires that the time rate of change of the BOD mass in reach i, $V_i(dB_i/dt)$, equals the net amount of BOD transported by the flow $Q_{i-1,i}$ from reach $i-1$ to reach i and the flow $Q_{i+1,i}$ from reach $i+1$ to i.

$$Q_{i-1,i}[\alpha B_{i-1} + (1-\alpha)B_i] + Q_{i+1,i}[\alpha B_{i+1} + (1-\alpha)B_i]$$

plus the net amount of BOD transported by dispersion,

$$E'_{i-1,i}(B_{i-1} - B_i) + E'_{i+1,i}(B_{i+1} - B_i)$$

less the BOD that is decayed [at a rate of K_i (T^{-1})] or that settles to the bottom [at a rate of K_S (T^{-1})],

$$V_i(K_i + K_S)B_i$$

plus the BOD load BR_i imposed on the reach by runoff, scour, and all the controlled discharges into the reach. Thus

$$\begin{aligned} V_i\frac{dB_i}{dt} &= Q_{i-1,i}[\alpha B_{i-1} + (1-\alpha)B_i] \\ &\quad + Q_{i+1,i}[\alpha B_{i+1} + (1-\alpha)B_i] \\ &\quad + E'_{i-1,i}(B_{i-1} - B_i) \\ &\quad + E'_{i+1,i}(B_{i+1} - B_i) \\ &\quad - V_i(K_i + K_S)B_i + BR_i \end{aligned} \tag{10.45}$$

where V_i is the volume (L^3) in reach i, α is the advection factor that could also be denoted as α_{ij} between each pair of reaches i and j, and E'_{ij} is the exchange coefficient from reach i to reach j (L^3T^{-1}). The eddy exchange coefficient between two successive reaches is as defined for equation 9.28. The BOD variables are concentrations (ML^{-3}) and the BR_i variable is a mass input rate (MT^{-1}) that may be partially controlled.

$$BR_i = W_i(1 - P_i) \tag{10.46}$$

To convert equation 10.45 to a steady-state equation, one requires that, for the time interval involved, $dB_i/dt = 0$ and BR_i is constant. Equation 10.45

for each of n reaches can then be written

$$a_{i,i-1}\mathrm{B}_{i-1} + a_{ii}\mathrm{B}_i + a_{i,i+1}\mathrm{B}_{i+1} = -\mathrm{BR}_i \tag{10.47}$$

where

$$\begin{aligned} a_{i,i-1} &= \alpha Q_{i-1,i} + E'_{i-1,i} \\ a_{ii} &= (1-\alpha)(Q_{i-1,i} + Q_{i+1,i}) - E'_{i-1,i} - E'_{i+1,i} \\ &\quad - (K_i + K_S)V_i \\ a_{i,i+1} &= \alpha Q_{i+1,i} + E'_{i+1,i} \end{aligned} \tag{10.48}$$

Equation 10.47 represents a set of simultaneous equations, the number of unknowns B_i equaling the number of equations.

To solve for the steady-state dissolved oxygen concentration DO_i in each reach i, an equation similar to equation 10.45 (or equation 9.28) can be derived:

$$\begin{aligned} V_i\frac{d\mathrm{DO}_i}{dt} = 0 = {} & Q_{i-1,i}(\alpha\mathrm{DO}_{i-1} + (1-\alpha)\mathrm{DO}_i) \\ & + Q_{i+1,i}(\alpha\mathrm{DO}_{i+1} + (1-\alpha)\mathrm{DO}_i) \\ & + E'_{i-1,i}(\mathrm{DO}_{i-1} - \mathrm{DO}_i) \\ & + E'_{i+1,i}(\mathrm{DO}_{i+1} - \mathrm{DO}_i) \\ & - V_iK_i^C\mathrm{B}_i^C - V_iK_i^N\mathrm{B}_i^N \\ & + V_iK_{ai}(\mathrm{DO}_i^S - \mathrm{DO}_i) + \mathrm{OR}_i \end{aligned} \tag{10.49}$$

where each term, except OR_i, is as previously defined. The oxygen addition OR_i, in reach i, is expressed as a mass rate (MT^{-1}), whereas the dissolved oxygen and BOD variables are concentrations (ML^{-3}). The use of these very simple predictive equations (either 10.33 to 10.36 or 10.45 and 10.49) provides an estimate of the average BOD and DO concentrations for the design flow conditions, and hence an estimate of the required treatment plant efficiencies P_i^C and P_i^N. The indicated efficiencies should be adjusted to the nearest discrete commercially feasible efficiency, and then these efficiencies should be simulated using more detailed simulation models of the river system. The application of these models to the Delaware estuary (Figure 10.4) is illustrated in Figure 10.5.

Some of the more detailed nonlinear multi-parameter water quality constituent simulation models discussed in Chapter 9 can also be converted into an economic management model. For example, the application of finite-difference or finite-element techniques permits the conversion of the QUAL II simulation model to an optimization or management model (see Section 9.5.6 and Table 9.1). In this case a larger number of water quality constituent concentrations can be considered.

The literature contains many examples of water quality management models which use various optimization algorithms. Dynamic programming,

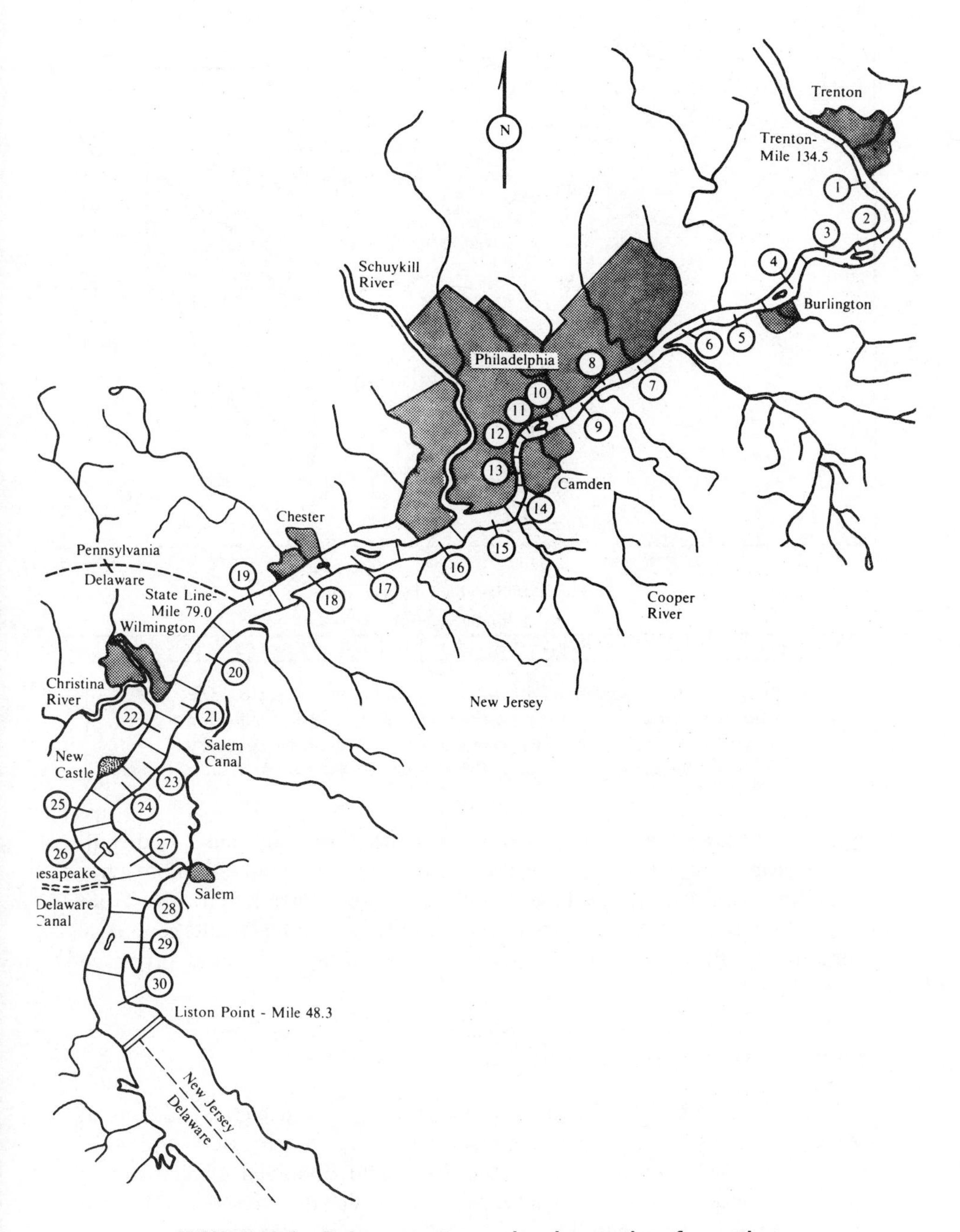

FIGURE 10.4. Delaware estuary, showing sections for mathematical model.

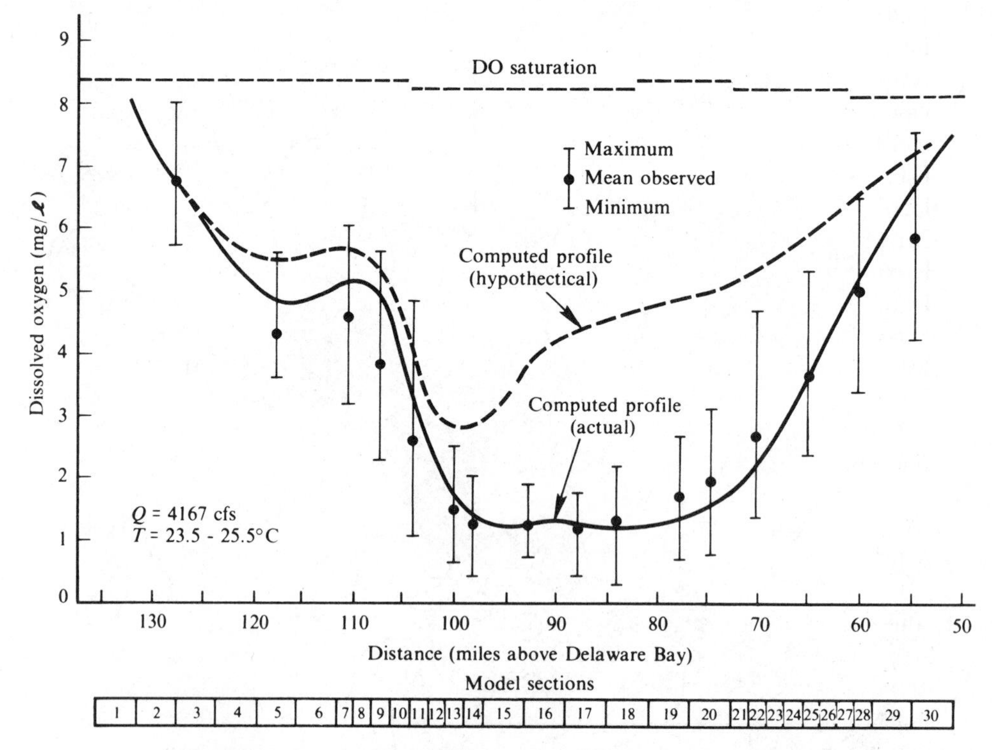

FIGURE 10.5. Computed dissolved oxygen concentration profiles for actual conditions (Aug.–Oct. 1970) and hypothetical (increased BOD removal) conditions in Delaware estuary. (U.S. EPA Delaware Estuary Water Quality Standards Study, August 1973.)

along with linear programming, are perhaps the most commonly used optimization approaches. For example, a very comprehensive study, using dynamic programming, was reported by Newsome [22] and Warn [35] for the Trent River in England. Others have used nonlinear programming [1] and geometric programming [6], to name only some of the nonlinear techniques proposed and applied.

10.4.4 Flow Augmentation

The models discussed above are all based on a critical design flow, denoted as Q_j at each quality site j. Given a constant discharge rate for BOD and other wastes from one or more sources, the waste and dissolved oxygen deficit concentrations at various sites may either decrease or increase, depending on the magnitude of the streamflow.

Increasing streamflows have three primary effects on water quality. First, increasing flow increases the volume of water in a water body. If the increased streamflow is of higher quality than the base flow, the increased flow dilutes the waste constituent concentrations and increases the minimum dissolved oxygen concentration. Second, increasing flows increase the water velocity, which in turn usually affects the reaeration rate and lengthens the distance over which an oxygen-demanding pollutant causes an oxygen deficit. Third, if base flows are augmented with cooler water, the deoxygenation rate decreases and the saturation concentration of dissolved oxygen increases. Conversely, if higher temperature waters are used for augmentation, deoxygenation rates increase and saturation dissolved oxygen concentrations decrease. Finally, increased streamflows may increase the BOD addition due to runoff and scour of benthal deposits. All these factors may well result in flow augmentation being beneficial at some sites and detrimental at others [12, 16, 26].

For example, consider a single waste source site upstream from some quality sites. If the dissolved oxygen deficit and BOD concentrations of the streamflow upstream of the waste discharge site are less than the respective dissolved oxygen deficit and BOD concentrations in the discharged wastewater, and if the increased streamflows do not pick up too much additional BOD from increased runoff and scour of bottom deposits, the additional dilution of the wastewater flow will decrease the BOD concentration and therefore will increase the minimum dissolved oxygen concentration downstream from the waste discharge site. While the minimum dissolved oxygen concentration is increased, these same conditions may lower the actual dissolved oxygen concentration at one or more specific quality sites, as illustrated in Figure 10.6.

In Figure 10.6 each pair of functions corresponds to the oxygen "sag" concentration and the BOD concentration resulting from a single waste source at site $i = 0$ given three streamflows $Q^1 < Q^2 < Q^3$. For the conditions stated above, the minimum dissolved oxygen concentration increases as the streamflow increases. At a site $j = 1$, an increase in the streamflow results in a decrease in the dissolved oxygen deficits and BOD concentrations (i.e., the water quality improves with increasing streamflows). The opposite may occur at quality sites $j = 2$ and 3. At these sites the quality decreases as the streamflow increases, at least up to a certain quantity. Hence there can exist situations, as illustrated in Figure 10.6, in which the waste removal efficiencies at upstream wastewater treatment facilities designed to meet both dissolved oxygen and BOD stream quality standards at low design streamflow conditions (e.g., the minimum average 7-day consecutive streamflow expected once in 10 years) are not sufficient to meet these same quality standards at higher streamflows. This was illustrated very clearly in an optimization-simulation study of the water quality of the Saint John River in Canada. As seen in Figure 10.7, daily flows typically higher than the 7-day, 10-year low-design-flow condition often resulted in lower predicted DO concentrations.

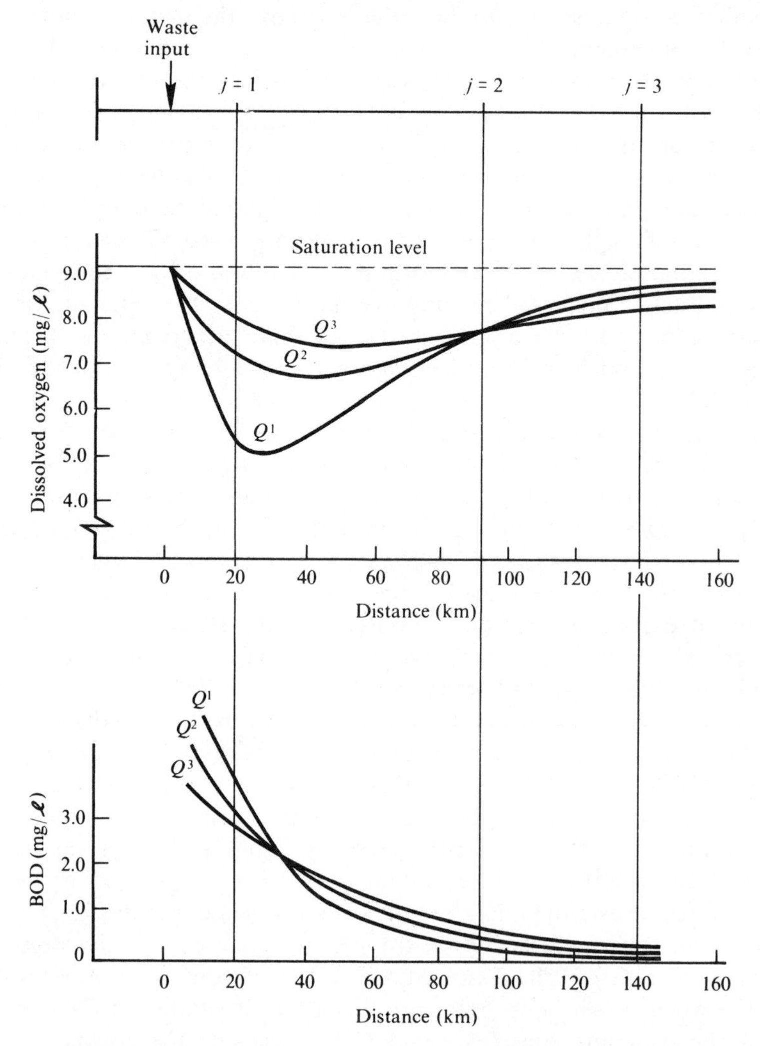

FIGURE 10.6. Dissolved oxygen and BOD concentrations downstream from a single waste source under increasing flow conditions ($Q^1 < Q^2 < Q^3$).

Since streamflows in excess of the commonly chosen low "design" flow occur much more frequently, there may be a considerable portion of the time when the stream quality at particular downstream sites is less than that which would occur given a critical design low flow condition. The selection of the critical "design" flow Q_j therefore may become an important consideration in the determination of treatment-facility waste removal efficiencies.

The critical streamflows Q_j can be augmented or increased by releasing additional waters from reservoirs or by reducing water withdrawals. The

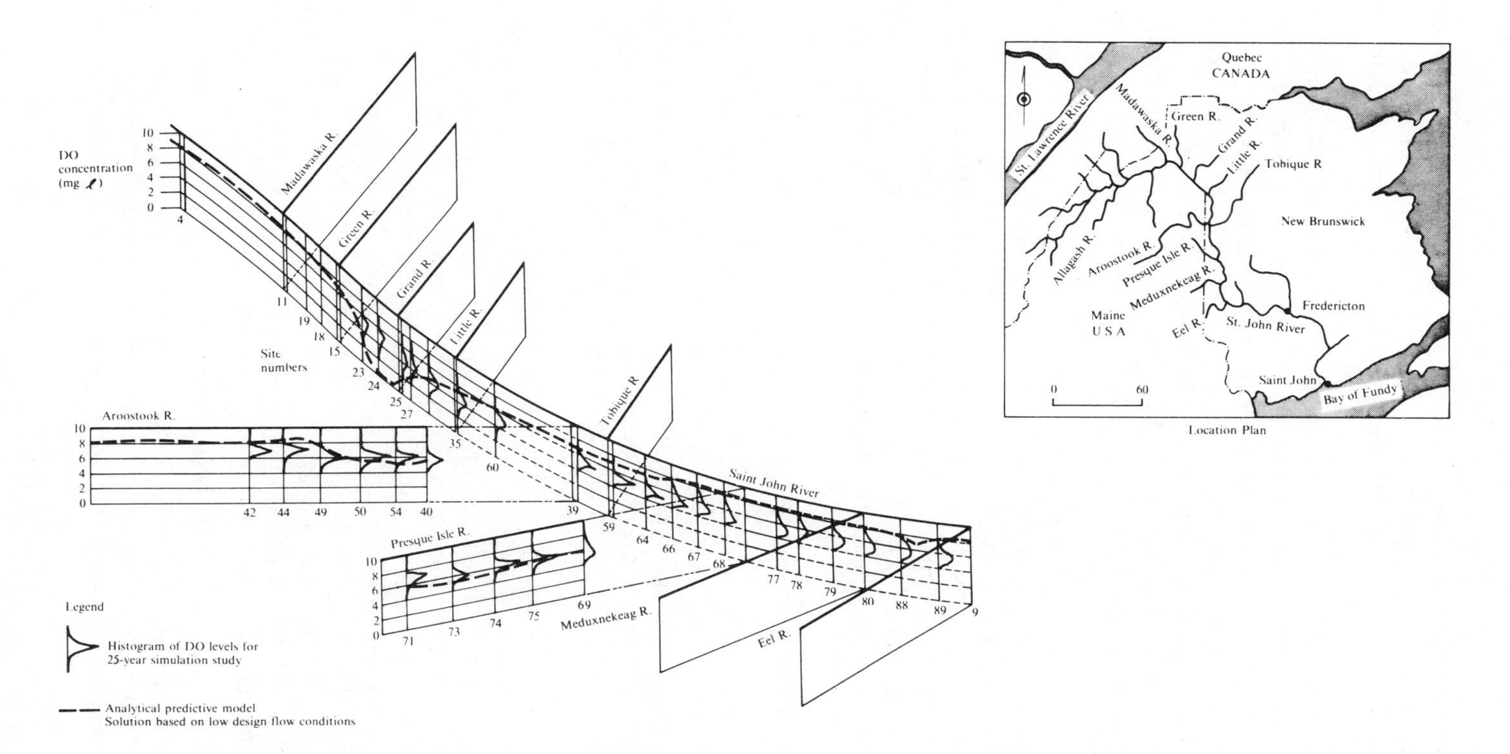

FIGURE 10.7. Comparison of dissolved oxygen concentrations predicted by analytical model for low-flow conditions and by a daily simulation model for more typical flow conditions in the Saint John River in the U.S. and Canada. (H. G. Acres, Ltd., Water Quality Management Methodology and Its Application to the Saint John River, Appendix G, Niagara Falls, Ontario, August 1971.)

increased flows may be a means of improving the critical flow conditions (e.g., increasing the flow in the stream or river, reducing temperatures, increasing the reaeration rates, etc.), thereby reducing the required treatment capacities.

The net costs of flow augmentation can be defined as the minimum reduction in net benefits from all other uses that is necessary to increase the flows during the period of low flows for the sole purpose of water quality management. In other words, the cost of augmentation $C_q(\Delta Q)$ is the reduction in the maximum net benefits that could be obtained from all other uses if flow augmentation requirements ΔQ are imposed. This net cost $C_q(\Delta Q)$ of the flow increase ΔQ can then be included in a cost-minimization objective function [16].

If the temperature and velocity of the flow between any pair of sites i and j change due to flow augmentation, the parameters and variables that define each of the transfer coefficients, b_{ij} and d_{ij} and $\hat{B}_j$ and $\hat{D}_j$, will change. Hence these transfer coefficients, $b_{ij}(Q_{ij})$ and $d_{ij}(Q_{ij})$, and $\hat{B}_j(Q_j)$ and $\hat{D}_j(Q_j)$, are functions of the streamflow Q_{ij} or Q_j. These functions would take the place of the constants b_{ij}, d_{ij}, $\hat{B}_j$, and $\hat{D}_j$; in addition, $Q_j + \Delta Q$ would replace the term Q_j in equations 9.8a, 9.13a, 10.41, and 10.42 [16, 26].

10.4.5 Artificial Instream Aeration

Instream aeration is another method of increasing the dissolved oxygen concentration in rivers. This is usually accomplished by injecting air into water through a network of perforated pipes or by rotating devices that cause surface turbulence, thereby increasing the area over which oxygen transfer can occur. These methods may be particularly efficient for the temporary improvement of near anaerobic conditions (i.e., at sites where the dissolved oxygen deficits are relatively high), but they require energy and may cause excessive noise.

The oxygen transfer rate from artificial aeration varies depending on the oxygen deficit concentration D, the water quality, and the temperature. For flowing rivers with negligible dispersion, the dissolved oxygen O_j (MT^{-1}) added at any site j will decrease the deficit immediately downstream, but not the deficit D_j (ML^{-3}) at that site. The oxygen transfer rate is a function of the deficit D_j [23, 26]. The higher the deficit, the higher the mass of oxygen that can be transferred per unit of aerator power capacity. If P_j is the aerator power capacity (LMT^{-1}) at site j, then

$$O_j = O_j(D_j, P_j) \qquad \forall j \tag{10.50}$$

The power capacity is usually limited so that for the design conditions (the critical flows, temperatures, and BOD concentrations) the dissolved oxygen concentration increase O_j/Q_j does not exceed the deficit concentration

D_j. Otherwise supersaturation results, which while possible, is generally not cost-effective.

Dividing any stream into homogenous reaches and numbering all reach junctions and potential aerator sites in successive order in the downstream direction, the deficit concentration D_j just upstream of site j, prior to any artificial addition of oxygen at that site, is a function of the BOD mass B_{j-1}, the initial deficit mass $D_{j-1}Q_{j-1}$, and the mass of added oxygen O_{j-1} at the immediate upstream site $j - 1$.

$$D_j = B_{j-1}d_{j-1,j} + (D_{j-1}Q_{j-1} - O_{j-1})b^a_{j-1,j} \tag{10.51}$$

Recall from equation 9.15 that the parameter $d_{j-1,j}$ is the dissolved oxygen deficit concentration at site j resulting per unit of BOD mass at site $j - 1$, and the parameter $b^a_{j-1,j}$ is the dissolved oxygen deficit concentration at site j resulting from a unit deficit mass at site $j - 1$.

Dissolved oxygen stream standards would apply to each D_j before any artificial aeration. Given the dissolved oxygen saturation concentration DO^S_j at site j,

$$DO^S_j - D_j \geq DO^{min}_j \qquad \forall j \tag{10.52}$$

where DO^{min}_j is the minimum allowable dissolved oxygen concentration at site j.

Aeration devices can be the most cost-effective means of meeting dissolved oxygen standards that would otherwise be violated during rarely encountered extreme low-flow, high-temperature conditions [26]. The typical dissolved oxygen control problem consists of determining the least-cost number of aeration units to be used, their location along the stream, and their design capacity measured in units of power.

The cost of aerators operated at capacity can be expressed as a function of their capacity. If the aerators are not operated at capacity, additional equations will be needed to define separately the annual capacity and the operating costs [23]. Let $C_j(P_j)$ be the cost of each aerator unit of power capacity P_j at location j along a stream having a specified extreme design flow of Q_j and known upstream oxygen demanding waste discharges. These costs, together with equations for predicting the oxygen deficit and oxygen addition from artificial aeration, can be used to develop an optimization model for finding the optimal capacities P_j at various sites j along the river. Assuming that the objective is one of total cost minimization, this can be written

$$\text{minimize} \sum_j C_j(P_j) \tag{10.53}$$

To estimate the annual cost functions $C_j(P_j)$, some estimates of operating times are needed. These estimates have to be based on an analysis of the hydrologic record of streamflow conditions, some assumptions regarding the oxygen demanding waste discharges during critical flow conditions, and the cost of energy.

The control problem defined by equations 10.50 to 10.53 can be structured for solution by discrete dynamic programming in which the stages are the possible locations of the aerators, and the states are the deficit concentrations D_j.

10.5 LAKE QUALITY MANAGEMENT

There are a number of alternatives for managing lake water quality. For man-made lakes these include selection of the impoundment site if it is not already fixed; design of the outlet structure and release policy; control of constituents in the inflow; artificial destratification, such as from diffused air or mechanical pumping; dredging; and other means of altering the normal physical, biological, and chemical processes that affect water quality [31]. Simulation models (such as those outlined and referred to in Chapter 9) that are able to predict with reasonable accuracy the water quality impact of any of these management alternatives are indeed just beginning to appear. With the exception of fully mixed impoundments, optimization models that incorporate water quality prediction together with various management alternatives for lakes and reservoirs have not yet been developed. However, just as multi-parameter water quality simulation models such as QUAL II (Section 9.5.6) are being adapted for optimization (management) modeling, so will the multi-parameter simulation models for stratified lakes and reservoirs sometime in the future.

10.5.1 Constant-Volume Well-Mixed Lakes

For lakes that are reasonably well mixed and of constant volume, the input–output models discussed in Chapter 9 (Section 9.6.1) can be extended to include as unknown decision variables the extent and cost of waste discharge reduction. Assuming a constant lake volume V (L^3), an outflow Q_t (L^3T^{-1}), and a nutrient or waste input of N_t (MT^{-1}) having a net decay rate constant of K (T^{-1}), the concentration C_t (ML^{-3}) at the end of period t, given an initial concentration C_{t-1}, can be found from integrating equation 9.54:

$$\frac{dC}{dt} = \frac{N_t}{V} - \frac{Q_t C}{V} - KC \tag{10.54}$$

which yields

$$C_t = \frac{N_t}{Q_t + \mathrm{KV}} - \left(\frac{N_t}{Q_t + \mathrm{KV}} - C_{t-1}\right) \exp\left[-\Delta t\left(\frac{Q_t}{V} + K\right)\right] \tag{10.55}$$

In the equation above, t represents a discrete time period and C_t is the concentration at the end of the time period. The term Δt is a particular interval of time within period t. Hence, if Δt is 0, the concentration is C_{t-1} (i.e., what it was at the end of period $t-1$ or at the beginning of period t). If the inflow rates N_t and Q_t are constant for a long time (i.e., as Δt approaches infinity), the concentration C approaches the equilibrium concentration, $N_t/(Q_t + \mathrm{KV})$.

To illustrate the use of equation 10.55 in a lake quality management model, consider the discharge N_{ti} (MT^{-1}) in each period t at various sites i along a lake. The rate of discharge is assumed constant within each period t (having a time interval of Δt equal to 1), but the mass discharges can be altered by removing a fraction P_i at a cost of $C_i(P_i)$. Thus the mass discharge is $N_{ti}(1 - P_i)$ at each site i. Equation 10.55 can not be written for each period t within a year.

$$C_t = \frac{\sum_i N_{ti}(1 - P_i)}{Q_t + \mathrm{KV}}(1 - e_t) + C_{t-1}e_t \qquad \forall t \tag{10.56}$$

where each known e_t equals

$$e_t = \exp\left(-\frac{Q_t}{V} - K\right) \qquad \forall t \tag{10.57}$$

and the variables P_i are bounded:

$$P_i^{\min} \leq P_i \leq P_i^{\max} \qquad \forall i \tag{10.58}$$

The variables C_t and P_i are unknown. Assuming that the rates N_t and Q_t repeat themselves over each cycle of periods t, the initial concentration C_{O} can be set equal to the concentration C_{T} at the end of the final period T to yield a steady-state solution of concentrations C_t for all periods $t = 1, 2, \ldots, T$. This eliminates the need to specify arbitrary or observed initial concentrations C_{O}, unless of course a dynamic model is desired for a predefined number of periods t. In addition, the foregoing model could be made more realistic by the inclusion of changes in the decay rates K_t in each period t, possibly resulting from changes in water temperature.

A simple economic objective might be to minimize total cost:

$$\text{minimize} \sum_i C_i(P_i) \tag{10.59}$$

subject to all constraint equations 10.56 and 10.58 and to constraints ensuring that the concentrations C_t do not exceed some maximum acceptable level $C^{\max}$:

$$C_t \leq C^{\max} \qquad \forall t \tag{10.60}$$

Once again it should be emphasized that these models are extremely simplified for most lake quality management problems. Nevertheless, with some judgment as to the appropriate values of Q_t and V, they can sometimes be used as a preliminary means of establishing average nutrient or other pollutant concentrations in well-mixed lakes and of identifying the various mix of water quality control alternatives that should be further analyzed using more detailed simulation models.

10.5.2 Predicting Algal Bloom Potential

One of the major concerns of lake management in nutrient rich areas is the potential for algal blooms. Large algal populations or blooms can seriously depress dissolved oxygen concentrations at night when algal respiration continues but oxygen production has stopped. Particularly disastrous can be the sudden collapse of a bloom, resulting in a tremendous oxygen demand for decomposition of the dead algal cells. In either case the depression of dissolved oxygen levels can kill desirable fish and other aquatic life and produce bad odors and appearance due to the resulting anaerobic conditions. In addition, large algal populations can clog filters in water supply systems and often produce substances toxic to fish, shellfish, and animals that drink the water.

A set of optimization models can be used to predict the potential for algal blooms and thus can be used to help estimate the effectiveness of alternative management strategies (diversion of nutrient bearing waste and thermal discharges or lake destratification). This is done by determining the largest probable bloom that could occur given the constraints placed on algal growth by temperature and by nutrient and light availability.

Such models were developed and applied by Bigelow et al. [2] to saltwater lakes and estuaries in The Netherlands. For any given temperature, the maximum potential algal bloom is limited by nutrient and light availability. The model's nutrient constraints restrict the total concentration of critical nutrients to that actually available. Let x_j be the unknown concentration of living algae in species j (ML^{-3}), a_{ij} the mass of nutrient i contained in each unit of species j (MM^{-1}), y_i the concentration (ML^{-3}) of nutrient i temporarily tied up in nonliving matter (dead algae and decomposable organic matter) which will become available for algal growth, and w_i the unknown concentration (ML^{-3}) of dissolved nutrient i immediately available for algal growth. The total concentration N_i (ML^{-3}) of nutrient i, which must be specified, potentially available for algal growth is

$$N_i = w_i + y_i + \sum_j a_{ij}x_j \tag{10.61}$$

Upon the death of an algal cell, the nutrients in the cell enter the pool y_i of nutrients bound up in nonliving material. Upon decay of the nonliving material, the nutrients become available for further algal growth. The time rate of change of the nutrient concentration in the nonliving material, dy_i/dt, is the difference between the rate at which algae die and the rate at which nutrient i in nonliving material is released into the water column and joins the available pool w_i. Let D_j be the death rate (T^{-1}) of species j and u_i the rate of release of nutrient i from nonliving material (the mineralization rate

T^{-1}); then

$$\frac{dy_i}{dt} = \sum_j a_{ij} D_j x_j - u_i y_i \tag{10.62}$$

If algal blooms build up rather slowly compared to the rate at which nutrients are released from nonliving material, the concentration of nutrients y_i will be reasonably close to equilibrium ($dy/dt \simeq 0$) at the algal bloom's peak. Thus a reasonable estimate of y_i is the equilibrium concentration y_i^* obtained by setting equation 10.62 to zero to obtain

$$y_i^* = \frac{1}{u_i} \sum_j a_{ij} D_j x_j \tag{10.63}$$

From equations 10.61 to 10.63, it follows that the equilibrium concentration of nutrient i, N_i^*, is

$$N_i^* = w_i + \frac{1}{u_i} \sum_j a_{ij} D_j x_j + \sum_j a_{ij} x_j = w_i + \sum_j a_{ij}\left(1 + \frac{D_j}{u_i}\right) x_i \tag{10.64}$$

This relationship specifies how the total biomass in all algal species is constrained by the available nutrients if equation 10.63 applies.

In warm summer months, algal blooms can build up very rapidly such that the concentration of nonliving material lags behind the buildup of the algal populations. If at the peak of the bloom, the amount of nutrient i tied up in nonliving matter is approximately $k_i y_i^*$ for some empirically determined k_i, $0 \leq k_i \leq 1$, then

$$N_i = w_i + \sum_j a_{ij}\left(1 + k_i \frac{D_j}{u_i}\right) x_i \tag{10.65}$$

Of course, $k_i = 0$, corresponding to $y_i = 0$, would result in the restriction that the total mass of nutrient i in all living algae must be less than N_i, the total nutrient available. Although this places an upper bound on algal biomass, it may not be sufficiently restrictive to allow determination of the maximum probable or realistically achievable algal population densities. The latter is the more important quantity from the point of view of lake management.

The constraints on light availability relate an algal species' ability to grow and compete with other algal species. The amount of light algae receive depends on the incident solar radiation throughout the day $I_0(t)$, the fraction of the radiation transmitted across the air–water interface β, and the light extinction rate η (L^{-1}) in the water column. In general, the light intensity at depth z (L) is given by

$$I_z(t) = \beta I_0(t) e^{-\eta z} \tag{10.66}$$

An algae population can grow if its individuals receive light of an adequate intensity throughout the day as they move throughout the epilimnion or upper layers of a lake or estuary so that production exceeds respiration and mortality. If $P_j[I_z(t)]$ is algal species j production rate at light intensity $I_z(t)$, for species j to participate in a bloom the average production rate of algal biomass throughout the day cannot be less than the minimum required for growth, $P_j^{\min}$. In symbols,

$$\frac{1}{24\text{ hr}}\int_0^{24\text{ hr}}\left\{\frac{1}{z_m}\int_0^{z_m} P_j[\beta I_0(t)e^{-\eta z}]\,dz\right\}dt \geq P_j^{\min} \tag{10.67}$$

where z_m is the maximum mixing depth and $P_j^{\min}$ the minimum average production rate species j requires to maintain itself. For any temperature, an algal species's production rate $P_j[I]$ peaks at some characteristic value for which it is well adapted and decreases above and below that value. Thus for each species j, there will be a feasible range of the extinction coefficient

$$\eta_j^{\min} \leq \eta \leq \eta_j^{\max} \tag{10.68}$$

within which species j can thrive and outside of which the average light intensity is either too low or too high.

The extinction coefficient η depends on the natural color or turbidity of the water, resulting in a background extinction rate η_0, and on the density of algae (self-shading) and nonliving matter in the water column. If the contribution to the total extinction coefficient of a unit concentration of algal species j is η_j ($L^{-1}/ML^{-3} = L^2M^{-1}$), the extinction coefficient resulting from background and living algae is

$$\eta_0 + \sum_j \eta_j x_j \tag{10.69}$$

To this must be added the contribution of nonliving algae which is primarily due to the chlorophyll remaining in dead algae, because reflectance is not nearly as important as the absorbance of light. If the decay rate (T^{-1}) of the light-absorbing capacity of dead algae is ν, the differential equation describing the time behavior of the contribution η_d of dead algae to the total extinction coefficient η is

$$\frac{d}{dt}(\eta_d) = \sum_j \eta_j D_j x_j - \nu(\eta_d) \tag{10.70}$$

Again, if the bloom grows slowly so that the concentration of chlorophyll in dead algae and hence its contribution to the extinction coefficient η_d stays in relative equilibrium, then $d/dt(\eta_d) \simeq 0$, and the equilibrium extinction coefficient contribution from dead algae is approximately

$$\eta_d^* \simeq \sum_j \frac{\eta_j D_j}{\nu} x_j \tag{10.71}$$

However, if blooms grow rapidly, η_d may lag behind the value given in equation 10.71 and the total extinction coefficient at the bloom's peak is best given by

$$\begin{aligned}\eta &= \eta_0 + \sum_j \eta_j x_j + k\eta_d^* \\ &= \eta_0 + \sum_j \eta_j\left(1 + \frac{kD_j}{\nu}\right)x_j\end{aligned} \tag{10.72}$$

for some empirically determined k between 0 and 1.

For incorporation into a linear programming model, the acceptable extinction coefficient intervals $(\eta_j^{\min}, \eta_j^{\max})$ for each species are used to define a set of subintervals $(\eta_s^{\min}, \eta_s^{\max})$. Typically, a number of intervals s will be contained within any particular species interval $(\eta_j^{\min}, \eta_j^{\max})$. Each subinterval s between $\eta_s^{\min}$ and $\eta_s^{\max}$ will be values of η that are acceptable extinction coefficient values for a set S_s of algal species j.

A set of linear programming models for finding the maximum probable or reasonable biomass or chlorophyll concentration of an algal bloom can be developed using these nutrient and light availability constraints. To find the maximum potential biomass or chlorophyll concentration, one would maximize for each s either

biomass: $$B^s = \sum_{j \in S_s} x_j \tag{10.73}$$

or

chlorophyll: $$C^s = \sum_{j \in S_s} C_j x_j \tag{10.74}$$

where C_j is the chlorophyll content (MM^{-1}) per unit of algal species j. This maximization would be subject to the nutrient constraints

$$N_i = w_i + \sum_{j \in S_s} a_{ij}\left(1 + k_i \frac{D_j}{u_i}\right) x_j \tag{10.75}$$

and the light limitations

$$\eta_s^{\min} \le \eta_0 + \sum_{j \in S_s} \eta_j\left(1 + k \frac{D_j}{v}\right) x_j \le \eta_s^{\max} \tag{10.76}$$

and nonnegativity

$$x_j \ge 0, \qquad w_i \ge 0 \tag{10.77}$$

Solving the model for each set S_s with the corresponding extinction coefficient interval gives a set of potential biomass $\{B^s\}$ and chlorophyll concentrations $\{C^s\}$. The largest of these,

$$B^{\max} = \max_s B^s \tag{10.78}$$

$$C^{\max} = \max_s C^s \tag{10.79}$$

provides an estimate of the potential or maximum probable values of these parameters.

The peak biomass of a bloom could be estimated with a dynamic ecosystem simulation model. This requires considerably more information than that needed for this simple optimization model and the ability to identify those environmental conditions that would give rise to the maximum possible bloom.

This optimization model makes no attempt to describe the time dynamics of the aquatic system. Rather, it uses the available nutrient constraints and information about algal species's light requirements and self-shading characteristics to predict the maximum biomass or chlorophyll concentration that might reasonably be achieved during a particular week or month. Using a management period of this length should ensure that sufficient time is available for potential blooms to materialize given that algal populations are already present in reasonable densities in natural environments.

There may be no feasible solution to the foregoing model for some s because for the given temperature and corresponding light conditions an algal bloom may be unable to develop. In winter and early spring, because of the cold water temperatures, there may be no feasible solution for any value of s.

Feasibility could also be affected by small changes in some of the parameters. From the dual variables of the nutrient availability constraints, one can obtain some indication of the impact that changes in the available nutrient concentrations N_i and algal extinction coefficient limits will have on the bloom potential.

As formulated, the model considers only a bloom's impact in terms of total biomass or chlorophyll. The oxygen demand for decomposition of dead algal cells, which would occur over a short period of time were the maximum bloom to suddenly collapse and all the algae die, can also be determined. The effect of zooplankton has been omitted from the model but could be incorporated in an approximate way. The model is relatively simple and serves as an approximate description of a very complex system. Its potential value lies in its simplicity, its ability to estimate answers to some important management questions, and its relatively modest data requirements, compared to more complex and realistic time-dynamic algal models.

10.6 CONCLUDING REMARKS

It would be ideal if one could say that water quality models such as those discussed in this chapter could be used directly to identify optimal, or at least improved, solutions to water quality management problems. Perhaps they can in situations where there exist sufficient data and modeling expertise to develop, calibrate, and verify models and their solutions, and where the objectives of those responsible for water quality planning and management are clearly defined. In those situations where these ideal conditions do not exist, use of these models can serve another purpose.

Water quality management models such as those outlined in this chapter are most commonly used for developing an understanding of the relative water quality impacts of alternative management practices and for determining the significance or importance of having more accurate or more detailed data. These insights can guide the development of effective plans and decisions, if not actually identify the best plan or decision.

Water quality modeling, if done well, can give an understanding of why some management alternatives are better than others for a particular river basin. Modeling can provide one with estimates of how the river system will respond, at least in a relative sense, to different waste discharges. In addition, models can be used to help identify preferred management plans given various management objectives and assumptions concerning future resource costs, technology, and social and legal requirements.

In acknowledging the role that water quality models can and should play in the planning process, one must recognize the inherent limitations of models as representatives of any real problem. The input data, including management objectives and assumptions concerning the physical, biological, and chemical reactions that take place in the river, may be controversial or uncertain. Of course, the input affects the output. Although the input data and model may be the best available, one's knowledge about the prototype (the actual water body), and about how future events that may alter its behavior, will always be limited. In addition, since public water quality objectives change, water quality models must be viewed as flexible tools adaptable to changing circumstances as they are perceived and to changing data as they become available.

EXERCISES

10-1. Using the simulation procedure outlined in Section 10.4.1, identify three alternative sets (feasible solutions) of storage lagoon volume capacities V and corresponding land application areas A and irrigation volumes Q_{2t} in each month t within a year that satisfy a 10 mg/ℓ maximum NO_3–N content in the drainage water of a land disposal

Month, t	*Storage Pond Influent, Q_{1t} (10^3 m^3)*	*Organic Nitrogen Mineralization Rate m_t*	*Precipitation, P_t (cm)*	*Evapotranspiration, E_t (cm)*	*Nitrogen Removal Rate, k_t ($months^{-1}$)*
1 (Apr)	580	0.002	7.7	0	0.49
2 (May)	610	0.008	9.4	10.4	0.60
3 (Jun)	670	0.010	8.9	13.5	0.70
4 (Jul)	670	0.012	9.5	15.0	0.75
5 (Aug)	670	0.012	9.8	12.7	0.74
6 (Sep)	640	0.009	7.9	8.9	0.66
7 (Oct)	580	0.005	8.4	5.8	0.55
8 (Nov)	580	0.001	7.4	0	0.42
9 (Dec)	580	0	5.8	0	0.29
10 (Jan)	580	0	5.2	0	0.26
11 (Feb)	580	0	6.1	0	0.27
12 (Mar)	580	0.001	6.8	0	0.36

system. In addition to the data listed below, assume that the influent nitrogen n_{1t} is 50 mg/ℓ each month, with 10% ($\alpha = 0.1$) of the nitrogen in organic form. Also assume that the soil is a well-drained silt loam containing 4500 kg/ha of organic nitrogen in the soil above the drains. The soil has a monthly drainage capacity d of 60 cm and has a field capacity moisture content M of 10 cm. Maximum plant nitrogen uptake values N_t^{max} are 35 kg/ha during April through October, and 70 kg/ha during May through September. Finally, assume that because of cold temperatures, no wastewater irrigation is permitted during November through March. December, January, and February's precipitation is in the form of snow and will melt and be added to the soil moisture inventory in March.

10-2. Verify the optimal solution shown in Table 10.2 for the problem defined in Figure 10.2, using mixed-integer programming. Discuss the problems in solving models of regional treatment systems if the characteristics of the wastes generated throughout the region are different at different locations. How could the model defined for the problem illustrated in Figure 10.2 be extended to include stream quality as well as effluent quality standards, and how might this inclusion alter the final solution?

10-3. Consider the problem of estimating the minimum total cost of waste treatment in order to satisfy quality standards within a stream. Let the stream contain seven homogenous reaches r, reach $r = 1$ being at the upstream end and reach $r = 7$ at the downstream end. Reaches $r = 2$ and 4 are tributaries entering the mainstream at the beginning of reaches 3 and 5, respectively; hence the mainstream consists of reaches 1, 3, 5, 6, and 7. Point sources of BOD enter the stream at the beginning of reaches 1, 2, 3, 4, 6, and 7. Assuming that at least 60% BOD removal is required at each discharge site, solve for the least-cost solution given the data in the accompanying table. Can you identify more than one type of model to solve this problem? How would this model be expanded to specifically include both carbonaceous BOD and nitrogenous BOD and nonpoint waste discharges?

Reach No.	*Design BOD Load (mg/ℓ)*	*Present % Removal Load*	ANNUAL COSTS OF VARIOUS DESIGN BOD REMOVALS			
			60%	*75%*	*85%*	*90%*
1	248	67	0	22,100	77,500	120,600
2	408	30	630,000	780,000	987,000	1,170,000
3	240	30	210,000	277,500	323,000	378,000
4	1440	30	413,000	523,000	626,000	698,000
6	2180	30	500,000	638,000	790,000	900,000
7	279	30	840,000	1,072,000	1,232,500	1,350,000

Reach No.	*Time of Flow (days)*	*Waste water Discharge (10^6 m^3/day)*	*Entering Reach Flow (10^6 m^3/day)*	*Total Reach Flow (10^6 m^3/day)*	*DO Saturation Concentration (mg/ℓ)*	*Maximum Allowable DO Deficit (mg/ℓ)*	*DO Deficit of Wastewater (mg/ℓ)*	*DO Conc at Beginning of Reach (mg/ℓ)*	*BOD Conc. at Beginning Reach (mg/ℓ)*	*Average Deoxygenation Rate Constant for Reach ($days^{-1}$)*	*Reaeration-Rate Constant ($days^{-1}$)*
1	0.235	19	5,129	5,148	10.20	3.20	1.0	9.50	1.66	0.31	1.02
2	1.330	140	4,883	5,023	9.95	2.45	1.0	8.00	0.68	0.41	0.60
3	1.087	30	10,171	10,201	9.00	2.00	1.0	?	?	0.36	0.63
4	2.067	53	1,120	1,173	9.70	3.54	1.0	9.54	1.0	0.35	0.09
5	0.306	0	11,374	11,374	9.00	2.50	—	?	?	0.34	0.72
6	1.050	98	11,374	11,472	8.35	2.35	1.0	—	—	0.35	0.14
7	6.130	155	11,472	11,627	8.17	4.17	1.0	—	—	0.30	0.02

10-4. Indicate in matrix notation how you would solve equations 10.47 and 10.49. Define the elements in each matrix and vector.

10-5. Discuss what would be required to analyze flow augmentation alternatives in Exercise 10.4. How would the costs of flow augmentation be defined and how would you modify the model(s) developed in Exercise 10.4 to include flow augmentation alternatives?

10-6. Develop a dynamic programming model to estimate the least-cost number, capacity, and location of artificial aerators to ensure meeting minimum allowable dissolved oxygen standards where they would otherwise be violated during an extreme low-flow design condition in a nonbranching section of a stream. Show how wastewater treatment alternatives, and their costs, could also be included in the dynamic programming model.

10-7. Using equations 10.56 to 10.60 as required, and the data provided, find the steady-state concentrations C_t of a constituent in a well-mixed lake of constant volume 30×10^6 m^3. The production N_{ti} of the constituent occurs at three sites i, and is constant in each of four seasons in the year. The required fractions of constituent removal P_i at each site i are to be set so that they are equal at all sites i and the maximum concentration in the lake in each period t must not exceed 20 mg/ℓ.

Period, t	*Days in Period*	*Flow, Q_t (10^3 m^3/day)*	*Constituent Decay Rate, Constant K_t (days^{-1})*
1	100	90	0.02
2	80	150	0.03
3	90	200	0.05
4	95	120	0.04

Constituent Discharge Site, i	*Constituent Production (kg/day)*
1	38000
2	25000
3	47000

10-8. Suppose that the solution of a model such as that used in Exercise 10.7, or measured data, indicated that for a well-mixed portion of a saltwater lake, the concentrations of nitrogen ($i = 1$), phosphorus ($i = 2$), and silicon ($i = 3$) in a particular period t were 1.1, 0.1, and 0.8 mg/ℓ, respectively. Assume that all other nutrients required for algal growth are in abundance. The algal species of concern are three in number and are denoted by $j = 1, 2$, and 3. The data required to estimate the probable maximum algal bloom biomass concentration using equations 10.73 to 10.79, as appropriate, are given in the accompanying table. Compute this bloom potential for all k_i and k equal to 0, 0.8, and 1.0.

Algal Bloom Model Parameter Values

Parameter (*Algae Species Index j*)	PARAMETER VALUE *1*	*2*	*3*
a_{1j} = mg N/mg dry wt of algae j	0.04	0.01	0.20
a_{2j} = mg P/mg dry wt of algae j	0.06	0.02	0.10
a_{3j} = mg Si/mg dry wt of algae j	0.08	0.01	0.03
D_j = mortality and grazing rate constant (days^{-1})	0.6	0.4	0.2
d_j = mortality rate constant, (days^{-1})	0.3	0.1	0.1
v = extinction reduction rate constant for dead algae, (days^{-1})	0.07	0.07	0.07
η_j^{max} = max. extinction coef. (m^{-1})	0.07	0.07	0.10
η_j^{min} = min. extinction coef. (m^{-1})	0.01	0.03	0.03
η_j = increase in extinction coef. per unit increase in mg/ℓ (g/m^3) of dry wt of species j (m^2/g)	0.05	0.164	0.04

Nutrient Index i:	*1*	*2*	*3*
Nutrient:	N	P	Si
u_i = mineralization rate constant, (days^{-1}):	0.02	0.69	0.62

Extinction Subinterval s:	*1*	*2*	*3*
Extinction coefficient range:	0.01–0.03	0.03–0.07	0.07–0.10
Algae species $j \in S_s$:	1	1, 2, 3	3

Extinction coefficient without algae = η_0 = 0.01 m^{-1}

Note: Since there are three extinction subintervals, there are three models to solve for each value of $k_i = k$.

REFERENCES

1. Bayer, M. B., A Modelling Method for Evaluating Water Quality Policies in Nonserial River Systems, *Water Resources Bulletin*, Vol. 13, No. 6, 1977, pp. 1141–1151.

2. Bigelow, J. H., J. G. Bolten and J. C. DeHaven, *Protecting an Estuary from Floods—A Policy Analysis of the Oosterschelde*, Vol. 4, *Assessment of Algal Blooms, A Potential Ecological Disturbance*, Rand Corporation, R-2121/4-NETH, Santa Monica, Calif., April 1977.

3. CANALE, R. P., *Modeling Biochemical Processes in Aquatic Ecosystems*, Ann Arbor Science, Publishers, Inc., Ann Arbor, Mich., 1976.
4. CHI, T., Waste Water Conveyance Models, Chap. 8 in *Models for Managing Regional Water Quality*, R. Dorfman, H. D. Jacoby, and H. A. Thomas, Jr. (eds.), Harvard University Press, Cambridge, Mass., 1972.
5. DORFMAN, R., H. D. JACOBY, and H. A. THOMAS, JR., *Models for Managing Regional W ter Quality*, Harvard University Press, Cambridge, Mass., 1972.
6. ECKER, J. G., "A Geometric Programming Model for Optimal Allocation of Stream Dissolved Oxygen," *Management Science*, Vol. 21, 1975, pp. 658–668.
7. FAN, L. T., M. T. KUO, and L. E. ERICKSON, Effect of Suspended Wastes on System Design, *Journal of the Environmental Engineering Division, ASCE*, Vol. 100, No. EE6, 1974.
8. GRADY, C. P. L., JR., Simplified Optimization of Activated Sludge Process, *Journal of the Environmental Engineering Division, ASCE*, Vol. 103, No. EE3, 1977.
9. HAITH, D. A., Optimal Control of Nitrogen Losses from Land Disposal Areas, *Journal of the Environmental Engineering Division, ASCE*, Vol. 99, No. EE6, 1973.
10. HAITH, D. A., A. KOENIG, and D. P. LOUCKS, Preliminary Design of Waste Water Land Application Systems, *Journal of Water Pollution Control Federation*, Vol. 49, No. 12, 1977.
11. JAMES, A. (ed.), *Mathematical Models in Water Pollution Control*, John Wiley & Sons, Inc., New York, 1978.
12. JAWORSKI, N. A., W. J. WEEBER, and R. A. DEININGER, Optimal Reservoir Releases for Water Quality Control, *Journal of the Sanitary Engineering Division, ASCE*, Vol. 96, No. 3, 1970, pp. 727–742.
13. KOENIG, A., and D. P. LOUCKS, A Management Model for Waste Water Disposal on Land, *Journal of the Environmental Engineering Division, ASCE*, Vol. 103, No. EE2, 1977.
14. LAWRENCE, A. W., and P. L. MCCARTY, Unified Basis for Biological Treatment Design and Operation, *Journal of the Sanitary Engineering Division, ASCE*, Vol. 96, No. SA3, 1970.
15. LIEBMAN, J. C., and W. R. LYNN, The Optimal Allocation of Stream Dissolved Oxygen, *Water Resources Research*, Vol. 2, No. 3, 1966, pp. 581–591.
16. LOUCKS, D. P., and H. D. JACOBY, Flow Regulation for Water Quality Management, Chap. 9 in *Models for Managing Regional Water Quality*, R. Dorfman, H. D. Jacoby, and H. A. Thomas, Jr. (eds.), Harvard University Press, Cambridge, Mass., 1972.
17. LOUCKS, D. P., et al., A Review of the Literature in Water Pollution Control: Systems Analysis, *Journal of the Water Pollution Control Federation*, Vols. 41–47, No. 6, 1970–1978.
18. LOUCKS, D. P., Risk Evaluation in Sewage Treatment Plant Design, *Journal of the Sanitary Engineering Division, ASCE*, Vol. 93, No. 1, 1967, pp. 25–39.
19. LOUCKS, D. P., C. S. REVELLE, and W. R. LYNN, Linear Programming Models for Water Pollution Control, *Management Science*, Vol. 14, No. 4, 1967, pp. B–166 to B–181.
20. MIDDLETON, A. C., and A. W. LAWRENCE, Cost Optimization of Activated Sludge Systems, *Biotechnology and Bioengineering*, Vol. 16, 1974, pp. 807–826.

21. MIDDLETON, A. C., and A. W. LAWRENCE, Least Cost Design of Activated Sludge Systems, *Journal of Water Pollution Control Federation*, Vol. 48, No. 5, 1976, pp. 889–905.
22. NEWSOME, D. H., "The Trent River Model—An Aid to Management," in *Modelling of Water Resource Systems*, Vol. 2, A. K. Biswas (ed.), Harvest House, Montreal, 1972, pp. 490–509.
23. ORTOLANO, L., Artificial Aeration as a Substitute for Waste Water Treatment, Chap. 7 in *Models for Managing Regional Water Quality*, R. Dorfman, H. D. Jacoby, and H. A. Thomas, Jr. (eds.), Harvard University Press, Cambridge, Mass., 1972.
24. PATTERSON, W. L. and R. F. BANKER, Estimating Costs and Manpower Requirements for Conventional Waste Water Treatment Facilities, Environmental Protection Agency Water Pollution Control Research Series, No. 17090 DAN, Washington, D.C., October 1971.
25. PAVONI, J. L. (ed.), *Handbook of Water Quality Management Planning*, Van Nostrand Reinhold Company, New York, 1977.
26. RINALDI, S., R. SONCINI-SESSA, H. STEHFEST, and H. TAMURA, *Modeling and Control of River Quality*, McGraw-Hill Book Company, New York, 1978.
27. RUSSELL, C. S. (ed.), *Ecological Modeling in a Resource Management Framework*, RFF Working Paper QE-1, Johns Hopkins University Press, Baltimore, Md., July 1975.
28. *Sewage Treatment Plant Design*, Manual No. 36, by a Joint Committee of the American Society of Civil Engineers and the Water Pollution Control Federation, ASCE, 1959.
29. SHIH, C. S., System Optimization for River Basin Water Quality Management, *Journal of Water Pollution Control Federation*, Vol. 42, No. 10, 1970, pp. 1792–1804.
30. SOBEL, M. J., Water Quality Improvement Programming Problems, *Water Resources Research*, Vol. 1, No. 4, 1965, pp. 477–487.
31. SYMONS, J. M., et al., *Water Quality Behavior in Reservoirs*, U.S. Department of Health, Education and Welfare, Public Health Service Publication 1930, Cincinnati, Ohio, 1969.
32. THOMANN, R. V., and M. J. SOBEL, Estuarine Water Quality Management and Forecasting, *Journal of the Sanitary Engineering Division, ASCE*, Vol. 90, No. 5, 1964, pp. 9–36.
33. THOMANN, R. V., *Systems Analysis and Water Quality Management*, Environmental Research and Applications, Inc., New York, 1972.
34. VANNOTE, R. H., et al., *A Guide to the Selection of Cost-Effective Wastewater Treatment Systems*, Technical Report EPA 430/9-75-002, U.S. Environmental Protection Agency, Washington, D.C., July 1975.
35. WARN, A. E., "The Trent Mathematical Model," Chap. 7 in *Mathematical Models in Water Pollution Control*, A. James (ed.), John Wiley & Sons, Inc., New York, 1978.

Index

D

J

K

L

M

N

Y

Z